FORMAL TECHNIQUES FOR NETWORKED AND DISTRIBUTED SYSTEMS

IFIP - The International Federation for Information Processing

IFIP was founded in 1960 under the auspices of UNESCO, following the First World Computer Congress held in Paris the previous year. An umbrella organization for societies working in information processing, IFIP's aim is two-fold: to support information processing within its member countries and to encourage technology transfer to developing nations. As its mission statement clearly states,

IFIP's mission is to be the leading, truly international, apolitical organization which encourages and assists in the development, exploitation and application of information technology for the benefit of all people.

IFIP is a non-profitmaking organization, run almost solely by 2500 volunteers. It operates through a number of technical committees, which organize events and publications. IFIP's events range from an international congress to local seminars, but the most important are:

- The IFIP World Computer Congress, held every second year;
- open conferences;
- working conferences.

The flagship event is the IFIP World Computer Congress, at which both invited and contributed papers are presented. Contributed papers are rigorously refereed and the rejection rate is high.

As with the Congress, participation in the open conferences is open to all and papers may be invited or submitted. Again, submitted papers are stringently refereed.

The working conferences are structured differently. They are usually run by a working group and attendance is small and by invitation only. Their purpose is to create an atmosphere conducive to innovation and development. Refereeing is less rigorous and papers are subjected to extensive group discussion.

Publications arising from IFIP events vary. The papers presented at the IFIP World Computer Congress and at open conferences are published as conference proceedings, while the results of the working conferences are often published as collections of selected and edited papers.

Any national society whose primary activity is in information may apply to become a full member of IFIP, although full membership is restricted to one society per country. Full members are entitled to vote at the annual General Assembly, National societies preferring a less committed involvement may apply for associate or corresponding membership. Associate members enjoy the same benefits as full members, but without voting rights. Corresponding members are not represented in IFIP bodies. Affiliated membership is open to non-national societies, and individual and honorary membership schemes are also offered.

FORMAL TECHNIQUES FOR NETWORKED AND DISTRIBUTED SYSTEMS

FORTE 2001
IFIP TC6/WG6.1 — 21st International Conference on Formal Techniques for Networked and Distributed Systems August 28–31, 2001, Cheju Island, Korea

Edited by

Myungchul Kim
Information and Communications University (ICU)
Korea

Byoungmoon Chin
Telecommunications Technology Association (TTA)
Korea

Sungwon Kang
Korea Telecom (KT)
Korea

Danhyung Lee
Korea IT Industry Promotion Agency (KIPA)
Korea

SPRINGER SCIENCE+BUSINESS MEDIA, LLC

ISBN 978-1-4757-8825-9 ISBN 978-0-306-47003-5 (eBook)
DOI 10.1007/978-0-306-47003-5

Electronic Services <http://www.wkap.nl>

Library of Congress Cataloging-in-Publication Data

A C.I.P. Catalogue record for this book is available from the Library of Congress.

Originally published by Kluwer Academic Publishers in 2001
Softcover reprint of the hardcover 1st edition 2001

Printed on acid-free paper.

Contents

Preface

FORTE 2001, formerly FORTE/PSTV conference, is a combined conference of FORTE (Formal Description Techniques for Distributed Systems and Communication Protocols) and PSTV (Protocol Specification, Testing and Verification) conferences. This year the conference has a new name FORTE (Formal Techniques for Networked and Distributed Systems). The previous FORTE began in 1989 and the PSTV conference in 1981. Therefore the new FORTE conference actually has a long history of 21 years.

The purpose of this conference is to introduce theories and formal techniques applicable to various engineering stages of networked and distributed systems and to share applications and experiences of them. This FORTE 2001 conference proceedings contains 24 refereed papers and 4 invited papers on the subjects. We regret that many good papers submitted could not be published in this volume due to the lack of space.

FORTE 2001 was organized under the auspices of IFIP WG 6.1 by Information and Communications University of Korea. It was financially supported by Ministry of Information and Communication of Korea.

We would like to thank every author who submitted a paper to FORTE 2001 and thank the reviewers who generously spent their time on reviewing. Special thanks are due to the reviewers who kindly conducted additional reviews for rigorous review process within a very short time frame. We would like to thank Prof. Guy Leduc, the chairman of IFIP WG 6.1, who made valuable suggestions and shared his experiences for conference organization.

This year we have seen exceptionally concerted efforts of the program committee to make FORTE 2001 a successful conference. We thank each one of the program committee for their contribution and cooperation. The enthusiasm and dedication the program committee showed has made us believe that FORTE will remain a prestigious conference with quality and distinction for a long time in the future.

Myungchul Kim
Byoungmoon Chin
Sungwon Kang
Danhyung Lee

Cheju Island, Korea

Program Committee and Reviewers

Program Committee Co-Chairmen

Myungchul Kim (ICU, Korea)
Byoungmoon Chin (TTA, Korea)
Sungwon Kang (KT, Korea)
Danhyung Lee (KIPA, Korea)

Program Committee

G. v. Bochmann (Univ. of Ottawa, Canada)
T. Bolognesi (IEI, Italy)
H. Bowman (Univ. of Kent, UK)
E. Brinksma (Univ. of Twente, Netherlands)
S. Budkowski (INT, France)
A. Cavalli (INT, France)
S. D. Cha (KAIST, Korea)
S. Chanson (Hong Kong Univ. of Science and Technology, China)
E. H. Choi (Korea Telecom, Korea)
J. Y. Choi (Korea Univ., Korea)
J. P. Courtiat (LAAS-CNRS, France)
R. Dssouli (Univ. of Montreal, Canada)
D. Frutos-Escrig (Universidad Complutense - Madrid, Spain)
S. Fischer (International Univ., Germany)
R. Groz (France Telecom R&D, France)
R. Gotzhein (Univ. of Kaiserslautern, Germany)
T. Higashino (Osaka Univ., Japan)
D. Hogrefe (Univ. of Luebeck, Germany)
G. J. Holzmann (Bell Labs, USA)
J. K. Kim (ETRI, Korea)
S. U. Kim (Pukyung Univ., Korea)
H. Koenig (BTU, Germany)
R. Lai (La Trobe Univ., Australia)
D. Latella (C.N.R., lst. CNUCE, Italy)
G. Leduc (Univ. of Liege, Belgium)
D. Lee (Bell Labs, USA)
D. I. Lee (KJIST, Korea)
J. Y. Lee (Yonsei Univ., Korea)
L. Logrippo (Univ. of Ottawa, Canada)
E. Najm (ENST, France)

D. Peled (Bell Labs, USA)
A. Petrenko (CRIM, Canada)
O. Rafiq (Univ. of Pau, France)
K. Suzuki (Advanced Comm. Co., Japan)
K. Tarnay (Nokia, Hungary)
R. Tenney (Univ. of Massachusetts, USA)
K. Turner (Univ. of Stirling, UK)
U. Ultes-Nitsche (Univ. of Southampton, UK)
S. Vuong (Univ. of British Columbia, Canada)
J. Wu (Tsinghua Univ., China)
N. Yevtushenko (Tomsk Univ., Russia)

Further Reviewers

D. Amyot
R. Andrade
A. Bailly
A. Bertolino
S. Boroday
L. Cacciari
O. Catrina
D. Chen
R. G. Clark
M. Ebner
A. En-Nouaary
A. Fantechi
A. Fontaine
N. De Francesco
C. Gervy
S. Gnesi
J. Grabowski
P. Gradit
S. Gruner
R. Hao
M. Heiner
H. Hermanns
M. Higuchi
I. Hwang
A. Idoue
P. Janowski
T. Kato
R. Langerak
C. Liu
L. Llana
S. Maag
E. Magill
J.' C. Maldonado
M. Massink
J.- F. Monin
M. Nakamura
A. Nakata
H. Neukirchen
S. Prokopenko
R. Pugliese
Y.- M. Quemener
C. Rinderknecht
A. Salah
J. Sincennes
A. Verdejo
F. Vernadat
H. Yamaguchi
G. Yang
H. Yokota
J. Zhu

Part One

Formal Methods in Software Development I

AUTOMATED DERIVATION OF ILP IMPLEMENTATIONS FROM SDL SPECIFICATIONS

Sven Twarok[1], Peter Langendoerfer[2], Hartmut Koenig
Brandenburg University of Technology at Cottbus, Department of Computer Science, PF 101344, 03013 Cottbus, Germany, Tel./Fax: +49-355-69-2236 / 2127, email: twarok@unilab.de, langendoerfer@ihp-microelectronics.com, koenig@informatik.tu-cottbus.de

Abstract In this paper we present a mapping strategy for an important manual protocol implementation technique: the Integrated Layer Processing (ILP) which is used for protocol stack implementations with complex data manipulation operations. The Integrated Layer Processing was implemented beside the server and the activity thread model as third code generation procedure in the configurable SDL compiler COCOS. We describe the chosen transformation principle and discuss in particular the semantic problems which have to be solved for this mapping. Finally we present first measurements which show the effect of the transformation. Performance gains up to 20 per cent compared to COCOS server implementations and of about 300 per cent compared to the SDT *Cadvanced* tool were measured.

Keywords: Formal description techniques, integrated layer processing, automated protocol implementation, configurable FDT compiler, SDL

1. MOTIVATION

The increasing range of specialized communication solutions, in particular in mobile communication [Lang01], will increase the importance of the deployment of automatically generated code in real-life protocol implementations. Automated code generation can make the implementation process rather efficient and can fast adapt to changes in the design. The

[1] Sven Twarok is now with UniLab Software AG Braunschweig
[2] Peter Langendoerfer is now with IHP Microelectronics Frankfurt (Oder)

acceptance of automated implementation techniques depends on the efficiency of the generated code. Automated implementation techniques will be only then applied if the efficiency of the generated code comes close to that of manual implementations. The research of the last 10 years in this area has shown that there is still a large potential for efficiency increase by introducing optimizing techniques in the code generation process as discussed in [Lang99]. A remarkable efficiency gain can be achieved by adapting implementation techniques which are used for manual protocol implementations [Koen00]. In particular two approaches have been pursued:

- the integration of advanced implementation techniques and
- a flexible adaption to the implementation context.

The solution proposed for this is that of a configurable FDT compiler. The SDL compiler COCOS (*Configurable Compiler for SDL*), developed in our group, is an example for this approach. It possesses a variable runtime system which can be configured according to the requirements of the implementation context and the applied implementation technique. COCOS supports different implementation techniques as the server model, activity threads, the integrated layer processing, and techniques for avoiding data copying within a protocol stack. The basic principles of COCOS and the applied mapping strategies were presented in [Lang99]. In this paper we describe the generation of implementations according to the Integrated Layer Processing technique which has not been reported so far.

The Integrated Layer Processing (ILP) is an efficient implementation technique for protocol stacks with complex data manipulation operations [Clar90]. It in particular aims at protocol implementations with high data throughput requirements such as multimedia transmissions. The Integrated Layer Processing is especially applied to implement applications which follow the Application Level Framing principle. The Application Level Framing (ALF) approach supports a layer-independent processing of the data units. It aims among others at an application oriented definition of the data units, so called application data units, e.g. video frames, and the avoidance of protocol functions like fragmentation and multiplexing. The idea of the Integrated Layer Processing is to reduce memory accesses which may become a performance bottleneck in high performance communication when handling complex data units. This is achieved by a combined execution of complex data manipulation operations (encryption, decryption, checksum calculation, syntax transformations) in one or two memory access cycles independently of the layer the protocols belong to.

In this paper we show how ILP implementations can be automatically derived from SDL specifications. In Section 2 we give a short introduction to the Integrated Layer Processing implementation technique. Section 3 gives a short overview of related work. Section 4 describes the applied mapping principle for SDL. Section 5 discusses semantic constraints. Section 6 sketches the code generation process. In Section 7 we present measurements to show the effect of the approach. The concluding remarks give an outlook on further required research steps.

2. PRINCIPLE OF INTEGRATED LAYER PROCESSING

The Integrated Layer Processing (ILP) represents an advanced protocol implementation technique. It was proposed by Clark and Tennenhouse [Clar90]. It is dedicated to the efficient implementation of data manipulation operations in protocols with a high data throughput. Typical examples for such data manipulation operations are the calculation of the checksum for error control, data encryption/decryption, data marshalling/unmarshalling and the compression/decompression of data streams. The approach is based on the observation that the execution of data manipulation operations over a protocol data unit (PDU) consumes more time for reading and writing of the data from and into the memory than for the proper execution of the operation. Since these data manipulation operations due to their functionality belong to different protocol layers many memory accesses are required in a traditional protocol stack implementation. Memory access times decrease at a slower rate compared to increase of the processor execution speed. For that reason memory access has become a bottleneck in high performance communication.

The Integrated Layer Processing aims at reducing the number of memory accesses by a combined execution of all data manipulation operations (DMOs) of a protocol stack. The DMOs are grouped into the so called ILP loop (see Figure 1). The DMOs can belong to different layers. They can even be located in different address spaces. The execution of the ILP loop does not influence the protocol execution, because the ILP loop combines operations which otherwise would be executed sequentially when the PDUs pass the different protocol layers. For execution, the DMOs are nested. They pass the data from one operation to the other whereby only fragments of the PDUs are processed each time. The DMOs aim at possibly small data fragments which fit into the register set of the processor. Thus only one read and one write operation per PDU are required for executing the ILP loop. The DMOs often use a different number of bytes as basis for their calculation. The calculation of the checksum is processed byte-wise, while an RSA encryption needs 16 byte blocks. To adjust the PDU fragments in an ILP loop word filters were proposed in [Abbo93] which concatenates bytes according to the length needed for the next DMO.

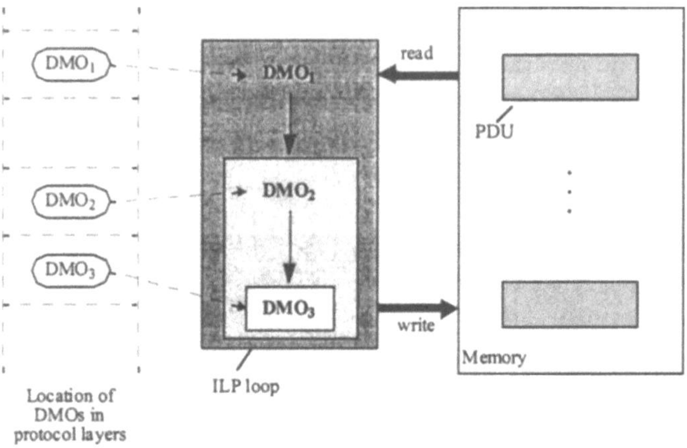

Figure 1: Principle of the Integrated Layer Processing

The ILP approach aims at the execution of the correct protocol paths, also called common paths [Leue96]. If an protocol error occurs in one of the protocols another execution path must be pursued to handle this error. In this case the results of the already executed DMOs of the higher layers are not needed any more and will be discarded. Low protocol error rates do not much influence the performance gain of the approach. If the error rate increases other implementation techniques will be more appropriate. We also discuss this effect later in this paper.

The Integrated Layer Processing can bring remarkable performance gains. For simple data manipulation operations like the calculation of the checksum, performance gains of up to 50 per cent were observed [Clar90], [Abbo93]. Further experiments revealed that the performance increase depends on the complexity of the data manipulation operations. It is less for more complex operations [Brau95]. In [Ahlg96] it was shown that ILP can even in part lead to efficiency losses compared to traditional implementation techniques if all data and state information of the data manipulation operation to execute cannot be stored in the register set of the processor. These data are moved to the cache or the memory. If the data are held in the cache the decrease may not be that strong but if the required information is neither found in the registers nor in the cache the performance loss may be dramatic. The deployment of the technique therefore requires a careful preparation to use its benefits.

3. RELATED WORK

The use of ILP has been investigated for formal description techniques (FDTs). For SDL, an approach for deriving the common paths was proposed in [Leue96], but no mechanisms for error recovery were discussed. The treatment of semantic constraints arising, for instance, from transitions with `save` statements also has not been solved. To our knowledge a compiler using ILP for one of the standardized FDTs has not been implemented yet. Automated derivations of ILP implementations from formal description techniques have been only reported for Esterel so far [Brau96], but the paper scarcely gives hints on the handling of semantic constraints. It is reported that the generated code achieves the efficiency of hand-coded implementations. Esterel though is more implementation oriented compared to the standardized FDTs. The problems for applying the technique to SDL arise from semantic constraints like the asynchronous communication between the processes. A practicable solution for SDL was proposed in [Twar00]. It is presented in the following.

4. MAPPING ON ILP IMPLEMENTATIONS

The mapping of SDL specifications on ILP implementations is not straight-forward. There are several problems to solve. First it is necessary to have an SDL specification of several protocol layers or of the whole protocol stack, respectively. We assume in the following that this specification is given. Furthermore, data manipulation operations are usually not entirely specified using SDL statements. In addition, SDL exchanges only complete signals between processes. The specification therefore does not provide sufficient support for a byte-wise exchange of the signals between DMOs. The compiler cannot always nest the DMOs. In order to use nevertheless the benefits of ILP implementations we distinguish two kinds of ILP loops:

- the simple ILP loop and
- the advanced ILP loop.

The loops differ in the nesting degree of the DMOs, but both variants lead to optimized implementations.

Simple ILP loop

In *simple ILP loops* the DMOs are not nested. Instead of this they are mapped into a common procedure and executed sequentially. Figure 2 shows the principle. The transitions which are involved in the loop remain preserved in the implementation. The PDUs of the different layers can be uniquely identified. Protocol sequences which cannot be mapped on this loop are implemented in the usual way. No additional measures are applied.

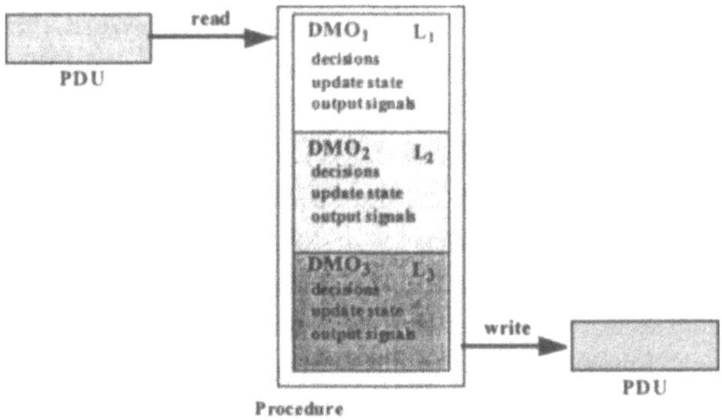

Figure 2: Structure of the simple ILP loop

The performance gain of simple ILP loops results from the avoidance of data copy operations between the DMOs. Besides no runtime support is needed for the process change. This is implicitly done by the ILP loop. The overhead needed for the signal exchange and the execution of the respective operation is avoided.

Advanced ILP loop

The advanced ILP loop fully implements the ILP concept as proposed in [Clar90]. The DMOs are nested and enforce a byte-wise processing of the data units. Because of this the PDUs of the different layers cannot be identified any more. All decisions concerning the correct transmission of the PDUs have to be delayed until the processing of the data unit is completed (see Figure 3). After that the decisions are evaluated. If the PDUs were correctly transmitted the respective protocol state is updated and the `output` statements are executed. For this, the transitions belonging to the loop must be restructured. This can be done automatically based on the following rules:

1. Move all decisions which influence the protocol flow behind the DMO execution,
2. Postpone outputs until these decisions are made,
3. Suppress all outputs if a protocol error occurs and trigger the error handling procedure,
4. Include word filters if the DMOs use different data formats (see below).

The performance gain of the *advanced ILP loop* is of course higher than that of the *simple loop* because only registers are used for the data exchange between DMOs.

Specification of ILP loops and word filters

Another important problem is the specification of the DMOs which belong to a certain ILP loop. The SDL specification does not contain any information on this. This information has to be included into the implementation specification which is the starting point for automated code generation. The implementation specification refines the protocol specification for the given target systems and includes all information for automatically deriving an implementation. The reasons for the use of an implementation specification are given in [Lang99], [Koen00].

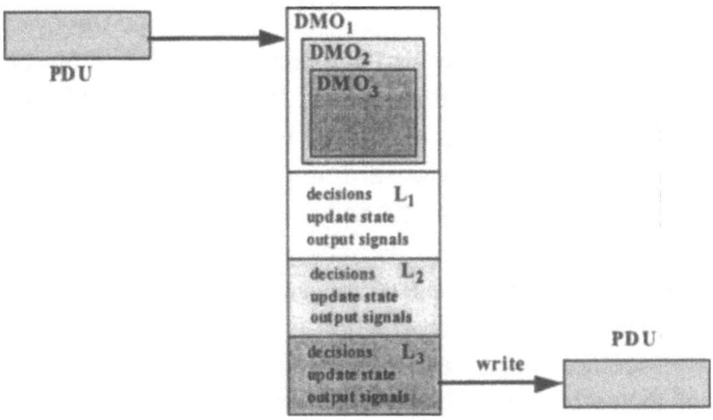

Figure 3: Structure of the advanced ILP loop

In our approach implementation oriented information is included by means of iSDL statements [Lang99]. iSDL is an implementation oriented annotation to control the code generation process. The statements are included as comments into the SDL specification and translated together with the specification. iSDL provides statements for specifying ILP loops and word filters.

An ILP loop is specified by means of the statements *define simple/advanced ILP loop*, e.g.

```
/*{$ define simple ILP loop L13 }*/
      . . .
   input sig1   /*${ belongs to L13}*/
      . . .
   input sig3   /*${ belongs to L13}*/
```

The statement indicates the transitions containing the DMOs in the order of their appearance. Note that the *belongs to* statement is attached as comment to the respective transition in the SDL specification.

A word filter for *advanced ILP loops* is defined as follows:

```
/*{$ define word filter 2_to_4 }*/.
```

The use of the filter is specified at places where it is required, e. g.

```
/*{$ P1 2_to_4 P2 }*/.
```

5. SEMANTIC CONSTRAINTS

There are two possibilities to implement the ILP loop execution: a separate and an integrated implementation. In the first variant the ILP loop is executed separately; the results are stored until they are needed for protocol execution. This is a straight-forward implementation of the concept. It, however, requires an additional memory access. From the point of view of automated code generation this variant is less advantageous, because it requires separating the DMOs and the actualization of state information. Therefore, the integrated implementation is preferred, in which the ILP loop is executed together with the protocol. This, however, requires that the protocols or the respective SDL processes are in states in which they accept the respective PDUs. There may be situations in which this is not given. The ILP loop cannot be executed then. Another problem is the detection of transmission errors. They cannot be detected before executing the loop. The handling of these situations is discussed in the following.

State errors

The execution of a PDU or a signal, respectively, may not be possible due to the current states of the SDL processes involved in the ILP loop. If a signal is executed then all processes must be in a state in which they await the PDUs contained in this signal. If these processes accept only signals exchanged within the loop and/or if they only possess one state, as supposed in [Leue96], the ILP loop can be always executed. The DMOs are processed consecutively. This is not allowed in situations when, for instance, a signal is saved by a `save` statement or when a signal is read from the input queue and discarded. Figure 4 depicts such a situation. The process *encrypt* may only encrypt video frames when it got a sending credit. Otherwise it waits in the state *wait_for_credit*. Incoming frames are saved until new credit is available. If the ILP loop would force the encryption of incoming frames then the frames will be lost because the receiver would not accept them without credit. Besides the state would change in DMO *encryption* although it should not change at all. This can influence the protocol execution. Similar problems appear when signals are discarded in a state. In such cases the implementation has to switch to the conventional protocol execution (see below).

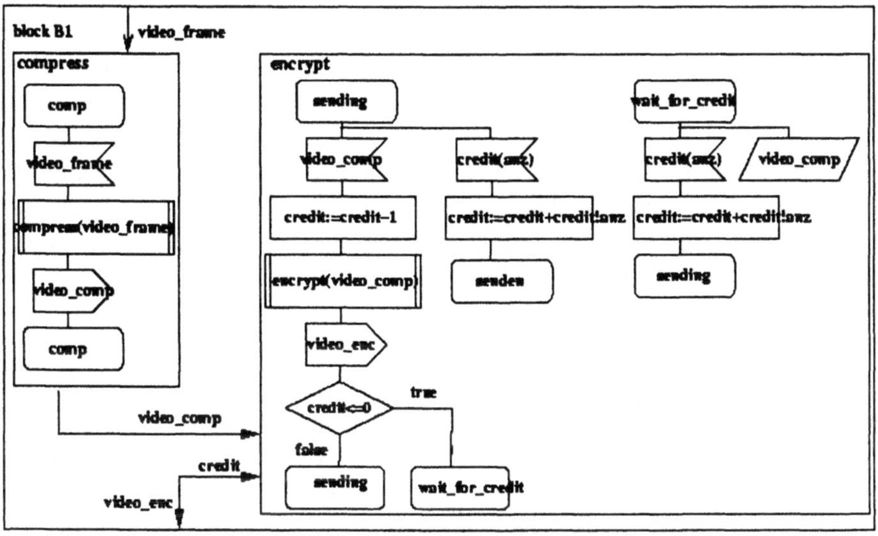

Figure 4: Refinement of the DMOs *compress* and *encrypt*

Transmission errors

Transmission errors can cause execution paths other than the common path the ILP loop aims at. In a *simple ILP loop* the error is indicated after proving the checksum, because the DMOs are processed PDU-wise. The execution can immediately branch off to another execution path to handle the error. In an *advanced ILP loop* with a byte-wise processing of the signals, errors are only detected after all DMOs are completed. At that point the signals are already converted into a representation which cannot used any more for error correction, e.g. for forward error control.

Solutions

In order to avoid these situations appropriate measures have to be taken. Mostly a double implementation is used. Another possibility is a recovery mechanism.

Double implementations mean that beside the ILP implementation a conventional implementation, either a server model or a activity thread implementation, is generated. Before the processing of the PDU the states of all protocols are checked. If a protocol is not in the appropriate state then the implementation switches to the conventional execution path. A special treatment is required for handling transmission errors when applying the *advanced ILP loop*. To ensure the correct protocol behaviour the PDU and all state information (variables, timers, states, sequence numbers) have to be stored before the ILP loop execution. When a transmission error is indicated the PDU and the original protocol state have to be restored before the conventional execution starts. Thus the signal is processed a second time to determine the kind of the error and to react appropriately.

Recovery mechanisms are based on log files and reset points. The actualization requires rather large efforts. Since transmission errors in fibre optics network with error rates of 10^{-15} are very rare the efforts for recovery mechanisms are not justified. In particular for automated code generation double implementations should be preferred, because all required information already exists and can be used by the code generator. Recovery mechanisms have to be supplemented manually in the implementation.

Overtaking of signals

An overtaking of signals in SDL can only take place if two or more signals are sent over the same `signalroute-channel` connection. ILP loops do not cause such a behaviour. All signals are sequentially executed within an ILP loop so that the order of the signals is preserved. However, double implementations, i.e. the combination of ILP with server model or activity thread implementations, can lead to signal overtaking. This can be avoided by applying mechanisms like the transition reordering described in [Lang99].

6. CODE GENERATION PROCESS

The generation of ILP implementations in COCOS is principally the same as for the other implementation techniques. It comprises two steps. First the implementation specification, i.e. the SDL specification together with the iSDL annotation, is transformed into an *intermediated representation* which consists of connected lists of the processes, of the states and of the transitions. The intermediate representation is used for various checks and optimizations [Lang99]. It is also the input for the code generation process which is carried out by the *code synthesis* component (see Figure 5). The *synthesizer* inspects the intermediate format to identify the iSDL statements which control the code generation by indicating the mapping strategy and the respective parameters. The selected implementation model identifies the respective code segments for inclusion into the generated code sequences. The *code segments* contain the implementations of the SDL statements in the target language, currently in C. They may be different for certain statements, e.g. the input statement, due to the different implementation techniques. The code segments can be parametrized. The actual parameters are included during code synthesis, e.g. the signal types for each `output` statement. The *synthesizer* further accesses the *code repository*. It contains user implemented functions such as word filters for the *advanced ILP loop* as well as segments for inlining manual coded sequences. The *code synthesis component* also determines parameters for the initialization functions of the runtime environment, e.g. the maximum length of the common input for all SDL processes. If an ILP implementation is supposed to be generated then the ILP loop is integrated either in a server model or an activity thread implementation as shown in Figure 5.

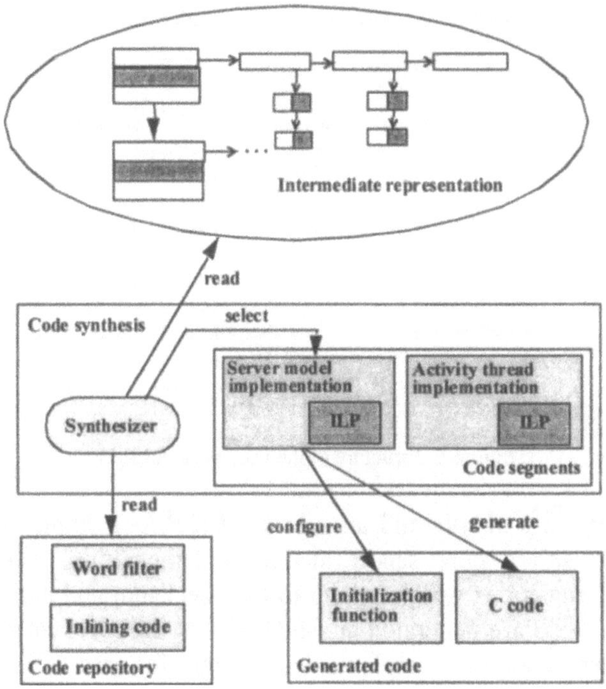

Figure 5: Code generation in COCOS

7. MEASUREMENTS

In order to evaluate the performance of the transformation we used the protocol stack depicted in Figure 6. The protocol stack is based on the XDT (*eXample Data Transfer*) protocol [Koen97]. It is an example protocol used for teaching protocol engineering. XDT is a connection-oriented data transfer protocol which applies the *go back N* error correction principle. For ILP, some data manipulation operations were added. They are located in the application layer as shown in Figure 6. Two DMOs are applied, one for calculating the checksum and one for coding/decoding the PDUs. The coding DMO applied for this experiment is very simple. It only byte-wise changes the payload data of the PDUs. The SDL/PR specification of the whole protocol stack has 991 lines. This corresponds 21 Kbyte.

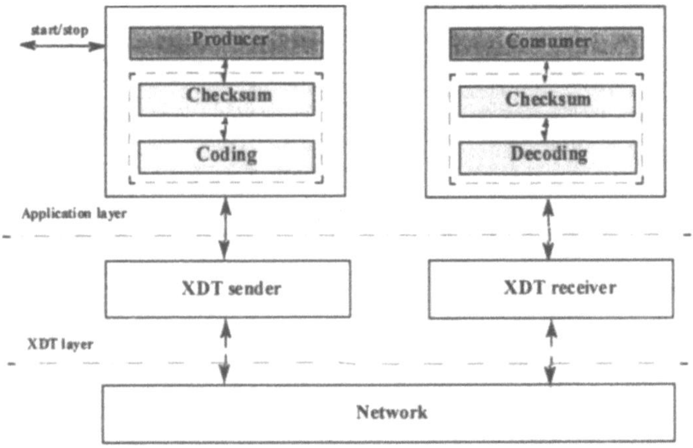

Figure 6: Structure of the used protocol stack

We generated a *simple* and an *advanced ILP loop* implementation and compared it with a pure server model implementation. The ILP loops comprise all transitions which belong to the checksum and coding/decoding DMOs. The loops are generated at sender and at the receiver side. The rest of the implementation is implemented straight-forward according to the server model [Lang99]. The measurement distance is the time between the sending of a PDU and the reception of the acknowledgment at sender side. It is marked by the signals *start* and *stop* in Figure 6. In each measurement cycle 10000 PDUs with a payload of 25 bytes were transferred. The measurements were repeated 50 times. Figure 7 shows the results the measurements for 3 workstations with varying performance parameters. Workstation A is a HYPER SPARC station with one processor (140 MHz) and 512 Kbyte cache, workstation B is a SPARC 20 with four processors

(125 MHz) and a cache of 256 Kbyte, whereas workstation C represents a ULTRA SPARC station with two processors and a cache of 2 Mbyte.

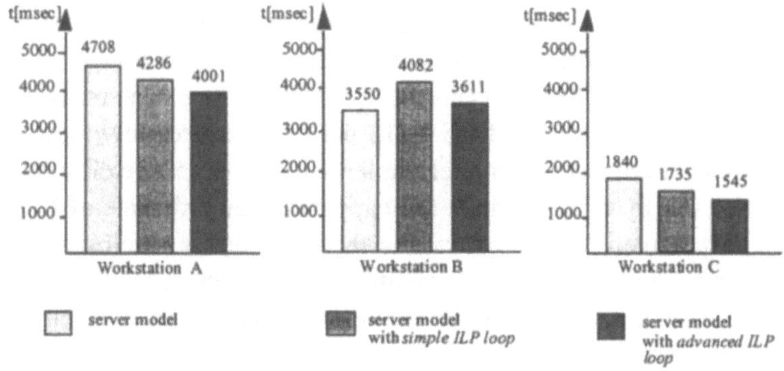

Figure 7: Measurement results

A comparison of the measurements at workstations A and C show that *simple loop* implementations bring a performance increase of about 5 to 9 per cent, while the *advanced ILP loop* implementations still add 7 to 11 to a total gain of almost 20 per cent. This shows that ILP may significantly improve the performance of protocol stack implementations. We further have to take the fact into account that our configuration is pretty simple. The two DMOs are adjacent and located in the same layer. Thus the cache reduces memory accesses. Several DMOs in different layers would more strongly demonstrate the effect of the approach. The phenomenon of workstation B that the server model implementation was about 1.8 per cent faster can be explained by the small cache of that workstation. The variation of the PDU execution time (measured in a separate experiment) was considerably higher at workstation B than for the other ones. This explains the fact that needed data and state information are not any more available in the cache (see also Section 2). When using larger PDUs of 200 byte payload this effect disappears (at least for low transmission error rates), because there are more memory accesses in the server model implementation (see below).

Another important aspect in this context is the influence of the error rate. Transmission errors force ILP loop results to be discarded and to start an error recovery procedure or continue with the conventional implementation. In order to determine the influence of transmission errors on the efficiency of ILP implementations we investigated the efficiency loss depending on the error rate. For this, we changed the order of the DMOs at receiver side, i.e. the checksum was calculated before the PDUs were decoded. Due to the combined execution of the operations all data are also decoded in error case. Unlike a non-ILP implementation the conventional implementation must be re-established before error handling. We compared the modified *advanced ILP loop* implementation with a server model and an activity thread implementation using various error rates. This was done by falsifying a corresponding number of signals during transmission. The measurements were carried out at workstation B with a signal payload of 200 bytes and 10000 transmissions. The results are depicted in Figure 8. The measurements show that from an error rate of 6.5 or 10 per cent, respectively, upwards the activity thread and the server model implementation are more efficient. With a payload of 100 bytes this change only appears at an error rate of 13.5 or 19 per cent, respectively. The error rate is therefore not a conclusive measure for the performance of ILP implementations. The latter depends on the point where the execution stops when an error occurs and the number of operations to be repeated.

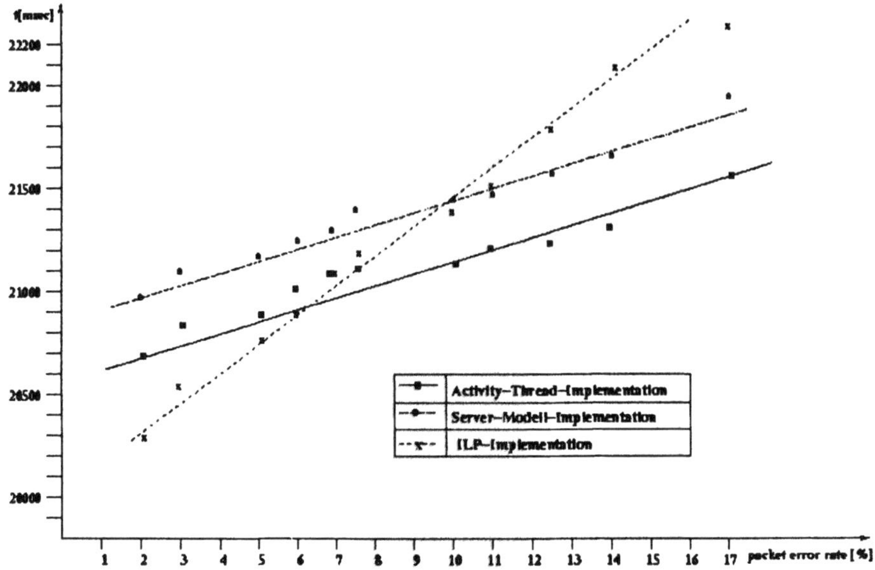

Figure 8: Influence of transmission errors on the efficiency of ILP implementations

The same experiment was also carried out with the *Cadvanced* code generator of the SDT tool version 3.2 [Tele98]. The execution times for error rates of 0 and 10 per cent were 65196 msec and 65342 msec, respectively, i.e. a performance gain of about 300 per cent could be achieved by using ILP compared to conventional, commercially available code generation tools.

8. FINAL REMARKS

In this paper we have presented a mapping approach for the automated derivation of ILP implementations from SDL specifications. We have discussed the semantic constraints and the code generation process. The measurements confirmed the performance increase reported from manual implementations. In addition we compared the *advanced ILP loop* implementation described in Section 6 with a manually coded server model implementation of the XDT protocol stack under the same conditions (200 byte payload, 10000 measurements). It showed that the manual implementation is only by a factor of 2.5 faster. Although this is only a conditional comparison, it points out that the integration of advanced manual implementation techniques like Integrated Layer Processing, activity threads or the offset technique [Lang99] may considerably improve the efficiency of automated code generation from formal descriptions.

Further research is needed to refine the techniques and to reveal places which slow down the execution speed of the generated code. One of such points is the coding/decoding of PDUs. Automated implementations of these procedures are often less efficient than manually coded ones. An inlining of manual codings in the automated code generation process might be appropriate support in this case. Moreover, it pointed out that the decision which implementation technique is selected is not straight-forward. Several experiments proved that the possible performance increase depends on the structure of the specification and on the target hardware. The better the specification supports the used implementation technique the better is the performance gain to be expected. The experiments further revealed that the different implementation techniques are not orthogonal to each other. The combination of the techniques does not bring a linear performance increase, since some optimizations are implicitly contained in other techniques. Besides the effect of certain techniques can be also achieved by combining of techniques, e.g. activity threads in combination with the offset technique can achieve similar performance parameters as ILP implementations. The selection of the appropriate techniques therefore requires a careful analysis of the specification and the implementation context. Performance predictions based on the implementation specification are a valuable means to support this process. Such a component is currently being integrated into the COCOS compiler.

An issue for a broad application of automated implementation techniques as well as other FDT based methods is the lack of complex formal specifications of real-life protocols. If larger specifications are available they are often not fully specified, especially PDU coding and decoding operations. The completion of the specification is still a time consuming process which takes several man-months. This mostly cannot be done in an academic environment. It hinders a realistic comparison of automatically generated protocol implementations with manually coded ones in a large scale. The availability of such specifications is also decisive prerequisite for a broad and thorough application of FDT based design, validation and implementation technologies.

REFERENCES

[Abbo93]Abbott, M.; Peterson, L.: Increasing Network Throughput by Integrated Protocol Layers. IEEE/ACM Transaction on Networking 1(1993)5

[Ahlg96] Ahlgren, B.; Gunningberg, P.G.; Moldeklev, K.: Increasing Communication Performance with a Minimal-Copy Data Path Supporting ILP and ALF. Journal of High Speed Network 5(1996) 2, 203-214

[Brau95] Braun, T.; Diot, C.: Protocol Implementation Using Integrated Layer Processing. Proc. ACM SIGCOMM, 1995.

[Brau96] Braun, T. et al.: ALFred - An ALF/ILP Protocol Compiler for Distributed Application Automated Design. Rapp. de Recherche No. 2786, INRIA, 1996.

[Clar90] Clark, D. D.; Tennenhouse, D. L.: Architectural Considerations for a New Generation of Protocols. Proc. ACM SIGCOMM '90, pp. 200 - 208.

[Koen97]Koenig, H.: eXample Data Transfer Protocol (XDT). Technical Report I-18/1997, BTU Cottbus, Department of Computer Science, 1997

[Koen00]Koenig, H.; Langendoerfer, P.; Krumm, H.: Improving the Efficiency of Automated Protocol Implementations Using a Configurable FDT Compiler. Computer Communications 23 (2000) 12, 1179-1195

[Lang97] Langendoerfer, P.; König, H.: Increasing the Efficiency of automatically generated code using implementation specific annotations. Proc. 3rd Workshop High Performance Protocol Architectures (HIPPARCH), P. Gunningberg, C. Huitema (eds.), Uppsalla, 1997.

[Lang99] Langendoerfer, P.; Koenig, H.: Deriving Activity Thread Implementations from Formal Descriptions Using Transition Reordering. In Wu, J.; Chanson, S.T.; Gao, Q. (eds.): Formal Methods for Protocol Engineering and Distributed Systems. Kluwer Academic Publishers, 1999, pp. 169 - 184.

[Lang01] Langendoerfer, P.; Koenig, H.; Kraemer, R.: Evaluation of well-known Implementation Techniques for Application in Mobile Networks. Accepted for Special Issue of the Journal of Supercomputing, Kluwer Academic Publishers

[Leue96] Leue, S:, Oechslin, P.: On Parallelizing and Optimizing the Implementation of Communication Protocols. IEEE/ACM Trans. on Networking 4 (1996) 1

[Tele98] Telelogic Malmoe AB: SDT 3.4 User's Guide. SDT 3.4 Reference Manual, 1998

[Twar00] Twarok, S.: Extension of the COCOS code generators for deriving ILP implementations. Master thesis, BTU Cottbus, Department of Computer Scvience, 2000 (in German)

STEPWISE DESIGN WITH MESSAGE SEQUENCE CHARTS*

Ferhat Khendek[1], Stephan Bourduas[1], Daniel Vincent[2]

[1]Department of Electrical and Computer Engineering, Concordia University
1455, de Maisonnneuve W., Montréal (P.Q.), Canada H3G 1M8
E-mail: {khendek, s_bourdu}@ece.concordia.ca
[2]France Telecom R & D, Lannion, France

Abstract Use cases are useful in various stages of the software process. They are very often described using text that has to be interpreted by system designers. This could lead to implementation errors. Another drawback of using such informal notations is that automating the process of moving from use cases to design specification is difficult, if not impossible. It would be beneficial to represent use cases in an unambiguous way, thereby reducing the probability of misunderstanding and allowing for automation of various activities in the software process. Message Sequence Charts (MSC) is a formal language and widely used in telecommunications for the specification of the required behaviors. In this paper, we use MSC for describing use cases and we propose an approach for stepwise refinement from high-level use cases in MSC to design MSCs that contain more details about the internal components of the system and their interactions. The refinement steps are done by the designer and guided by the system architecture. For each step, the newly obtained MSC is validated automatically against the previous MSC using a conformance relation between MSCs.

1. INTRODUCTION

Distributed software systems, like any software system, go through requirement, design, implementation and testing phases. Ensuring the quality of such software systems from the initial stages is a challenging task.

UML use cases are becoming the standard form for requirements specification. An UML use case describes some functionality offered by a system as perceived by the user or an external actor of the system [1, 2, 9]. The user sees the system as a black box that responds to inputs with a specified output. Use cases are very often specified using a combination of text and diagrams that must be interpreted by the system designers and translated into a more concrete representation. It would be beneficial to represent use cases in an unambiguous way from the start, thereby reducing the probability of misunderstanding, and enabling the use of tools. MSC [3, 4, 11] is an excellent candidate as discussed in [10].

* Partially supported by France Telecom R&D.

MSC and SDL (Specification and Description Language) [5, 6] are widely used for telecommunication software engineering. In a previous work [7, 8] we have developed an approach and a tool for generating SDL specifications from a MSC and a given architecture. The given MSC is seen as a design MSC where the internal behavior of the components of the system is given. In this paper, we introduce a new approach for refining use cases specified with MSC into design MSC in a stepwise manner. The resulting MSC is then used as input for our MSC to SDL translation approach.

In our approach, a use case model is developed through interactions with the system designers and the customers as described in [1, 2, 9]. The result of this represents the functional requirements of the system under construction. Once the use cases have been agreed on, the designers must specify the architecture of the system. Guided by this architecture, the use cases are then refined in a stepwise manner. We distinguish between horizontal and vertical refinements. Horizontal refinement consists of adding new messages or actions to existing MSC axes. Vertical refinement consists of decomposing an axis into at least two other axes following the architecture of the system. The enriched MSC must conform to the previous (or parent) MSC. We therefore define a conformance relation between MSCs that is used to validate refinement steps made by the designer.

The rest of this paper is organized as follows. In Section 2, we discuss the modeling of use cases with MSC. Section 3 introduces the notion of conformance between MSCs and discusses the related work. In Section 4, Subsection 4.1 describes our stepwise refinement approach for bMSCs, while Subsection 4.2 extends this approach to HMSCs. In Section 5, we apply our approach to the ATM (Automatic Teller Machine) example, before concluding in Section 6.

2. DESCRIBING USE CASES WITH MSC

Jacobson introduced the concept of use cases in the Object-Oriented Software Engineering (OOSE) method [9]. Use cases are now part of the UML standard. A use case is an abstract description of some desired functionality provided to the user of a system. The use case description sees the system as a black box that generates some output for a given input. The fact that they are not formalized allows for flexibility in the early stages of development. This same advantage can turn into a liability if inexperienced designers are expected to interpret the use cases and then produce a design specification. In [10], the authors propose a method for formalizing use cases using MSCs. We follow this approach for the modeling of use cases.

2.1 bMSC Use Cases

bMSCs are simple diagrams that capture the interactions between system components, and between system components and the environment. For use case modeling, bMSCs are used to show the interactions between the system and the environment only as illustrated in Figure 1. The user sees a "black box" represented by a single process instance in the MSC. The axis represents the system boundary that interacts with the user. In this example, the user sends message "a" to the system, and the system reacts by sending message "b" to the user.

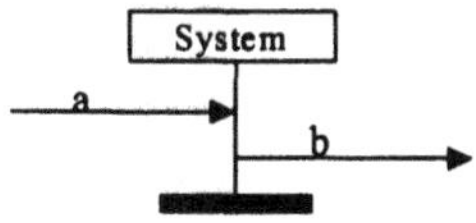

Figure 1. A bMSC use case.

2.2 HMSC Use Cases

HMSC allows for a more abstract view of the system [10, 11]. HMSCs improve the readability of the system by hiding low-level details and showing graphically how a set of MSCs can be combined. They are viewed as roadmaps composed of MSCs using sequential, alternative and parallel operators. HMSCs are seen as directed graphs where the nodes are: start symbol, end symbol, MSC reference, a condition, a connection point, or a parallel frame.

For simple use cases, a single bMSC is sufficient to show the behavior, but complex behaviors such as conditional executions or alternatives can lead to large and unreadable bMSCs. For these complex use cases, it is easier to use several bMSCs and then combine them using a HMSC use case.

A use case describes one possible usage of a system. Most systems will have many functions that will result in many use cases. In order to be able to organize many use cases and to allow a designer to view and navigate them with ease, a HMSC that references the HMSC use cases could be used to show an even higher level of abstraction. Using three level of abstraction allows us to specify the use cases for an entire system. Figure 2 shows how the various levels relate to each other.

The system level expresses the functional view, as a set of use cases, of the system by using a top-level HMSC. The structure level describes each use case using one HMSC without going into details. The basic level shows the interactions between the system and the environment using bMSCs. A bMSC at that level cannot refer to a HMSC as allowed in the standard [3]. A bMSC at the basic level does not contain alternatives or optional behaviors. Any alternative or optional behavior can be represented using HMSC at the structure level. In this paper, we restrain the HMSC language to the weak sequencing operator, alternative operator and iteration. The parallel operator is not taken into consideration.

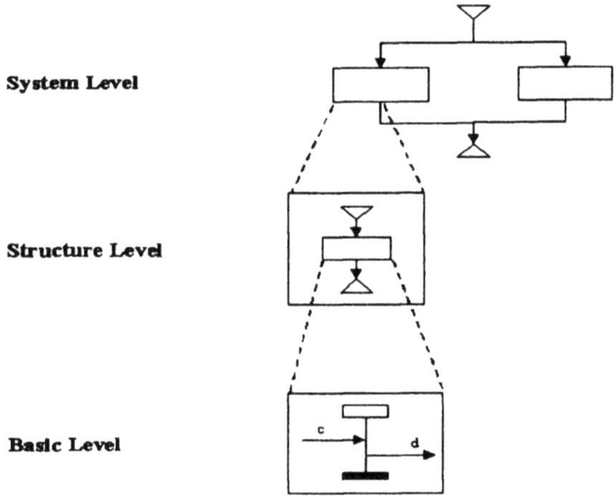

Figure 2. Three levels of abstraction for use case specification.

3. CONFORMANCE RELATION FOR MSC

In this section, we first introduce the definition of conformance between bMSCs, and then we extend this definition to HMSC.

3.1 bMSC

The bMSC refinement process consists of several steps. We refer by M_k to the bMSC obtained at step k. In order to preserve, throughout the refinement process, the semantics of the original use case defined by the user, each newly obtained bMSC should preserve the semantics of its parent bMSC. In other words, bMSC M_n must preserve the behavior described by bMSC M_{n-1}. M_n must have the events of M_{n-1} and for these events M_n preserves the orders defined in M_{n-1}. The new messages and events introduced in M_n are not constrained. Informally, we say that a bMSC M_2 preserves the behavior of (or *conforms to*) a bMSC M_1, if and only if for each axis A_1 in M_1 there is a corresponding set of axes $\{A_{21}, A_{22}, \ldots, A_{2n}\}$ in M_2 with all events of A_1 included in $\{A_{21}, \ldots, A_{2n}\}$, and all the orders defined between events in M_1 are preserved in M_2.

For a formal definition of the conformance relation, we need a formal definition of a bMSC. For this, we follow the definitions in [12, 13].

Definition 1 (bMSC). A bMSC is a tuple $\langle V, \ll, P, M, L, T, N, m \rangle$, where

- V is a finite set of events,
- $\ll \subseteq V \times V$: is a transitive and acyclic relation,
- P is a set of processes,
- M is a set of message names,
- L: V → P is a mapping that associates each event with a process,
- T: V → {send, receive, local} defines the types of each event as a send, receive or local,
- N: V → M maps every event to a name
- m: a partial function that pairs up send and receive events.

The relation $\ll$ is defined between the sending event (s_m) and the receiving event (r_m) of message "m" as $s_m \ll r_m$, and between events e_1 and e_2 in the same axis, $e_1 \ll e_2$, if e_1 appears before e_2.

Definition 2 (Conformance for bMSCs). A *bMSC M2* = $\langle V_2, \ll_2, P_2, M_2, L_2, T_2, N_2, m_2 \rangle$ *conforms to a bMSC M1* = $\langle V_1, \ll_1, P_1, M_1, L_1, T_1, N_1, m_1 \rangle$, if and only if there exist an injective mapping Γ: $V_1 \rightarrow V_2$ and a surjective function ϕ: $P_2 \rightarrow P_1$ such that:

- $L_1(e) = \phi(L_2(\Gamma(e)))$
- $T_1(e) = T_2(\Gamma(e))$
- $N_1(e) = N_2(\Gamma(e))$
- if $m_1(e) = f$ then $m_2(\Gamma(e)) = \Gamma(f)$
- if $e \ll_1 f$ then $\Gamma(e) \ll_2 \Gamma(f)$

Our conformance relation is similar to the matching relation defined in [12, 13]. However, there are two differences. The first one is related to the MSC semantics. In fact, we do not distinguish between the visual order and the enforced order as it is done in [12, 13]. The visual order being the transitive and reflexive closure $\ll^*$ of $\ll$, while the enforced order depends on the communication architecture where some visual orders may not hold, because of race conditions [14]. In the formal semantics of MSC (for bMSC [15]), the visual order has to be enforced. The question "if an architecture allows for that order or not" is another issue. The second difference as mentioned earlier in our conformance relation, an axis in bMSC Mn may correspond to a set of axes in bMSC Mn+1. In other words, we allow for one-to-many relationship between axes, contrary of the one-to-one relation in [12, 13].

Mauw and Reniers have proposed a refinement relation for interworkings [16]. This refinement consists of decomposing an instance into its constituents and adding internal messages between these constituents. Beside the semantic issues, our conformance relation can be seen as a combination of the refinement relation in [16] and the matching relation in [12, 13]. In this paper, the terms "refinement" and "conformance" are not synonymous. With the term "refinement" we only mean developing further a MSC, without automatically ensuring conformance.

For the illustration of the conformance relation, let us consider the bMSCs in Figure 3. A mapping Γ between M1 and M2 that associates each event to itself and a function ϕ that associates each process to itself satisfy the conditions in Definition 2. bMSC M2 conforms to bMSC M1. We have more messages and events in M2, but these events preserve the orders defined in M1. The bMSC M3 also conforms to M1. In this case, the function ϕ associates P21 and P22 to P2. In the case of bMSC M4, we cannot find a mapping of events that preserves the orders defined in M1. In fact, reception of message "x" by P21 and reception of "y" by P22 are not ordered as specified in P2 in M1.

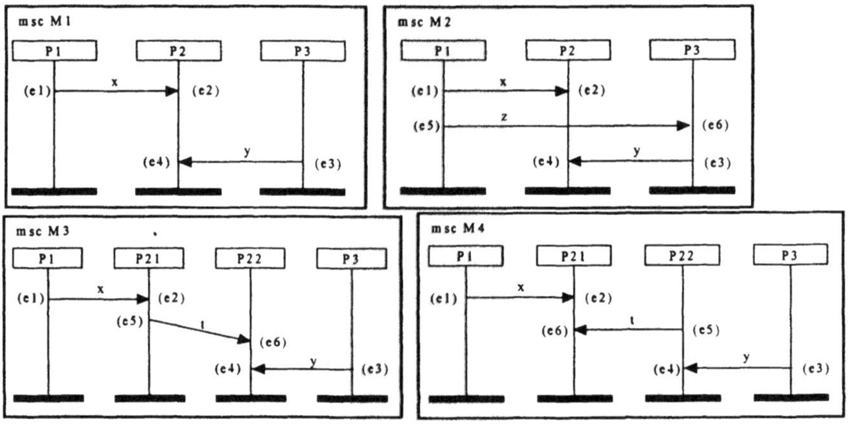

Figure 3. Examples of conformance and non-conformance between bMSCs.

3.2 HMSC

In order to build complex behaviors from simple behaviors, the MSC standard [3] has defined weak sequential, iteration, alternative and parallel compositions of MSCs. Informally, weak composition of two bMSCs can be seen as a concatenation of the axes of the common processes in the bMSCs.

Definition 3 (Weak composition of bMSCs). Given two bMSCs, $M1 = \langle V_1, \ll_1, P_1, M_1, L_1, T_1, N_1, m_1 \rangle$ and $M2 = \langle V_2, \ll_2, P_2, M_2, L_2, T_2, N_2, m_2 \rangle$, with $V_1 \cap V_2 = \varnothing$, $M1 \circ M2 = \langle V_1 \cup V_2, \ll, P_1 \cup P_2, M_1 \cup M_2, L_1 \cup L_2, T_1 \cup T_2, N_1 \cup N_2, m_1 \cup m_2 \rangle$, where $\ll \; = \; \ll_1 \cup \ll_2 \cup \{(e_1, e_2)$, such that $L_1(e_1) = L_2(e_2)$, $e_1 \in V_1$ and $e_2 \in V_2\}$.

From the syntactical point of view, a HMSC defines a roadmap where MSCs are combined using weak sequential, iteration, alternative and parallel compositions. As mentioned in Section 2, we do not take into account the parallel operator in this paper. From a semantic point of view, we define a HMSC as a potentially infinite set of alternatives of (infinite) weak sequential composition of bMSCs.

Definition 4 (HMSC). A HMSC H1 is a set of sequences seq_i (or bMSCs), with $seq_i = Mi1oMi2oMi3o\ldots oMiq$, where Mij is a bMSC, for j = 1, ..., q and i = 1, ..., n.

The set and the sequences could be infinite. The iteration is defined as the repetition, using weak sequential composition, of the same bMSC.

Informally, we say that a HMSC H1 conforms to HMSC H2, if and only if for each alternative bMSC M1 in H1 there is a bMSC M2 in H2, such that M2 conforms to M1. M2 preserves the behavior of M1. H2 may contain more alternatives than H1, but each alternative of H1 is preserved in at least one alternative of H2.

Definition 5 (Conformance for HMSCs). Given two HMSCs, H1 = $\{seq1_i, i = 1, \ldots, n\}$ and H2 = $\{seq2_j, j = 1, \ldots, m\}$, *H2 conforms to H1*, if and only if for each $seq1_i$ in H1 there exist $seq2_j$ in H2 such that *$seq2_j$ conforms to $seq1_i$*.

The conformance between $seq1_i$ and $seq2_j$ is given in Definition 2. Both conformance relations (for bMSCs and HMSCs) are transitive and reflexive.

4. FROM USE CASES TO DESIGN SPECIFICATION

Typically, designers will use their experience to generate design specifications from textually represented use cases. This method can lead to errors in implementation if the use cases are unclear. How these types of errors can be avoided is the goal of our approach. It enables a designer to specify an initial use case in MSC and refine it incrementally into a MSC design specification. The system architecture plays an important role in the stepwise refinement of MSC use cases. A refined MSC must conform to its parent MSC and follow the underlying system architecture. This system architecture can be represented using either UML or SDL. Our methodology for stepwise refinement of use cases is to be used in conjunction with existing tools for translation of a UML architecture into a SDL architecture [17], and the generation of a SDL specification from a given target architecture and MSCs [8].

4.1 bMSC Refinement Methodology

During the refinement process, we distinguish between vertical and horizontal refinements. Vertical refinement reflects the architectural decisions, while horizontal refinement allows to enrich the behavior of a bMSC. Both types of refinements are used together to incrementally refine a use case bMSC into a design specification. Figure 4 illustrates how both types of refinements are combined to reach a design specification.

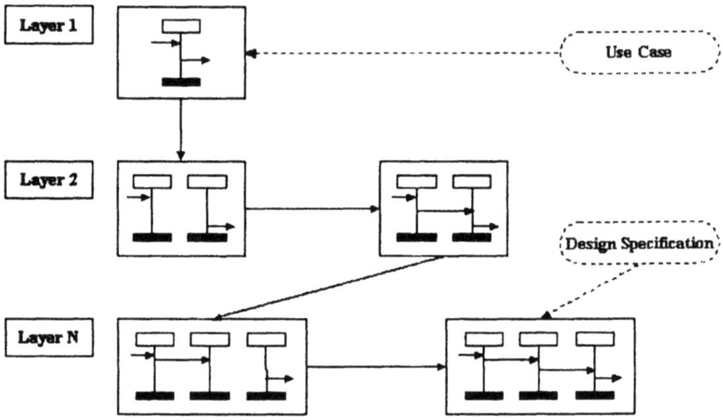

Figure 4. Overall refinement methodology.

The refinement process is dependent on the system architecture. The amount of layers used in the architecture limit the number of vertical refinements. There can only be one vertical refinement per layer. Each vertical refinement step splits a bMSC instance into several instances according to the architecture. The instances that are not decomposed keep the same names. When an instance is decomposed the designer has to distribute its events among the new instances. For each message "m", the names of the associated events (sending and receiving) are kept unchanged. The messages and events associated with instances that are not refined are kept unchanged and re-generated automatically. In general, a combination of vertical and at least one horizontal refinement is needed in order to generate a MSC that may conform to the previous level.

Horizontal refinement is concerned with adding messages, events and local actions to the bMSC. The designer can add new messages, sending and receiving events, as well as local actions. The messages that can be used in a horizontal refinement are specified in the architecture. A designer cannot introduce messages that are not specified in the architecture at the current level of refinement. Unlike vertical refinement, there is no limit to the number of horizontal refinements that can be performed at any given layer.

After each horizontal refinement, the designer can check automatically whether the refined bMSC conforms to the previous one. This verification of the conformance relation between bMSCs is implemented with the Event Order Tables (EOT) introduced in [7, 8]. As mentioned earlier, the bMSCs used in this paper do not contain alternative or optional behaviors.

Event Order Tables (EOT)

An EOT for a bMSC is a matrix that shows precedence relationships between events of the MSC. Figure 5 shows a bMSC and its corresponding EOT. Each message between two instances corresponds to two events: sending and receiving. Once the events on a MSC have been labeled appropriately, the EOT can be constructed. Following the MSC semantics and the relation <<, we use two rules to generate the EOT [8]:

- Each instance is totally ordered (except for co-regions where there is no order between events).
- A reception event happens after its matching sending event.

Inspection of each process axis in Figure 5 reveals that the first rule must be applied to P2, yielding the relation $e_2 << e_3$. Applying the second rule to the MSC of Figure 5 yields $\{e_1 << e_2, e_3 << e_4\}$. The transitive closure $<<^*$ of << allows for the construction of the EOT. For instance, $e_1 << e_2$ and $e_2 << e_3$ implies $e_1 << e_3$. A cell in the EOT is set to true if the row event occurs before the column event.

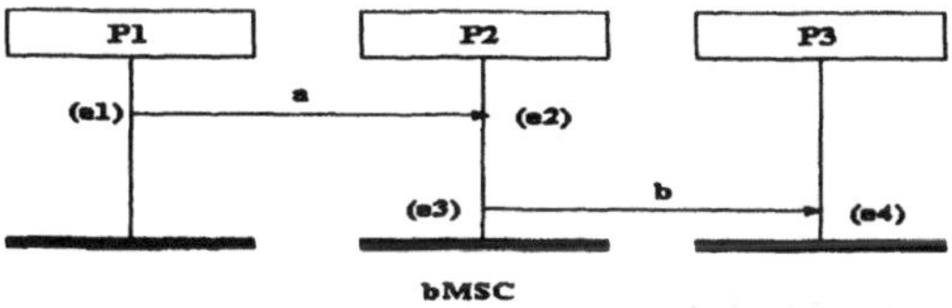

	e1	e2	e3	e4
e1		T	T	T
e2			T	T
e3				T
e4				

Event Order Table

Figure 5. A bMSC and its EOT.

Definition 6 (Inclusion for EOTs). We say that *EOT T_1 is included in EOT T_2*, if and only if

- all the events of T_1 are in T_2, and
- for each pair of events (e_i, e_j) if $e_i << e_j$ in T_1 then $e_i << e_j$ in T_2.

Provided the relation, enforced by the bMSC refinement approach (and tool), between instances and events in bMSC M_{k+1} and bMSC M_k, we say that M_{k+1} conforms to M_k if and only if EOT of M_k is included in EOT of M_{k+1}.

Proposition 1 (Conformance of bMSCs using EOTs). Given bMSCs, M_k and M_{k+1}, provided that events in M_k are mapped into themselves in M_{k+1}, M_{k+1} conforms to M_k if and only if EOT of M_k is included in the EOT of M_{k+1}.

Proof. The proof is straightforward. Our refinement approach maps events in M_k into themselves in M_{k+1}. The orders defined in M_k are preserved in M_{k+1} if and only if the EOT of M_k is included into the EOT of M_{k+1}. **[End]**

Figure 6 shows a bMSC that conforms to the bMSC in Figure 5, the message "c" sent from P2 to P3 has been added. In order to preserve the original event labels, the event labels e5 and e6 have been used for the new events.

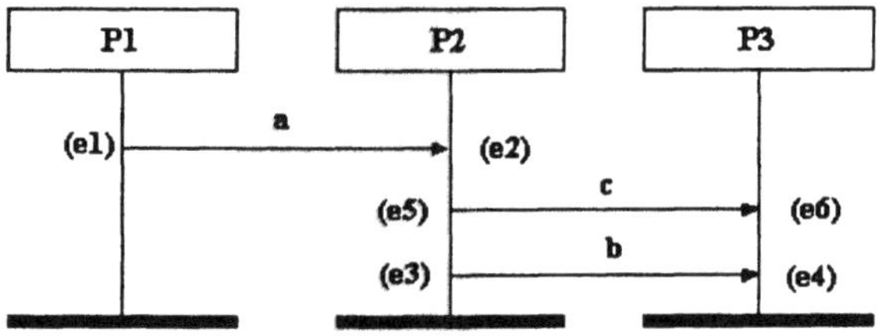

Refined bMSC

	e1	**e2**	**e3**	**e4**	**e5**	**e6**
e1		**T**	**T**	**T**	T	T
e2			**T**	**T**	T	T
e3				**T**		
e4						
e5			T			T
e6				T		

Event Order Table

Figure 6. Example of conformance verification using EOTs.

The EOT in Figure 5 specifies an ordering of events that is maintained by the refined bMSC. In fact, the EOT in Figure 5 is included in the EOT in Figure 6. The old relations are bolded in Figure 6; as long as these are not violated, the new bMSC conforms to its parent bMSC.

4.2 Stepwise Refinement of HMSCs

A HMSC is composed of a number of bMSCs following a certain roadmap. To enrich a use case HMSC, we keep the same roadmap and we refine the bMSCs. A HMSC is refined according to the following rules:

1. Keep the roadmap unchanged.
2. For vertical refinement, all bMSCs are vertically refined at the same time so that each bMSC has the same number of instances.
3. For horizontal refinement, we refine horizontally the referenced bMSCs. Each bMSC can be refined horizontally independently of the others.

Vertical refinements are done automatically according to the system architecture and the current level of abstraction. The user refines bMSCs horizontally using the method described previously and messages described in the architecture at the current level of abstraction. Unfortunately, the conformance of the refined bMSCs to their corresponding bMSCs does not lead automatically into the conformance of the refined HMSC to the original HMSC. In fact, orders between events in a given axis may not hold anymore when the axis is decomposed and the events distributed among the refining axes as shown in Figure 7. Indeed, in the HMSC M1 o M2, the sending of "x" always precedes the sending of "y", but this not the case in the HMSC M1' o M2', because of the distribution of these two events. In this example, M1' conforms to M1 and M2' conforms to M2, but M1' o M2' does not conform to M1 o M2.

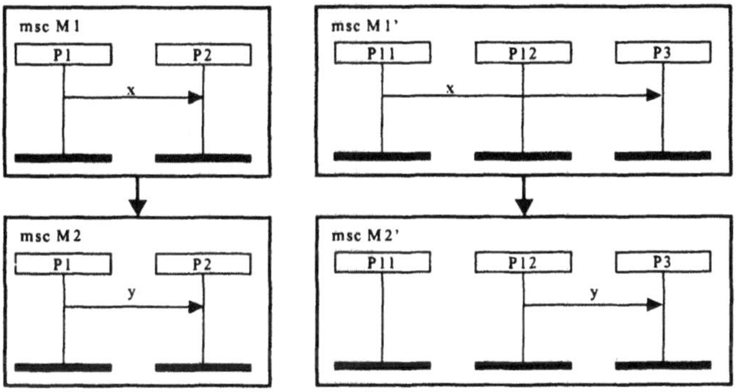

Figure 7. Non-conformance between HMSCs.

Theorem 1 (Conformance between HMSCs).
Given a HMSC H1 and its refinement H2, (with the same roadmap), if

- each bMSC in H2 conforms to its corresponding bMSC in H1, and
- in the HMSC H2, for each pair of bMSCs, M1' and M2', such as M1' o M2' in H2, for each set of axes {A11, ..., A1n} in M1' and M2' resulting from the decomposition of A1 (in M1 and M2 in H1), the order between the last event of A1 in M1 and the first event of A1 in M2 is preserved,

then H2 conforms to H1.

Proof. Consider two HSMCs, H1 and its refinement H2 obtained following the rules mentioned above. From a semantic point of view, each HMSC is a set of alternatives of concatenation of bMSCs. Roadmaps of H1 and H2 are identical, each bMSC in H1 has a corresponding bMSC in H2. Any sequence of concatenation seq1 in H1 has its corresponding sequence of concatenation seq2 in H2, using corresponding bMSCs in H2. In order to show that H1 conforms to H2, we have to show that seq2 conforms to seq1. For that, we only have to prove that if M1' conforms to M1, M2' conforms to M2, and the second condition of the theorem holds, M1' o M2' conforms to M1 o M2.

M1' conforms to M1 means that orders in M1 are preserved in M1'. M2' conforms to M2 also means that orders of M2 are preserved in M2'. M1 o M2 (before decomposition of axes) imposes orders on events in the same axes and these orders have to be preserved in M1' o M2'. The second condition of the theorem ensures the preservation of these orders in M1'o M2'. Therefore, M1' o M2' conforms to M1 o M2. **[End]**

During the refinement of HMSC use cases, we check the conformance between the new HMSC and its parent HMSC by checking the conformance between each pair bMSCs and the second condition of the theorem above. When a bMSC in the new HMSC does not conform to its corresponding bMSC in the parent HMSC, the refinement is not accepted. New refinements and modifications have to be applied to the new bMSC and the conformance checked again. In some cases, the designer has to add new messages in the refined HMSC to ensure that the second condition of the theorem holds. The tool can also add automatically "dummy" messages to satisfy this condition of the theorem. Notice that in our refinement approach the conformance of bMSCs is a sufficient condition and not a necessary one. However, in order to ensure the conformance between HSMCs, we enforce conformance between pairs of bMSCs. The study of all the possible necessary conditions is left for future investigations.

5. APPLICATION

Automation (tool) reduces the likelihood of human error and increases the confidence in the generated specifications.

5.1 Tool

We have developed a MSC refinement tool. The use cases are specified with HMSC and the system architecture is specified using UML (or SDL directly). The stepwise refinement of MSCs is used in conjunction with other methodologies that have been automated, namely:

- UML to SDL architectural translation [17].
- Generation of SDL specification from a given SDL architecture and MSC [7, 8]. The SDL specification preserves the behaviors specified in the MSC and is free of any design error such as deadlocks.

Figure 8 shows how the MSC refinement tool can be combined with existing tools. The shaded region represents the MSC refinement process. A tool guides the user through the refinement process and verifies that the resulting MSCs conform to the original use cases and follow the system architecture.

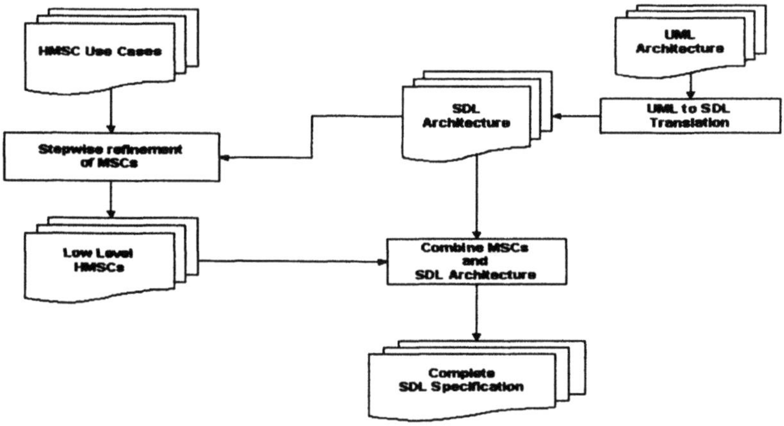

Figure 8. A toolkit for developing SDL and MSC specifications.

The MSC refinement approach (and tool) can also be used in the front-end of the SDL specification enrichment approach introduced in [18].

5.2 Example

This section illustrates our refinement approach through an example. Figure 9 shows a system architecture representing an ATM machine using SDL. There are two processes, namely "ATM" and "BANK". There is a channel that allows the user of the system to interact with the ATM process instance, and another channel allows for the two process instances to interact with each other. The signals these processes exchange with each other and their environment are specified in the architecture.

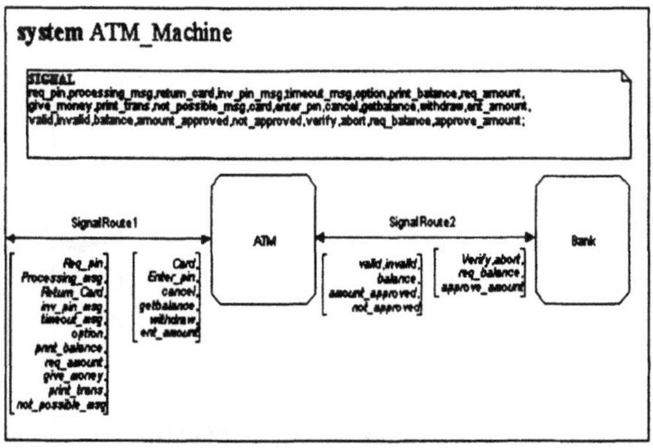

Figure 9. ATM system architecture in SDL.

Two functions provided by an ATM machine are the ability to withdraw and deposit money. Both these functions have a common starting point. The user must first insert a card and enter the correct PIN before a transaction can occur. Figure 10 shows the system and structure level HSMCs.

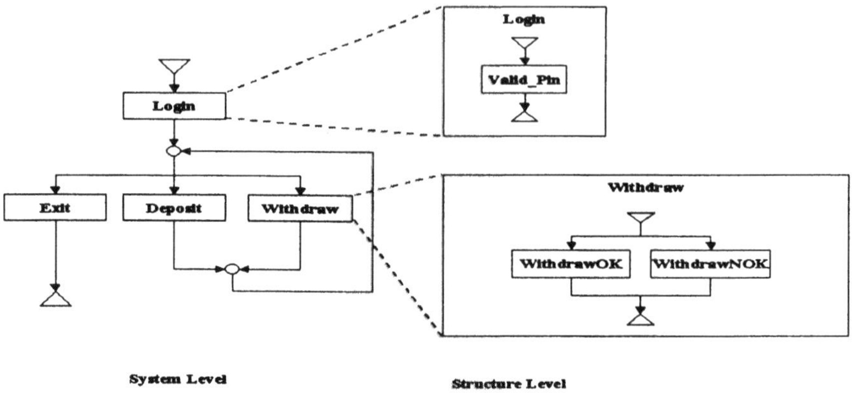

Figure 10. System and structure level HMSCs for the ATM example.

The system level HMSC shows three different use cases, "Exit", "Deposit" and "Withdraw". It also shows that each use case has the common behavior denoted by the "Login" block. The structure level shows more details. The "Login" HMSC has only one bMSC, "Valid_Pin". Later on, the system designer can decide to add an "Invalid_Pin" scenario to the "Login" HMSC that would not affect the rest of the system. The "Withdraw" HMSC shows that the withdraw use case can have two results, "WithdrawOK" and "WithdrawNOK". For illustration purpose, we will perform refinements on "WithDrawOK" only. The bMSC "WithdrawOK" in Figure 11.a is first refined vertically into the bMSC in Figure 11.b. Note that there is no new behavior added, only a splitting of the instance "ATM_Machine" to reflect the architecture of Figure 9.

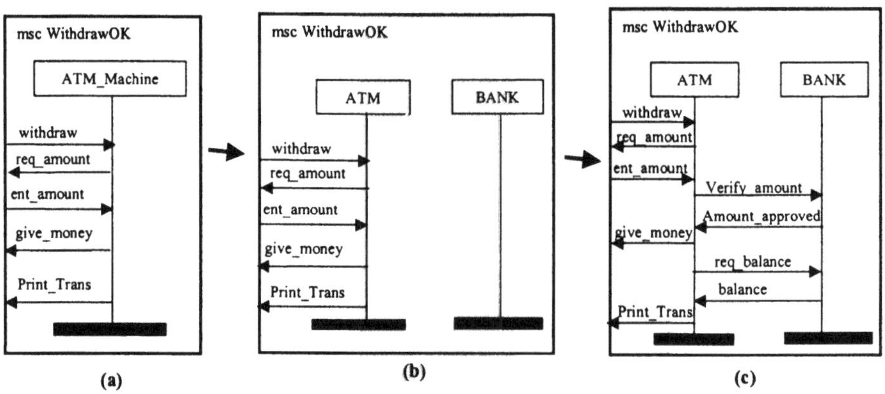

Figure 11. Vertical and horizontal refinements of "WithdrawOK".

Once the vertical refinement has been performed, the designer can now enrich the bMSC. A horizontal refinement is shown in Figure 11.c. The new MSC conforms to the original MSC in Figure 11.a.

Now suppose that the ATM block of Figure 9 was further decomposed into three separate processes: *Dialogue, MoneyHandler* and *Printer*. With our refinement process, the MSC of Figure 11.c is first refined vertically to show the decomposition of the ATM block. This refinement is followed by a horizontal refinement with the addition of messages. The resulting MSC, that conforms to the MSC in Figure 11.c and therefore to the initial use case in Figure 11.a, is shown in Figure 12.

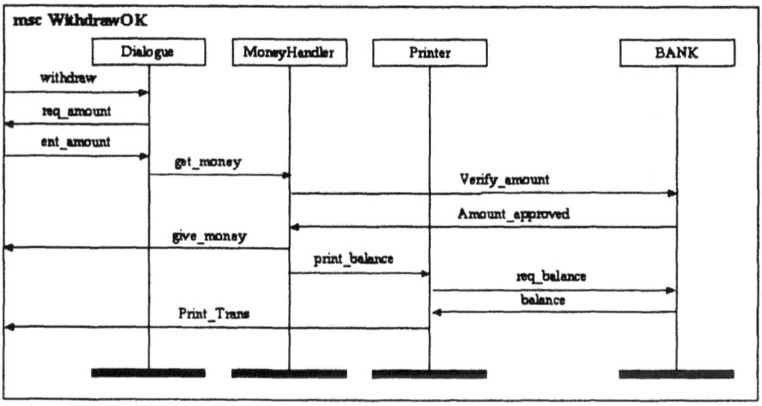

Figure 12. Another refinement of "WithdrawOK".

While horizontal refinements are being done separately for each bMSC of the HMSC. The designer can verify the conformance of each new bMSC (HMSC) against its corresponding bMSC (HMSC) in the previous stage.

6. CONCLUSION

Use cases are a good way to capture user requirements. They are abstract and are often described using textual documents that can be misinterpreted later on by designer. The reason for this is that designers apply their experience to translate use cases into a design and implementation. A structured way of going from use cases to specification would benefit the software development process by detecting and avoiding errors early on, and by enabling easier software comprehension. This will reduce development and maintenance costs. MSCs are well suited for this because they can show a system at a high level of abstraction, or they can be used for low-level specifications. They are intuitive, easy to learn and formally defined.

The goal of this project is to create a methodology and a CASE tool to semi-automate the process of refining a use case HMSC into a design specification. The MSC refinement tool is used as a front-end tool for an existing set of tools for translating UML architecture specifications to SDL architecture specifications, and from MSC specifications to SDL specifications. This set of tools aims at improving the quality of the software as well as shortening the development process.

Other issues such as conformance relations that allow for enrichment and modification of roadmaps are under consideration.

Acknowledgements: We would like to thank Loïc Helouët from France Telecom for his very helpful comments.

REFERENCES

1 OMG, "*UML version 1.3*", June 1999.

2 H. E. Eriksson and M. Penker, "*UML Toolkit*", John Wiley & Sons Inc., 1998.

3 ITU-T, Z.120, "*Message Sequence Charts – MSC'2000*", November 1999.

4 S. Mauw, M. A. Reniers and T.A.C. Willemse., "*Message Sequence Charts in the Software Engineering Process*", Report 00/12, Department of Computer Science, Eindhoven University of Technology, 2000.

5 ITU-T, Z.100, "*Specification and Description Language – SDL'2000*", November 1999.

6 J. Ellsberger, D. Hogrefe and A. Sarma, "*SDL: Formal Object Oriented Language for Communicating Systems*", Prentice-Hall, 1997.

7 M.M. Abdalla, F. Khendek and G. Butler, "*New Results on Deriving SDL Specifications from MSCs*", Proceedings of SDL Forum'99, Montréal, Canada, June 21-25, 1999.

8 G. Robert, F. Khendek and P. Grogono, "*Deriving an SDL Specification with a Given Architecture from a Set of MSCs*", Proceedings of SDL Forum'97, Evry, France, Sept. 1997.

9 I. Jacobson et al, "*Object-Oriented Software Engineering*", Addison-Wesley, 1992.

10 M. Andersson and J. Bergstrand, "*Formalizing Use Cases with Message Sequence Charts*", Masters Thesis, Lund University, Sweden, 1995.

11 E. Rudolph et al, "*Tutorial on Message Sequence Charts (MSC'96)*", Proceedings of FORTE/PSTV'96, Kaiserslautern, Germany, Oct. 1996.

12 A. Muscholl and D. Peled, "*Analyzing Message Sequence Charts*", Proceedings of SDL And MSC Workshop (SAM'2000), Grenoble, France, June 26-28.

13 A. Muscholl, D. Peled and Z. Su, "*Deciding properties of message sequence charts*", Proceedings of FoSSaCS'98, LNCS 1378, Springer, 1998.

14 R. Alur, G. Holzmann and D. Peled, "*An analyser for message sequence charts*", Software Concepts and Tools, 17(2), 1996, pp. 70-77.

15 S. Mauw and S. A. Reniers, "*An algebraic semantics for Basic Message Sequence Charts*", The Computer Journal, 37(4), 1994, pp. 269-277.

16 S. Mauw and M. Reniers, "*Refinement in interworkings*", Proceedings of CONCUR'96, LNCS 1119, Springer, 1996.

17 S. Bourduas, F. Khendek and D. Vincent, "*From MSC + UML to SDL*", Technical Report, Concordia University, November 2000.

18 F. Khendek and D. Vincent, "*Enriching SDL Specifications with MSCs*", Proceedings of SDL And MSC Workshop (SAM'2000), Grenoble, France, June 26-28, 2000.

FORMAL SYNTHESIS AND CONTROL OF SOFT EMBEDDED REAL-TIME SYSTEMS

Pao-Ann Hsiung
Department of Computer Science and Information Engineering
National Chung Cheng University, Chiayi–621, Taiwan, ROC
hpa@computer.org

Abstract Due to rapidly increasing system complexity, ever-shortening time-to-market, and growing demands for soft real-time, formal methods are becoming indispensable in the synthesis of embedded real-time systems. In this work, a formal method based on *Time Free-Choice Petri Nets* (TFCPN) is proposed for synthesizing and controlling *Soft Embedded Real-Time Systems* (SERTS). Technically, the proposed method employs quasi-static data scheduling for satisfying limited embedded memory requirements and controls firing interval bounds for satisfying soft real-time constraints. An application example is given to illustrate the feasibility of the formal method, which can also be used for code generation.

Keywords: embedded real-time systems, formal synthesis, control, quasi-static scheduling, Time Free-Choice Petri nets

1. INTRODUCTION

With the proliferation of *Soft Embedded Real-Time Systems* (SERTS) into our daily lives in the form of home appliances, internet appliances, personal assistants, wearable computers, telecommunication gadgets, and transportation facilities, we are now faced with a growing escalation of system complexity, ever-shortening time-to-market, and growing demands for soft real-time applications. All these factors have propelled the need for practical formal methods that can be used to synthesize and control SERTS. A large portion of formal methods is mainly devoted to hard real-time system analysis. In contrast, we show how formal methods can also be applied to *soft* real-time systems that have flexible ranges of acceptable behaviors.

Our target systems are soft real-time systems such as multimedia servers, communication networks, telecommunication devices, home electric appliances, and information appliances. These systems can tolerate deadline misses up to a certain threshold value. Not every deadline needs to be met, only most of the deadlines need be met. Further, deadlines themselves can be specified as a time interval (α, β) such that if a system task completes execution no earlier than α and no later than β, then the task does not miss its deadline, where α and β are integers representing some points in the time-line.

Informally, our target problem is to synthesize an embedded real-time system starting from an initial set of loose specifications into a final set of strict specifications such that the final specification satisfies all user-given real-time constraints such as response times, deadline, and periods. A specification ψ_1 is said to be looser than another specification ψ_2 if all the behaviors given by ψ_2 are implied by ψ_1. In plain terms, our solution is to restrict loose specifications into stricter ones such that given constraints are met.

The two main issues involved in the design of SERTS are as follows:

- *Bounded Memory Execution*: A processor cannot have infinite amount of memory space for the execution of any software process. This fact is even more emphasized in an embedded system, which generally has only a few hundreds of kilobytes memory installed. Thus, a SERTS must utilize as less memory as possible.

- *Soft Real-Time Constraints*: A processor may have to execute several concurrent tasks with precedence and temporal constraints. Thus, a SERTS is generally composed of several soft concurrent real-time tasks.

Our formal model is based on the recently proposed *Time Free-Choice Petri Nets* (TFCPN) [15], which is a sub-class of Time Petri Nets. In solution to the above two issues, our proposed method consists of the following two phases:

- *Quasi-Static Data Scheduling* (QSDS): This scheduling phase ensures that embedded real-time applications do not require an unbounded amount of memory for execution, since embedded real-time systems have limited amount of embedded memory,

- *Firing-Interval Bound Synthesis* (FIBS): This synthesis phase ensures that an embedded real-time system meets all soft real-time constraints, which are generally modeled as action or firing time intervals. This phase is also called *Controller Synthesis* since controllers can be synthesized for soft real-time systems using this method.

Software code can also be generated for soft real-time systems by applying our synthesis method to a recently proposed code generation scheme, which was for hard real-time systems [15]. Due to page-limits, we will not delve into this part of the work in this article.

This article is organized as follows. Section 2 gives some previous work related to SERTS synthesis. Section 3 will formulate, model, and solve the SERTS synthesis problem. Section 4 will illustrate the proposed method through an application example. Section 5 will conclude the article with some research directions for future work.

2. PREVIOUS WORK

Currently, *synthesis of soft real-time systems* is a hot topic of research in the field of hardware-software codesign of embedded systems [11]. Previously, a large effort was directed towards synthesis of hard real-time systems, especially in the application of formal methods. Synthesis was mainly carried out for communication protocols [19], plant controllers [4, 18, 5], and real-time schedulers [25, 1] because they generally exhibited regular behaviors. Only recently has there been some work on automatically generating code for embedded systems [17, 16, 23, 26, 6]. In the following, we will briefly survey the existing works on the synthesis of non real-time software and controller synthesis, on which our work is based.

Lin [16, 17] proposed an algorithm that generates a software program from a concurrent process specification through intermediate Petri-Net representation. This approach is based on the assumption that the Petri-Nets are safe, *i.e.*, buffers can store at most one data unit, which implies that it is always schedulable. The proposed method applies *quasi-static scheduling* to a set of safe Petri-Nets to produce a set of corresponding state machines, which are then mapped syntactically to the final software code. Later, Zhu and Lin [26] proposed a compositional version of the synthesis method that reduced the generated code size and was thus more efficient.

A software synthesis method was proposed for a more general Petri-Net framework by Sgroi et al. [23]. A quasi-static scheduling algorithm was proposed for *Free-Choice Petri Nets* (FCPN) [23]. A necessary and sufficient condition was given for a FCPN to be schedulable. Schedulability was first tested for a FCPN and then a valid schedule generated by decomposing a FCPN into a set of *Conflict-Free* (CF) components which were then individually and statically scheduled. Code was finally generated from the valid schedule.

Balarin et al. [6] proposed a software synthesis procedure for reactive embedded systems in the *Codesign Finite State Machine* (CFSM) [7] framework with the POLIS hardware-software codesign tool [7]. This work cannot be easily extended to other more general frameworks.

Besides synthesis, there are also some recent work on the verification of software in an embedded system such as the *Schedule-Verify-Map* method [12], linear hybrid automata techniques [10, 13], and mapping strategy [8].

Controller synthesis for plants (also called supervisor synthesis) was mainly performed in the discrete time domain, with a large portion of classical work done by Ramadge and Wonham [21, 22]. Around 1994, when timed automata was proposed as a dense-time model for real-time systems [3], controller synthesis was extended to dense real-time systems [4, 18, 25] as well as to hybrid systems [24]. Recently, the same technique was further extended to multimedia scheduler synthesis [1]. Given a dense real-time system modeled by timed

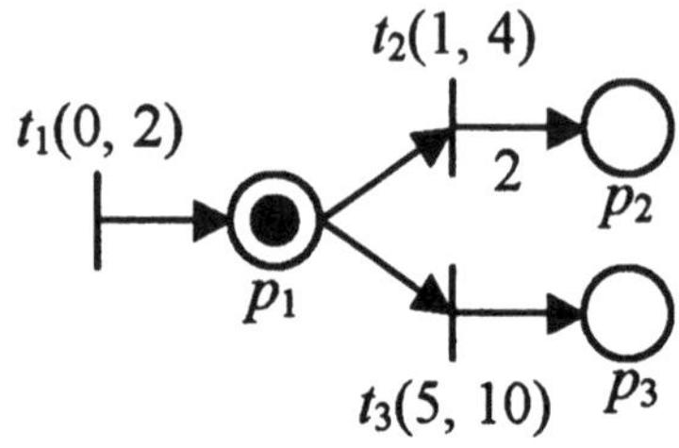

Figure 1. **A Time Free-Choice Petri Net**

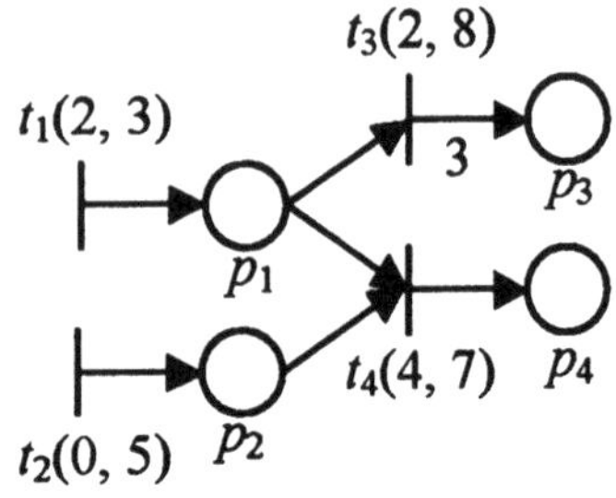

Figure 2. **Not A TFCPN (Confusion)**

automata and a (temporal) property given as a formula in *Timed Computation Tree Logic* (TCTL) [2, 9], a controller is synthesized such that it restricts the behavior of the system for satisfying the property. This is the *controller synthesis problem*. Recently, system parameters have also been taken into consideration for real-time controller synthesis [14].

3. FORMAL SYNTHESIS AND CONTROL

A formal synthesis method for soft embedded real-time systems is presented in this section. Its basic features are that the synthesized system executes in *bounded memory* and satisfies all user-given *soft real-time constraints*. Before going into details, the system model and related terminologies are presented.

A soft embedded real-time system is specified as a set of *Time Free-Choice Petri Nets* (TFCPN), which are time extensions of Free-Choice Petri Nets (FCPN) [23]. As mentioned in Section 2, FCPN was used for the quasi-static scheduling of embedded real-time software. But, there was no concept of time in the FCPN model, which makes it an inconvincing model for *real-time* systems. FCPN was recently extended to include time just as Merlin and Farber's *Time Petri Nets* (TPN) [20] are time extension of standard Petri Nets. The extended version called TFCPN was introduced in [15] and is presented here.

In the rest of this section, we first define TFCPN, give its properties, and explain why TFCPN are used for modeling SERTS. Then, the problem to be solved is formulated. Finally, our proposed synthesis algorithm is described.

3.1 System Model

Definition 1 : <u>**Time Free-Choice Petri Nets**</u> (TFCPN)
A *Time Free-Choice Petri Net* is a 5-tuple (P, T, F, M_0, τ), where: P is a finite set of places, T is a finite set of transitions, $P \cup T \neq \emptyset$, and $P \cap T = \emptyset$, $F : (P \times T) \cup (T \times P) \rightarrow \mathcal{N}$ is a weighted flow relation between places and transitions, represented by arcs, such that every arc from a place is either a unique outgoing arc or a unique incoming arc to a transition (this is called

Free-Choice), where $\mathcal{N}$ is a set of nonnegative integers, $M_0 : P \to N$ is the initial marking (assignment of tokens to places), and $\tau : T \to Q^* \times (Q^* \cup \infty)$, i.e., $\tau(t) = (\alpha, \beta)$, where $t \in T$, α is the *earliest firing time* (EFT), and β is *latest firing time* (LFT); together they are called *firing interval bounds* (FIB).

Graphically, a TFCPN can be depicted as in Fig. 1, where circles represent places, vertical bars represent transitions, arrows represent arcs, black dots represent tokens, and integers labeled over arcs represent the weights as defined by F. Here, $F(x, y) > 0$ implies there is an arc from x to y with a weight of $F(x, y)$, where x and y can be a place or a transition. *Conflicts* are allowed in a TFCPN, where a conflict occurs when there is a token in a place with more than one outgoing arc such that only one enabled transition can fire, thus consuming the token and disabling all other transitions. For example, t_2 and t_3 are conflicting transitions in Fig. 1. But, *confusions* are not allowed in TFCPN, where a confusion is a result of coexistence of concurrency and conflict. An example of confusion is given in Fig. 2. Transitions t_1 and t_2 are concurrent and t_3 and t_4 are in conflict.

By disallowing confusions, a system modeled by TFCPN can be easily analyzed and synthesized because conflicts can be resolved through net decomposition of a TFCPN into conflict-free components. Though the free-choice restriction disables a designer from describing systems that have coexisting concurrency and conflicts (i.e. synchronization with conflict as in Fig. 2), yet the net decomposition approach can be extended to general Petri nets, which will be part of our future research.

Semantically, the behavior of a TFCPN is given by a sequence of *markings*, where a marking is an assignment of tokens to places. Formally, a marking is a vector $M = \langle m_1, m_2, \ldots, m_{|P|} \rangle$, where m_i is the non-negative number of tokens in place $p_i \in P$. Starting from an initial marking M_0, a TFCPN may transit to another marking through the firing of an enabled transition and re-assignment of tokens. A transition is said to be *enabled* when all its input places have the required number of tokens for the required amount of time, where the required number of tokens is the weight as defined by the flow relation F and the required amount of time is the earliest starting time α as defined by τ. An enabled transition upon firing, the required number of tokens are removed from all the input places and the specified number of tokens are placed in the output places, where the specified number of tokens is that specified by the flow relation F on the connecting arcs. An enabled transition may not fire later than the latest firing time β defined by τ.

SERTS have both data-dependent executions as well as time-dependent specifications. Both of these characteristics are well-captured by TFCPN. TFCPN can distinguish clearly between *choice* and *concurrency*, hence they are good models of data-dependent and concurrent computations. Further, TFCPN can also distinguish clearly between data-dependent and time-dependent choices, thus TFCPN are well-defined models for our target SERTS.

Some properties of Petri Nets (PN) can be defined as follows. *Reachability*: a marking M' is reachable from a marking M if there exists a firing sequence σ starting at marking M and finishing at M'. *Boundedness*: a PN is said to be k-bounded if the number of tokens in every place of a reachable marking does not exceed a finite number k. A safe PN is one that is 1-bounded. *Deadlock-free*: a PN is deadlock-free if there is at least one enabled transition in every reachable marking. *Liveness*: a PN is live if for every reachable marking and every transition t it is possible to reach a marking that enables t.

3.2 Problem Formulation

In multimedia presentations, network computing, distance learning, and other soft real-time systems, the real-time behavior can be *controlled*, that is, restricted such that the system satisfies some pre-defined specification. For example, if the tolerable network lag in some kind of network computing is pre-specified as 10 seconds, then the behavior of the network computing environment could be controlled such that under all circumstances a maximum of 10 seconds network lag is encountered during computation.

To model the above soft real-time behavior, we define a new simplified linear temporal logic, which a controller is supposed to enforce in a SERTS.

Definition 2 : Timed Reachability Specification
A *Timed Reachability Specification* (TRS) for a TFCPN $A = (P, T, F, M_0, \tau)$ has the following syntax:

$$\phi ::= \exists\Diamond_{\sim c}\vec{p} \mid \forall\Box_{\sim c}\vec{p} \mid \phi_1 \wedge \phi_2 \tag{1}$$

where $\sim \in \{<, \leq, =, \geq, >\}$, $\vec{p}$ is a non-negative integer vector of $|P|$ elements, and ϕ_1 and ϕ_2 are TRS formulae. ||

Semantically, $\exists\Diamond_{\sim c}\vec{p}$ means eventually and obeying the timing restriction $\sim c$ there exists a TFCPN marking M such that $M = \vec{p}$, where $\vec{p}$ is a *token assignment* represented by a non-negative integer vector of $|P|$ elements such that each element represents the amount of tokens that must reside in the corresponding place. This definition is the same as a marking, but we do not call it a marking because $\vec{p}$ might not be reachable from the initial marking. Further, $\forall\Box_{\sim c}\vec{p}$ means for all reachable markings M, while obeying the timing restriction $\sim c$, $M = \vec{p}$. Thus, a TRS gives a linear temporal condition that a TFCPN must satisfy. Since we consider a single microprocessor (executing software) in our soft embedded real-time systems, linear temporal logic in the above TRS form (Equation 1) is sufficient for expressing all reachability properties such as safeness, deadlines, boundedness, deadlock-free, and starvation.

Table 1. Soft Embedded Real-Time System Synthesis Algorithm

SERTS_Synthesize(S, μ, ϕ)
set of TFCPN $S = \{A_i \mid A_i = (P_i, T_i, F_i, M_{i0}, \tau_i), i = 1, 2, \ldots, n\}$;
integer μ; // Maximum memory
TRS ϕ; // Specification
{
  // *Quasi-Static Data Scheduling* (QSDS)
  for each A_i in S { (1)
    $B_i =$ **CF_Generate**(A_i); // B_i: set of CF components (2)
    for each CF component A_{ij} in B_i { (3)
      $QSS_{ij} =$ **Quasi_Static_Schedule**(A_{ij}, μ); // QSS: schedules (4)
      if QSS_{ij} = NULL { (5)
        print "QSDS failed for A_{ij}"; (6)
        return $QSDS\_Error$; } (7)
      else $QSS_i = QSS_i \cup \{QSS_{ij}\}$; } } (8)
  // *Firing Interval Bound Synthesis* (FIBS)
  if **Controller_Synthesize**$(S, QSS_1, \ldots, QSS_n, \phi) = NULL$
    return $FIBS\_Error$; (9)
  else return $Synthesized$; (10)
}

Other properties which are not as important for SERTS such as liveness cannot be specified using TRS.

Given a system model TFCPN (Definition 1) and a specification logic (Definition 2), we are now ready to formulate our problem as follows.

Definition 3 : **Soft Embedded Real-Time System Synthesis**
Given a system modeled by a set of TFCPN $S = \{A_i \mid A_i = (P_i, T_i, F_i, M_{i0}, \tau_i), i = 1, 2, \ldots, n\}$ and a specification ϕ in TRS, the system description S is to be synthesized by scheduling and by modifying firing interval bounds such that S is made to satisfy ϕ. ||

3.3 Synthesis Algorithm

As introduced in Section 1 and formulated in Definition 3, there are two objectives for our SERTS synthesis algorithm, namely bounded memory execution and soft real-time constraints satisfaction. Thus, the algorithm **SERTS_Synthesize()** proposed in Table 1 is intuitively divided into two phases corresponding to the two objectives.

As shown in Table 1, given a set of TFCPNs $S = \{A_i \mid A_i = (P_i, T_i, F_i, M_{i0}, \tau_i), i = 1, 2, \ldots, n\}$, a maximum bound on memory μ, and a TRS ϕ, a system is synthesized upon completion of the following two phases.

3.3.1 Quasi-Static Data Scheduling (QSDS). The basic concept here is to employ net decomposition such that firing choices that exists in a TFCPN are segregated into individual *Conflict-Free* (CF) components. This is done by a procedure **CF_Generate()** as in Step (2) for each A_i, which results in a set B_i of CF components corresponding to A_i. The CF components are not distinct decompositions as a transition may occur in more than one component. As in Step (4), each CF component of each TFCPN is quasi-static scheduled, that is, starting from an initial marking for each component, a *finite complete cycle* is constructed, where a finite complete cycle is a sequence of transition firings that returns the net to its initial marking. A CF component is said to be *schedulable* if a finite complete cycle can be found for it and if it is deadlock-free. Once all CF components of a TFCPN are scheduled, a *valid quasi-static data schedule* QSS_i for the TFCPN A_i can be generated as a set of the finite complete cycles. The reason why this set is a valid schedule is that since each component always returns to its initial marking, no tokens can get collected at any place. Details of this procedure can be found in [23].

We have extended the quasi-static scheduling approach given in [23] to consider timing constraints on transition firings during the scheduling process. A quasi-static schedule is said to be feasible only if all transition firing intervals are satisfied. Satisfaction of memory bound can be checked by observing if the memory space represented by the maximum number of tokens in any marking does not exceed the bound. Here, each token represents some amount of buffer space (i.e., memory) required after a computation (transition firing). Hence, the total amount of actual memory required is the memory space represented by the maximum number of tokens that can get collected in any marking, which results from the transition firings in a quasi-static data schedule.

3.3.2 Firing Interval Bound Synthesis (FIBS). This phase consists of a procedure **Controller_Synthesize()** as in Step (9) of Table 1, which synthesizes a controller for system S with quasi-static schedules QSS_1, $\ldots, QSS_n$ to satisfy a TRS ϕ.

Some embedded soft real-time systems, such as multimedia and networks, can tolerate *latencies* that occur due to network lags, inferior display technologies, weak processing power, and limited memory bandwidth. In order to control such systems, normally a *controller* is needed to ensure quality of service (QOS), predictability, and reliability. The two main issues involved in the design of a controller for embedded soft real-time systems are as follows:

- *Synchronization Wait*: A software task, upon completion of its scheduled jobs, may have to wait for a period of time to synchronize with another software task or with the hardware.

- *Real-Time Specification*: In order to satisfy some given real-time specification, such as deadlines, a software task must finish execution of its scheduled jobs earlier than system-permitted deadlines.

Solving the above two issues, a synthesis method must generate a controller that ensures all synchronizations and real-time specifications are met. In our proposed method, the above two issues are solved as follows. Here, each software task T is associated with a time interval (α, β), where α is the earliest start time of T and β is the latest finish time of T.

- *Postpone Release Time*: For synchronization to be feasible and for predictable behavior, a software task that needs to wait for some other tasks, should have its earliest start time α changed into $\alpha + \delta_w$, where $\delta_w > 0$ is the amount of wait time required.

- *Advance Finish Time*: For satisfaction of real-time specifications, the deadline of a software task is advanced from β to $\beta - \delta_h$, where $\delta_h > 0$ is the difference in the user-specified and system-permitted deadlines.

As shown in Table 2, a solution to FIBS is proposed as an algorithm **Controller_Synthesize()**, which consists of three nested for-loops spanning over each TFCPN (Step (1)), over each schedule of a TFCPN (Step (2)), and over each transition in a schedule (Step (3)). *Firing Interval Bound Synthesis* or *Controller Synthesis* mainly restricts some transition firing interval $\tau(t) = (\alpha, \beta)$ into a smaller interval (α', β'), where $\alpha' \geq \alpha$ and $\beta' \leq \beta$, such that a given TRS formula is satisfied. In the above case, (α, β) is said to be *less restricted* than (α', β').

The conditions given in Step (3) of Table 2 specifies that we consider only each prefix $\vec{t} = \langle t_0, t_1, \ldots, t_k \rangle$ of a schedule v_{ij} that leads to a possible token assignment specified in some component of ϕ. A transition in in_trans(p) is an incoming transition of place p and the function token$_{\phi_i}(p)$ gives the number of tokens at place p specified in the ith component ϕ_i of ϕ. First, as in Step (4) the aggregate delay interval τ is calculated for a schedule prefix $\vec{t}$ by summing up all the EFT α_i and all the LFT β_i of transitions t_i in $\vec{t}$. Then, a full set of new interval bounds (New_IBS$_i$) is constructed by procedure **IBS_Synthesize()** in Step (5), with details in Table 3.

Corresponding to the two kinds of path-formulae in a TRS ϕ, there are two ways for incorporating the new set of interval bounds in S.

1 $\phi = \exists \Diamond_{\sim c} \vec{p_i}$:
A variable Min_IBS$_i$ keeps track of the set of minimally restricted transition firing intervals of A_i for satisfying ϕ (Steps (6) and (7)). A solution

Table 2. Controller Synthesis Algorithm for TFCPN/TRS

Controller_Synthesize$(S, QSS_1, \ldots, QSS_n, \phi)$
set of TFCPN $S = \{A_i \mid A_i = (P_i, T_i, F_i, M_{i0}, \tau_i), i = 1, 2, \ldots, n\}$;
TRS $\phi = M_1\vec{p_1} \wedge M_2\vec{p_2} \wedge \ldots, M_n\vec{p_n}$, where $\phi_i = M_i\vec{p_i}$,
$M_i \in \{\exists\Diamond_{\sim c}, \forall\Box_{\sim c}\}, \vec{p_i} = \langle x_1, x_2, \ldots, x_{|P_i|}\rangle, x_i \in \mathcal{N}_{\geq 0}$;
{
  for $i = 1, \ldots, n$ { (1)
    for each schedule $v_{ij} \in QSS_i$ { (2)
      for each $t_k \in v_{ij}$, $t_k \in$ in_trans(p) and token$_{\phi_i}(p) > 0, p \in P_i$ { (3)
        $\tau = (\Sigma_{i=0}^{k}\alpha_i, \Sigma_{i=0}^{k}\beta_i)$; // $\langle t_0, t_1, \ldots, t_k\rangle$ is a prefix of v_{ij} (4)
        New_IBS$_i$ = **IBS_Synthesize**$(v_{ij}, t_k, \tau, \phi_i)$; (5)
      if $M_i = \exists\Diamond_{\sim c}$ and New_IBS$_i$ > Min_IBS$_i$ (6)
        Min_IBS$_i$ = New_IBS$_i$; (7)
      if $M_i = \forall\Box_{\sim c}$ Old_IBS$_i$ = Old_IBS$_i$ $\cap$ New_IBS$_i$; } } (8)
    if $M_i = \exists\Diamond_{\sim c}$ and Min_IBS$_i$ $\neq NULL$
      IBS_assign(Min_IBS$_i$); // modify τ (9)
    else if $M_i = \forall\Box_{\sim c}$ and Old_IBS$_i$ $\neq NULL$
      IBS_assign(Old_IBS$_i$); // modify τ (10)
    else return NULL; } (11)
  return τ; (12)
}

consisting of a set of *minimally restricted* intervals is sought because such a solution contains the *maximal behavior* of the original system S that satisfies specification ϕ.

2 $\phi = \forall\Box_{\sim c}\vec{p_i}$:
A variable Old_IBS$_i$ records the intersection of all sets of restricted transition firing intervals of A_i for satisfying ϕ (Step (8)). A set intersection is performed by individual intersections of each pair of intervals $\tau_1 = (\alpha_1, \beta_1)$ and $\tau_2 = (\alpha_2, \beta_2)$, that is, $\tau_1 \cap \tau_2 = (\alpha', \beta')$, where $\alpha' = \max\{\alpha_1, \alpha_2\}$ and $\beta' = \min\{\beta_1, \beta_2\}$.

IBS_assign() in Steps (9) and (10) assigns the final set of interval bounds to the system S.

IBS_Synthesize() in Table 3 synthesizes (modifies) the firing interval bounds for a sequence of transition firings, which is a prefix of a schedule $v_{ij} = \langle t_0, t_1, \ldots, t_k, \ldots\rangle$, such that the modified system satisfies both ϕ and the aggregate delay interval τ. The switch-case statement in Step (1) to Step (7) first decides what is the least restriction on $\tau = (\alpha, \beta)$ by calculating $\tau' = (\alpha', \beta')$ such that ϕ is satisfied. Then, depending on whether the calculated restriction

Table 3. Synthesis of Interval Bounds Set

IBS_Synthesize(v_{ij}, t, τ, ϕ)
schedule $v_{ij} = \langle t_{i0}, \ldots, t_{ik}, \ldots \rangle$; // $\tau(t_{ij}) = (\alpha_{ij}, \beta_{ij})$
transition $t = t_k \in v_{ij}$;
FIB $\tau = (\alpha, \beta)$;
TRS $\phi = M\vec{p}, M \in \{\exists\Diamond_{\sim c}, \forall\Box_{\sim c}\}, \vec{p} = \langle x_1, \ldots, x_{|P_i|}\rangle$;
{
  switch "$\sim$" { (1)
    case $<$: if $(c \leq \beta)$ $\tau' = (\alpha, c-1)$; break; (2)
    case $\leq$: if $(c \leq \beta)$ $\tau' = (\alpha, c)$; break; (3)
    case $=$: $\tau' = (c, c)$; break; (4)
    case $\geq$: if $(c \geq \alpha)$ $\tau' = (c, \beta)$; break; (5)
    case $>$: if $(c \geq \alpha)$ $\tau' = (c+1, \beta)$; break; (6)
  } // let $\tau' = (\alpha', \beta')$ (7)
  if $\alpha' > \alpha$ { (8)
    for $j = k, \ldots, 0$ do { (9)
      α_{ij} += $\min\{\alpha' - \alpha, \beta_{ij} - \alpha_{ij}\}$; (10)
      if $(\alpha' - \alpha \leq \beta_{ij} - \alpha_{ij})$ break; (11)
      else α' −= $\beta_{ij} - \alpha_{ij}$; (12)
      if $j = 0$ return *Unsynthesizable*; } } (13)
  if $\beta' < \beta$ do { (14)
    for $j = k, \ldots, 0$ do { (15)
      β_{ij} −= $\min\{\beta - \beta', \beta_{ij} - \alpha_{ij}\}$; (16)
      if $(\beta - \beta' \leq \beta_{ij} - \alpha_{ij})$ break; (17)
      else β' += $\beta_{ij} - \alpha_{ij}$; (18)
      if $j = 0$ return *Unsynthesizable*; } } (19)
  return $\{(\alpha_{ij}, \beta_{ij}) : j = 0, \ldots, k\}$; (20)
}

is on EFT (Step (8)) or LFT (Step (14)), there is a loop for modifying the firing interval bounds $(\alpha_{ij}, \beta_{ij})$ of transitions starting from the kth one. If even after all transitions have firing intervals modified and ϕ is still not satisfied then an error is returned (Steps (13) and (19)). Otherwise, the set of modified firing intervals is returned (Step (20)).

After applying the controller synthesis algorithm (Table 2) to a system S, some transition firing intervals of the TFCPNs in S are restricted into smaller intervals such that the restricted (controlled) system S' satisfies a given specification TRS ϕ and there is no other lesser restricted system S'' that can satisfy ϕ. An example will be given in Section 4.

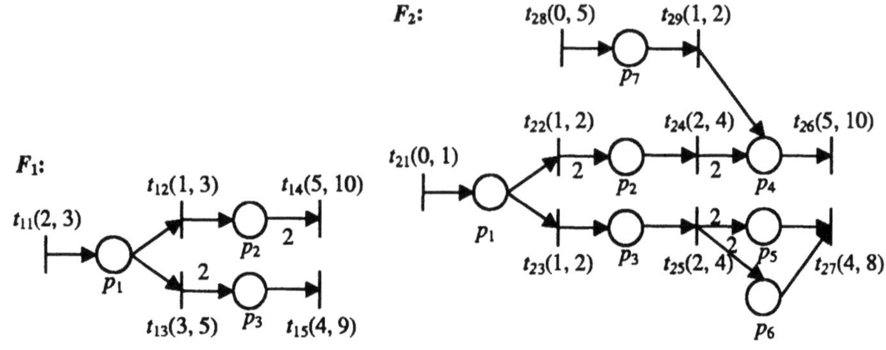

Figure 3. **Application Example $S = (F_1, F_2)$**

4. APPLICATION EXAMPLE

A 2-process system example is given in this section to illustrate the proposed SERTS synthesis algorithm. It consists of two TFCPN (F_1 and F_2) as shown in Fig. 3 and a *Timed Reachability Specification* (TRS) formula as below:

$$\phi : \exists\Diamond_{\leq 7}\langle 002\rangle \wedge \exists\Diamond_{\geq 30}\langle 0000001\rangle \tag{2}$$

According to our proposed algorithm (Table 1), we apply quasi-static data scheduling and controller synthesis to the given system.

QSDS for F_1: Since t_{12} and t_{13} are conflicting transitions, two CF components (R_{11} and R_{12} in Fig. 4) are derived, which are then individually scheduled, resulting in the following two schedules, with their associated execution time intervals.

$$v_{11} = (t_{11}t_{12}t_{11}t_{12}t_{14}), \qquad 11 \leq \tau(v_{11}) \leq 22 \tag{3}$$

$$v_{12} = (t_{11}t_{13}t_{15}t_{15}), \qquad 13 \leq \tau(v_{12}) \leq 26 \tag{4}$$

There can be two sets of valid schedules for this TFCPN as given below.

$$\Sigma_1 = \quad \{(t_{11}t_{12}t_{11}t_{12}t_{14}), (t_{11}t_{13}t_{15}t_{15})\} \tag{5}$$

$$\Sigma_2 = \quad \{(t_{11}t_{13}t_{15}t_{15}), (t_{11}t_{12}(t_{11}t_{13}t_{15}t_{15})^k t_{11}t_{12}t_{14}, \quad k \in \mathcal{N})\} \tag{6}$$

QSDS for F_2: Since t_{22} and t_{23} are conflicting transitions, two CF components (R_{21} and R_{22} in Fig. 5) are derived, which are then individually scheduled, resulting in the following two schedules, with their associated execution time

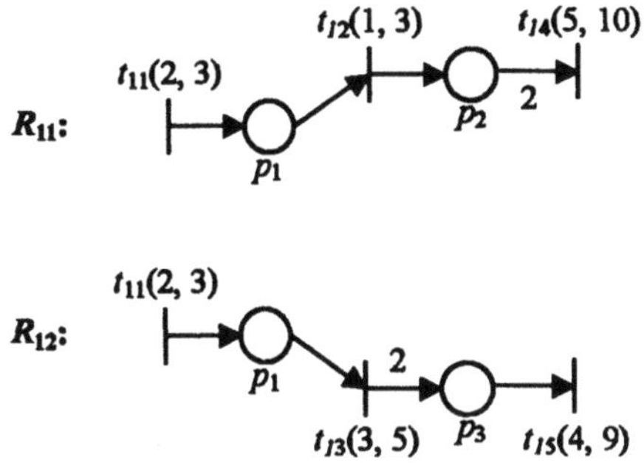

Figure 4. **Conflict-Free Components of F_1**

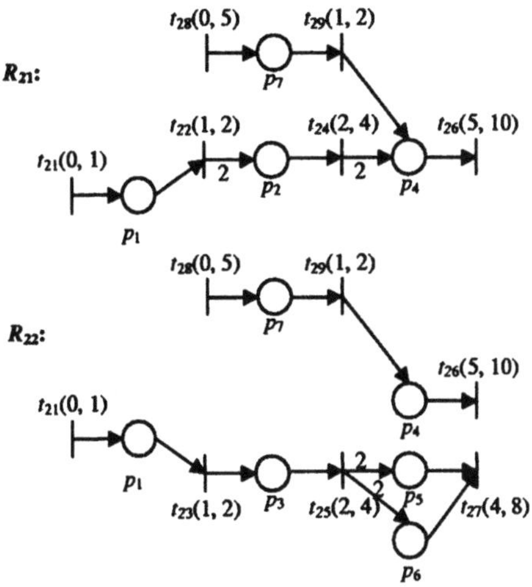

Figure 5. **Conflict-Free Components of F_2**

intervals.

$$v_{21} = (t_{21}t_{22}t_{24}t_{24}t_{26}t_{26}t_{26}t_{26}t_{28}t_{29}t_{26}), \qquad 31 \le \tau(v_{21}) \le 68 \qquad (7)$$

$$v_{22} = (t_{21}t_{23}t_{25}t_{27}t_{27}t_{28}t_{29}t_{26}), \qquad 15 \le \tau(v_{22}) \le 36 \qquad (8)$$

The set of valid schedules for this TFCPN is as given below.

$$\Sigma_3 = \{(t_{21}t_{22}t_{24}t_{24}t_{26}t_{26}t_{26}t_{26}t_{28}t_{29}t_{26}), (t_{21}t_{23}t_{25}t_{27}t_{27}t_{28}t_{29}t_{26})\} \qquad (9)$$

Controller Synthesis: In Equation (2), the first conjunct in ϕ corresponds to F_1 and specifies that the TFCPN F_1 reaches a marking within less than or equal to 7 time units such that there are no tokens in places p_1 and p_2 and there are two tokens in p_3. The second conjunct corresponds to F_2 and specifies that the TFCPN F_2 reaches a marking after 30 time units, inclusive, such that there are no tokens in any of the first six places $(p_1, \ldots, p_6)$ and there is one token in place p_7. Applying the controller synthesis algorithm from Table 2, we have the following results.

FIBS for F_1: First, consider the conjunct in ϕ that corresponds to F_1, that is, $\exists\Diamond_{\le 7}\langle 002\rangle$. Since there is only one schedule $(v_{12} = (t_{11}t_{13}t_{15}t_{15}))$ in Σ_1 (Equation (4)) that results in p_3 having tokens. We calculate the time required by the prefix of the schedule that leads to 2 tokens in p_3 as follows:

$$\begin{aligned} 2+3 &\le \tau(t_{11}) + \tau(t_{13}) \le 3+5 \\ 5 &\le \tau(t_{11}) + \tau(t_{13}) \le 8 \end{aligned}$$

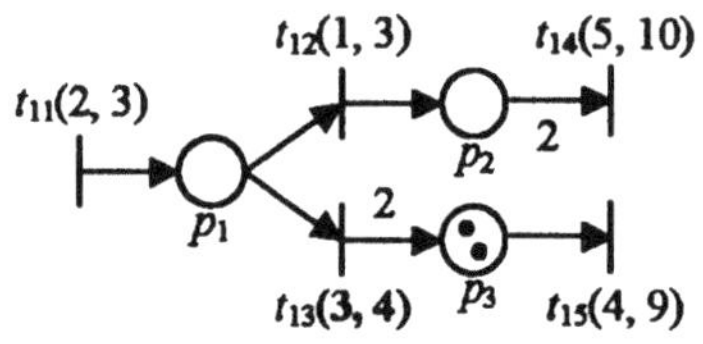

Figure 6. **Controlled TFCPN F_1**

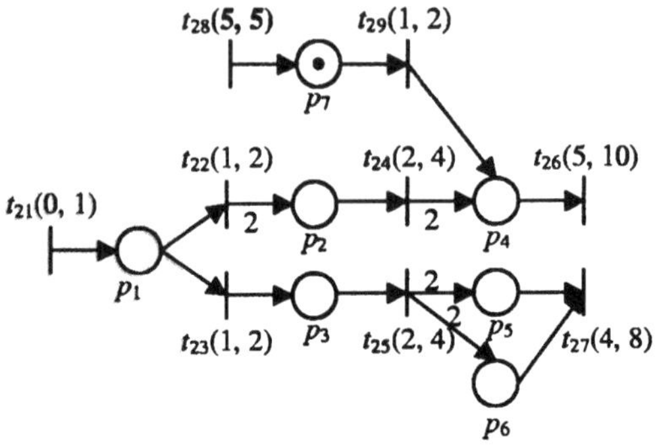

Figure 7. **Controlled TFCPN F_2**

Thus applying the IBS synthesis algorithm from Table 3, for the time spent on the schedule prefix $(t_{11}t_{13})$ to satisfy (≤ 7) constraint, the firing interval of t_{13} is modified as follows.

$$\tau(t_{13}) = (3, 4) \tag{10}$$

FIBS for F_2: For TFCPN F_2, we must consider both the schedules v_{21} and v_{22} from Equations (7) and (8) because both the schedules have prefixes that lead to a token in place p_7. First, the aggregate delay interval is calculated for a prefix of v_{21} as follows.

$$25 \leq \tau(t_{21}t_{22}t_{24}t_{24}t_{26}t_{26}t_{26}t_{26}t_{28}) \leq 56$$

Thus, to satisfy the constraint of ≥ 30, the firing interval of t_8 is modified as follows.

$$\tau(t_{28}) = (5, 5) \tag{11}$$

When we consider a prefix of schedule v_{22}, as shown below it is impossible to modify any firing interval of transitions in T_2 to satisfy the ≥ 30 constraint because the maximum firing delay is only 28.

$$11 \leq \tau(t_{21}t_{23}t_{25}t_{27}t_{27}t_{28}) \leq 28$$

After modifying the firing intervals of transitions t_3 and t_8, we get the two controlled TFCPN as illustrated in Figs. 6 and 7.

As a last note on controller synthesis for this example, let us suppose the TRS specification is changed to the following.

$$\phi' : \exists\Diamond_{\leq 7}\langle 002\rangle \wedge \exists\Diamond_{\geq 60}\langle 0000001\rangle \tag{12}$$

Then, there is no modification of any firing interval of any transition that can make the system S to satisfy ϕ'. In this case, the controller is *unsynthesizable.*

5. CONCLUSION

Instead of ad-hoc, trial-and-error methods that engineers use in developing *Soft Embedded Real-Time Systems* (SERTS), it has been proposed in this work how SERTS can be developed by a formal automatic synthesis method. To satisfy the limited memory space and processor power requirements of a soft real-time embedded system, two phases, namely *Quasi-Static Data Scheduling* (QSDS) and *Firing Interval Bound Synthesis* (FIBS) are performed before code generation. Engineers will benefit from our work when he/she applies the proposed method to automatically and formally synthesize a system specification modeled as a set of TFCPNs. Future research directions include the extension of system models (TFCPN) to more general ones such that a larger domain of system can be synthesized.

REFERENCES

[1] K. Altisen, G. Gobler, A. Pneuli, J. Sifakis, S. Tripakis, and S. Yovine. A framework for scheduler synthesis. In *Proceedings of the Real-Time System Symposium (RTSS'99).* IEEE Computer Society Press, 1999.

[2] R. Alur, C. Courcoubetis, N. Halbwachs, and D. Dill. Modeling checking for real-time systems. In *Proceedings of the IEEE International Conference on Logics in Computer Science (LICS'90)*, 1990.

[3] R. Alur and D.L. Dill. A theory of timed automata. *Theoretical Computer Science*, 126:183 – 235, 1994.

[4] E. Asarin, O. Maler, and A. Pneuli. Symbolic controller synthesis for discrete and timed systems. In P. Antsaklis, W. Kohn, A. Nerode, and S. Sastry, editors, *Hybrid Systems II*, volume 999, pages 1 – 20. Lecture Notes in Computer Science, Springer Verlag, 1995.

[5] E. Asarin, O. Maler, A. Pneuli, and J. Sifakis. Controller synthesis for timed automata. In *Proceedings of System Structure and Control.* IFAC, Elsevier, July 1998.

[6] F. Balarin and M. Chiodo. Software synthesis for complex reactive embedded systems. In *Proceedings of International Conference on Computer Design (ICCD'99)*, pages 634 – 639. IEEE CS Press, October 1999.

[7] F. Balarin and et al. *Hardware-software Co-design of Embedded Systems: the POLIS approach.* Kluwer Academic Publishers, 1997.

[8] J.-M. Fu, T.-Y. Lee, P.-A. Hsiung, and S.-J. Chen. Hardware-software timing coverification of distributed embedded systems. *IEICE Transactions on Information and Systems*, E83-D(9):1731–1740, September 2000.

[9] T.A. Henzinger, X. Nicollin, J. Sifakis, and S. Yovine. Symbolic model checking for real-time systems. In *Proceedings of the IEEE International Conference on Logics in Computer Science (LICS'92)*, 1992.

[10] P.-A. Hsiung. Timing coverification of concurrent embedded real-time systems. In *Proceedings of the 7th IEEE/ACM International Workshop on Hardware Software Codesign (CODES'99, Rome, Italy)*, pages 110 – 114. ACM Press, May 1999.

[11] P.-A. Hsiung. CMAPS: A cosynthesis methodology for application-oriented parallel systems. *ACM Transactions on Design Automation of Electronic Systems*, 5(1):51–81, January 2000.

[12] P.-A. Hsiung. Embedded software verification in hardware-software codesign. *Journal of Systems Architecture — the Euromicro Journal*, 46(15):1435–1450, December 2000.

[13] P.-A. Hsiung. Hardware-software timing coverification of concurrent embedded real-time systems. *IEE Proceedings — Computers and Digital Techniques*, 147(2):81–90, March 2000.

[14] P.-A. Hsiung. Synthesis of parametric embedded real-time systems. In *Proceedings of the International Computer Symposium (ICS'00), Workshop on Computer Architecture (ISBN 957-02-7308-9)*, pages 144–151, December 2000.

[15] P.-A. Hsiung. Formal synthesis and code generation of embedded real-time software. In *International Symposium on Hardware/Software Codesign (CODES'01, Copenhagen, Denmark)*, pages 208–213. ACM Press, New York, USA, April 2001.

[16] B. Lin. Efficient compilation of process-based concurrent programs without run-time scheduling. In *Proceedings of Design Automation and Test Europe (DATE'98)*, pages 211 – 217. ACM Press, February 1998.

[17] B. Lin. Software synthesis of process-based concurrent programs. In *Proceedings of IEEE/ACM Design Automation Conference (DAC'98)*, pages 502 – 505. ACM Press, June 1998.

[18] O. Maler, A. Pnueli, and J. Sifakis. On the synthesis of discrete controllers for timed systems. In *12th Annual Symposium on Theoretical Aspects of Computer Science (STACS'95)*, volume 900, pages 229 – 242. Lecture Notes in Computer Science, Springer Verlag, March 1995.

[19] P. Merlin and G.V. Bochman. On the construction of submodule specifications and communication protocols. *ACM Transactions on Programming Languages and Systems*, 5(1):1 – 25, January 1983.

[20] P. Merlin and D. Farber. Recoverability of communication protocols – implication of a theoretical study. *IEEE Transactions on Communications*, September 1976.

[21] P.J. Ramadge and W.M. Wonham. Supervisory control of a class of discrete event processes. *SIAM Journal of Control and Optimization*, 25:206 – 230, 1987.

[22] P.J. Ramadge and W.M. Wonham. The control of discrete event systems. *Proceedings of the IEEE*, 77:81 – 98, 1989.

[23] M. Sgroi, L. Lavagno, Y. Watanabe, and A. Sangiovanni-Vincentelli. Synthesis of embedded software using free-choice petri nets. In *Proceedings IEEE/ACM Design Automation Conference (DAC'99)*. ACM Press, June 1999.

[24] H. Wong-Toi. The synthesis of controllers for linear hybrid automata. In *Proceedings of the International Conference CDC'97*, 1997.

[25] H. Wong-Toi and G. Hoffman. The control of dense real-time discrete event systems. Technical Report STAN-CS-92-1411, Stanford University, 1992.

[26] X. Zhu and B. Lin. Compositional software synthesis of communicating processes. In *Proceedings of International Conference on Computer Design (ICCD'99)*, pages 646 – 651. IEEE CS Press, October 1999.

Part Two

Distributed Systems Testing

TOWARDS A FORMAL FRAMEWORK FOR INTEROPERABILITY TESTING*

César Viho, Sébastien Barbin and Lénaïck Tanguy
IRISA/IFSIC - Université de Rennes I, Campus de Beaulieu, 35042 Rennes, France
{viho,ltanguy,sbarbin}@irisa.fr

Abstract This paper shows how the existing concepts of conformance testing can be used to define a formal framework for interoperability testing. First, the different possible interoperability testing architectures are discussed. Then, we define several ***interoperability relations*** based on the existing relations defined for conformance testing. A comparison of these relations is given, in terms of their power to detect non-interoperability. Some guidelines are given to help in generating interoperability tests.

Keywords: interoperability, test, architecture, relation, protocol, conformance

1. INTRODUCTION

In the context of distributed systems, there are basically two approaches to testing implementations to ensure that they will work effectively together. Conformance testing determines whether a single implementation under test (IUT) conforms to its specification (generally a standard) or not. In contrast, interoperability testing determines the ability of two or more implementations to work together in a real operational environment. As we can see, the purposes of these two kinds of tests are not really the same. While conformance testing evaluates an implementation in terms of its correspondence to a specific standard, interoperability testing compares an implementation with other products.

The words "interoperability", "interwork", "interoperate" are often used in the descriptions of computer systems. Different needs of interoperability testing can be identified. The most basic is to put together two implementations built by different vendors and verify that they "interwork" correctly. Another common situation is the so called "one against N" interoperability testing in which there already exist N (N≥1)

*This work has been supported by the CELAR/TCOM Contract 9842.561 MTI (Méthodologie de test d'interopérabilité), the European ITEA Project 99011 RTIPA, and the CNRS/Programme Télécommunications 99N36/0029

systems working together. One may wonder if a new system S will interoperate with the N existing systems. In other words, the question here is "will the N+1 systems still work together ?". There exist many other situations which require interoperability. Thus, it is difficult to give a unique definition of this notion. Indeed, several definitions of interoperability exist [1, 2, 3, 4]. However, the functional meaning is clear: the components of the system under test (SUT) have to communicate with each other correctly and provide the expected services.

A lot of work have been done in the area of conformance testing [5] leading to a precise definition of conformance [6]. The so called conformance/implementation relations in [7, 8] give formal characterizations of conditions under which an IUT can be considered as conformant to its specification. This allows the automatic generation of conformance tests [9]. Unfortunately, such work is lacking for interoperability testing [3]. There are several reasons for this situation. The first reason is because conformance of implementations has been considered a prerequisite for achieving interoperability. Thus, a very important effort has been concentrated on conformance testing. Another reason is that interoperability testing is considered a pragmatic and practical requirement, as it relates to implementations; contrary to conformance testing where the specification serves as reference.

The fact that both conformance and interoperability testing concerns the same objects (specification, IUT, etc) is evident. Thus, most of the concepts, methodologies and theory developed for conformance testing can be adjusted for interoperability testing. This simple observation suggests to adapt the existing formal conformance testing framework for interoperability testing. Following this idea, the work presented in this paper proposes a formal framework (architectures, interoperability relations, tests generation) for interoperability testing.

This paper is structured as follows. Section 2 presents the different classes of interoperability testing architectures. For each class, contexts in which an architecture can be used are explained. We give an indication of how the existing architectures can be positioned in regard to these classes. The model and notations used are described in Section 3. Section 4 gives formal definitions of interoperability relations, that are based on the existing relations defined for conformance testing. Each interoperability relation specifies formally the conditions to be satisfied by two implementations in order to be considered interoperable. These relations are compared in terms of their power to detect non interoperability. In Section 5, we give guidelines to generate interoperability tests for the proposed architectures and relations. Conclusion and future work are in Section 6.

2. INTEROPERABILITY ARCHITECTURES

Depending on how the implementations to be tested are interconnected, and depending on the degree to which we can observe their interaction, different interoperability testing architectures can be used. Indeed, a lot of architectures have been proposed [1, 2, 4, 10].

2.1. Definition of testing architectures

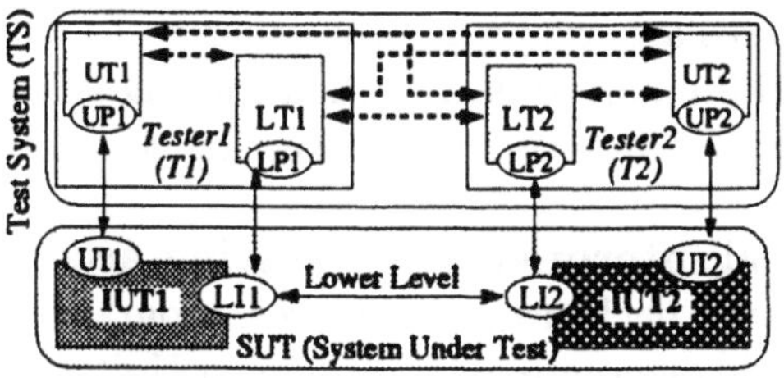

Figure 1. General architecture of interoperability testing

Let us consider the general architecture of figure 1 where the SUT is composed of two IUTs (IUT_1 and IUT_2). Each IUT is a black-box. The **Upper Interface** UI_i (resp. **Lower Interface** LI_i) is the interface of the IUT_i, through which it communicates with its upper (resp. lower) layer. The expected services are furnished through UI_i while LI_i is used by IUT_i to interact with the peer entities in the same layer.

Depending on the environment in which the interoperability testing will be done, different levels of control and observation of the SUT are possible. Thus, the **Test System** (TS) in charge of testing, via its PCOs (points of control and observation), the interoperability of this SUT can consist of some or all of these following elements. The **Upper Tester** i UT_i (resp. **Lower Tester** i resp.LT_i) is part of TS in charge of the control and/or the observation of UI_i (resp. LI_i), via the **Upper PCO** UP_i (resp. the **Lower PCO** LP_i). The **Tester** T_i (composed by UT_i and/or LT_i) is the part of TS in charge of the control and/or observation of IUT_i .

The different possible compositions of both TS and SUT induce different kinds of interoperability testing architectures as described below. The reader can see that they are easily extendible to the general case where TS and/or SUT are composed by more than two components.

2.1.1 Unilateral interoperability testing architectures.

These architectures correspond to the situation in which the control and/or observation of only one component (IUT_1 or IUT_2) of the SUT is possible. Thus, the decision of interoperability is based on these partial observations done by only one tester T_1 or T_2. The tester T_i may contain only LT_i (resp. UT_i). This can happen if the upper (resp. lower) interface is embedded under an upper (resp. lower) layer protocol. In this case, the architecture is called Unilateral **Lower** (resp. **Upper**) Interoperability Testing Architecture. The Unilateral **Total** Interoperability Testing Architecture corresponds to the case where the tester T_i contains both UT_i and LT_i. In this case, a test coordination procedure (TCP) may exist between UT_i and LT_i.

These architectures are used when the other IUTs which compose the SUT are embedded (i.e., all or part of their interfaces are not accessible) or in a "one against N" interoperability testing context (see section 1).

2.1.2 Bilateral interoperability testing architectures.

In these architectures, each tester T_i realizes separately the control and observation of the corresponding IUT_i, using a Unilateral Interoperability Testing Architecture. Depending on the composition of the testers, the terms **Upper**, **Lower** and **Total** apply also here.

This architecture is interesting in the unusual situations where the designers of the IUTs require their own control and opinion of how their respective IUT interact with the others. This happens for example when a certain level of confidentiality is still required by each of the vendors [4].

2.1.3 Global interoperability testing architectures.

In these architectures, the control and observation are done globally by the two testers T_1 and T_2 on both sides (IUT_1 and IUT_2) of the SUT. They are called **global** because the decision of interoperability is achieved according to a global view (simultaneously on both sides) of the SUT. This is the main difference with the previous architectures, as a test coordination procedure exists between the testers. In this sense, it is the more complex architecture. Global upper, lower, and total architectures are defined as for bilateral architectures.

2.1.4 Hybrid interoperability testing architectures.

Some situations of interoperability combine previously defined architectures. All the architectures which are not strictly unilateral, bilateral or global architectures are called **hybrid** architectures. As an example, the architecture in which we have only the upper tester in tester T_1 and upper and lower testers of the tester T_2, is typically a hybrid architecture. This kind of architecture can be encountered in the "testing against reference" context where one of the implementations serves as the reference, the upper tester over the tested IUT is used as a responder.

2.2. Positioning the existing architectures

Many interoperability architectures and associated terminologies can be found in the literature [1, 2, 4, 10]. In the following, we give elements which help in understanding why there is so much confusion in the state of the art of interoperability testing architectures. We show that each of the existing architectures can be associated with one of the canonical architectures described above.

Black-box versus grey-box testing
It is well-known that conformance testing is done in a black-box context. But there is still a debate about considering interoperability testing as black-box or grey-box testing [2]. The contested point comes from the possibility for the interoperability testing architecture to observe or not (and/or control) the interactions between the implementations which compose the SUT. Our purpose is not to close the debate, but one can see that "grey-box" architectures correspond to architectures which include lower testers. In contrast, the so called "black-box" architectures are those which do not contain any lower tester.

Protocol versus service interoperability testing
The definition of interoperability can be "layer-oriented" [4, 11]. For example in [4], they consider four levels of interoperability called respectively protocol, service, application and user-perceived interoperability architectures. It is easy to see that protocol interoperability architectures correspond to a lower interoperability testing architecture, while service (application or user perceived) interoperability architectures can be definitely classified as upper interoperability testing architectures.

Active versus passive testing architectures
If the test system has the possibility to give stimuli (control) and/or to corrupt events in the SUT, the corresponding testing architectures are called active [1]. If not, they are called passive architectures [10, 11]. Thus, any possible architecture (including those proposed in this paper) can be considered either as active or passive.

Monitor versus arbiter
In these architectures [10], a test component is placed between the implementations to control (monitor) or observe (arbiter) their communication. To be monitor or arbiter depends on the active/passive behavior of the tester. Thus, any architecture proposed in this paper that contains at least one lower tester can be used as monitor or arbiter. Arbiters are used to identify faulty implementations [12].

3. MODELS AND NOTATIONS

In order to give formal definitions of the notion of interoperability, we need a model which allows the formal description of all the elements involved in the interoperability testing activity. As said before, most of the terms and components in interoperability testing are similar to those used in conformance testing. We choose the commonly used model of the IOLTS (Input-Output Labeled Transition System) [13] to model specifications as well as implementations.

3.1. IOLTS, definitions and notations

Definition 3.1 *An IOLTS is a tuple $M = (Q^M, \Sigma^M, \Delta^M, q_0^M)$ where*

- *Q^M is the set of states of the system and $q_0^M \in Q^M$ is the initial state.*
- *Σ^M denotes the set of observable (input and/or output) events on the interaction points (with the environment) of the system. Thus, we have : $\Sigma^M \subseteq P^M \times \{?,!\} \times A^M$ where P^M is the finite set of interaction points (ports) through which the system communicates with its environment (lower or upper layer, or other systems), "?" and "!" respectively denote a input and an output of message, A^M is the alphabet of input-output messages exchanged by the system through its ports.*
- *$\Delta^M \subseteq Q^M \times (\Sigma^M \cup \tau) \times Q^M$ is the transition relation, where $\tau \notin A^M$ denotes an internal event. We note $q \xrightarrow{\alpha}_M q'$ for $(q, \alpha, q') \in \Delta^M$.*

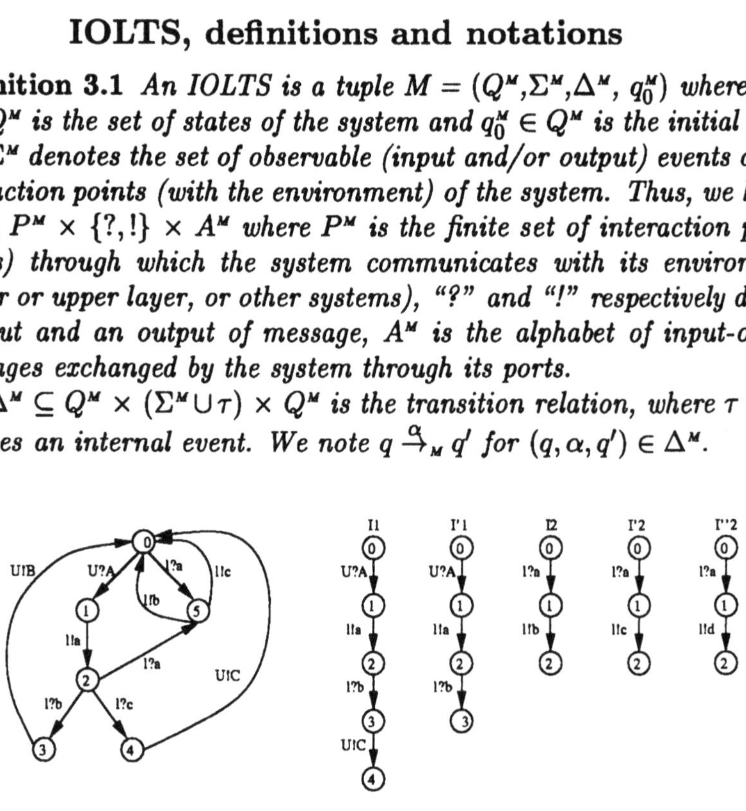

Figure 2. A specification S and possible implementations I_1, I_1', I_2, I_2', I_2''

Σ^M can be decomposed as follow: $\Sigma^M = \Sigma_U^M \cup \Sigma_L^M$, where Σ_U^M (resp. Σ_L^M) is the set of messages exchanged on the upper (resp. lower) interface. Σ^M can be also decomposed in order to distinguish input messages from output messages. $\Sigma^M = \Sigma_I^M \cup \Sigma_O^M$, where Σ_I^M (resp. Σ_O^M) is the finite set of input (resp. output) messages.

In the following we denote the set of all input/output labeled transition systems by $\mathcal{IOLTS}$. Let us consider an IOLTS $M \in \mathcal{IOLTS}$, and let $\alpha \in \Sigma^M$ with $\alpha = p.\{?,!\}.m$, $\mu_i \in \Sigma^M \cup \tau$, $\sigma \in (\Sigma^M)^*$, $q, q', q_i \in Q^M$:

- $q \overset{\mu_1 \dots \mu_n}{\longrightarrow}_M q' =_\Delta \exists\, q_0 = q, q_1 \dots, q_n = q', \forall i \in [1,n], q_{i-1} \overset{\mu_i}{\rightarrow}_M q_i$.
- $q \overset{\epsilon}{\Rightarrow}_M q' =_\Delta q = q'$ or $q \overset{\tau \dots \tau}{\longrightarrow}_M q'$.
- $q \overset{\alpha}{\Rightarrow}_M q' =_\Delta \exists\, q_1, q_2, q \overset{\epsilon}{\Rightarrow}_M q_1 \overset{\alpha}{\rightarrow}_M q_2 \overset{\epsilon}{\Rightarrow}_M q'$.
- $q \overset{\sigma}{\Rightarrow}_M q' =_\Delta q \overset{\mu_1 \dots \mu_n}{\Longrightarrow}_M q' =_\Delta \exists\, q_0 = q, q_1 \dots, q_n = q', \forall i \in [1,n], q_{i-1} \overset{\mu_i}{\Rightarrow}_M q_i, \sigma = \mu_1 \cdots \mu_n$.
- $out(q) =_\Delta \{\alpha \in \Sigma^M_O \mid \exists\, q'$ and $q \overset{\alpha}{\rightarrow}_M q'\}$ is the set of outputs from q.
- q *after* $\sigma =_\Delta \{q' \in Q^M \mid q \overset{\sigma}{\Rightarrow}_M q'\}$ is the set of states which can be reached from q by the sequence of actions σ. By extension, all the states reached from the initial state of the IOLTS M is (q_0^M *after* σ) and will be noted by (M *after* σ). In the same way, $Out(M, \sigma) =_\Delta out(M$ *after* $\sigma)$.
- $Traces(q) =_\Delta \{\sigma \in (\Sigma^M)^* \mid q$ *after* $\sigma \neq \emptyset\}$ is the set of possible observable traces from q. And, $Traces(M) =_\Delta Traces(q_0^M)$.

In interoperability testing, we usually need to observe some specific events among all the possible traces of a SUT. These traces, reduced to the expected messages, can be obtained by a **projection** of those traces on a set representing criteria used to select the expected events.

Definition 3.2 *Let us consider an IOLTS M, a trace $\sigma \in (\Sigma^M)^*$, $\alpha \in \Sigma^M$, and a set X. The projection of σ on X is noted by σ/X and is defined by: $\epsilon/X = \epsilon, (\alpha.\sigma)/X = \sigma/X$ if $\alpha \notin X$, and $(\alpha.\sigma)/X = \alpha.(\sigma/X)$ if $\alpha \in X$.*

For example, consider a trace $\sigma = U?A.l!a.l?c.U!B \in (\Sigma^M)^*$, then $\sigma/\{l!a, U?D\} = l!a$, $\sigma/\Sigma^M_U = U?A.U!B$, $\sigma/\Sigma^M_I = U?A.l?c$. Notice that $\sigma/X \in (\Sigma^M)^*$ but we do not have necessarily $\sigma/X \in Traces(M)$.

Definition 3.3 *Let us consider an IOLTS M, a trace $\sigma \in (\Sigma^M)^*$ and a set X, then $Out_X(M,\sigma) =_\Delta Out(M,\sigma) \cap X$ is the set of outputs from (M after σ), reduced to the outputs that belong to X.*

3.2. Synchronous and asynchronous interaction

Definition 3.4 (Synchronous composition $\|_S$) *The synchronous composition of two IOLTS M_1 and M_2 is noted $M_1\|_S M_2 = (Q^{M_1} \times Q^{M_2}, \Sigma^{M_1\|_S M_2}, \Delta^{M_1\|_S M_2}, (q_0^{M_1}, q_0^{M_2}))$ where $\Sigma^{M_1\|_S M_2} \subseteq \Sigma^{M_1} \cup \Sigma^{M_2}$, and the transition relation $\Delta^{M_1\|_S M_2}$ is obtained as follow, $\forall (q_1, q_2) \in Q^{M_1} \times Q^{M_2}$,*

$$\frac{(q_1, a, q_1') \in \Delta^{M_1}, a \in \Sigma_U^{M_1} \cup \{\tau\}}{((q_1, q_2), a, (q_1', q_2)) \in \Delta^{M_1\|_S M_2}}, \frac{(q_2, a, q_2') \in \Delta^{M_2}, a \in \Sigma_U^{M_2} \cup \{\tau\}}{((q_1, q_2), a, (q_1, q_2')) \in \Delta^{M_1\|_S M_2}} \quad (1)$$

$$\frac{(q_1, a, q_1') \in \Delta^{M_1}, (q_2, a, q_2') \in \Delta^{M_2}, a \in \Sigma_L^{M_1} \cap \Sigma_L^{M_2}}{((q_1, q_2), a, (q_1', q_2')) \in \Delta^{M_1\|_S M_2}} \quad (2)$$

The interaction between components of a distributed system is done through an environment. This interaction can be either synchronous, or asynchronous. As in [13], we will suppose that the (synchronous or asynchronous) environment can be modeled by an IOLTS $\mathcal{E}$. Thus, the interaction of an IOLTS S_i with another IOLTS S_j through an environment $\mathcal{E}$ is obtained by the synchronous composition $S_i\|_s\mathcal{E}\|_sS_j$. This will be noted by $S_i\|S_j$ and corresponds to the reachability graph in [1, 2].

4. INTEROPERABILITY RELATIONS

In this section, we give a formal definition of the notion of interoperability based on the model of IOLTS introduced in section 3.1. To a certain extent, testing the interoperability between two implementations consists of a kind of conformance testing of their interaction with the expected behavior or service. Thus, it becomes natural to study how these conformance relations [7, 8] can be adapted to obtain interoperability relations. We will consider the most commonly used **ioconf** conformance relation [9, 13]. It states that an implementation I is conformant to its specification S if after a trace of S, outputs of I are foreseen in S.

Definition 4.1 I **ioconf** $S =_\Delta \forall\sigma \in Traces(S),\ Out(I,\sigma) \subseteq Out(S,\sigma)$

4.1. The notion of interoperability relation

An interoperability relation formally specifies conditions to be satisfied by two implementations in order to be considered interoperable.

Definition 4.2 *An interoperability relation is a relation $\mathcal{R}$ between two implementations I_1 and I_2: $\mathcal{R}(I_1,I_2)$ means that I_1 interoperates with I_2.*

Interoperability relations are not transitive: the fact that an implementation I_1 interoperates with I_2 and that I_2 interoperates with I_3, does not allow one to draw a conclusion regarding interoperability between I_1 and I_3. The main difference between an interoperability relation and a conformance relation is that: a conformance relation connects an implementation with its specification while an interoperability relation concerns implementations. More precisely, two implementations can interoperate even though their interaction does not correspond to anything allowed in their respective specification. Thus, an interoperability relation can take into account or not the specifications on which the implementations are based. This leads to two main classes of interoperability relations called in the following **"Specification-less" interoperability relations** and **"Specification-based" interoperability relations**.

As the "Specification-less" interoperability relations do not use any specification, it is impossible to generate *a priori* tests based on these relations. Verifying these relations can be done preliminary to other tests using specification-based relations. This can explain why it is called interconnectivity testing in [4]. As a consequence, we will no longer discuss these relations in the sequel.

4.2. Specification-based relations

Specification-based interoperability relations are relations which refers to the specification of at least one component of the SUT.

4.2.1 Specification-based lower interoperability relations.

In this section, we focus on the lower layer interfaces of the implementations. We consider the different possible situations and we propose the associated interoperability relations which can be used.

The relation $\mathcal{R}_1$ considers the situation where we have the specification of only one of the two implementations. Let us consider that S_1 is the specification used to develop the implementation I_1. Now, let us consider that only the lower interface of I_1 is accessible. During the interaction between I_1 and I_2, the least we can expect from the implementation I_1 is to behave as foreseen in its specification S_1. This is described in the relation $\mathcal{R}_1$. It considers the traces observed ($Traces(I_1 \| I_2)$) during the interaction between I_1 and I_2. It states that, after any corresponding trace in the specification S_1 of the unilaterally considered implementation (here I_1), all the outputs (sent to the other IUT on the lower layer) are allowed in S_1.

Definition 4.3 (Unilateral Lower Interoperability Relation $\mathcal{R}_1$)
$\mathcal{R}_1(I_1, I_2) =_\Delta \forall \sigma_1 \in Traces(S_1), \forall \sigma \in Traces(I_1 \| I_2), \sigma/\Sigma^{I_1} = \sigma_1 \Rightarrow Out_{\Sigma_L^{I_1}}(I_1 \| I_2, \sigma) \subseteq Out_{\Sigma_L^{S_1}}(S_1, \sigma_1)$

Remarks:
• In figure 2, $\mathcal{R}_1(I_1, I_2'')$ but $\neg\, \mathcal{R}_1(I_2'', I_1)$.
• Notice also that there is a subtle difference with the **ioconf** relation: $\mathcal{R}_1(I_x, I_y)$ does not mean necessarily that I_x **ioconf** S_x.

The relation $\mathcal{R}_1$ can be applied independently for I_2 (based on the specification S_2). When we have both $\mathcal{R}_1(I_1, I_2)$ and $\mathcal{R}_1(I_2, I_1)$, this corresponds to the interoperability relation $\mathcal{R}_2$ defined below.

Definition 4.4 (Bilateral Lower Interoperability Relation $\mathcal{R}_2$)

$$\mathcal{R}_2(I_1, I_2) =_\Delta \mathcal{R}_1(I_1, I_2) \wedge \mathcal{R}_1(I_2, I_1)$$

In figure 2, $\mathcal{R}_2(I_1, I_2)$ but $\neg\, \mathcal{R}_2(I_1, I_2'')$. In the two relations $\mathcal{R}_1$ and $\mathcal{R}_2$, only lower layer outputs of each IUT are considered. Typically, we can have $\mathcal{R}_2(I_1, I_2)$, whatever happens in their interaction with upper layers.

The next relation $\mathcal{R}_3$ is called **global** because it is based on the global behavior of the interactions between respectively the two specifications ($S_1 \| S_2$) and the two implementations ($I_1 \| I_2$). The relation $\mathcal{R}_3$ states that: after a trace included in the composition of the specifications $S_1 \| S_2$ and observed in the interaction between the two implementations, all outputs observed on the lower layer are allowed in $S_1 \| S_2$.

Definition 4.5 (Global Lower Interoperability Relation $\mathcal{R}_3$)
$\mathcal{R}_3(I_1, I_2) =_\Delta$
$\forall \sigma \in Traces(S_1 \| S_2) \Rightarrow Out_{\Sigma_L^{I_1 \| I_2}}(I_1 \| I_2, \sigma) \subseteq Out_{\Sigma_L^{S_1 \| S_2}}(S_1 \| S_2, \sigma)$.
In figure 2, $\mathcal{R}_3(I_1, I_2)$ but $\neg\, \mathcal{R}_3(I_1, I_2'')$. $\mathcal{R}_3$ is a kind of reduced conformance relation in the sense that only a subset of outputs is considered.

4.2.2 Specification-based upper interoperability relations.

The following three relations are similar to the previous ones. The difference is that only observations on the upper (instead of lower) layer are used. Consequently, similar explanations and remarks apply for the three relations which are enumerated below.

◇ The **Unilateral Upper Interoperability relation** $\mathcal{R}_4$ can be defined as follows: $\mathcal{R}_4(I_1, I_2) =_\Delta \forall \sigma_1 \in Traces(S_1), \forall \sigma \in Traces(I_1 \| I_2)$, $\sigma / \Sigma^{I_1} = \sigma_1 \Rightarrow Out_{\Sigma_U^{I_1}}(I_1 \| I_2, \sigma) \subseteq Out_{\Sigma_U^{S_1}}(S_1, \sigma_1)$.

◇ For the **Bilateral Upper Interoperability relation** $\mathcal{R}_5$, we have: $\mathcal{R}_5(I_1, I_2) =_\Delta \mathcal{R}_4(I_1, I_2) \wedge \mathcal{R}_4(I_2, I_1)$.

◇ Finally, the **Global Upper Interoperability relation** $\mathcal{R}_6$ is defined by: $\mathcal{R}_6(I_1, I_2) =_\Delta \forall \sigma \in Traces(S_1 \| S_2)$, $Out_{\Sigma_U^{I_1 \| I_2}}(I_1 \| I_2, \sigma) \subseteq Out_{\Sigma_U^{S_1 \| S_2}}(S_1 \| S_2, \sigma)$.

On the example of figure 2, we have:
- $\mathcal{R}_4(I_1, I_2'')$, $\mathcal{R}_4(I_2, I_1)$, $\mathcal{R}_4(I_2'', I_1)$ but $\neg \mathcal{R}_4(I_1, I_2)$.
- $\mathcal{R}_5(I_1, I_2'')$, $\mathcal{R}_5(I_1, I_2')$ but $\neg \mathcal{R}_5(I_1, I_2)$.
- $\mathcal{R}_6(I_1, I_2'')$, $\mathcal{R}_6(I_1', I_2)$, $\mathcal{R}_6(I_1, I_2')$ $\mathcal{R}_6(I_1', I_2')$, but $\neg \mathcal{R}_6(I_1, I_2)$.

4.2.3 Specification-based total interoperability relations.

The following relations are qualified **total** because the decision of interoperability uses observations on both upper and lower layers. These observations can be done locally/unilaterally, bilaterally or globally. As in the two sections above, we obtain the three following relations.

◇ The **Unilateral Total Interoperability relation** $\mathcal{R}_7$ is defined as follows: $\mathcal{R}_7(I_1, I_2) =_\Delta \forall \sigma_1 \in Traces(S_1)$, $\forall \sigma \in Traces(I_1 \| I_2), \sigma / \Sigma^{S_1} = \sigma_1 \Rightarrow Out_{\Sigma^{I_1}}(I_1 \| I_2, \sigma) \subseteq Out(S_1, \sigma_1)$.

◇ It is easy to obtain the **Bilateral Total Interoperability relation** $\mathcal{R}_8$. It is defined by: $\mathcal{R}_8(I_1, I_2) =_\Delta \mathcal{R}_7(I_1, I_2) \wedge \mathcal{R}_7(I_2, I_1)$.

◇ The definition of the **Global Total Interoperability relation** $\mathcal{R}_9$ is also straightforward. $\mathcal{R}_9(I_1, I_2) =_\Delta \forall \sigma \in Traces(S_1 \| S_2),\ Out(I_1 \ \| I_2, \sigma) \subseteq Out(S_1 \| S_2, \sigma)$.

Remarks:

• In figure 2, we have $\mathcal{R}_7(I_1, I_2'')$ but $\neg\ \mathcal{R}_7(I_1, I_2)$, $\mathcal{R}_8(I_1, I_2')$ but $\neg\ \mathcal{R}_8(I_1, I_2'')$ and $\neg\ \mathcal{R}_8(I_1, I_2)$

• One can notice that $\mathcal{R}_7(I_1, I_2) = \mathcal{R}_1(I_1, I_2) \wedge \mathcal{R}_4(I_1, I_2)$, $\mathcal{R}_8(I_1, I_2) = \mathcal{R}_2(I_1, I_2) \wedge \mathcal{R}_5(I_1, I_2)$, and $\mathcal{R}_9(I_1, I_2) = \mathcal{R}_3(I_1, I_2) \wedge \mathcal{R}_6(I_1, I_2)$.

• $\mathcal{R}_9(I_x, I_y)$ corresponds to $I_x \| I_y$ **ioconf** $S_x \| S_y$. We will discuss this point later in Section 5.

4.3. Comparison of interoperability relations

According to the rigor required for interoperability testing, we need to know which interoperability relation to use. In this section, we give a comparison between the specification-based interoperability relations defined in Sections 4.2.1, 4.2.2 and 4.2.3, in terms of their power of non-interoperability detection. Let us call the set of these relations $\mathcal{SBIR}$, then the formalization of the comparison of the relations leads to the well-known theory of testing equivalence [6, 13] and the related preorder expressed here by a relation $\sqsubseteq_{\mathcal{R}} \subseteq \mathcal{SBIR} \times \mathcal{SBIR}$.

Definition 4.6 $\forall\, \mathcal{R}_x, \mathcal{R}_y \in \mathcal{SBIR}$,
$\mathcal{R}_x \sqsubseteq_{\mathcal{R}} \mathcal{R}_y =_{def} \forall I_1, I_2 \in \mathcal{IOLTS},\ \mathcal{R}_x(I_1, I_2) \Rightarrow \mathcal{R}_y(I_1, I_2)$

So, $\mathcal{R}_x \sqsubseteq_{\mathcal{R}} \mathcal{R}_y$ means that interoperability tests based on $\mathcal{R}_x$ detect more non-interoperable implementations than $\mathcal{R}_y$. By extension, the testing equivalence between two relations will be denoted by $\cong_{\mathcal{R}}$. We note $\mathcal{R}_x \not\sim_{\mathcal{R}} \mathcal{R}_y$ to say that $\mathcal{R}_x$ and $\mathcal{R}_y$ are not comparable. It is represented by "-" in the Figure 3 which synthesizes all the comparison of the specification-based interoperability relations. Proofs are given below.

	$\mathcal{R}_1$	$\mathcal{R}_2$, $\mathcal{R}_3$	$\mathcal{R}_4$	$\mathcal{R}_5$, $\mathcal{R}_6$	$\mathcal{R}_7$
$\mathcal{R}_8$ ($\cong_{\mathcal{R}}$ $\mathcal{R}_9$)	$\sqsubseteq_{\mathcal{R}}$	$\sqsubseteq_{\mathcal{R}}$	$\sqsubseteq_{\mathcal{R}}$	$\sqsubseteq_{\mathcal{R}}$	$\sqsubseteq_{\mathcal{R}}$
$\mathcal{R}_7$	$\sqsubseteq_{\mathcal{R}}$	-	$\sqsubseteq_{\mathcal{R}}$	-	
$\mathcal{R}_5$ ($\cong_{\mathcal{R}}$ $\mathcal{R}_6$)	-	-	$\sqsubseteq_{\mathcal{R}}$		
$\mathcal{R}_4$	-	-			
$\mathcal{R}_2$ ($\cong_{\mathcal{R}}$ $\mathcal{R}_3$)	$\sqsubseteq_{\mathcal{R}}$				

Figure 3. Comparison of specification-based interoperability relations

4.3.1 Some proofs. The decision of interoperability relies on observations done on the available interfaces. So the set of interfaces can be used to compare interoperability relations. "Lower-oriented" and "upper-oriented" relations are not comparable because their sets of interfaces are not comparable. Total interoperability relations are stronger than the corresponding upper or lower interoperability relations because their set of observations is greater. Bilateral and Global relations are stronger than their corresponding unilateral relations. Most of the proofs are based on these remarks. Nevertheless, we have observed that complete formal proofs of $\mathcal{R}_2 \cong_{\mathcal{R}} \mathcal{R}_3$, $\mathcal{R}_5 \cong_{\mathcal{R}} \mathcal{R}_6$, and $\mathcal{R}_8 \cong_{\mathcal{R}} \mathcal{R}_9$ need some intermediate lemmas and propositions. We give them in the following.

Lemma 4.1 *Let $M_1, M_2 \in \mathcal{IOLTS}$, and $\sigma \in Traces(M_1 \| M_2)$,*
$$Out(M_1\|M_2,\sigma) = Out(M_1,\sigma/\Sigma^{M_1}) \cup Out(M_2,\sigma/\Sigma^{M_2}).$$

Proof: Let $(q_1, q_2) \in ((M_1\|M_2) \text{ after } \sigma)$ and $a \in Out(M_1\|M_2,\sigma)$. Rules of definition 3.4 apply for (q_1, q_2). Thus, we have either $a \in \Sigma^{M_1}$ or $a \in \Sigma^{M_2}$ which means that $a \in Out(M_1,\sigma/\Sigma^{M_1}) \cup Out(M_2,\sigma/\Sigma^{M_2})$. In the other sense, it is easy to see that $Out(M_1,\sigma/\Sigma^{M_1}) \cup Out(M_2,\sigma/\Sigma^{M_2}) \subseteq Out(M_1\|M_2,\sigma)$. Thus, we have $Out(M_1\|M_2,\sigma) = Out(M_1,\sigma/\Sigma^{M_1}) \cup Out(M_2,\sigma/\Sigma^{M_2})$. $\diamond$

Lemma 4.2 *Let $M_1, M_2 \in \mathcal{IOLTS}$, and $\sigma \in Traces(M_1 \| M_2)$,*
$$Out_{\Sigma_L^{M_1\|M_2}}(M_1\|M_2,\sigma) = Out_{\Sigma_L^{M_1} \cup \Sigma_L^{M_2}}(M_1\|M_2,\sigma) \quad .$$

Proof: Using the definition 3.4 of the composition $\|$, we have $\Sigma_L^{M_1\|M_2} \subseteq \Sigma_L^{M_1} \cup \Sigma_L^{M_2} \Rightarrow Out_{\Sigma_L^{M_1\|M_2}}(M_1\|M_2,\sigma) \subseteq Out_{\Sigma_L^{M_1} \cup \Sigma_L^{M_2}}(M_1\|M_2,\sigma)$. Now, let $a \in Out_{\Sigma_L^{M_1} \cup \Sigma_L^{M_2}}(M_1\|M_2,\sigma)$, then $\sigma.a \in Traces(M_1\|M_2)$. Thus, $a \in \Sigma_L^{M_1\|M_2}$ and $Out_{\Sigma_L^{M_1} \cup \Sigma_L^{M_2}}(M_1\|M_2,\sigma) \subseteq Out_{\Sigma_L^{M_1\|M_2}}(M_1\|M_2,\sigma)$. $\diamond$

Proposition 4.1 $\mathcal{R}_2 \sqsubseteq_{\mathcal{R}} \mathcal{R}_3$.

Proof: Let I_1, I_2, S_1, $S_2 \in \mathcal{IOLTS}$ such that $\mathcal{R}_2(I_1, I_2)$. Let $\sigma \in Traces(S_1\|S_2)$ such that $\sigma \in Traces(I_1\|I_2)$. Let us consider $\sigma_1 = \sigma/\Sigma^{I_1}$ and $\sigma_2 = \sigma/\Sigma^{I_2}$, we can notice that $\sigma_1 \in Traces(S_1)$ and $\sigma_2 \in Traces(S_2)$. Using the definition of $\mathcal{R}_2(I_1, I_2)$, we have $Out_{\Sigma_L^{I_1}}(I_1\|I_2,\sigma) \subseteq Out_{\Sigma_L^{S_1}}(S_1,\sigma_1)$ and $Out_{\Sigma_L^{I_2}}(I_1\|I_2,\sigma) \subseteq Out_{\Sigma_L^{S_2}}(S_2,\sigma_2)$. Thus, with the definition 3.3, and the classical properties of $\cup$ and $\cap$, we have :
$$Out_{\Sigma_L^{I_1} \cup \Sigma_L^{I_2}}(I_1\|I_2,\sigma) \subseteq Out_{\Sigma_L^{S_1} \cup \Sigma_L^{S_2}}(S_1,\sigma_1) \cup Out_{\Sigma_L^{S_1} \cup \Sigma_L^{S_2}}(S_2,\sigma_2).$$
Using the definition of σ_1 and σ_2, we have :
$Out_{\Sigma_L^{I_1} \cup \Sigma_L^{I_2}}(I_1\|I_2,\sigma) \subseteq Out_{\Sigma_L^{S_1} \cup \Sigma_L^{S_2}}(S_1,\sigma/\Sigma^{I_1}) \cup Out_{\Sigma_L^{S_1} \cup \Sigma_L^{S_2}}(S_2,\sigma/\Sigma^{I_2})$
$\Leftrightarrow Out_{\Sigma_L^{I_1} \cup \Sigma_L^{I_2}}(I_1\|I_2,\sigma) \subseteq Out_{\Sigma_L^{S_1} \cup \Sigma_L^{S_2}}(S_1\|S_2,\sigma)$ (lemma 4.1)
$\Leftrightarrow Out_{\Sigma_L^{I_1\|I_2}}(I_1\|I_2,\sigma) \subseteq Out_{\Sigma_L^{S_1\|S_2}}(S_1\|S_2,\sigma)$ (lemma 4.2)
Thus, $\forall \sigma \in Traces(S_1\|S_2)$, $Out_{\Sigma_L^{I_1\|I_2}}(I_1\|I_2,\sigma) \subseteq Out_{\Sigma_L^{S_1\|S_2}}(S_1\|S_2,\sigma)$ which proves: $\mathcal{R}_2 \sqsubseteq_{\mathcal{R}} \mathcal{R}_3$. $\diamond$

Proposition 4.2 $\mathcal{R}_3 \sqsubseteq_{\mathcal{R}} \mathcal{R}_2$.

Proof: Let I_1, I_2, S_1, $S_2 \in \mathcal{IOLTS}$ such that $\mathcal{R}_3(I_1, I_2)$ and let us prove that it implies $\mathcal{R}_1(I_1, I_2)$. As $\mathcal{R}_1(I_1, I_2)$ does not refer to the specification of I_2, we can consider that $S_2 = I_2$. Let $\sigma_1 \in Traces(S_1)$, and $\sigma \in Traces(I_1||I_2), \sigma/\Sigma^{I_1} = \sigma_1$. As $\sigma_1 \in Traces(S_1)$ and $S_2 = I_2$, we have $\sigma \in Traces(S_1||S_2)$. Using the fact that $\mathcal{R}_3(I_1, I_2)$ gives $Out_{\Sigma^{I_1||I_2}}(I_1||I_2, \sigma) \subseteq Out_{\Sigma^{S_1||S_2}}(S_1||S_2, \sigma)$. We want to prove that $Out_{\Sigma^{I_1}}(I_1||I_2, \sigma) \subseteq Out_{\Sigma^{S_1}}(S_1, \sigma_1)$.

As $\Sigma^{I_1} \cap \Sigma^{I_2} = \emptyset$ and $\Sigma^{S_1} \cap \Sigma^{S_2} = \emptyset$ and considering only the outputs of I_1, we have: $Out_{\Sigma^{I_1}}(I_1||I_2, \sigma) \subseteq Out_{\Sigma^{S_1}}(S_1||S_2, \sigma)$. Using the definition 3.3, lemmas 4.1 and 4.2 gives the following equations $Out_{\Sigma^{S_1}}(S_1||S_2, \sigma) = Out_{\Sigma^{S_1}}(S_1, \sigma/\Sigma^{S_1}) \cup Out_{\Sigma^{S_1}}(S_2, \sigma/\Sigma^{S_2}) = Out_{\Sigma^{S_1}}(S_1, \sigma/\Sigma^{S_1}) = Out_{\Sigma^{S_1}}(S_1, \sigma_1)$. Thus, $Out_{\Sigma^{I_1}}(I_1||I_2, \sigma) \subseteq Out_{\Sigma^{S_1}}(S_1, \sigma_1)$. So, $\forall \sigma_1 \in Traces(S_1), \forall \sigma \in Traces(I_1||I_2)$, $\sigma/\Sigma^{I_1} = \sigma_1$, $Out_{\Sigma^{I_1}}(I_1||I_2, \sigma) \subseteq Out_{\Sigma^{S_1}}(S_1, \sigma_1)$ and $\mathcal{R}_3(I_1, I_2) \Rightarrow \mathcal{R}_1(I_1, I_2)$.

Using the fact that $\mathcal{R}_3$ is symmetrical, we have $\mathcal{R}_3(I_1, I_2) \Rightarrow \mathcal{R}_1(I_2, I_1)$. Thus, $\mathcal{R}_3(I_1, I_2) \Rightarrow (\mathcal{R}_1(I_1, I_2) \wedge \mathcal{R}_1(I_2, I_1)) = \mathcal{R}_2(I_1, I_2)$. $\diamond$

Theorem 4.1 *$\mathcal{R}_3 \cong_{\mathcal{R}} \mathcal{R}_2$, $\mathcal{R}_6 \cong_{\mathcal{R}} \mathcal{R}_5$, and $\mathcal{R}_9 \cong_{\mathcal{R}} \mathcal{R}_8$*

Proof: The first proof is achieved with the proposition 4.1 and the proposition 4.2. The two other proofs are achieved with similar lemmas adapted to upper and total contexts. This means that the global and bilateral interoperability relations are equivalent. $\diamond$

5. INTEROPERABILITY TESTS GENERATION

Several methods have been proposed to generate interoperability tests [2, 14, 15, 16]. Now that we have identified the interoperability architectures and we have defined possible interoperability relations, we give in this section some guidelines to derive interoperability tests.

How to choose interoperability relations?
Depending on several parameters (accessibility of interfaces of the implementations, problems of confidentiality, etc), different architectures can be used for interoperability testing. In Section 2.1, we have identified the four possible classes of interoperability testing architectures. An architecture determines the possible interoperability relations which can be used. As proved in Section 4.3, the power of the generated tests depends on the chosen relation. In the following, we give elements which help in choosing appropriate interoperability relations.

▷ For unilateral testing architectures, the choice of interoperability relations depends on available interfaces (upper, lower or both).

▷ As bilateral and global interoperability relations are equivalent (see theorem 4.1), it can be more interesting to use a bilateral interoperability relation in a global testing architecture. Indeed, this choice avoids the often difficult and error-prone task of designing test coordination procedures which are required when using global interoperability relations.

▷ Hybrid architectures can be tested by generalizing interoperability relations to hybrid cases, or by using a combination of existing relations. For example, architectures where one lower interface is unavailable can be tested using either a combination of $\mathcal{R}_4$ (unilateral upper) relation and $\mathcal{R}_7$ (unilateral total) relation, or a combination of $\mathcal{R}_6$ and $\mathcal{R}_1$ (global upper and unilateral lower) relations.

Can conformance tests be used for interoperability?

There is still considerable debate about the value of conformance testing as a means of achieving interoperability [1, 16]. Let us consider the **ioconf** relation. Proposition 5.1 suggests that an implementation tested with conformance relation **ioconf** does not need to be tested again when the $\mathcal{R}_7$ interoperability relation is used.

Proposition 5.1 *Let* $I_1, I_2 \in \mathcal{IOLTS}$, I_1 **ioconf** $S_1 \Rightarrow \mathcal{R}_7(I_1, I_2)$

Proof: Let us consider $I_1, I_2 \in \mathcal{IOLTS}$ and $S_1 \in \mathcal{IOLTS}$ the specification on which I_1 is based. I_1 **ioconf** S_1 implies $\forall \sigma_1 \in Traces(S_1)$, $Out(I_1, \sigma_1) \subseteq Out(S_1, \sigma_1)$. Let us consider a trace $\sigma \in Traces(I_1 \| I_2)$ such that $\sigma/\Sigma^{I_1} = \sigma_1$. Using lemma 4.1 and properties of the projection, we have $Out_{\Sigma^{I_1}}(I_1 \| I_2, \sigma) = Out(I_1, \sigma_1)$. Thus, as $Out(I_1, \sigma_1) \subseteq Out(S_1, \sigma_1)$, we have I_1 **ioconf** $S_1 \Rightarrow \mathcal{R}_7(I_1, I_2)$. ◇

The problem is that conformance testing is not exhaustive. Thus, the result of Proposition 5.1 is "theoretical". On the other hand, proposition 5.2 says that an implementation can be considered $\mathcal{R}_7$-interoperable with another one without being conformant to its specification.

Proposition 5.2 *Let* $I_1, I_2 \in \mathcal{IOLTS}$, $\mathcal{R}_7(I_1, I_2) \not\Rightarrow I_1$**ioconf** S_1

Proof: It is based on the fact that : $Traces(I_1 \| I_2)/\Sigma^{I_1} \subseteq Traces(I_1)$. Thus, it can exist a trace $\sigma_1' \in Traces(I_1) \setminus Traces(I_1 \| I_2)/\Sigma^{I_1}$, such that $\mathcal{R}_7(I_1, I_2)$ but not I_1**ioconf** S_1. ◇

We have similar results for other interoperability relations. Given the complexity of standards and the limits on exhaustive conformance testing, interoperability testing is seen to be a practical requirement. In particular, it is used to uncover incompatibilities (such as incompatible options, coding, etc) even when both implementations have successfully undergone (whatever rigorous) conformance tests. It is a matter of increasing the confidence in the real interoperability between implementations.

How to generate interoperability tests
Interoperability relations defined in Section 4.2 can be written in the shape of "parameterized" conformance relations. Indeed, interoperability relations consider the interaction between implementations, rather than a separately considered implementation in conformance relations. Thus, parameters to be introduced in conformance relations (like **ioconf**) in order to obtain interoperability relations are implementations (resp. specifications) with which the considered implementation (resp. specification) has to interact and the available interfaces. Thus, one may wonder if existing automatic test generation tools like TGV [9] can be used and/or adapted for automatically generating interoperability tests. The general answer to this question needs some further

Theorems 4.1 which prove that bilateral and global interoperability relations are equivalent hints that we may avoid the construction of the composition of the specifications. It is well-known that this construction (even if it is done on-the-fly like in TGV) is a bottleneck for most of the existing tools. This is our current work and first results are promising.

6. CONCLUSION AND FUTURE WORK

In this paper, we have proposed a formal framework (architectures, interoperability relations, test generation techniques) for interoperability testing. We show that the existing concepts of conformance testing can be used to give a formal definition of the notion of interoperability. Different interoperability testing architectures are proposed and discussed. We have defined several interoperability relations based on the existing implementation relations defined for conformance testing. A comparison of these relations is given, in terms of their power to detect non-interoperability. This comparison suggests that the generation of interoperability tests do not necessarily need to build the whole interaction between the specifications of the implementations. This paper ends with some guidelines to help in generating interoperability tests.

As future work, we will focus on quiescence management in the interoperability relations. Based on the defined interoperability relations and the results of their comparison presented in this paper, we will investigate more deeply the different methods and associated algorithms in order to automatically and efficiently generate interoperability tests.

REFERENCES

[1] O. Rafiq and R. Castanet. From conformance testing to interoperability testing. In *Protocol Test Systems*, volume III, pages 371–385, North-Holland, 1991. IFIP, Elsevier sciences publishers B. V.

[2] R. Castanet and O. Koné. Deriving coordinated testers for interoperability. In O. Rafiq, editor, *Protocol Test Systems*, volume VI C-19, pages 331–345, Pau-France, 1994. IFIP, Elsevier Science B.V.

[3] T. Walter and B. Plattner. Conformance and interoperability a critical assessment. Technical Report 9, Computer engineering and networks laboratory (TIK), Swiss federal institute of technology Zurich, 1994.

[4] J.P. Baconnet, C. Betteridge, G. Bonnes, F. Van den Berghe, and T. Hopkinson. Scoping further EWOS activity for interoperability testing. Technical Report EGCT/96/130 R1, EWOS, 1996.

[5] ISO. Information Technology - Open Systems Interconnection Conformance Testing Methodology and Framework - Parts 1-7. *International Standard ISO/IEC 9646/1-7*, 1992.

[6] E. Brinksma, R. Alderden, J. Langerak, R. Van de Lagemaat, and J. Tretmans. A Formal Approach to Conformance Testing. In J. De Meer, L. Mackert, and W. Effelsberg, editors, *Second International Workshop on Protocol Test Systems*, pages 349–363, North Holland, 1990.

[7] M. Phalippou. *Relations d'implantations et Hypothèses de test sur les automates à entres et sorties*. PhD thesis, Université de Bordeaux, France, 1994.

[8] J. Tretmans. *A formal approach to conformance testing*. PhD thesis, University of Twente, Enschede, The Netherlands, 1992.

[9] J.-C. Fernandez, C. Jard, T. Jéron, and C. Viho. An experiment in automatic generation of test suites for protocols with verification technology. *Science of Computer Programming - Special Issue on Industrial Relevant Applications of Formal Analysis Techniques*, 1997.

[10] T. Walter, I. Schieferdecker, and J. Grabowski. Test architectures for distributed systems : state of the art and beyond. In Petrenko and Yevtushenko, editors, *Testing of Communicating Systems*, vol. 11, pages 149–174. IFIP, Kap, 1998.

[11] J. Gadre, C. Rohrer, C. Summers, and S. Symington. A COS study of OSI interoperability. *Computer standards and interfaces*, 9(3):217–237, 1990.

[12] G. Bochmann, R. Dssouli, and J. Zhao. Trace analysis for conformance and arbitration testing. *IEEE Trans. on Software Eng.*, 15(11):1347–1356, 1989.

[13] L. Verhaard, J. Tretmans, P. Kars, and E. Brinksma. On Asychronous Testing. In G. Von Bochman, R. Dssouli, and A. Das, editors, *Fifth International Workshop on Protocol Test Systems*, pages 1–13, North Holland, 1993. IFIP Transactions.

[14] N. Arakawa, M. Phalippou, N. Risser, and T. Soneoka. Combination of conformance and interoperability testing. *FORTE'92*, V(C-10):397–412, 1992.

[15] S. Kang and M. Kim. Test sequence generation for adaptive interoperability testing. In A. Cavally and S. Budkowski, editors, 8th *International Workshop on Protocol Test Systems*, pages 193–206, Evry, France, 1995. IFIP.

[16] K. Myungchul, K. Gyuhyeong, and D.C. Yoon. Interoperability testing methodology and guidelines. In *Digital Audio-Visual Council, system integration TC*, volume DAVIC/TC/SYS/96/06/006, New York, 1996.

Acknowledgments

Authors wish to thank their colleagues D. Clarke for carefully reading this paper, and C. Jard & T. Jéron for their constructive remarks.

DISTRIBUTED TEST USING LOGICAL CLOCK

Young Joon Choi[1], Hee Yong Youn[2], Soonuk Seol[1], and Sang Jo Yoo[3]
[1]*School of Engineering, Information and Communications University*
{cyjnice, suseol}@icu.ac.kr
[2]*School of ECE, Sungkyunkwan University, Suwon, Korea*
youn@ece.skku.ac.kr
[3]*The Graduate School of Information Technology & Telecommunications, Inha University, Inchon, Korea*
sjyoo@inha.ac.kr

Abstract It is difficult to test a distributed system because of the task of controlling concurrent events. Existing works do not propose the test sequence generation algorithm in a formal way and the amount of message is large due to synchronization. In this paper, we propose a formal test sequence generation algorithm using logical clock to control concurrent events. It can solve the control-observation problem and makes the test results reproducible. It also provides a generic solution such that the algorithm can be used for any possible communication paradigm. In distributed test, the number of channels among the testers increases non-linearly with the number of distributed objects. We propose a new remote test architecture for solving this problem. SDL Tool is used to verify the correctness of the proposed algorithm, and it is applied to the message exchange for the establishment of Q.2971 point-to-multipoint call/connection as a case study.

Keywords: concurrent events, control-observation problem, distributed testing, logical clock, output-shifting faults, test sequence

1. INTRODUCTION

Distributed objects have been implemented by Open Distributed Processing (ODP) [11] and others. ODP is an ISO/ITU standard, offering a generic framework for the development of distributed system. In this paper, distributed system developed based on ODP is taken as the target [1] to test.

Benattou et al. [1] proposed a method to test distributed system as a black box using a distributed test method. Here extra signals were proposed to be used among the testers through coordination channels, called multicast channel, for synchronizing them. The method has a merit of making the events occurred at the distributed objects be totally ordered with special signals, but has a drawback of possible output-shifting faults [8]. A method in [2] is to use synchronization messages for controlling concurrent events as a grey box. However, it has a difficulty in choosing a way to observe the internal status of implementation under test (IUT) and needs other messages in addition to TS. Luo et al. [3] analyzed existing models based on synchronized TS and surveyed fault coverage in terms of output, transfer, and hybrid. However, it did not propose any Test Sequence (TS) generation algorithm in a formal way.

In overall, existing works for controlling concurrent events in the test of distributed system do not propose the algorithm in a formal way and increase the amount of message within IUT due to synchronization. To overcome the drawbacks, we propose a TS generation algorithm using logical clock that formally generates signals to control concurrent events by numerically labeling the events of TS with logical clock values and comparing them. It can also solve the control-observation problem [1,3]. The proposed TS generation algorithm based on logical clock allow the test results to be reproduced since a totally ordered TS can be achieved with the controlling of concurrent events.

In distributed test, several testers are used and synchronized by exchanging coordination messages. While traditional distributed test method is effective in testing IUT of multiple ports, it has a shortcoming that the number of channels among the testers increases non-linearly with the number of distributed objects. To overcome this drawback, we also propose an approach of remote test architecture. Specification and Description Language (SDL) Tool [14] is used to verify the correctness of the proposed algorithm, and it is applied to the message exchange for the establishment of Q.2971 point-to-multipoint call/connection as a case study.

The rest of the paper is structured as follows. Section 2 explains the background on testing distributed system. The proposed TS generation algorithm based on logical clock is proposed in Section 3. Section 4 applies the algorithm to the remote test architecture, and finally Section 5 concludes the paper with some future research work.

2. BACKGROUND

When a distributed system is tested, some of the events executed by the distributed objects do not always have casual relationship but possibly concurrent relationship with each other. If the events are concurrent, the following **control-observation problem** may take place [1, 3].

- When concurrent inputs are applied to a black box of a form of IUT, the test results depend on the order of input process in the IUT.
- In testing a black box IUT, even though the mapping of an input and its output is incorrect, test results may be correct.

The events occurred at each distributed object, which is an element of a distributed system, may have concurrent relationship with other events. In this case, the test results can be different from each other when an IUT is tested with the same TS repeatedly since all the events are not totally ordered. Therefore we need a mechanism to make concurrent events have causal relation in order to reproduce the same test results. Causal relation is also called "happened before" relation, represented by '→'. Relation 'a' → 'b' should satisfy one of the following rules to be 'happened before' events [5, 6].

Rule 1

(1) 'a' and 'b' are in the same distributed object and 'a' → 'b'.
(2) 'a' is sender and 'b' is receiver. However, in case of synchronous communication, both send and receive events occur at the same time.
(3) 'a' → 'c' and 'c' → 'b'.

Logical clocking is a time-stamping mechanism, which can be represented with linear time, vector time, and matrix time [5]. In this paper, we adopt vector time describing logical clock in one-dimensional matrix whose field is assigned to each object [5]. The value of a field of logical clock denotes the time the corresponding event occurs at a distributed object. With the vector time, the following rule should be satisfied so that two events can be causally related.

Rule 2

If the following relation is satisfied between event 'X' and 'Y' in an IUT consisting of distributed objects a, b and c, the two events are said to have a causal relation "a → b" except a case as ($X_a = Y_a, X_b = Y_b, X_c = Y_c$). The logical clock values of events 'X' and 'Y' are (X_a, X_b, X_c) and (Y_a, Y_b, Y_c), respectively.

$$(X_a \leq Y_a, X_b \leq Y_b, X_c \leq Y_c)$$

A stable state means a global state, which is not changed into another state without any input [7]. In testing a black box IUT, it is difficult to observe the transient states and only the stable states are visible.

Figure 1. Stable states and transient states.

Figure 1 represents a finite state machine (FSM), which reaches stable state-3 via transient state-1 and 2 after input 'a' is applied at stable state-0. A tester can infer the test result by observing state-0 before the input 'a' was applied and state-3 after the output 'x' and 'z' are obtained. This allow to reduce the number of states which should be managed in the test and also avoid the difficulty of observing transient states.

3. THE PROPOSED ALGORITHM

In this section we first analyze the existing method proposed for controlling concurrent events in a black box IUT. We then propose a TS generation algorithm based on logical clock.

3.1 The Problems in Existing Test Method

As a case study, we consider a typical test method reported in [1]. Figure 2(a) represents a FSM M, where each state is stable. The FSM M is a 3-port FSM whose inputs and outputs at each port are shown in Table 1. Figure 2(b) is faulty FSM M derived from Figure 2(a). Here each port is connected to a tester whose identifier is the same as the port.

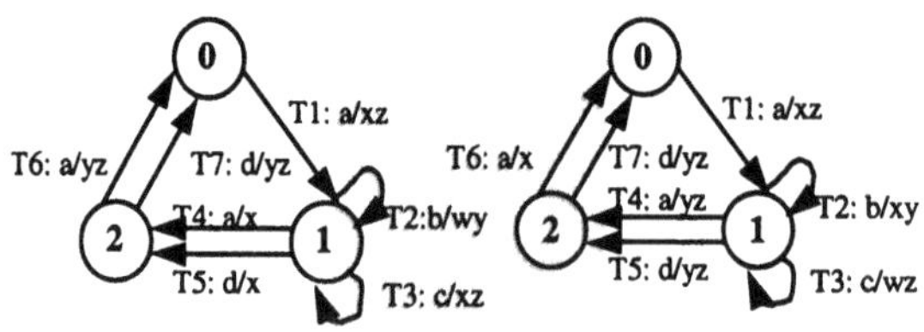

(a) FSM M (b) Faulty FSM M

Figure 2. 3-port FSM.

Table 1. Inputs and outputs at each port of Figure 2.

	port 1	port 2	port 3
Input	a	b	c, d
Output	x, w	y	z

Definition 1: Transition T is represented by $(s_i, s_j, x/y)$, where $x \in X$, $y \in Y$, $\{s_i, s_j\} \subset S$, $s_j = \delta(s_i, x)$, and $y = \lambda(s_i, x)$. X is a finite set of input symbols, Y is a finite set of output symbols, S is a finite set of states, δ is a state transfer function, and λ is an output function.

If an IUT implemented with a faulty FSM M is tested by a TS (T1 → T3 → T2 → T5 → T6), the control-observation problem may occur in the subsequence T3 → T2 and T5 → T6. None of the testers can detect the faults if 'b' is read first by the IUT when the input 'c' of transition T3 and the input 'b' of transition T2 are applied to the IUT concurrently in state-1. It is because the order of the outputs observed by the testers is correct even though the mapping of the input and its output is incorrect in the subsequence T5 → T6.

The TS of a distributed object usually consists of inputs and outputs, and signal 'C' and 'O' controlling concurrent events. The signal 'C' gives casual relation to concurrent inputs applied to IUT, while signal 'O' gives causal relation between concurrent input and output. The following steps produce signal 'C' and 'O'. A new TS for each distributed object including signal 'C' and 'O' is called Local Test Sequence (LTS).

1. Label logical clock values for the events of TS.
2. Classify the events into either causal or concurrent by comparing the logical clock values of the events.
3. If there exist concurrent events, generate signal 'C' and 'O' which make the concurrent events be causally related and insert them to the TS of each distributed object.

To solve the control-observation problem, [1] proposed coordination channels whose function is to send or receive signal 'C' and 'O' among the testers. However, output–shifting fault which is a consequence of the control-observation problem can occur.

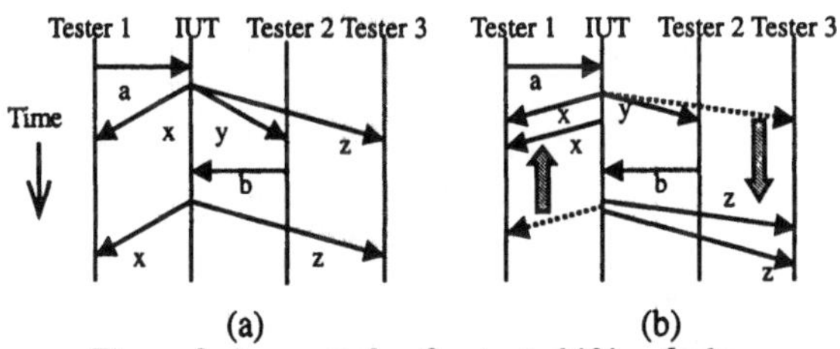

Figure 3. An example of output-shifting faults.

Refer to Figure 3(a) where an IUT is tested by TS (a/xyz → b/xz). The signal 'C' and 'O' are not generated by the method proposed in [1]. As a result, none of the testers can detect the faults even if the output 'z' of "a/xyz" is shifted forward and the output 'x' of "b/xz" is shifted backward as shown in Figure 3(b). This is because the order of the input and output observed by each tester is the same as Figure 3(a). This is called **output-shifting faults** [8]. Output-shifting faults are the case that the test results are correct even though the mappings of input and output are incorrect. This occurs since it cannot properly control the testers receiving the outputs of the previous transition which has concurrency relation with the new input. A new formal way for signal generation is thus needed to be developed.

3.2 The Proposed Local Test Sequence Generation Algorithm

Here we propose an LTS generation algorithm which can solve the control-observation problem including output-shifting faults in a formal way using logical clock. We make assumptions as follows:

- IUT is a black box.
- The logical clock vector is represented as a format (tester 1, IUT, tester 2, tester 3). When an event occurs at an element, the value of the corresponding field is increased.
- The output receive delay for a transmission is zero.
- Logical clock is increased by the unit time.

Figure 4 shows TS1 (T1 → T3 → T2 → T4 → T7) which tests the IUT implemented with FSM M. Observe that each event is labeled with a logical clock value. To control concurrent events classified by Rule 2, extra signals are needed. Table 2 lists the comparisons of logical clock values labeled with the events of subsequence T1 → T3 and T3 → T2. Here Compare_LC(x, y) is a function that returns `true' if event 'x' and 'y' satisfies Rule 2. Otherwise, it returns `false'.

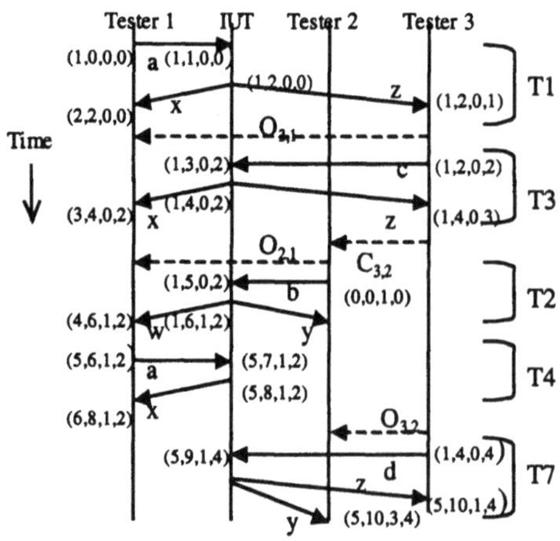

Figure 4. Logical clock labeling at the events of Test Sequence 1 (TS1).

Table 2. Comparison of logical clock values.

T1 (0, 1, a/xz) →T3 (1, 1, c/xz)	T3 (1, 1, c/xz) →T2 (1, 1, b/wy)
Compare_LC(a, c) = true	Compare_LC(c, b) = false
Compare_LC(x, c) = false	Compare_LC(x, b) = false
Compare_LC(z, c) = true	Compare_LC(z, b) = false

According to Table 2, only output ‘x’ of transition T1 and input ‘c’ of transition T3 are concurrent in subsequence T1 → T3. In this case, a signal making input ‘c’ and output ‘x’ be casually related is needed. This signal is called signal ‘O’. Accordingly, Tester 3 that provides input ‘c’ of transition T3 to IUT sends signal ‘$O_{3,1}$’ to Tester 1, which receives output ‘x’ of transition T1. Similarly, input ‘b’ of transition T2 is concurrent with all the inputs and outputs of transition T3 in subsequence T3 → T2. Therefore a signal is needed to let input ‘b’ be applied to the IUT after transition T3 occurs. This signal is called signal ‘C’. In this case, Tester 3 that is one of the testers receiving the outputs of transition T3 sends signal ‘$C_{3,2}$’ to Tester 2, which applies input ‘b’ of transition T2 to the IUT. The signal ‘$C_{3,2}$’ suggests that Tester 3 had already received output ‘z’. However, Tester 2 sends additional signal ‘$O_{2,1}$’ to Tester 1 since input ‘b’ is still concurrent with output ‘x’ of transition T3.

None of the testers can detect faults even though output ‘y’ is backward-shifted since the test scenario of Tester 2 is the same as that for correctly implemented FSM being tested by subsequence T4 → T7. To solve this problem, Tester 3 sends signal ‘$O_{3,2}$’ to Tester 2 before Tester 3 applies input ‘d’ to the IUT.

As aforementioned, the output receive delay is assumed to be zero, and thus each tester receives the output of a transition instantly. Therefore, signal ‘O’ controlling concurrent outputs is generated by the tester which sends input to the IUT. This assumption was also adopted in [1]. With this assumption, though, the TS generation algorithm has some limitations in testing real distributed system.

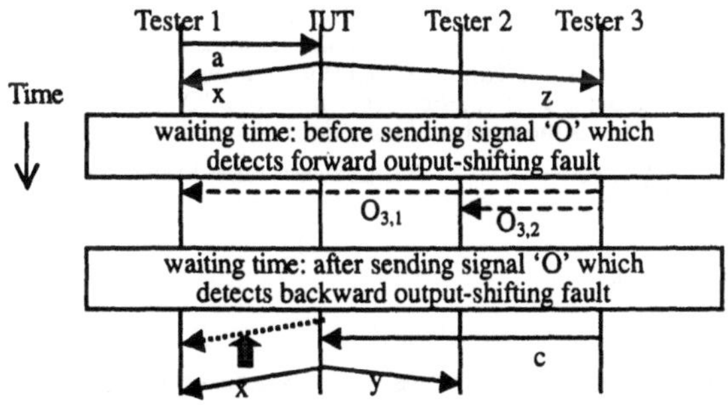

Figure 5. Waiting time and over-waiting time problem.

For example, refer to Figure 5. Tester 1 decides that the IUT is faulty when it receives output ‘x’ in “a/xz” after Tester 1 eceives signal ‘$O_{3,1}$’ even if the IUT is correctly implemented. Also Tester 2 cannot detect faults in case that output ‘y’ in “c/xy” is backward-shifted. This is because synchronization among the testers is done through synchronization signal, but IUT cannot be directly synchronized with the testers. Accordingly, the testers should have sufficient waiting time to synchronize with the IUT.

However, if Tester 1 receives backward-shifted output 'x' in "c/xy" after some waiting time, it is impossible to detect the fault. Let us call this **over-waiting problem**. In order to prevent the problem, the waiting time and receive delay time of outputs for a transmission are assumed to be zero. The related definitions and the proposed algorithm are presented as follows.

Definition 2

- In transition T = (si, sj, x/y), "x/y" stands for "!x?y".
- '-' means that signal is sent to the tester and '+' means that signal is received from other testers.
- Port(x) is a function that returns 'port id' generating input or output 'x'.
- PortS(X) = {Port(x1), Port(x2), ..., Port(xn)}, where X = {x1, x2, ..., xn} and xi is an event.
- Compare_LCS(x, Y) is a function that returns false if Compare_LC(x, a) = false for all a ∈ Y, and return true otherwise where Y is a set of input or output.
- Label_LC(x, k) is a function that updates logical clock value of port k and records logical clock value of x where x is an input or output.
- Label_LCS(X, PortS(X)) = {Label_LC(x1, Port(x1)), Label_LC(x2, Port(x2)), ..., Label_LC(xn, Port(Xn))}. Append(X, Y) is a function concatenating string X and Y.
- Append_In(Port(x), x) = Append(LTSPort(x), Append("!", x)) where LTSPort(x) is an LTS of the tester connected to Port(x).
- Append_Out(PortS(X), X) = {Append(LTSPort(X1), Append("?", "x1")), Append(LTSPort(X2), Append("?", "x2")), ..., Append(LTSPort(Xn), Append("?", "xn"))}.

```
ALGORITHM_LTS /* LTS GENERATING ALGORITHM */
INPUT: TS = !s1?Y1...!sm?Ym /*The elements of Yi can be more than zero and is represented with an array */
OUTPUT: local test sequences = (LTS1, ... ,LTSn) /* n is the number of distributed objects/
Send_Set = null /* Send_Set is a set of testers receiving signal 'O' */
for(i = 1; i ≤ m; i++)   /* m is the number of transitions */
 Label_LC(si, Port(si ));        <------   /* Logical clock values labeling at input si and outputs Yi */
 Label_LCS(Yi, PortS(Yi));
end for
for(i = 1; i ≤ m; i++)
 k = Port(si); h = Port(si+1);
 Append_In(Port(si), si );
 if i < m
  case |Yi| ≠ 0:
   Append_Out(PortS(Yi), Yi);  <--  /* Comparison of logical clock values for generating signal 'C' */
   if (Compare_LCS(si+1,Yi) = false) and (Compare_LC( si+1, si) = false)
    select a sender (tester) which generates signal 'C' where sender ∈ PortS(Yi);
    Append(LTSsender, Append("-", "Csender,k")); Append(LTSk, Append("+", "Csender,k"));
   end if
   for(j = 0; j < |Yi|; j++)   <---  /* Comparison of logical clock values for generating
    if (Compare_LC(si+1, Yi[j]) = false) and (Port(Yi[j])) ≠ sender)   signal 'O' to protect forward
     Send_Set ← Send_Set ∪ Port(Yi[j]);                                   output-shifting faults */
    end if
   end for                          /* Comparison of logical clock values for generating
   for(j = 0; j < |Yi+1|; j++)  <---  signal 'O' to protect backward output-shifting faults */
    if [(Compare_LCS(Yi+1[j], Yi) = true) and (Port(Yi+1[j]) ∉ PortS(Yi)) and
    (Port(Yi+1[j])≠ h)]  or [(Compare_LCS(Yi+1[j], Yi) = false) and (Port(Yi+1[j]) ≠ h)]
     Send_Set ← Send_Set ∪ Port(Yi+1[j]);
    end if
```

```
27:   end for
28:  case |Y_i| = 0:         <---   /* Comparison of logical clock value for generating signal 'C' */
29:   if Compare_LC( s_i, s_i+1) = false
30:    Append(LTS_k, Append("-", "C_k,h"); Append(LTS_h, Append("+", "C_k,h"));
31:   end if
32:   for(i = 0; i < |Y_i+1|; j++)
33:    if Port(Y_i+1[j]) ≠ h  <---------------   /* Port comparison for generating signal 'O' to
34:     Send_Set ← Send_Set ∪ Port(Y_i+1[j]);       protect backward output-shifting faults */
35:    end if
36:   end for
37:  end case
38:  for ∀p ∈ Send_Set
39:   Append(LTS_h, Append("-", "O_h,p"); Append(LTS_p, Append("+", O_h,p));
40:  end for
41: else if (i = m) and (|Y_i| ≠ 0)
42:  Append_Out(Port(Y_i), Y_i);
43: end if
44: end for
end Algorithm_LTS
```

In the algorithm above, Lines from 4 to 7 are to label logical clock values for the events of TS. Whenever input or output is generated from the IUT, Label_LC (or Label_LCS) function updates the logical clock value of the distributed object where the input or output is generated. Lines from 14 to 40 are to generate signal 'C' and 'O' in the transitions except the final transition. When the output of a transition is not null, the logical clock values of the events between the current transition and the input events of the next transition are compared using Line 19. If they are concurrent, signal 'C' is generated. Lines 19 and 24 are the statements comparing logical clock values for controlling forward output-shifting faults and backward output-shifting faults, respectively. When the output of a transition is null, the logical clock values between the current and next input are compared using Line 29. If they are concurrent, signal 'C' is generated. Line 33 is to solve backward output-shifting faults in case of null output by comparing the clock values. Figure 6 is a test scenario obtained as a result of applying Algorithm_LTS to TS1.

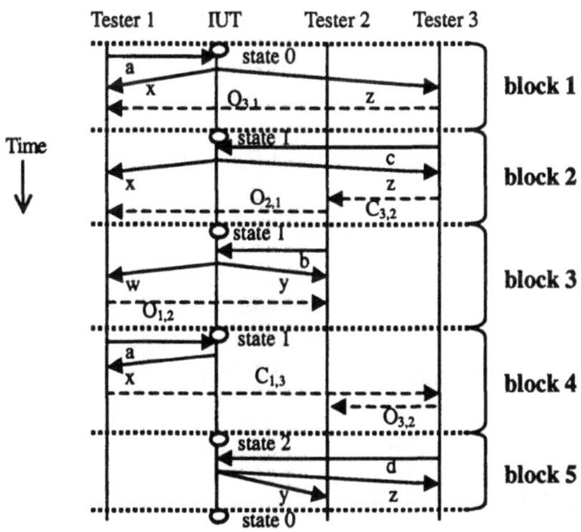

Figure 6. Applying Algorithm_LTS to TS1.

Note that state 0, 1 and 2 in Figure 6 are stable states since they are not changed to other states before a new input is applied to the IUT. If the area between the stable states is represented as a block, happened-before relationships also exist between the blocks since the events in each block are totally ordered with the events in other blocks by signal 'C' and 'O'.

3.3 Extension of the Proposed Local Test Sequence Generation Algorithm

The LTS generation algorithm [1] and the proposed Algorithm_LTS assume that the output receive delay is zero. If the assumption is not satisfied, test results may not be able to be reproduced. For example, in Figure 6, even though IUT is correct, the test results can be faulty when Tester 3 receives output 'z' and then sends signal '$O_{3,1}$' to Tester 1 before output 'x' is received by Tester 1 in the first block. To solve the problem, LTS for Tester 3 should be updated so that input 'c' can be sent to the IUT after Tester 1 sends signal '$O_{3,1}$' to Tester 3 and Tester 3 receives both signal '$O_{3,1}$' and output 'z'.

Let us call the LTS generation algorithm for which output receive delay is not zero Algorithm_LTS_DELAY. Append_RD(X, Y) is a function attaching Y to X while keeping the order of the elements of them. For example, Append_RD("xy", "z") means that 'xyz', 'xzy', or 'zxy'.

ALGORITHM_LTS_DELAY

```
INPUT: TS = !s1?Y1...!sm?Ym /*The elements for Yi can be more than zero and is represented by an array */
OUTPUT: local test sequences = (LTS1, ... ,LTSn) /*n is the number of distributed objects */
Send_Set = null /* Send_Set is a set of testers which should receive signal 'O' */
for(i = 1; i ≤ m; i++)  /* m is the number of transitions */
 Label LC(si, Port(si ));
 Label_LCS(Yi, PortS(Yi));   <--- /* Logical clock values labeling at input si and outputs Yi */
end for
for(i = 1; i ≤ m; i++)
 k = Port(si); h = Port(si+1);      /* Case 1: comparison of logical clock values for
 Append_In(k, si);                     generating signals 'C' and 'O' */
 if i < m
  case |Yi| ≠ 0:
   if (Compare_LCS(si+1, Yi) = false) and (Compare_LC(si+1, si) = false)
    Append_Out(PortS(Yi), Yi);
    select a sender (tester) which generates signal 'C' where sender ∈ PortS(Yi);
    Append(LTSsender, Append("-", "Csender,h"));
    Append_RD(temp, Append("+", "Csender,h"));
    Temp_Set ← PortS(Yi) \ sender;  /* / is a difference set */
   else if (Compare_LCS(si+1, Yi) = false) and (Compare_LC(si+1, si) = true)
    Append_Out(PortS(Yi), Yi);          <--- /* Case 2: comparison of logical clock
    Temp_Set ← PortS(Yi);                         values for generating signal 'O' */
   else if Compare_LCS(si+1, Yi) = true
    for(j = 0; j < |Yi|; j++)           <--- /* Case 3: comparison of logical clock
     if Compare_LC(si+1, Yi[j]) = false           values for generating signal 'O' */
      Temp_Set ← Temp_Set ∪ Port(Yi[j]);
      Append(LTSa, Append("?", Yi[j]));
     else Append(temp, Append("?", Yi[j]));
     end if
    end for
```

```
30:  end if
31:  for ∀p ∈ Temp_Set
32:   Append(LTS_p Append("-", "O_p,h")); Append_RD(temp, Append("+", "O_p,h"));
33:  end for
34:  Append(LTS_h, temp);               /* Comparison of logical clock values for generating signal
35:  for(i = 0; i < |Y_i+1|; i++)  <----- 'O' to protect backward output-shifting faults */
36:   if [(Compare_LCS(Y_i+1[j], Y_i) = true) and (Port(Y_i+1[j]) ∉ PortS(Y_i)) and
      (Port(Y_i+1[j]) ≠ h)] or [(Compare_LCS(Y_i+1[j], Y_i) = false) and (Port(Y_i+1[j]) ≠ h)]
37:    Send_Set ← Send_Set ∪ Port(Y_i+1[j]);
38:   end if
39:  end for
40: case |Y_i| = 0:
41:  if Compare_LC( s_i, s_i+1) = false  <----- /* Comparison of logical clock value for generating signal 'C' */
42:   Append(LTS_k, Append("-", "C_k,h"); Append(LTS_h, Append("+", "C_k,h"));
43:  end if
44:  for(i = 0; i < |Y_i+1|; j++)   ------ /* Port comparison for generating signal 'O' to protect backward
45:   if Port(Y_i+1[j]) ≠ h <------             output-shifting faults*/
46:    Send_Set ← Send_Set ∪ Port(Y_i+1[j]);
47:   end if
48:  end for
49: end case
50: for ∀p ∈ Send_Set
51:  Append(LTS_k, Append("-", "O_k,p"); Append(LTS_p, Append("+", O_k,p));
52: end for
53: else if (i = m) and (|Y_i| ≠ 0)
54:  Append_Out(Port(Y_i), Y_i);
55: end if
56: end for
end Algorithm_LTS_DELAY
```

The lines from 13 to 34 are modified from Algorithm_LTS. Line 13 (Case 1), Line 19 (Case 2), and Line 22 (Case 3) compare logical clock values for controlling forward output-shifting faults in case that the outputs are not null. Case 1 is applied when all the events of the current transition and an input of the subsequent transition are concurrent, Case 2 is applied when outputs of the current transition and an input of the subsequent transition are concurrent, and Case 3 is applied when some outputs of the current transition and an input of subsequent transition are causally related, respectively. The signal 'O' generated by each case is sent from the tester (Temp_Set), which receives the outputs of the current transition, to a tester which sends an input of the subsequent transition to the IUT. Therefore, the LTS for input-side testers should manage all combinations of the receive order of the input events. Generating signal 'O' that controls backward output-shifting faults is the same as the one in Algorithm_LTS.

Figure 7 is a sample test scenario. Here Specification and Description Language (SDL) Tool [14] is used to verify the correctness of Algorithm_LTS_DELAY. An FSM representing the IUT and the LTS obtained by Algorithm_LTS_DELAY for TS1 are described in SDL processes, respectively. Note that the output receive delay is not zero.

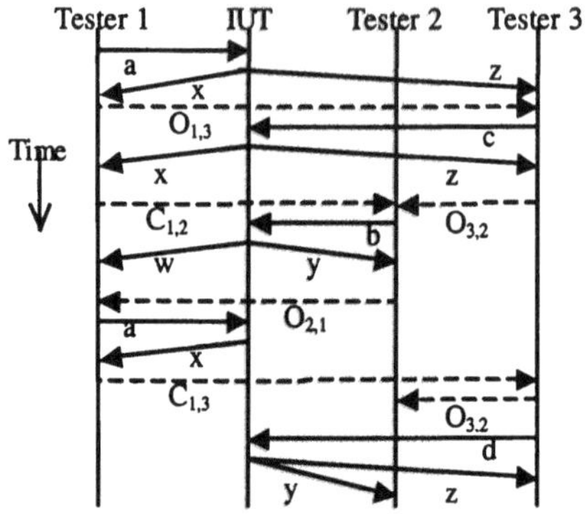

Figure 7. The LTS obtained by Algorithm_LTS_DELAY for TS1.

We analyze the results of Message Sequence Chart (MSC) Trace which is the test scenario for Figure 7 using reachability tree. Figure 8 shows the reachability tree obtained by the SDL Tool. The states represented at each point are global ones showing the states of Tester 1, IUT, Tester 2 and Tester 3, respectively.

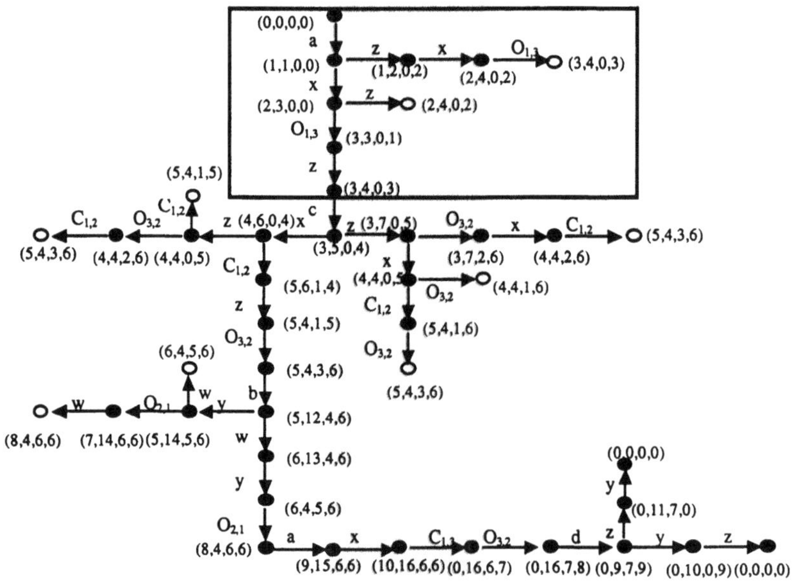

Figure 8. The reachability tree for Figure 9.

The box shown in Figure 8 represents the reachability tree for transition T1. To send input 'c' of transition T3 to the IUT, the logical clock value should be (3, 4, 0, 3). In other words, input 'c' is applied to the IUT only at stable state (3, 4, 0, 3). This demonstrates that Algorithm_LTS_DELAY can reproduce the test results by controlling concurrent events.

We apply Algorithm_LTS and Algorithm_LTS_DELAY to the message exchange for the establishment of Q.2971 point-to-multipoint call/connection. Figure 9 illustrates a message exchange where call initiator is S_A and call receivers are R_A and R_B [12, 13]. In here, the IUT has 3 ports connected to S_A, R_A and R_B.

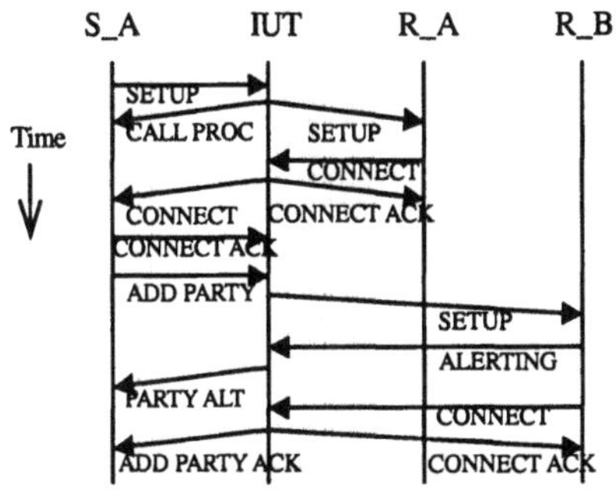

Figure 9. Message exchange for the establishment of Q.2971 network-side npoint-to-multipoint call/connection.

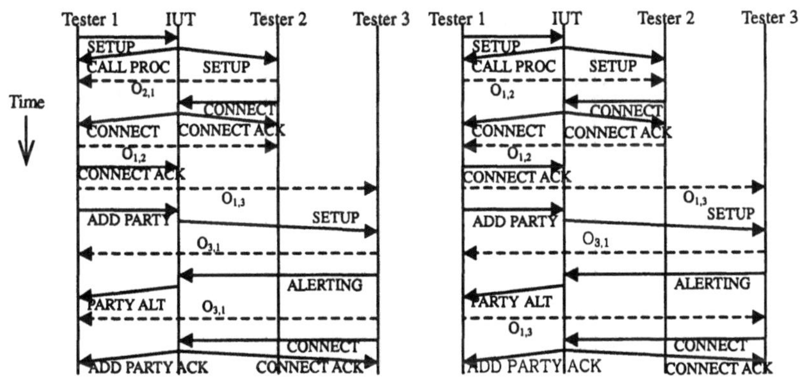

(a) Algorithm_LTS (b) Algorithm_LTS_DELAY

Figure 10. The LTS obtained by the two algorithms.

In Figure 10(a), the testers must receive the outputs without delay. However, message exchange in communication networks such as Q.2971 cannot satisfy such condition. The test scenario of Figure 10(b) can be said to be much more general. Table 3 compares the LTS generation algorithms: the algorithm in [1], Algorithm_LTS, and Algorithm_LTS_DELAY.

Table 3. Comparison of Local Test Sequence generation algorithms.

	[1]	Algorithm_LTS	LTS_DELAY
Output receive delay	No	No	Yes
Backward output-shifting faults	Yes	No	No
Forward output-shifting faults	Yes	No	No
Over-waiting time problem exists	-	Yes	No
Channel complexity	$O(n^2)$	$O(n^2)$	$O(n^2)$

In Table 3, Algorithm_LTS prevents output-shifting faults except over-waiting time problem. With Algorithm_LTS_DELAY, appropriate waiting time is allowed for preventing over-waiting time problem. The channel complexity of the algorithms is $O(n^2)$ since they employ a test method of mesh-structure. The numbers of signal 'C' and 'O' generated by the proposed algorithms are same, while it is slightly larger than that of [1].

4. TEST ARCHITECTURE

In this section, we propose a method for testing distributed system using Algorithm_LTS_DELAY with no coordination channels.

4.1 The Proposed Test Architecture

The number of coordination channels among the testers in distributed test environment exponentially increases since the channels are connected as a mesh. So, the diagnostic power of a tester diminishes due to the communication overhead. In order to solve the problem, a test architecture having O(n) channel complexity is proposed as Figure 11. Here Management and Control Entity (MCE) receives signal 'O' from the testers and sends the signal to them.

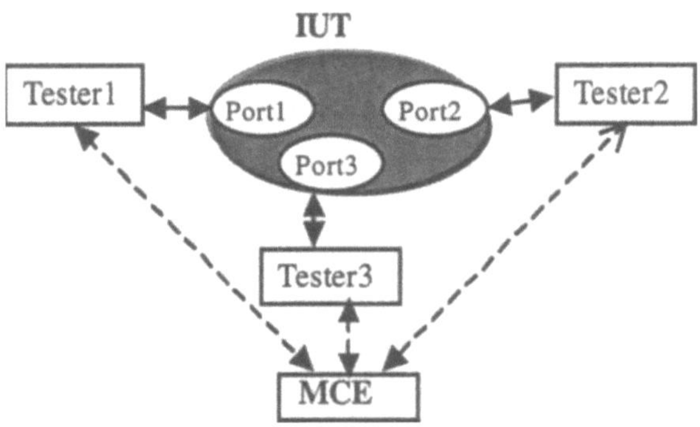

Figure 11. Proposed test architecture.

Here the transitions (state a, state b, x, y) and (state b, state c, x , y), and the tester connected to Port(x) should receive synchronization signals except cases that Port(y) = null, Port(x) = Port(x), and Port(x) = Port(y). This is because the testers are synchronized with other testers via only MCE. Therefore, only signal 'O' is used to synchronize the testers. Figure 12 shows the result obtained by applying the proposed algorithm to the Figure 9.

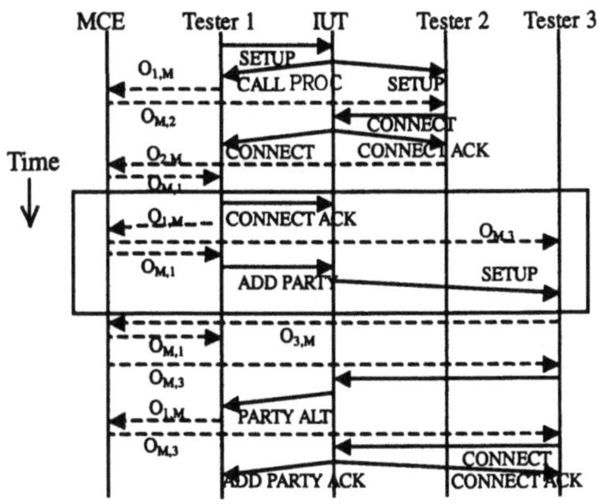

Figure 12. An LTS for message exchange of the establishment of Q.2971.

To apply the proposed algorithm, extra signals are needed between the testers and an MCE in case of the transitions having concurrent events or no output for any input. To generate signal '$O_{M,3}$' to prevent backward output-shifting faults for output 'SETUP' in "ADD PARTY/SETUP", extra signal '$O_{1,M}$' is generated as shown in a box of Figure 12. Note that only MCE can transfer signal 'O'. Here, signal '$O_{M,1}$' is synchronization signal of Tester 1, while Tester 1 applies input 'ADD PARTY' to the IUT.

The test architecture suggested in Figure 11 is more effective than conventional distributed test architecture for testing Q.2971. It is because modification of test architecture is minimal and test errors caused by channel faults becomes less while the number of call receives increases. However, there may be a bottleneck in MCE due to the increase of signal 'O's.

5. CONCLUSION AND FUTURE WORK

In this paper, a LTS generation algorithm has been proposed to test a distributed system using logical clock. Proposed algorithm classifies the events into concurrent or causal by labeling and comparing the logical clock values of the events of TS. Based on that, it generates additional signals 'C' and 'O' to control concurrent events and inserts them to the TS for each distributed object.

We have also proposed a variation of the LTS generation algorithm for which the output receive delay can be nonzero in distributed test architecture with channels among testers. It can automatically produce LTS onto already generated TS. Analyzing test scenario given with the proposed algorithm using reachability tree demonstrates that the proposed algorithm can solve the control-observation problem including output-shifting faults in a formal way.

In conventional distributed test architecture, the number of channel increases non-linearly since channels among testers are connected as a mesh. To overcome this drawback, a test architecture having O(n) channel complexity is proposed where the testers are not directly connected with each other. With this, diagnostic power of the testers can increase owing to the simplified test architecture.

As future work, over-waiting problem and appropriate ways to test distributed system using a dynamic testing method will be studied.

REFERENCES

[1] M. Benattou, L. Cacciari, R. Pasini, and O. Rafiq, "Principles and Tools for Testing Open Distributed Systems," Int'l Workshop on Testing of Communicating Systems, pp.77-92, Budapest, Hungary, September 1999.

[2] A. Ulrich and H. Konig, "Architectures for Testing Distributed Systems," Int'l Workshop on Testing of Communicating Systems, pp.93-107, Budapest, Hungary, September 1999.

[3] G. Luo, R. Dssouli, G.v. Bochmann, P. Venkataram, and A. Ghedamsi, "Test Generation With Respect To Distributed Interfaces," Computer Standards and Interfaces, pp.119-132, 1994.

[4] K. Tai, R. Carver, and E. Obaid, "Debugging Concurrent Ada Programs by Deterministic Execution," IEEE Trans. Software Engineering, Vol. 17, No.1, pp.45-63, January 1991.

[5] M. Kim, S. T. Chanson, S. Kang, and J. Shin, "An Enhanced Model for Testing Asynchronous Communicating Systems," FORTE/PSTV'99, June 1999.

[6] C. Fidge, "Logical Time in Distributed Computing Systems," IEEE Computer, pp.28-33, August 1991.

[7] G. Luo, G. v. Bochman, and A. Petrenko, "Test Selection Based on Communicating Nondeterministic Finite-State Machines using a Generalized Wp-Method," IEEE Trans. Software Engineering, Vol. 20, No. 2, pp.149-162, February 1994.

[8] Y. C. Young and K. C. Tai, "Observational Inaccuracy in Conformance Testing with Multiple Testers," IEEE 1st Workshop on Application-specific Software Engineering and Technology, pp.80-85, 1998.

[9] T. V. Gioles, I. Schieferdecker, M. Born, M. Winkler, and M. Li, "Configuration and Execution Support for Distributed Tests," Int'l Workshop on Testing of Communicating Systems, pp.61-76, Budapest, Hungary, September 1999.

[10] G. Coulouris, J. Dollimore, and T. Kindberg, "Distributed Systems, Concepts and Design," Second Edition, Addison-Wesley, 1994.

[11] H. Herzog and K. Sunderhaft, "General Framework for fault tolerance from ISO/ITU Reference Model for Open Distributed Processing (RM-ODP)," Object-Oriented Real-Time Dependable Systems, pp. 111-118, 1999.

[12] Y. Jung and J. Lee, "Experiences with Generation of Conformance Test Suite for Q.2971 Network-side Testing," Information Networking, pp. 286 – 289, 1998.

[13] ITU-T Draft Recommendation, Q.2971 B-ISDN Digital Subscriber Signaling No. 2 (DSS2) – User Network Interface layer 3 Specification for Point-to-Multipoint Call/Connection Control, 1995.

[14] Telelogic SDT3.2: Getting Started, Part1: Tutorials on SDT Tools, Telelogic, September 1997.

DIAGNOSING MULTIPLE FAULTS IN COMMUNICATING FINITE STATE MACHINES

Khaled El-Fakih+, Nina Yevtushenko++ and Gregor v. Bochmann+
+School of Information Technology and Engineering, University of Ottawa, ON, K1N 6N5, Canada. {kelfakih, bochmann}@ site.uottawa.ca
++Tomsk State University, 36 Lenin Str., Tomsk, 634050, Russia, yevtushenko.rff@elefot.tsu.ru

Abstract In this paper, we propose a method for diagnostic test derivation when the system specification and implementation are given in the form of two communicating finite state machines and at most a single component machine can be faulty. The method enables to decide if it is possible to identify the faulty machine in the system, once faults have been detected in a system implementation. If this is possible, it also provides tests for locating the faulty component machine. Two examples are used to demonstrate the different steps of the method. The method can also be used for locating faults within a machine when the system specification and implementation are given in the form of a single FSM.

Keywords: Diagnostics, conformance testing, communicating finite state machines

1. INTRODUCTION

The purpose of conformance testing is to check whether an implementation conforms to its specification. Usually a conforming implementation is required to have the same I/O behavior. An interesting complementary yet more complex problem is to locate the differences between a specification and its implementation when the implementation is found to be nonconforming [7]. A solution to this problem has various applications. For example, it makes easy the job of correcting the implementation so that it conforms to its specification [7].

In the software domain where a system may be represented as an FSM, some work has already been done for the *diagnostic* and *fault localization* problems [2][4][7]. However, the work done for distributed systems represented as two Communicating FSMs [3] makes some important simplifying assumptions. In [2][7] and [3] the difference between the system specification and its implementation is located under the assumption of a single fault in the implementation. In [4] the differences can be located for multiple faults under the assumption that each of the faults is reachable through non-faulty transitions.

In this paper, we consider a system consisting of two communicating FSMs, called components. One component, called *context machine,* communicates with the environment and the other component, called *embedded machine*. The interaction between these two components is assumed to be hidden, that is, unobservable. We assume that multiple output or transfer faults may occur in at most one component [5]. None of these faults will increase the number of states in the implementation of the component. It is not always possible to locate the faulty component of the given system when faults have been detected in a system implementation (SUT). This happens when certain faults in an implementation of the context and other faults in the implementation of the embedded component may cause the same observable behavior of the SUT. We present a novel approach for diagnostic test derivation. The approach enables us to decide whether it is possible to identify the faulty component in the given system, and if this is possible then tests for locating the fault are derived. We use a non-deterministic FSM for the compact representation of possible faulty transitions. The same technique can also be used for locating multiple faults when the system specification and implementation are given in the form of a single FSM.

This paper is organized as follows. Section 2 comprises necessary definitions for communicating FSMs. In Section 3, the diagnostic problem is discussed, while the method to solve the problem is presented in Section 5. In Section 4, the different steps of the method are presented illustrated by a working example. Future work is described in Section 6.

2. FINITE STATE MACHINES

A non-deterministic finite state machine (FSM) is an initialized non-deterministic Mealy machine which can be formally defined as follows [9]. A *non-deterministic finite state machine A* is a 6-tuple $\langle S, I, O, h, D_A, s_0\rangle$, where S is a finite nonempty set of n states with s_0 as the initial state; I and O are input and output alphabets; D_A is a specification domain which is a subset of SxI; and h: $D_A \rightarrow 2^{S\times O}\backslash\varnothing$ is a behavior function where $2^{S\times O}$ is the set of all subsets of the set $S\times O$. The behavior function defines the possible transitions of the machine. Given present state s_i and input symbol i, each pair $(s_j,o)\in h(s_i,i)$ represents a possible transition to the next state s_j with the output o. This is also written as a transition of the form $s_i \xrightarrow{i/o} s_j$. If $D_A = S\times I$ then A is said to be a *complete* FSM; otherwise, it is called a *partial* FSM. In the complete FSM we omit the specification domain D_A, i.e. complete FSM is 5-tuple $A = \langle S, I, O, h, s_0\rangle$. If for each pair $si\in D_A$ it holds that $|h(s,i)| =1$ then FSM A is said to be *deterministic*. In the deterministic FSM A instead of behavior function h we use two functions, transition function δ: $S\times I \rightarrow S$ and output function λ: $S\times I \rightarrow O$. FSM $B = (S',I,O,g,s_0)$, $S'\subseteq S$, is a *sub-machine* of complete FSM A if for each pair $si\in S'\times I$, it holds that $g(s,i) \subseteq h(s,i)$. Similar to [6], we further denote $Sub(A)$ the set of all deterministic sub-machines of FSM A. We also use the definition of a deterministic path in FSM A since only such paths can occur in its deterministic sub-machines. A *deterministic path P* of FSM A starts at the initial state and has no transitions with different next states and/or outputs for the same state-input combination.

As usual, function h can be extended to the set I^* of finite input sequences. Given state $s\in S$ and input sequence $\alpha=i_1i_2...i_k\in I^*$, output sequence $o_1o_2...o_k\in h(s,\alpha)$ if there exist states $s_1=s$, s_2, ... , s_k, s_{k+1} such that $(s_{j+1},o_j)\in h(s_j,i_j)$, $j = 1,\ldots,k$. We let the set $h^o(s,\alpha) = \{\gamma| \exists s' \in S [(s', \gamma)\in h(s, \alpha)]\}$ denote the *output projection* of h, while denoting $h^s(s,\alpha) = \{s'| \exists\gamma \in Y^* [(s', \gamma)\in h(s, \alpha)]\}$ the *state projection* of h. Input/Output sequence $i_1o_1i_2o_2...i_ko_k$ is called a *trace* of A if $o_1o_2...o_k\in h^o(s_0,i_1i_2...i_k)$. We also use the notation $h^s_\gamma(s,\alpha)$ to denote the set $\{s'| (s', \gamma) \in h(s, \alpha)]\ \}$of all states where the sequence α can take FSM A from the initial state with the output response γ. For appropriate (s, γ) the set $h^s_\gamma(s,\alpha)$ can be empty. If the set $h^s_\gamma(s,\alpha)$ has the unique state s then we say the state s is *observably reachable* from the initial state via the trace α/γ.

Given states s_1 and s_2 of complete FSM A, states s_1 and s_2 are *equivalent*, written $s_1 \cong s_2$, if for each input sequence $i_1i_2...i_k\in I^*$, it holds that $h(s_1,i_1i_2...i_k) = h(s_2,i_1i_2...i_k)$. If states s_1 and s_2 are not equivalent then they are *distinguishable*, written $s_1 \neq s_2$. Given complete FSM A, sequence $\alpha\in I^*$ such that $h(s_1, \alpha) \neq h(s_2, \alpha)$ is said to *distinguish* states s_1 and s_2. FSM A with pair-wise distinguishable states is called a *reduced* FSM.

Complete FSMs $A = (S,I,O,h,s_0)$ and $B = (T,I,O,g,t_0)$ are *equivalent*, written $A \cong B$, if their sets of traces coincide. It is well known, given complete deterministic FSM A, there always exists a reduced FSM that is equivalent to A.

2.1 A System of Two Communicating FSMs

Many complex systems are typically specified as a collection of communicating components. We consider here a special case, where the system consists of two Communicating FSMs (*ComFSMs*), called embedded machine (or M_2) and context machine (or M_1). We let the alphabet X and Y represent the externally observable input/output actions of the system, while the U and Z alphabets represent the internal (hidden) input/output interactions between the two components. The two (deterministic) FSMs communicate asynchronously via bounded input queues where input/output messages are stored. An FSM produces an output in response to each input. We assume that the system at hand has at most one message in transit, i.e. the next external input is submitted to the system only after it produced an external output y to the previous input. Under these assumptions, the collective behavior of the two communicating FSMs can be described by *product machine.* The product machine $M_1 x M_2$ is a Labeled Transition System (LTS) that describes the joint behavior of the component machines in terms of all actions within the system. If the product machine has a cycle labeled only with internal actions from the alphabet $U \cup Z$ then the system falls into live-lock when an input sequence leading to this cycle is applied; i.e. the system will not produce any external output. Here we accept a catastrophic interpretation of live- locks and, similar to [9], say the composed FSM of M_1 and M_2 is not defined. Otherwise, a *composed machine*, denoted as *Reference System(RS)* = $M_1 \lozenge M_2$, can be obtained from the product machine by hiding all internal actions in the product machine, and pairing input with output actions [9].

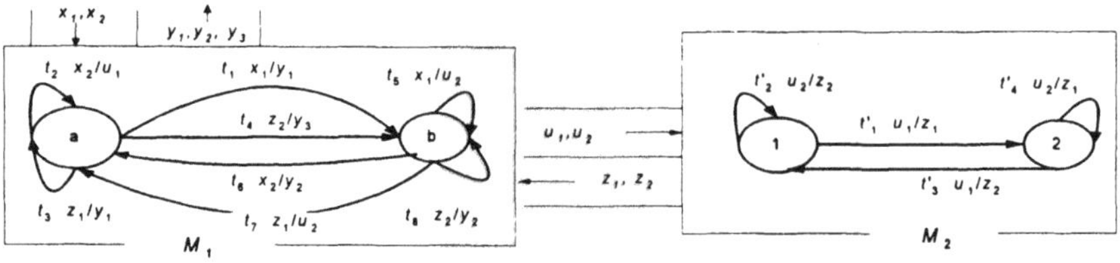

Figure 1. A system of two ComFSMs M_1 and M_2

As an example, we consider the two machines M_1 and M_2 shown in Fig.1. The set of external inputs is $X=\{x_1,x_2\}$, the set of external outputs is $Y=\{y_1, y_2, y_3\}$, the set of internal inputs is $U=\{u_1, u_2\}$, and the set of internal outputs is $Z = \{z_1, z_2\}$. Their corresponding reference system $RS = M_1 \lozenge M_2$ is shown in Fig. 2.

A tester, implements a given test by executing external input sequences (test cases) simultaneously against both the SUT consisting of the implementation of M_1 and M_2, and the reference system in order to generate the observed and expected outputs. If for each test case, the sequences of the observed and expected outputs coincide then the system is said to *pass* the test suite.

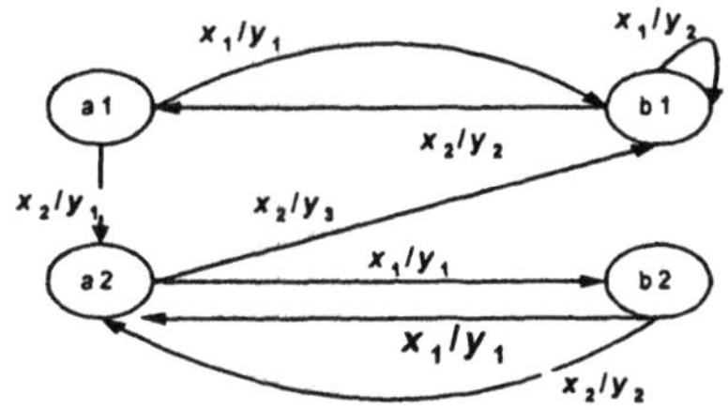

Figure 2. Reference System ($M_1 \lozenge M_2$) of the M_1 and M_2 of Fig. 1

2.2 A Fault Model for the System of Communicating FSMs

We consider a fault model based on output and transfer faults of a deterministic FSM [2]. Given complete deterministic specification and implementation FSMs $\langle S, I, O, \delta, \lambda, s_0\rangle$ and $\langle S, I, O, \Delta, \Lambda, s_0\rangle$, we say a transition (s,i) has an *output fault* if $\delta(s,i)=\Delta(s,i)$ while $\lambda(s,i)\neq\Lambda(s,i)$. An implementation has a *single output fault* if one and only one of its transitions has an output fault. We say that a transition has a *transfer fault* if $\delta(s,i)\neq\Delta(s,i)$, i.e. given state s and input i, the implementation enters a different state than that specified by the next-state function of the specification. An implementation has a *single transfer fault* if there is no output fault in the implementation and one and only one of its transitions has a transfer fault. We say that the implementation under test of the given system may have *multiple faults* if several transitions have output or transfer faults. Here we notice the fault model based on output and transfer faults is often used for diagnosis of a system decomposed into components, where only one component may be faulty [5]. In our context, the specification is a decomposed system; i.e. its implementation is a system of two *ComFSMs*, where at most one of these machines is faulty.

3. DIAGNOSIS PROBLEM

3.1 Diagnosis Problem Statement

Let M_1 and M_2 be complete deterministic FSMs representing the specifications of the components of the given system while M'_1 and M'_2 are their corresponding implementations. We propose an adaptive method for diagnostic test derivation. If the implementation system does not pass a given test suite, our algorithm enables to decide whether it is possible to identify a faulty component of the given system under the assumption that multiple faults may occur only in one of the implementations M'_1 or M'_2. Furthermore, if this is possible, the algorithm enables to decide whether it is possible to locate the faulty transitions within the faulty component, and if possible it locates them. Moreover, the algorithm draws the conclusion "*Faults cannot be captured by the assumed fault model*" if it has been detected that the implementation at hand has faults that cannot be captured by the assumed fault model.

The diagnostic method can be used for the case where multiple transfer or output faults can occur in one of the implementation component FSMs M'_1 or M'_2. However, without loss of generality and for simplicity of presentation, hereafter, we assume that only output faults may occur.

3.2 An Overview of the Diagnostic Approach

Let $RS = M_1 \Diamond M_2$ be a specification system while *TS* is a conformance test suite. If the composition $M'_1 \Diamond M'_2$ of the implementations M'_1 and M'_2 of the given system produces unexpected output responses to the given test suite *TS*, then the composition $M'_1 \Diamond M'_2$ is not equivalent to *RS*, i.e. either M'_1 or M'_2 is a faulty implementation. Our objective is to determine whether M'_1 is not equivalent to M_1 while M'_2 and M_2 are equivalent, or vice versa.

In order to determine whether the output responses of *TS* can be produced when FSM M'_2 has output faults and M'_1 is equivalent to its specification, we derive the FSM $(M_2)_a^{out}$ by adding new (faulty) transitions with all possible outputs to each transition of M_2, and then we combine the obtained non-deterministic FSM with M_1. We call the obtained non-deterministic machine, $M_1 \Diamond (M_2)_a^{out}$, the Fault Function (FF) of the embedded component (or *FF-Embedded*). Similarly, we derive the FF of the context M_1, *FF-Context* = $(M_1)_a^{out} \Diamond M_2$, to determine whether the output responses of *TS* can be produced when FSM M'_1 has output faults and M'_2 is equivalent to its specification. Fault functions were introduced in [8] to represent in a concise way all mutants of a given FSM with a given type of implementation errors.

The set of all deterministic sub-machines of $M_1 \Diamond (M_2)_a^{out}$ (or $(M_1)_a^{out} \Diamond M_2$) includes each implementation $M_1 \Diamond M'_2$ (or $M'_1 \Diamond M_2$), where output faults are possible in the implementation of component FSM M_2 (or M_1). However, the set also includes superfluous sub-machines that do not correspond to a composition of any possible deterministic component machines. This is due to the fact that while deriving the composition, we do not take into consideration that for a specific state-input combination (s,i), one and only one output is possible in a deterministic implementation.

Since our implementation system is deterministic we do not take into account non-deterministic paths of machines $M1\Diamond(M2)_a^{out}$ and $(M1)_a^{out}\Diamond M2$. Moreover, we also remove from $M1\Diamond(M2)_a^{out}$ and $(M1)_a^{out}\Diamond M2$ a behavior that does not agree with the observed outputs to the applied test suite *TS*. In Section 4, we describe the algorithm that removes from machine $M1\Diamond(M2)_a^{out}$ (or $(M1)_a^{out}\Diamond M2$) sub-machines whose output responses to the test suite do not agree with those obtained by applying the test suite *TS* to the SUT.

If the SUT is equivalent to a deterministic sub-machine of $M1\Diamond(M2)_a^{out}$ then the faulty machine is $M2$, and if it is equivalent to a deterministic sub-machine of $(M1)_a^{out}\Diamond M2$, then the faulty machine is $M1$. However, if the SUT is equivalent to a deterministic sub-machine of $M1\Diamond(M2)_a^{out}$ and to a deterministic sub-machine of $(M1)_a^{out}\Diamond M2$, then the faulty machine cannot be identified. This is due to the fact that there are some possible faults in $M1$ and some possible faults in $M2$ that cause the same observable behavior of the SUT. If none of the above cases applies, we conclude that the implementation has faults that are not captured by the considered fault model.

In order to draw one of the above conclusions, we should have test cases such that by observing the output responses of the SUT to these test cases, we can distinguish the SUT and sub-machines of $(M1)_a^{out}\Diamond M2$ and of $M1\Diamond(M2)_a^{out}$. If the machines $(M1)_a^{out}\Diamond M2$ and $M1\Diamond(M2)_a^{out}$ have no equivalent deterministic sub-machines then there exists a so-called *distinguishing set* of input sequences [6] such that given the set of deterministic output responses to these input sequences, we always can determine whether the machine under test is a sub-machine of $(M1)_a^{out}\Diamond M2$ or $M1\Diamond(M2)_a^{out}$. The algorithm for deriving such a distinguishing set is proposed in [6]. We illustrate this algorithm in Section 4. In other cases, it may happen that there are sub-machines of $(M1)_a^{out}\Diamond M2$ and others of $(M1)_a^{out}\Diamond M2$ that are equivalent, but these sub-machines are not equivalent to the SUT. In this case, the observed outputs to the given test suite are insufficient to distinguish between these sub-machines and the SUT. Therefore, more test cases must be generated and added to the test suite. This can be done by breaking *FF-Embedded* and *FF-Context* into sub-machines, and by comparing each pair of the obtained sub-machines. Each time when two obtained sub-machines become distinguishable, their distinguishing test is added to the test suite. This enables the elimination of all sub-machines whose output responses to a distinguishing test are different from those observed by applying this test to the SUT. We break the machines by fixing some transitions as deterministic transitions, i.e. by reducing the non-determinism of Fault Functions. In the worst case, we can come up with explicit enumeration of all faulty machines. However, as the considered examples show, we usually need to fix only a small number of transitions in order to draw a conclusion.

The details of the diagnostic method are presented in Section 5. Meanwhile, in the following section we present its constituent algorithms illustrated by a working example.

3.3 Working Example

Consider the two machines M_1 and M_2 given in Fig. 1, and their corresponding reference system $RS = M_1 \lozenge M_2$ shown in Fig. 2. A given complete test suite TS derived from RS is $TS = \{x_1x_1x_1, x_1x_2x_2, x_2x_1x_1x_2, x_2x_1x_2x_2, x_2x_2x_1\}$. TS is derived using the method presented in [1], and it detects any complete FSM $M'_1 \lozenge M'_2$ that is not equivalent to RS, under the assumption that FSMs M'_1 and M'_2 have up to two states. The set of expected output responses to the TS is as follows:

$\lambda(s_0, x_1x_1x_1) = y_1y_2y_2$; $\lambda(s_0, x_1x_2x_2) = y_1y_2y_1$; $\lambda(s_0, x_2x_1x_1x_2) = y_1y_1y_1y_3$; $\lambda(s_0, x_2x_1x_2x_2) = y_1y_1y_2y_3$; $\lambda(s_0, x_2x_2x_1) = y_1y_3y_2$.

Let us assume that the composition $M'_1 \lozenge M'_2$ of the implementation component FSMs M'_1 and M'_2 produces unexpected output responses

$y_1y_3y_3$ to $x_1x_1x_1$; $y_1y_1y_3y_2$ to $x_2x_1x_1x_2$ and $y_1y_3y_3$ to $x_2\,x_2\,x_1$ **(1)**

Thus, the composition $M'_1 \lozenge M'_2$ is not equivalent to RS, i.e. either M'_1 or M'_2 is a faulty component implementation.

Figure 3 includes the *FF-Embedded* machine, $M_1 \lozenge (M_2)_a^{out}$, and the *FF-Context* ,$(M_1)_a^{out} \lozenge M_2$, obtained as described in Section 3.2. In this example, the *FF-Context* is derived under an assumption that in the faulty context implementation, external outputs can only be replaced with external outputs and internal outputs can only be replaced with internal outputs, respectively. This is done in order to have a simple and more readable working example.

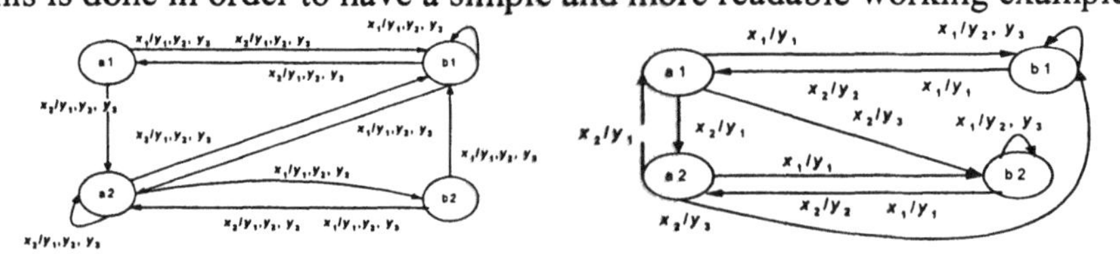

Figure 3. (a) Fault function of Embedded Component (b) Fault Function of Context

4. DISTINGUISHING NON-DETERMINISTIC FSMS

In this section, we present the different algorithms of the test suite derivation method, given in the subsequent section, for locating a faulty component FSM.

4.1 Removing Sub-machines of a Non-Deterministic FSM

The following algorithm is used to remove from $(M_1)_a^{out}\lozenge M_2$ (and from $M_1\lozenge(M_2)_a^{out}$) sub-machines whose output responses to the test cases of *TS* disagree with those obtained by applying these test cases to the SUT.

Given a complete non-deterministic FSM $A = (S,I,O,h,s_0)$ and a set V of deterministic sequences over alphabet $(IO)^*$, the algorithm returns a smallest sub-machine A^r of A which has the property that each sub-machine of A that comprises V as a subset of its traces is a sub-machine of A^r. We note that in our case, the input parts of the sequences of the initial set V are those of the given test suite *TS*, and the output parts are those observed by applying *TS* to the SUT.

Algorithm 4.1. Removing from FSM A sub-machines that do not match V

Input: A complete non-deterministic FSM $A = (S, I, O, h, s_0)$ and a set V of deterministic sequences over alphabet $(IO)^*$

Output: The smallest subFSM $A^r = (S, I, O, h, s_0)$ of A that contains each sub-machine of A, such that the set of its traces comprises V, if exists.

Step-1. Given FSM $A_1 = A$ and the set V of sequences over alphabet $(IO)^*$, we derive the tree $Tree_1$ of all deterministic paths through A_1 labeled with sequences of V. Assign i:=1 and go-to Step-2.

Step-2. If there exists a sequence in V such that no path in $Tree_i$ is labeled with this sequence, then there is no sub-machine in A such that V is a subset of the set of traces of that sub-machine (A^r does not exist, end of Algorithm 4.1). Otherwise, we build a machine A_{i+1} which is a sub-machine of A_i as follows. For each observably reachable node in $Tree_i$, we copy into the corresponding state in A_{i+1} the outgoing transitions from this node. If there are several observably reachable nodes in $Tree_i$ with the same label, we copy for the corresponding state in A_{i+1} only the matching transitions at these nodes (same input/output/next-state values). If there are no matching transitions in $Tree_i$, then there is no sub-machine in A such that V is a subset of the set of traces of that sub-machine (A^r does not exist, end of Algorithm 4.1). For each state in A_i, that labels only non-observably reachable nodes in $Tree_i$, we copy all the outgoing transitions from machine A_i into A_{i+1}. If no transition in the machine A_i has been changed (i.e. $A_{i+1}=A_i$), then we have $A^r = A_i$ (End of Algorithm 4.1). Otherwise, go-to Step-3.

Step-3. At this step we trim $Tree_i$ in order to obtain $Tree_{i+1}$ using the machine A_{i+1} obtained by Step-2 above. For each path in $Tree_i$ that has a node where the output and/or next node of the transition do not match machine A_{i+1}, remove this transition and its sub-tree. If all outgoing transitions from some node for an appropriate input have been removed, we remove the incoming transition to this node. If all transitions from the root node for an appropriate input have been removed, then there is no sub-machine in A such that V is a subset of the set of traces of that submachine (A^r does not exist, end of Algorithm 4.1). If the trees $Tree_i$ and $Tree_{i+1}$ coincide, then the machine $A^r = A_{i+1}$ is derived (End of Algorithm 4.1). Otherwise, increment i by 1 and go back to Step-2.

As an example, we consider $Tree_1$ in Fig. 4-a generated for the fault function of the context, i.e. $A_1 = (M_1)_a^{out} \Diamond M_2$ in Fig. 3-b, using Step-1 and *TS*. We do not include in $Tree_1$ the paths that do not match the observed output of *TS*. For example, for all test cases in the *TS* that start with the input x_2, the observed output for x_2 is y_1. Therefore, the paths of $(M_1)_a^{out} \Diamond M_2$ that start from the initial state a_1 by a transition labeled with a label other than x_2/y_1 are not included in $Tree_1$. Moreover, we do not include in $Tree_1$ any non-deterministic path. For example, we do not include the path $a_1 \xrightarrow{x_1/y_1} b_1 \xrightarrow{x_1/y_3} b_1 \xrightarrow{x_1/y_3} a_1$.

In $Tree_1$, the root node a_1 is observably reachable through the empty sequence. Therefore, in Step-2, we copy the outgoing transitions of that node into A_2 (shown in Fig. 4-b), i.e. transitions, $a_1 \xrightarrow{x_1/y_1} b_1$, $a_1 \xrightarrow{x_2/y_1} a_2$ and $a_1 \xrightarrow{x_2/y_1} b_1$. Moreover, in $Tree_1$, starting from the root node a_1, the node b_1 is observably reachable through the sequence x_1/y_1. Therefore, in Step-2, we copy the outgoing transitions of b_1 from $Tree_1$ into A_2. Nodes a_2 and b_2 in $Tree_1$ are only non-observably reachable. Therefore, we copy from A_1 in Fig. 3-b the outgoing transitions from states a_2 and b_2 into A_2.

Afterwards, using A_2 in Step-3, we consider in $Tree_1$, starting from the root node a_1, the outgoing transitions from the node b_1 that is reached through the sequence x_2/y_1. We notice that all its outgoing transitions do not match A_2. Therefore we trim the sub-tree of this node, and since all outgoing transitions for inputs x_1 and x_2 from this node are removed, we remove its incoming edge, and we get $Tree_2$, which is equal to $Tree_1$ except that the shaded area TRIM-1 is removed.

Back to Step-2, by considering $Tree_2$, the root node a_1 is observably reachable through the empty sequence. Moreover, starting from the root node a_1, the ending nodes a_1, b_1, and a_2 are observably reachable through the sequences $x_1/y_1x_2/y_2$, x_1/y_1, and x_2/y_1, respectively, and node b_2 is observably reachable through the sequence $x_2/y_1x_1/y_1$. Therefore, for these nodes, we copy their matching outgoing transitions from $Tree_2$ into A_3 in Fig.5-a.

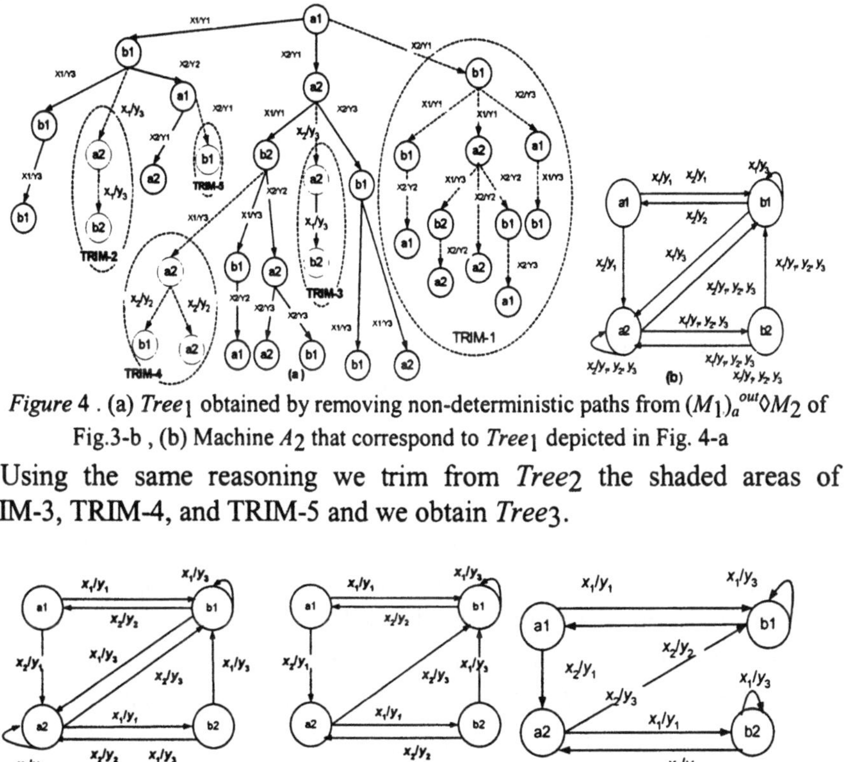

Figure 4 . (a) $Tree_1$ obtained by removing non-deterministic paths from $(M_1)_a^{out}\Diamond M_2$ of Fig.3-b , (b) Machine A_2 that correspond to $Tree_1$ depicted in Fig. 4-a

Using the same reasoning we trim from $Tree_2$ the shaded areas of TRIM-3, TRIM-4, and TRIM-5 and we obtain $Tree_3$.

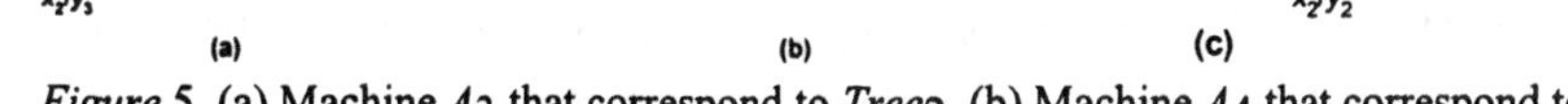

Figure 5. (a) Machine A_3 that correspond to $Tree_2$, (b) Machine A_4 that correspond to $Tree_3$, (c) Machine for the embedded component

Back to Step-2, by considering $Tree_3$, we notice that all nodes are observably reachable. We copy all the matching outgoing transitions of these nodes and obtain the final machine A_4 shown in Fig. 5-b.

We apply Algorithm 4.1. to the *FF-Embedded*, i.e. for $M_1\Diamond(M_2)_a^{out}$ of Fig. 3-a, and we obtain the machine shown in Fig. 5-c.

Theorem 4.1. Given a complete non-deterministic FSM $A = (S, I, O, h, s_0)$ and a set V of deterministic sequences over alphabet $(IO)^*$, if there exist a sub-machine B of A that has V as a subset of its traces, then the FSM A^r is derived by Algorithm 4.1 and it includes B. Otherwise, the FSM A^r does not exist.

Proof: In Step-2 of Algorithm 4.1, if there exists no deterministic path in $Tree_i$ labeled with some sequence $\alpha/\beta \in V$, then there is no deterministic sub-machine of A that can produce the observed output sequence β to the input sequence α, i.e. there is no sub-machine A^r in A such that V is a subset of traces of A^r. Moreover, In Step-3, if all the transitions from the root node in $Tree_i$ have been removed, then there is no deterministic sub-machine A^r in A such that V is subset of traces of A^r. The transitions of A are changed only during Step-2 for the states that label observably reachable nodes of the corresponding tree. For this reason, it is enough to show that each sub-machine B of A_i that has V as a subset of its traces, is a sub-machine of A_{i+1}. Let state s label an observably reachable node of $Tree_i$ that is reachable through some sequence $\alpha/\beta \in V$. Any sub-machine of A_i that has V as a subset of its traces must reach state s after applying the sequence α. Therefore, all transitions from the state s of B match transitions from this node, i.e. B is a sub-machine of A_{i+1}.

4.2 Distinguishing the Sets of Deterministic Sub-Machines of Two Non-Deterministic FSMs

If the two sub-machines of the FSMs *FF-Context* and *FF-Embedded* obtained after using the above procedure for removing behaviors that do not match observed outputs, do not have equivalent sub-machines then we can derive a distinguishing set Dis_{Set} that allows us to recognize which of the component FSMs is faulty. This set can be constructed using the algorithm given in [6]. In other words, let machines $M_1 \Diamond (M_2)_a^{out}$ and $(M_1)_a^{out} \Diamond M_2$ be distinguished with the distinguishing set Dis_{Set}. Let also P be a sub-machine of $M_1 \Diamond (M_2)_a^{out}$ or of $(M_1)_a^{out} \Diamond M_2$. Then by observing the output responses of P to sequences in the set Dis_{Set}, we can always conclude whether $P \in Sub(M_1 \Diamond (M_2)_a^{out})$ or $P \in Sub((M_1)_a^{out} \Diamond M_2)$, i.e. the diagnostic problem is always solvable. Therefore, we derive, using the algorithm given in [6], a distinguishing set Dis_{Set} for the FSMs *FF-Context* and *FF-Embedded* in order to recognize a sub-machine that corresponds to the faulty SUT.

As an example, we consider the machines *FF-Context* (machine A_4 of Fig. 5-b) and *FF-Embedded* (Fig. 5-c) obtained after deleting sub-machines with traces that do not match observed output responses. The sequence $x_2x_1x_1x_2x_2$ distinguishes these machines. Therefore, if we apply the input x_2 after the input sequence $x_2x_1x_1x_2$ and the implementation at hand produces the output response y_1 to the tail input x_2, then we conclude that M'_1 is the faulty implementation. If the output y_3 is produced to the tail input x_2, then we conclude that M'_2 is faulty. If the implementation produces the output different from y_3 and y_1, then the implementation at hand has faults that cannot be captured by the assumed fault model.

4.3 Determining a Superfluous Sub-machine

Due to the considered fault model, the SUT M is a sub-machine of $M_1\Diamond(M_2)_a^{out}$ or $(M_1)_a^{out} \Diamond M_2$. However, we mentioned above that not each sub-machine of $M_1\Diamond(M_2)_a^{out}$ (or $(M_1)_a^{out} \Diamond M_2$) can be obtained through output faults in the implementation of M_2 (or M_1). The reason is that we do not take into account deterministic and non-deterministic paths when combining the compact representation $(M_2)_a^{out}$ of all possible implementations of M_2 with M_1 (or the compact representation $(M_1)_a^{out}$ of all possible implementations of M_1 with M_2), and thus we may obtain superfluous sub-machines. Therefore, it may happen that the SUT M is equivalent to a sub-machine of $M_1\Diamond(M_2)_a^{out}$ and to a submachine of $(M_1)_a^{out}$ $\Diamond$ M_2, but there is no sub-machine $M'_2\in(M_2)_a^{out}$ such that $M_1\Diamond M'_2$ is equivalent to M. In this case, only the implementation of M_1 is faulty. Thus, we have the following problem.

Given FSM $M_1\Diamond(M_2)_a^{out}$ and its sub-machine M, we must check whether there exists FSM $M'_2 \in (M_2)_a^{out}$ such that $M=M_1\Diamond M'_2$. To solve the problem we can project sub-machine M onto the set of states of M_2 and input and output alphabets of M_2. There exists FSM $M'_2\in(M_2)_a^{out}$ such that $M=M_1\Diamond M'_2$ if and only if the obtained FSM is deterministic. A sub-machine M of $M_1\Diamond(M_2)_a^{out}$ for which there is no $M'_2\in(M_2)_a^{out}$ such that FSMs $M_1\Diamond M'_2$ and M are equivalent, is called a *superfluous* sub-machine.

5. FAULT DIAGNOSIS ALGORITHM

Algorithm 5.1. Recognizing a faulty component FSM

***Input*:** Composition $M \cong M_1\Diamond M_2$ of two FSMs M_1 and M_2, and the set $V=TS$ of sequences over alphabet $(IO)^*$,

***Output*:** Verdict "*Component FSM M_1* (or M_2) *is faulty*", or verdict "*Both M_1/M_2 could be faulty*" when there is a possible faulty implementation of M_1 and a possible faulty implementation of M_2 that cause the same observable behavior as the implementation at hand, or verdict "*Faults cannot be captured by the assumed fault model*" if it has been detected that the implementation at hand has faults that can not be captured by the assumed fault model.

Step-1. Derive machines, $A_1 = (M_1)_a^{out} \lozenge M_2$ (Fault Function of M_1) and $A_2 = M_1 \lozenge (M_2)_a^{out}$(Fault Function of M_2). Let the set $\mathfrak{R}_1$ be equal to $\{A_1\}$, and the set $\mathfrak{R}_2$ be equal to $\{A_2\}$.

Step-2. For each machine say A_k in the sets $\mathfrak{R}_1$ and $\mathfrak{R}_2$, call Algorithm 4.1. to obtain the smallest sub-machine A^r of A_k which includes all sub-machines of A_k that have V as a subset of their traces. If such an A^r exist, replace A_k by A^r. Otherwise, remove A_k from the corresponding set $\mathfrak{R}_1$ or $\mathfrak{R}_2$.

If the sets $\mathfrak{R}_1$ and $\mathfrak{R}_2$ are empty, then the implementation at hand has a fault that is not captured by the assumed fault model (End of diagnosis algorithm). If the set $\mathfrak{R}_1$ (or $\mathfrak{R}_2$) is empty, then we conclude the other machine M_2 (or M_1) is faulty (End of diagnosis algorithm).

Otherwise, check, as described in [6], whether there are two machines, say A_i in $\mathfrak{R}_1$ and A_j in $\mathfrak{R}_2$, that are distinguishable.

♦- If there exist such two machines, we obtain, using the algorithm given in [6], the $Dist_{set}$ that distinguish them, and we apply the input sequences of this set to the SUT.

--If $|\mathfrak{R}_1| = 1$ and $|\mathfrak{R}_2| = 1$, i.e. $\mathfrak{R}_1 = \{A_i\}$ and $\mathfrak{R}_2 = \{A_j\}$, then:

-If the output responses of the SUT to the $Dist_{set}$ are different from those expected by both machines A_i and A_j, then we conclude that the implementation at hand has faults that cannot be captured by the assumed fault model (End of diagnosis algorithm).

-Else, if the output responses of the SUT are different than those expected by A_j (or A_i), then we conclude that M_1 (or M_2) is the faulty machine (End of diagnosis Algorithm).

-- If $|\mathfrak{R}_1| > 1$ or $|\mathfrak{R}_2| > 1$, then after observing the output responses of the SUT to the sequences in $Dist_{set}$, we remove A_i (or A_j) from the set $\mathfrak{R}_1$ (or $\mathfrak{R}_2$) if these responses are different than those expected by A_i (or A_j). Then, we add the sequences in $Dist_{set}$ with the observed output responses to V, and we return return back to Step-2.

♦- If all the machines in $\mathfrak{R}_1$ are indistinguishable from those in $\mathfrak{R}_2$, then

--If $\mathfrak{R}_1$ and $\mathfrak{R}_2$ have only deterministic machines, then check whether all the machines in $\mathfrak{R}_1$ and in $\mathfrak{R}_2$ are superfluous as described above.

-If all the sub-machines in $\mathfrak{R}_1$ (or in $\mathfrak{R}_2$) are superfluous, then machine M_2 (or M_1) is faulty (End of diagnosis algorithm).

-Else, If there exist a sub-machine in $\Re_1$ and another in $\Re_2$ that are not superfluous, then we conclude that "Both_M_1/M_2 could_be_faulty". There is a possible faulty implementation of M_1 and a possible faulty implementation of M_2 that cause the same observable behavior as that of the implementation at hand

--Else, if the set $\Re_1$ (or $\Re_2$) has at least one non-deterministic FSM, then we break that machine into k machines by fixing one of its non-deterministic transitions. Then, we replace that machine in the set $\Re_1$ (or $\Re_2$) with the obtained sub-machines, and we go back to Step-2.

As another example, suppose that the implementation of the composed system $M'_1 \Diamond M'_2$ of specifications in Fig. 1 produces the unexpected output responses $y_1 y_1 y_2 y_2$ to the input sequence $x_2 x_1 x_1 x_2$ of *TS*. This can happen if M'_1 has an output fault; namely it produces y_2 instead of u_2 on executing transition t_5.

By applying the diagnostic method described above, we find that M'_1 can not be identified as the faulty implementation since there exists a faulty implementation M'_2 of M_2 such that $M_1 \Diamond M'_2$ and $M'_1 \Diamond M_2$ are equivalent. It is the case when M'_2 produces z_2 instead of z_1 on executing t'_4.

6. FURTHER RESEARCH WORK

In this paper we presented a method for diagnostic test derivation when the system specification and implementation are given in the form of two communicating finite state machines and at most a single component machine can be faulty. The algorithm can be extended for locating the faults within the faulty machine (if possible). The idea here is to locate the faulty implementation that corresponds to the faulty machine. This implementation is a sub-machine of $(M_1)_a^{out} \Diamond M_2$ when M_1 has a faulty implementation, or a sub-machine of $M_1 \Diamond (M_2)_a^{out}$ when M_2 has a faulty implementation. Therefore, the algorithm can be extended to decide which sub-machine of $(M_1)_a^{out} \Diamond M_2$ (or $M_1 \Diamond (M_2)_a^{out}$) is equivalent to the system at hand, i.e. comparison between sub-machines of a given FSM can be added to the algorithm. We note that sometimes it is not possible to locate the faulty sub-machine. It is the case when the SUT has the same observable behavior for different output faults of the faulty machine. Moreover, the algorithm can be extended for locating faults when the specification and implementation systems are given in the form of a single FSM

Currently, we are performing experiments to estimate the effectiveness and the complexity of the method. The preliminary results obtained for the case when only single output faults in a component are taken into account, show that almost always the faulty component can be identified.

REFERENCES

[1] S. Fujiwara, G. v. Bochmann, F. Khendek, M. Amalou and A. Ghedamsi.'Test selection based on finite state models', *IEEE Trans. SE*-17, No. 6, 1991, pp. 591-603.

[2] A. Ghedamsi and G. v. Bochmann.'Test result analysis and diagnostics for finite state machines', *Proc. of the 12-th ICDS*, Yokohama, Japan, 1992.

[3] A. Ghedamsi, G. v. Bochmann and R. Dssouli.'Diagnostic Tests for Communicating Finite State Machines', *Proc. of the 12th IEEE Internaitonal Phoenix Conference on Communications*, Scottsdale, USA, March 1993.

[4] A. Ghedamsi, G. v. Bochmann and R. Dssouli. 'Multiple fault diagnosis for finite state machines', *Proc. of IEEE INFOCOM'93*, 1993, pp.782-791.

[5] J. de. Kleer and B.c. Williams.'Diagnosing multiple faults', *Artificial Intelligence* 32(1), 1987, pp. 97-130.

[6] I. Koufareva. 'Using non-deterministic FSMs for test suite derivation', *Ph.D. Thesis*, Tomsk State University, Russia, 2000 (In Russian).

[7] D. Lee and K. Sabnani. 'Reverse engineering of communication protocols', *Proc. of ICNP*, October 1993, pp. 208-216.

[8] A. Petrenko and N. Yevtushenko.'Test suite generation for a FSM with a given type of implementation errors', *Proc. of the 12th IWPSTV*, 1992, pp. 229-243.

[9] A. Petrenko, N. Yevtushenko, G. v. Bochmann, and R. Dssouli. 'Testing in context: Framework and test derivation', *Computer Communications Journal*, Special issue on protocol engineering, 1996, pp. 1236-1249.

FROM ACTIVE TO PASSIVE: PROGRESS IN TESTING OF INTERNET ROUTING PROTOCOLS

Jianping Wu, Yixin Zhao and Xia Yin
Department of Computer Science, Tsinghua University, Beijing 100084
jianping@cernet.edu.cn, {zhaoyx,yxia}@csnet1.cs.tsinghua.edu.cn

Abstract Routing protocols play an important role in the Internet and the test requirements are mounting. To test routing protocols more efficiently, several enhanced techniques are applied in our Protocol Integrated Test System. The Implementation Under Test is modeled as a black box with windows. The test system is endowed with multiple channels and multiple ports to test distributed protocols. The test suite and other related aspects are also extended. Meanwhile, a passive testing is introduced to test, analyze and manage routing protocols in production field, capable of performing the conformance test, the interoperability test and the performance test. Necessary algorithms and techniques were worked out for the passive testing to cover the scope from the protocol state machine to the routing information manipulation. This paper presents our efforts to test Internet routing protocols with both the active testing and the passive testing.

Keywords: routing protocol, test structure, passive testing

1. INTRODUCTION

High performance and stable routing protocols dominate the performance of the Internet. Currently the most commonly used routing protocols are RIP [1], OSPF [2] and BGP [3]. There are other routing protocols like [4-6], but despite their importance, the test activity is still rather limited due to the fact that the original test theories, structures and systems were aimed at communication protocols. While routing protocols possess different features.

This work has been supported by National Natural Science Foundation of China under Grant No. 69682002 and No.69725003

Several deficiencies were encountered when we started our work on routing protocol test with the Protocol Integrated Test System – PITS [7].

Then the Implementation Under Test (IUT or module under test MUT) is modeled as a windowed black box and tested with a multiple channel test method. The test suite has evolved into a heterogeneous one, while the test data of different types can be described in their appropriate formats. Enhanced techniques mainly focus on the Reference Implementation (RI). In cooperation with the Enhanced RI (ERI), the packet extension technique helps to describe the whole test activity integrally and uniformly.

Some abnormalities will only appear in practice and/or over a long time, such as route oscillations [8], useless route advertisements [9] or even exhausted routers. Although the active testing helps a lot to uncover the deficiencies of the implementations of routing protocols [10][11], it can hardly deal with these problems. The passive testing [12] can be performed in production field over a long period of time without interference on the network. It is able to perform the conformance, interoperability and performance tests. Therefore, the passive testing has also been adopted in our work on routing protocol test. Corresponding techniques have also been worked out [13][16].

All the following methods are applied in the test activity. We have built an Enhanced Reference Implementation [19] for PITS An intelligent online BGP-4 analyzer[20] has been implemented and then a prototype of a more general online test system [21]. Another important advance is that the active testing and the passive testing are combined, which helps us in the effort to cover all features of routing protocols in our test. These techniques are applicable for RIP, OSPF and BGP. In this paper, we will mainly focus on BGP-4 as the example.

2. FEATURES AND TEST REQUIREMENTS OF ROUTING PROTOCOLS

Compared with communication protocols such as TCP/IP, routing protocols are quite different. First, the purpose of the routing protocol is to collect the information needed to do routing [1], which is a basic summary of routing protocol functions. Another extraordinary character is the nature of distributed systems, which can be found from both the interaction between different routers and the communication between the interfaces of one single router. No matter the designation or the implementation, there are several important databases to store the routing information. The information flow among these databases is connected by the internal processes which dominate the correctness and efficiency of routing protocol implementations.

For a routing protocol, there is no "upper level protocol" at all. However the interaction of routing protocols is much more complex. For example, BGP-4 will talk with neighbor peers while exchanging information with other routing protocols such as OSPF on the same router. A more detailed discussion can be found in [14]. For the testing of communication protocols, the protocol state machine is heavily relied upon. The state machines only describe a small part of the behavior of routing protocols. The majority of a routing protocol activities begins after it enters a "stable" state (e.g. established state for BGP-4).

Therefore, the testing of routing protocol should cover (1) the packet structure, (2) the protocol state machine, (3) the interoperability, (4) exchange and manipulation of routing information, (5) performance evaluation and (6) other features such as the policy-based routing and the route flap damping.

3. CURRENT TEST SYSTEM AND DEFICIENCIES

3.1 Structure of PITS

PITS is the second generation of the test system we developed. Figure 1 shows the structure of PITS. PITS exploits test suites in TTCN[15] format and adopts a concurrent interpretive test execution method for high efficiency. The test suite generation tool TUGEN generates a TTCN-formatted test suite from the formal description of the protocol. Test Execution (TE) is the key engine in the test process, responsible for test process control and the generation of all the information needed in the test report. PITS is a general tester, that is, TE works independently of any particular protocol. All the protocol-related information is reflected in the test suite and the corresponding RI.

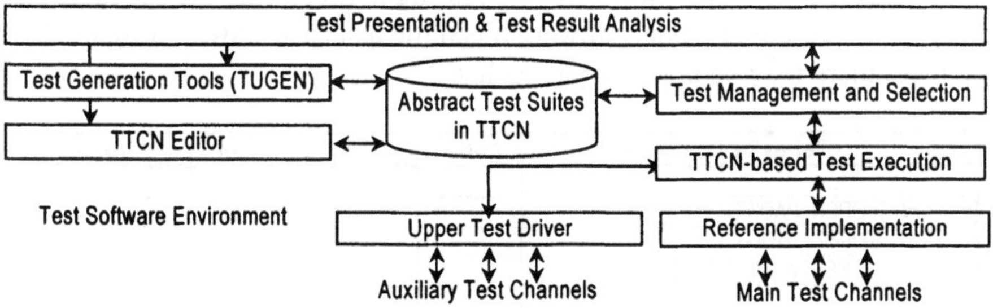

Figure 1 Protocol Integrated Test System

3.2 Deficiency and limitation of current test system

Since PITS was mainly aimed at the testing of communication protocols, some deficiencies were uncovered when testing routing protocols.

Originally, RI works in a very simple way, communicating with TE to get commands and data and sending/receiving data packets through the service provider. Test data and correct results are written in the test suite in advance and it is up to TE to accomplish the final comparison and judgement. Since there are large amounts of data transmission and routing information processing and TE depends on interpretive means to execute, TE can hardly accomplish the numerical calculation.

In traditional work, the test system only sets up a communication channel with the MUT and always exploits only one port. With such a test structure, only the packet structure and the protocol state machine can be tested. Since the routing protocol possesses features of distributed systems, the test system must be able to provide multiple ports. Only then can routing information exchange and processes be tested. Even if multiple ports are available, the entire scope of the testability is restricted to the data that the routing protocol module sends out. In addition, testing the correctness of the routing protocol under different conditions is also very important, especially for policy-based routing. Manual configurations are not only time-consuming but also error prone. On the other hand, the routing protocol is always implemented in the router while interacting with other modules. These modules provide the information and the access to the routing protocol module and the whole router through standard interfaces. In the original test structure, since the test system only connects with the MUT, this information and access can't be exploited at all.

There are further limitations regarding TTCN. It is difficult to describe certain cases with TTCN and it is almost impossible for TE to execute it. We believe that the latest TTCN-3 has greater ability and flexibility. However, the necessary editor, the executing environment, is under development.

The router always has a powerful processing ability while the speed of the test system is relatively much slower. When the test case shown in figure 2 is executed, there may be possible packet loss.

No	Behavior Description	Label	Constraints Ref	Verdict
	...			
1	T! OPEN_PACKET		C1_OPEN	
2	START T_HOLD			
3	T? KEEPALIVE_PACKET		C1_KEEP	
4	....			
5	? TIMEOUT			FAIL
6	T? OTHERWISE			FAIL
	...			

Figure 2 The test case with the possibility of packet loss

When step 1 is interpreted, the data and the command are sent to RI. RI then sends out the packet OPEN_PACKET. MUT will respond with a KEEPALIVE_PACKET. Since the speed of the router is rather fast, this packet is sent back immediately. At that time, if step 3 hasn't finished its interpretation RI will not accept the response packet. Further, when step 3 is done and RI is informed to receive some packet, what RI actually receives may be some other packet and step 6 will then be entered. The other scenario is that RI receives nothing at all. When the timer T_HOLD expires, step 5 will be entered.

4. ENHANCING TECHNIQUES

In order to solve these problems discussed above, the test system needs to be enhanced accordingly. After some analysis, almost all the enhancements are related to RI.

4.1 Functional sub-modules introduced into RI

In order to release TE from the heavy burden of the calculation, the dynamic calculation and the data generation sub-modules are introduced (shown in figure 3).

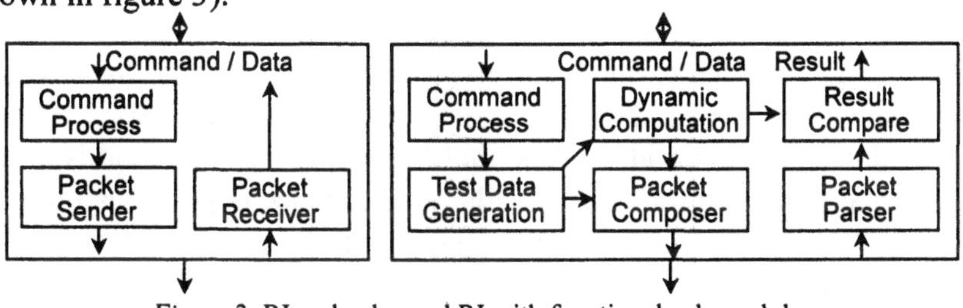

Figure 3 RI and enhanced RI with functional sub-modules

The test data generation sub-module in the ERI selects the test data or generates new test data with some algorithm according to the demand of TE. The packet composer sub-module forms the data packet and sends them out while the dynamic computation sub-module calculates the expected results. Response from the IUT (or MUT) is first processed by the packet parser sub-module, after which the results are compared with what are expected and finally presented to TE. Through these means, ERI accomplishes the calculation originally taken by TE, which improves the efficiency of the test execution.

4.2 Windowed black box model and multiple channels test method

In order to take full advantage of the information and access provided by the system (for example, a router) where the MUT resides, the windowed black box model and the multiple channel test method are adopted. The windowed black box takes into consideration not only the MUT but also other modules interacting with it. These modules offer standard interfaces and provide the tester with several windows to enhance the observation and the control of the MUT in the routing protocol test. Accordingly, the multiple channel test method is exploited. In traditional work, the test system only sets up a communication channel with the MUT, we call this channel a "basic channel", as in [10]. The principal difference of the multiple channel test method is that not only does the test system set up a communication channel with the MUT, it also sets up channels with other modules interacting with it (shown in figure 4).

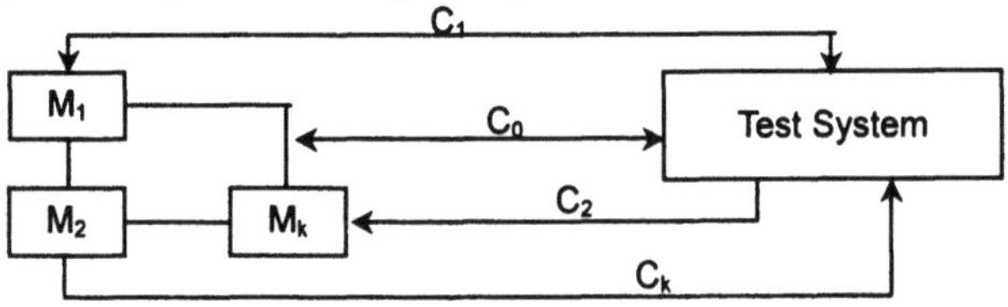

Figure 4. Multiple channels test method and test structure

The C_0 is the basic channel between the test system and the MUT. $M_1,M_2,\ldots,M_k$ are the k modules (will have been tested in advance) interacting with the MUT. The test system sets up channels $C_1,C_2,\ldots,C_k$ with these modules. The main idea of the multiple channel test method is to take full advantage of the information available to enhance the observability and the controllability, thus promoting the test efficiency and the coverage. The multiple ports needed to test distributed protocols can be easily implemented by expanding the channels.

4.3 Partition of acceptance sub layer and submission sub layer

In order to avoid possible packet loss in high-speed networks, RI is partitioned into two sub layers: the acceptance sub layer and the submission sub layer. The acceptance sub layer takes charge of receiving packets from the MUT, stores them in a buffer and maintains them. The submission sub layer fetches the packets and provides them to TE according to its demand. The acceptance sub layer makes RI independent of the speed of the lower network and the packet loss can be avoided by managing a proper interval between test case execution.

4.4 Relation of enhanced techniques

The structure of the ERI is shown in figure 5, which also explains the relation of these enhancing techniques. The acceptance and the submission sub layers reside at the bottom of the ERI. All the sub layers above can take advantage of them to make up for the speed gap. In the middle of the ERI structure, are basic and extended channels. Each channel can provide multiple ports in order to test distributed routing protocols. Meanwhile, each channel possesses functional sub-modules such as dynamic calculation and data generation (details can be referred to in figure 3) which can provide more processing ability and flexibility. The topside of the ERI contains the command/data dispensation and collection sub-module, communicating with TE, getting data/command, dispensing it to corresponding channels and returning the result to TE for the final verdict.

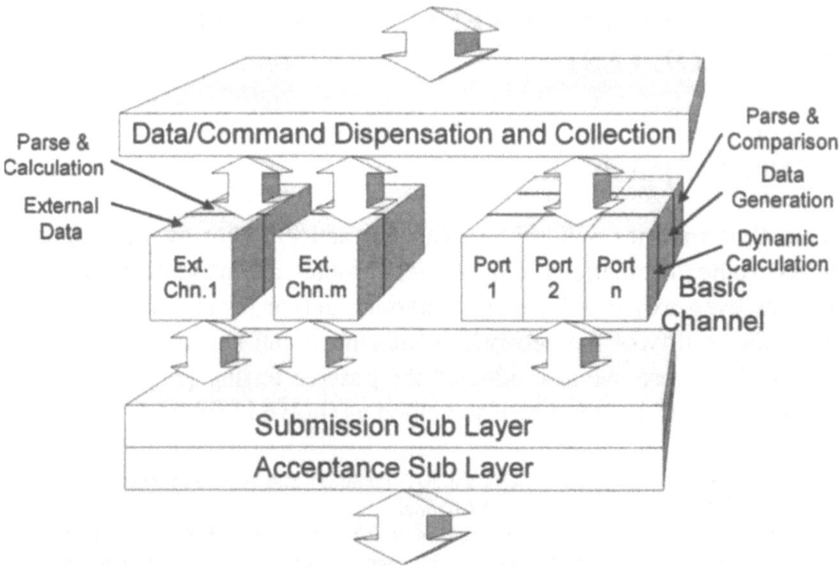

Figure 5 Structure of the Enhanced Reference Implementation

It can be seen that the ERI overcomes the problems discussed above and possesses these features: (1) functionally compatible with the former RI; (2) no effect on the syntax and semantic integrity of TTCN; (3) provides multiple channels to increase the observability and the controllability of the MUT; (4) multiple ports to test distributed routing protocols; (5) reduces the burden of TE and improves the flexibility of PITS. (6) enhanced functionality suitable for the routing protocol test.

4.5 Describing the test activity in the test suite integrally and uniformly

The multiple channel test method also brings new problems. One is how to describe the test activity uniformly and integrally. Therefore, the packet expansion technique is raised. For example, four types of BGP-4 packets are defined in [3] and we defined type 128-143 for the data sent through the configuration channel and 144-161 for the auxiliary information channel. This definition does not contradict current usage of packet types and is unlikely to cause future conflicts. However, there is no actual data in the date field of the packet with special type values. Instead, there is only an index. Corresponding channels will look up the data in their mapping tables. This form of representation is a natural extension of the PIXIT (Protocol Implementation eXtra Information) which can act as an external part of the test suite.

5. APPLIED PASSIVE TESTING IN ROUTING PROTOCOL TEST

5.1 Brief of passive testing

Some abnormalities will only appear in practice and/or over a long time, such as route oscillations [8], useless route advertisements [9] and sometimes even exhausted routers. Although active testing can do a lot to promote the quality of the network equipment, it can hardly deal with such problems. Therefore, we also adopted the passive testing [12] in our work. The passive testing is an online test method as shown in figure 6.

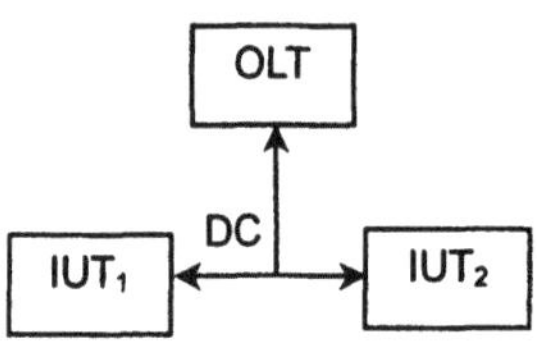

IUT: IUT_1 and IUT_2 communicate with each other.
DC: Data Channel
OLT: OnLine Tester. It monitors the sessions between IUT_1 and IUT_2 without sending our anything gives verdicts and provides information for the fault management periodically or on demand.

Figure 6. Model of the passive test

Compared with the active testing, the passive testing is simple in structure, while possessing more feasible and powerful abilities to perform the conformance and the interoperability tests and even analysis of the performance. Additionally, it allows more IUTs and OLTs, enabling distributed test methods.

5.2 Algorithms and methods for passive testing

Algorithms to test the protocol state machine can be found in [13]. As to routing information manipulation, one generic method is topology analysis. The main idea is that correct routing information must describe the network topology correctly. We construct the topology from the routes exchanged and compare it with the actual topology. If differences are found, then there may be some errors in the routing information manipulation. Since we can get the constructed network topology from the OSPF routing information base, we mainly focused on the topology analysis for BGP-4, which can be found in [16]. Other methods like the internal process simulation are also exploited. Different modules are implemented to process the routing information collected online similarly to what the processes in the router will do. The correct scene is calculated and compared with the actual result. For BGP-4, the internal process simulation covers the generic route selection, route selection with policy constraints, and the route flap damping, etc.

5.2.1 Testing of route selection with policy constraint

Route selection with policy constraints is an important feature of BGP-4. These policies directly serve the relations of the Internet Service Providers (ISP) and their customers. Three basic relations are classified (client-provider, provider-provider and customer with backup) and corresponding policies are presented. Policy conformance and violations are defined.

Definition 1. Policy conformance Conf = $<Conf_i, Conf_e>$, where $Conf_i$ is the import conformance which evaluates the process when receiving routing information. $Conf_e$ is the export conformance which evaluates the process when advertising routing information.

Suppose PS_r is the series of policies to be enforced on the received routes. $PS_r = [\ P_{r1}, P_{r2}, P_{r3}, \ \ldots\ldots, P_{rn}]\ \forall\ R_i \in R\_RCV,\ R_i' = P_{rn}(P_{r\ n-1}(\ldots\ldots P_{r1}(R_i)))$,denoted as $R_i' = PS_r(R_i)$.R_i' is route of import conformance: R_i' $Conf_I$ PS_r

Suppose PS_s is the series of policies to be enforced on the routes to be sent out. $PS_s = [\ P_{s1}, P_{s2}, P_{s3}, \ \ldots\ldots, P_{sm}]\ \forall\ R_j \in R\_ACP, R_j' = P_{sm}(P_{s\ m-1}(\ldots\ldots P_{s1}(S_j)))$,denoted as $R_j' = PS_s(R_j)$, R_j' is route of export conformance: R_j' $Conf_e$ PS_s.

Then two basic types of policy violations can be defined; other violations can be made out of these two.

Definition 2. Penetration: if $\exists R_k \in R\_ACP, \neg (\exists R_l \in R\_RCV, PS_r(R_l) = R_k)$, then "penetration" occurs, that is, the route is accepted after the route calculation and selection while it actually should not be.

Definition 3. Leak: if $\exists R_k \in R\_SNT, \neg (\exists R_l \in R\_ACP, PS_s(R_l) = R_k)$, then "leak" occurs, that is, the route is delivered while it actually should not be.

A separate module was implemented. With policy configurations (provided by the network operator), this module can find the basic policy violations and other subtle abnormalities.

5.2.2 Testing of route flap damping

Route flap is another serious problem of the Internet routing [9][17] and the "merit way" is the damping method now in use [18]. However, to our best knowledge, there is no test to evaluate the damping effect. Therefore, we worked out a simple but practical one (shown in figure 7). In this state machine, the flap of a single route, the interaction between two routes (replacement) and the condition with an alternative route are all covered. Since the route selection will finally go down to the comparison between two routes, this model is applicable for generic conditions.

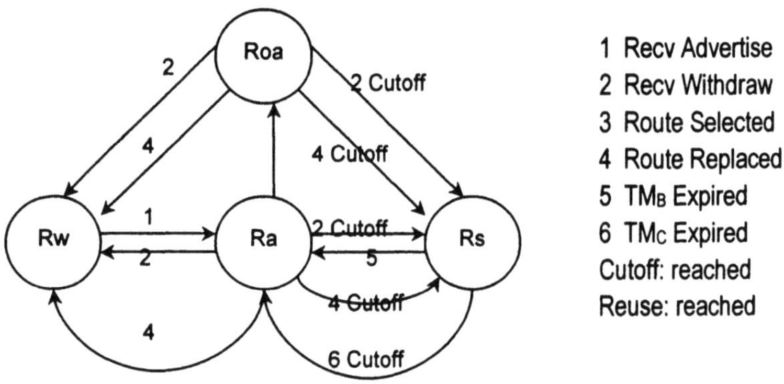

Figure 7. State machine of route flap damping

The flap module follows this state machine and performs the merit value calculation (increase and decay). Therefore, not only can the passive testing find the existence of the route flap and its severity, it can also evaluate the effect of the route flap damping.

6. IMPLEMENTATIONS AND TEST ACTIVITIES

6.1 PITS with ERI

The basic channel can support up to 8 ports and two extended channels: the auxiliary information and the configuration channels. The auxiliary information channel is set up to access the information needed in the routing protocol test in its session with the SNMP; the configuration channel connects through telnet and configures the routing protocol and the router with corresponding commands. A test case is shown in figure 8.

Dynamic Part				
Test Case Name:	UPDATE_ALL_IBGP			
Group:	bgp-4/IBGP/UPDATE_ALL_IBGP			
Purpose:	Test the information manipulation of IBGP			
Default:	NULL			
Comments:	NULL			
No	Behavior Description	Label	Constraints Ref	Verdict
1	T! TO_CONF		C_TO_CONF	
2	+ESTABLISHED			
3	T! UPDATE01		C_UPDATE01	
4	T! UPDATE01		C2_UPDATE01	
5	START T_HOLD			
6	T? FROM_AUX		C_FROM_AUX	
7	START T_KEEP			
8	T? KEEP_ALIVE01		C_KEEPALIVE01	PASS
9	? TIMEOUT			FAIL
10	T? OTHERWISE			FAIL
11	? TIMEOUT			FAIL
12	T? OTHERWISE			FAIL
Extended Comments: NULL				

Figure 8 An example of a test case

In this case, the configuration channel and the auxiliary information channel are exploited. This example also demonstrates how the test activity is described integrally and uniformly.

In the acceptance sub layer, a buffer of 1600 X 40 bytes is set aside. There is no packet loss in a 10M or 100M Ethernet environment.

6.2 Online intelligent analyzer

The intelligent online BGP-4 Analyzer [20] is our first application of the passive testing. Compared with earlier work, this analyzer offers more efficient means for both the protocol test and the network management. It can test the state machines of BGP peers', from which some conclusions on the conformance and the interoperability can be drawn. As to the routing information manipulation, the analyzer simulates internal processes, such as the route advertisement and route withdrawal, the policy monitoring and the flap damping. The analyzer can be set to focus on some specified routing information or on certain BGP speakers' behavior. It can also collect statistics and provide basic performance parameters regarding the frequency and the amount of the data interaction. Also, the analyzer can re-display the procedure, making it easy to analyze and eliminate faults. Further, it can generate a TTCN-formatted test suite with the recorded data. Besides, it has a friendly graphical user interface (GUI) running on Linux and can attend to multiple NICs simultaneously. Data is stored in a compressed form in consideration of working over a long period time.

6.3 Online test system

As a more generic system for all the routing protocols and other distributed protocols, we have implemented the prototype of an online test system [21]. The framework is shown in figure 9. There are five main components. The input component includes a protocol state machine description module, a protocol packet structure description module and a set condition module. The communication component includes a communication interface module and a condition check and filter module, which shields the difference among communication protocols of different layers. The condition check and filter module further processes the data collected and provides the test component with a uniform interface. The test component is the kernel of the OLTS, including the protocol state machine test module, the internal process simulation module and the statistics module. The cooperation component is used to organize multiple OLTS instances. The test result module, the LOG module and other modules belong to the conclusion and presentation component, which composes a test report, provides graphical presentations and generates the test suite.

For each protocol to test, a specific communication interface module is coded, descriptions of the protocol state machine and the protocol packet structure are input and the necessary internal process simulation module is applied or implemented. OLTS is implemented on a Sun Ultra 1 - Solaris and a x86-Linux separately in order to widen its application and verify its portability

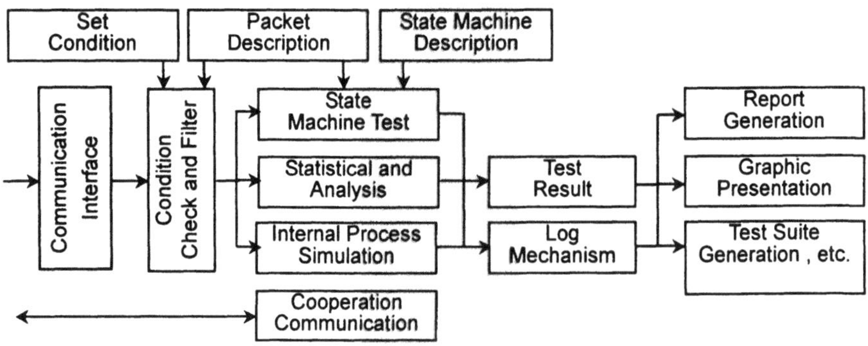

Figure 9. Framework of OLTS

Specific modules will be implemented for different communication media to shield the difference of the lower level. For example, the interface has been implemented for the most popular Ethernet, which is able to capture all the data of upper level protocols. On x86-Linux, the packet socket is exploited and DLPI (Data Link Provider Interface) [22] operating /dev/le on the Sun Ultra 1 - Solaris. For other communication media, there has been no difficulty encountered so far (even for a non-broadcast point to point link).

Based on this, the virtual network interface (VNI) is further implemented. VNI aims to provide a uniform interface that the main process can operate. When new communication media is introduced, the only requirement is to map its interface to the uniform interface.

The internal process simulation is done via a module stack; the modules can be pushed onto the stack for different protocols and test purposes.

In consideration of testing in a distributed environment, Simple Cooperation Protocol (SCP) [23] is defined for the cooperation component, which enables multiple instances of OLTS's to work together in different places of the network. We still have a tool set, including the packet composer. It can be controlled through a control channel by other systems such as PITS.

One of the most interesting aspects is that the active testing and the passive testing are combined. The active testing and the passive testing have both advantages and limitations of their own. The active testing can send out the desired packets; the passive testing can work in production field over a long time. By combining the active testing and the passive testing, we can expect a better efficiency.

An example is the erroneous session shown in figure 10. It is first captured by the analyzer. 166.111.69.91 is one port address of the commercial router and 166.111.69.75 is that of the Router II. This session shows that due to the incorrect implementation of BGP-4 of Router II, UPDATE packets are sent out repeatedly instead of OPEN after the TCP connection is established. But the response of the commercial router is not correct either. After it responds with a NOTIFICATION packet (line 2 in figure 10), it answers with the same UPDATE (line 5 in figure 10), which contradicts the protocol specification and leads to problems of the conformance and the interoperability along with the extra cost of the network bandwidth.

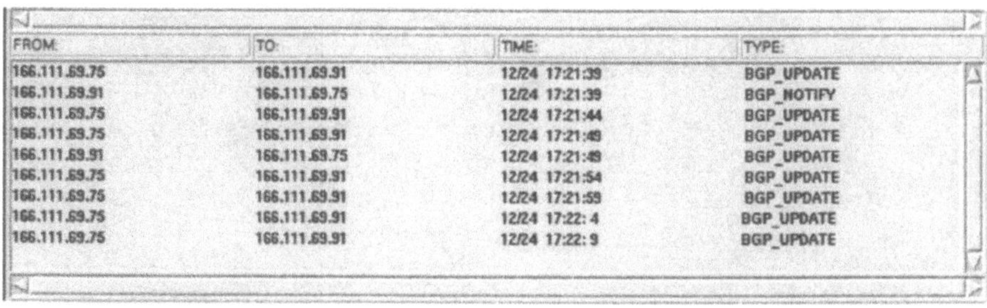

FROM:	TO:	TIME:	TYPE:
166.111.69.75	166.111.69.91	12/24 17:21:39	BGP_UPDATE
166.111.69.91	166.111.69.75	12/24 17:21:39	BGP_NOTIFY
166.111.69.75	166.111.69.91	12/24 17:21:44	BGP_UPDATE
166.111.69.75	166.111.69.91	12/24 17:21:49	BGP_UPDATE
166.111.69.91	166.111.69.75	12/24 17:21:49	BGP_UPDATE
166.111.69.75	166.111.69.91	12/24 17:21:54	BGP_UPDATE
166.111.69.75	166.111.69.91	12/24 17:21:59	BGP_UPDATE
166.111.69.75	166.111.69.91	12/24 17:22: 4	BGP_UPDATE
166.111.69.75	166.111.69.91	12/24 17:22: 9	BGP_UPDATE

Figure 10 An erroneous session recorded

Then this fault was confirmed with the active testing and several other similar faults were uncovered with this indication.

The test structure used in the policy-based routing test is shown in figure 11. BG_t is a real router, while W_1 and W_2 are two online test systems. N_1 and N_2 are two nodes, each of which may be a real router or an active test system. N_1 sets up an EBGP connection with BG_t and N_2, BG_t an IBGP connection. W_1 watches the session on EBGP and W_2 on IBGP. With different configurations, variant tests can be performed. In a production field, N_1 and N_2 are all real routers, online test systems W_1 and W_2 monitor if there is any leak or penetration. Under the constructed test environment, N_1 and N_2 are test systems and they can actively test policy-based routing to detect violations.

The test structure of the route flap damping (shown in figure 12) also covers both the active testing and the passive testing. There are three types of nodes in this structure: flap source nodes, router nodes and detector nodes. The router nodes construct a network. The flap source nodes inject flap into the network at different positions and the detector collects the information about the flap damping. The test is in an active one. However, if the network is a real network, it will generate flap itself. Then the set of flap source nodes can be empty and the test switches into a passive mode.

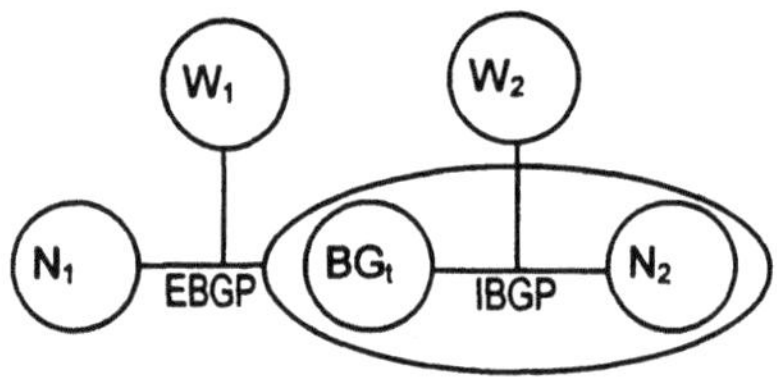

Figure 11 Test structure for policy based routing test

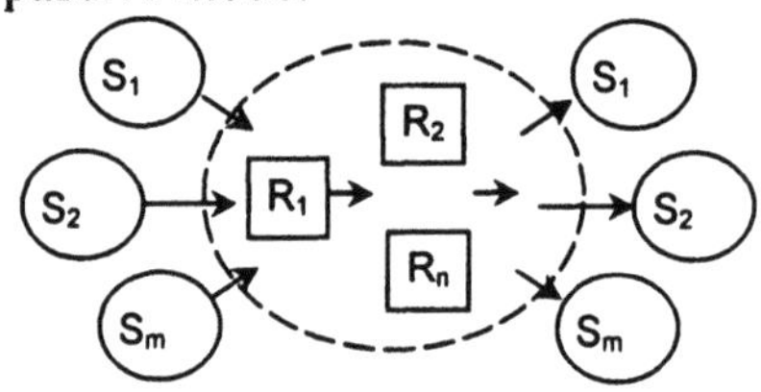

Figure 12 Test structure for route flap damping

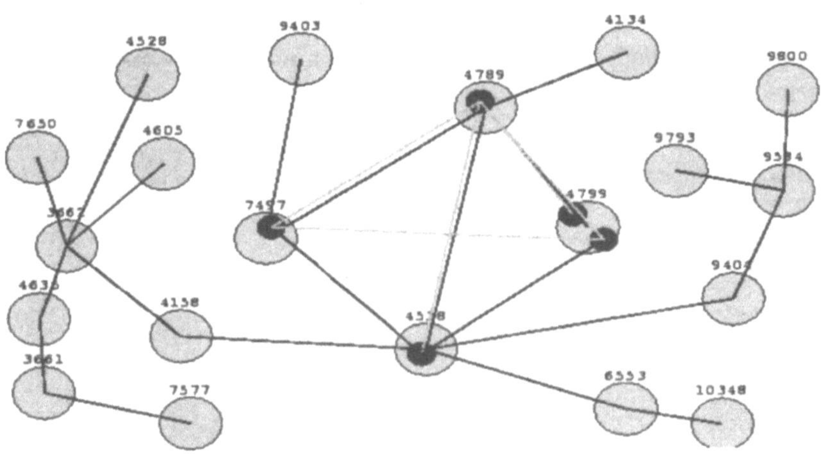

Figure 13 Topology constructed with the data collected from the backbone of CERNET

With both the active testing and the passive testing, a large number of test activities have been performed. We have tested some commercial routers in our lab and we have performed the test on the backbone of CERNET [24]. Figure 13 is the topology constructed with the data collected on CERNET. We compare this constructed topology with the actual one to discover errors in routing information manipulation. We also offer test support to the development of routers. Their pass ratio is considerably promoted and therefore the quality is improved accordingly.

7. CONCLUSIONS

Despite the great importance of routing protocols, test activity is still quite limited compared with the testing of communication protocols. In addition to our previous work, reference [11] discusses the testing of IP routing protocols in the environment generated by software tools. Craig Labovitz has collected the routing data for two years and published several papers [17]. Currently there are some commercial test systems, such as HP Router Tester [25] and NetCom Systems Smartbits [26], etc. They mainly focus on the performance of the system. They are all active testers and possess a high degree of industrialization. The price is rather expensive and they are always difficult to extend. For example, users can't develop test suites for new protocols.

This paper presents our efforts to cover important features of Internet routing protocols. However, there is still quite a lot to do. Routing protocols are still progressing, while models, methods and other techniques still need further refinement. At the same time, more efforts must be expended within the field of the protocol test, through which the validity of new methods can be proved and new problems can be discovered.

REFERENCES

[1] C. Hedrick. Routing Information Protocol. RFC 1058. 1988

[2] J. Moy. OSPF Version 2. RFC 2178. 1998

[3] Y. Rekhter, T. Li. A Border Gateway Protocol 4 (BGP-4). RFC 1771, 1995

[4] ISO,Open Systems Interconnection (OSI) Routing Protocol, 1985

[5] Cisco Systems, Inc. Interior Gateway Routing Protocol, 1987

[6] Cisco Systems, Inc., Enhanced Interior Gateway Routing Protocol, 1989

[7] Jianping Wu, Xiuhuan Chen, Ruibing Hao et al. PITS-The protocol integrated test system based on formal technology, Journal of Tsinghua Univ. 1998: 38(S1):26~29.

[8] T. G. Griffin, G. Wilfong, An Analysis of BGP Convergence Properties, Proc of Sigcomm 99, 1999

[9] C. Labovitz, G.R. Malan, F. Jahanian. Internet routing instability. IEEE/ACM Transactions on Networking Oct 1998 IEEE Piscataway NJ USA 1063-6692 IEANEP, 1998; 6: 515-527

[10] Yixin Zhao, Jianping Wu, Xia Yin and Bo Han. Test of BGP-4 based on the Protocol Integrated Test System. Proc of APCC2000. Korea. 2000

[11] Ruibing Hao, D. Lee, K. Rakesh. et al. Testing IP Routing Protocols - From Probabilistic Algorithms to a Software Tool. KLUWER ACADEMIC PUBLISHERS. Proc of FORTE/PSTV2000. Pisa, Italy. 2000.

[12] D. Lee, A.N. Netravali, K.K. Sabnani et al. Passive Testing and Applications to Network Management, Proc of ICNP 97 , 1997

[13] Yinin Zhao, Jianping Wu. Algorithm to determine states applied in online test. Chinese Journal of Electronics. 2000; 11: 33-40

[14] Yixin Zhao, Bo Han, Xia Yin and Jianping Wu. Study of Routing Protocols Test. Computer Engineering.2000; 26:769-774

[15] ISO. Information Technology, Open System Interconnection, Conformance Test Methodology and Framework. IS-9646, ISO, 1991

[16] Yixin Zhao, Xia Yin, Jianping Wu et al. Fault Management of Inter-domain Routing with Passive Test. Proc of ICT2001. Romania, Bucharest, 2001

[17] C. Labovitz, R. Wattenhofer, S. Venkatachary. The Impact of Internet Policy and Topology on Delayed Routing Convergence. Tech. Report. 2000

[18] C. Villamizar, R. Chandra, R. Govindan. BGP Route Flap Damping. RFC 2439 1998

[19] Yixin Zhao, Xia Yin, Bo Han and Jianping Wu. Enhanced Reference Implementation Applied in Routing Protocol Test. Proc of ICACT 2001, Korea, 2001

[20] Yixin Zhao, Jianping Wu and Bin Yu. Intelligent Online BGP-4 Analyzer. IEEE Computer Society. Proc of MASCOTS 2000. Los Vaqueros Circle: IEEE Computer Society, 2000

[21] Yixin Zhao, Xia Yin, Bo Han and Jianping Wu. The Online Test System Applied in the Test of Routing Protocols. High Technology Letters (In press).2000

[22] OSI. Data Link Provider Interface Specification, UNIX International OSI Work Group, revision 2.0.0. 1991

[23] Yixin Zhao. Study of BGP-4 conformance test. Master thesis. Tsinghua Univ. 2000

[24] China Education and Research Network CERNET. http://www.cernet.edu.cn

[25] Agilent Technologies. http://www.tm.agilent.com/. 1999

[26] Spirent Communications, Inc. http://www.netcomsystems.com. 2000

Part Three

Timed Automata

TIME AND ACTION LOCK FREEDOM PROPERTIES FOR TIMED AUTOMATA

Howard Bowman
Computing Lab, Univ of Kent at Canterbury, Canterbury, Kent, CT2 7NF, UK
H.Bowman@ukc.ac.uk

Abstract Time and action locks can arise freely in timed automata specification. We review techniques for avoiding the occurrence of timelocks, based on the Timed Automata with Deadlines framework. Then we present a notion of parallel composition which preserves action lock freeness, in the sense that, if any component automaton is action lock free, then the composition will also be action lock free.

Keywords: Timed Automata, Timed Systems, Timelocks.

1. INTRODUCTION

Deadlocks are the characteristic error situation arising in concurrent and distributed systems. In very general terms, they are states in which the system is unable to progress further.

Classically the term deadlock has been seen as synonymous with what we will call *action locks*. These are situations in which, how ever long time is allowed to progress, the system will never be able to perform an action. Such action locks often result from unmatchable action offers, e.g. when a component wishes to perform a synchronisation action, but is unable to because no other process can offer a matching synchronisation.

In the context of timed systems, new locking situations arise. In particular, in this paper, we will be working in an environment with two main types of locking situation. As a result of this, we have had to be careful with our choice of terminology. Thus, in this paper the term deadlock is the most general. It embraces action locks and the form of locking behaviour that comes with timed systems - timelocks.

Timelocks are situations in which, informally speaking, time is prevented from passing beyond a certain point. They are highly degenerate occurrences [4] because they yield a *global* blockage of the systems evolution. This is quite different from an action lock, which cannot affect the evolution of an independent process.

In particular, unless significant care is taken, the possible presence of timelocks is a major problem for the formal specification and analysis of time critical networks and distributed systems. This is especially the case when using timed automata, which are at the heart of current timed model checking technology, such as UPPAAL [1] and Kronos [7].

This problem was highlighted in [6] where a number of timelock errors were discovered in a timed automata model of a lip-synchronisation protocol. Furthermore, it was shown in [4] that, when using timed automata, even the simple task of defining a timeout in a communication protocol is hampered by the possible presence of timelocks.

In fact, the issue of whether timelocks are desirable or undesirable features of timed models remains a hotly debated topic. The standard argument in favour of models containing timelocks is that they represent specification inconsistencies (like logical contradictions) and that by discovering and eliminating them, specifications can be corrected. However, we take the contrary position for a number of reasons.

Firstly, detecting timelocks is, in fact, a difficult and expensive analysis task. The classic method is to show that a property such as the Kronos formula, $\texttt{init} \Rightarrow \Box\Diamond_{=1}(\texttt{true})$, holds over the specification. This is an unbounded liveness property which is one of the most difficult classes of formulae to check. Recent work by Tripakis [9] offers potential improvements in such analysis. However, his algorithm remains unimplemented and furthermore, such improvements will always be thwarted by systems with fundamentally large state spaces.

We are also strongly of the opinion that inconsistencies and contradictions fit in the domain of logical description, but are difficult to reconcile with behavioural specification techniques, such as Timed Automata (TA). Contradictions arise when conflicting *properties* are asserted / conjoined. However, although the mistake is frequently made, parallel composition of processes is not a property composition operator, rather its meaning is operational - two (or more) physical components are *run* in parallel. Error situations in behavioural techniques should have a behavioural / operational intuition that is justifiable in terms of real world behaviour. This is the case for action locks and livelocks. However, there is no real world counter-part for time stopping!

It is also important to emphasize that the situation with action locks and timelocks is, in this respect, a little different. In our opinion timelocks are highly counter-intuitive and thus we believe that constructively preventing the occurence of timelocks is essential. However, since action locks are not in the same way counter-intuitive, prevention is not in the same sense essential. Nonetheless investigating techniques which ensure action lock freeness is useful since it highlights forms of parallel composition that can be employed when building systems that are "correct by construction".

Although pleasingly simple, as previously implied, timed automata have the weakness that timelocks can freely arise and in a number of different ways. Perhaps most problematically they can arise through the interplay of urgency and synchronous interaction. We argue that urgency is given too strong an interpretation in timed automata, in the sense that an action can be forced (i.e. it becomes urgent) even if it is not possible (i.e. is not enabled).

The first part of this paper presents a re-interpretation of synchronisation that weakens the effect of urgency and thus limits the occurrence of timelocks. The approach uses ideas from the Timed Automata with Deadlines (TADs) framework of Bornot and Sifakis [2, 3]. However, we adapt the TADs definitions in the manner we introduced in [4].

The timed prioritised choice features offered by the TADs framework yield the possibility that the dynamic enabling of "competing" transitions can be defined statically. Hence notions of parallel composition that preserve different dynamic properties can be investigated. In this vein we also present a notion of parallel composition which preserves action lock freeness, in the sense that, if any of the component TADs is action lock free then the parallel composition will also be action lock free. Thus, in this paper we develop notions of parallel composition which are strongly compositional in the sense that if component processes are time and action lock free, then the parallel composition of these process will also be time and action lock free.

Structure of Paper. Section 2 introduces basic notation. Then we tackle the issue of timelocks in section 3 and action locks in section 4. Finally, section 5 presents concluding remarks.

2. BACKGROUND

Notation. $HA = \{x?, x! \mid x \in CA\}$ is a set of *half* (or *uncompleted*) actions. CA is a set of *completed* (or internal) actions. These give a simple CCS or UPPAAL [1] style point-to-point communication. Thus, two actions, $x?$ and $x!$ can synchronise and generate a completed action x. $\mathbb{A} = HA \cup CA$ is the set of *all* actions. We use a complementation notation over elements of $\mathbb{A}$: $\overline{x} = x$ if $x \in CA$, $\overline{x?} = x!$ and $\overline{x!} = x?$.

$\mathbb{R}^+$ denotes the positive reals without zero and $\mathbb{R}^{+0} = \mathbb{R}^+ \cup \{0\}$. $\mathbb{C}$ is the set of all clock variables, which take values in $\mathbb{R}^{+0}$. CC is a set of clock constraints. Also if $C \subseteq \mathbb{C}$ we write CC_C for the set of clock constraints generated from clocks in C.

$\mathbb{V}_C = C \rightarrow \mathbb{R}^{+0}$ is the space of clock valuations for clocks in C.

Timed Automata. An arbitrary TA has the form: (L, l_0, T, I, C), where, L is a finite set of locations; $l_0 \in L$ is a *start location*; and C is the set of clocks of the TA.

$T \subseteq L \times \mathbb{A} \times CC_C \times \mathbb{P}(C) \times L$ is a transition relation. A typical element of T would be, (l_1, e, g, r, l_2), where l_1, l_2 are locations; e labels the transition; g is a guard; and r is a reset set. $(l_1, e, g, r, l_2) \in T$ is typically written, $l_1 \xrightarrow{e,g,r} l_2$, stating that the automaton can evolve from location l_1 to l_2 if the (clock) guard g holds and in the process action e will be performed and all the clocks in r will be set to zero.

$I : L \rightarrow CC_C$ is a function which associates an invariant with every location. Intuitively, an automaton can only stay in a state while its invariant is satisfied.

It is important to understand the difference between the role of guards and of invariants. In this respect we can distinguish between *may* and *must* timing. Guards express may behaviour, i.e. they state that a transition is possible or in other words *may* be taken. However, guards cannot "force" transitions to be taken. In contrast, invariants define must behaviour. This must aspect corresponds to *urgency*, since an alternative expression is that when an invariant expires, outgoing transitions must be taken straightaway.

Semantics. TA are semantically interpreted over transition systems which are triples, $(S, s_0, \Rightarrow)$, where, $S \subseteq L \times \mathbb{V}_C$ is a set of states; $s_0 \in S$ is a start state; and $\Rightarrow \subseteq S \times Lab \times S$ is a transition relation, where $Lab = \mathbb{A} \cup \mathbb{R}^+$. Transitions can be of one of two types: *discrete transitions*, e.g. (s_1, e, s_2), where $e \in \mathbb{A}$ and *time transitions*, e.g. (s_1, d, s_2), where $d \in \mathbb{R}^+$ and denotes the passage of d time units. Transitions are written: $s_1 \stackrel{e}{\Longrightarrow} s_2$ respectively $s_1 \stackrel{d}{\Longrightarrow} s_2$.

For a clock valuation $v \in \mathbb{V}_C$ and a delay d, $v+d$ is the clock valuation such that $(v+d)(c) = v(c)+d$ for all $c \in C$. For a reset set r, we use $r(v)$ to denote the clock valuation v' such that $v'(c) = 0$ whenever $c \in r$ and $v'(c) = v(c)$ otherwise. v_0 is the clock valuation that assigns all clocks to the value zero.

The semantics of a TA $A = (L, l_0, T, I, C)$ is a transition system, $(S, s_0, \Rightarrow)$, where $S = \{ s' \in L \times \mathbb{V}_C \mid \exists s \in S, y \in Lab \, . \, s \stackrel{y}{\Longrightarrow} s' \} \cup \{ [l_0, v_0] \}$, $s_0 = [l_0, v_0]$ and $\Rightarrow$ is defined by the following inference rules:-

$$\frac{l \xrightarrow{e,g,r} l' \quad g(v)}{[l, v] \stackrel{e}{\Longrightarrow} [l', r(v)]} \qquad \frac{\forall d' \leq d \, . \, I(l)(v + d')}{[l, v] \stackrel{d}{\Longrightarrow} [l, v + d]}$$

The semantic map which generates transition systems from TA is written $[\![\,]\!]$. Also, notice that our construction ensures that only reachable states are in S.

Parallel Composition. We assume our system is described as a vector of TA, denoted, $|A = |\langle A[1], ..., A[n]\rangle$ where $A[i]$ is a timed automaton. In addition, we let u, u' etc, range over the set $\mathbb{U}$ of vectors of locations, which are written, $\langle u[1], ..., u[n]\rangle$. We use a substitution notation as follows: $\langle u[1], ..., u[j], ..., u[n]\rangle[u[j]'/u[j]] = \langle u[1], ..., u[j]', ..., u[n]\rangle$ and we write $[u[j]'/u[j]]$ as $[j'/j]$ and $u[i_1'/i_1]...[i_m'/i_m]$ as $u[i_1'/i_1, ..., i_m'/i_m]$.

If $\forall i(1 \leq i \leq n) \,.\, A[i] = (L_i, l_{i,0}, T_i, I_i, C_i)$ then the product automaton, which characterises the behaviour of $|\langle A[1], ..., A[n]\rangle$ is given by, (L, l_0, T, I, C) where $L = \{\, |u \mid u \in L_1 \times ... \times L_n \,\}$, $l_0 = |\langle l_{1,0}, ..., l_{1,n}\rangle$, T is as defined by the following two inference rules, $I(|\langle u[1], ..., u[n]\rangle) = I_1(u[1]) \wedge ... \wedge I_n(u[n])$ and $C = C_1 \cup ... \cup C_n$.

$$\frac{u[i] \xrightarrow{x?,g_i,r_i} u[i]' \quad u[j] \xrightarrow{x!,g_j,r_j} u[j]'}{|u \xrightarrow{x,g_i \wedge g_j, r_i \cup r_j} |u[i'/i, j'/j]} \qquad \frac{u[i] \xrightarrow{x,g,r} u[i]' \quad x \in CA}{|u \xrightarrow{x,g,r} |u[i'/i]}$$

where $1 \leq i \neq j \leq |u|$. Note, we write $x \leq k \neq r \leq y$ in place of $x \leq k \leq y \ \wedge \ x \leq r \leq y \wedge \ k \neq r$.

Timelocks. We can formulate the notion of a timelock in terms of a testing process. Consider, if we take our system which we denote `System` and compose it in parallel with the timed automaton, `Tester`, shown in figure 1, where, since it is completed, the **zzz** action is independent of all actions in `System`. Then for any d$\in \mathbb{R}^+$, if the composition `|<Tester(d),System>` can evolve to a state from which it cannot perform **zzz**, then the system contains a timelock at time d. Notice that we are not saying that *all* executions of `System` will timelock, but rather that `System` can timelock.

This illustration indicates why timelocks represent such degenerate situations - even though the `Tester` is in all respects independent of the system, e.g. it could be that `Tester` is executed on the Moon and `System` is executed on Earth without any co-operation, the fact that the system cannot pass time prevents the tester from passing time as well. Thus, *time really does stop* and it stops everywhere because of a degenerate piece of *local* behaviour.

We can also give a semantic definition of the notion, which is similar to definitions given in [9]. However, we first need a little notation. A trace of a timed automaton A has the form, $\rho = s_0\, y_1\, s_1 \dots s_{n-1}\, y_n\, s_n$, where, $\forall i(0 \leq i \leq n) \,.\, s_i \in [\![A]\!].1$ (throughout the paper we use the notation $t.i$ to access the ith element of a tuple); $s_0 = [l_0, v_0]$; $y_i \in \mathbb{A} \cup \mathbb{R}^+$; and

$\forall i (0 \leq i \leq n-1) \,.\, s_i \xrightarrow{y_i+1} s_{i+1}$. We let $Tr(A)$ denote the set of all traces of A and we define the function *delay* as,

$$delay(\rho) = \Sigma\{\, y_i \,|\, 1 \leq i \leq n \;\wedge\; y_i \in \mathbb{R}^+ \}$$

Now we say that A can timelock at time d iff

$$\exists \rho \in Tr(A) \,.\, (delay(\rho) < d \wedge \forall \sigma \in Tr(A) \,.\, (\rho \; pref \; \sigma \implies delay(\sigma) < d))$$

where $\rho_1 \; pref \; \rho_2$ if and only if ρ_1 is a prefix of ρ_2. Intuitively this expresses that there is a state reachable before d time units has passed, from which it is not possible for time to elapse beyond d. Notice this definition does not preclude the system evolving "while timelocked".

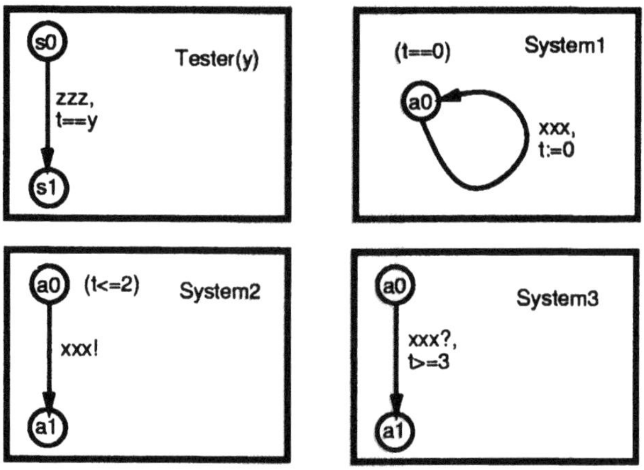

Figure 1. A Tester and Timelock Illustrations

There are two different forms of timelock:-

1 *Zeno Timelocks.* These arise when the system has an infinite behaviour but time cannot pass beyond a certain point. In other terms, an infinite number of discrete transitions are performed in a finite period of time. An example of such a specification is `System1` (see figure 1).

2 *Time Action Locks.* These are situations in which a state is reached from which neither time or action transitions can be performed. An example of such a lock is a trivial TA with no transitions and one location with invariant *false*. However, more problematically, time action locks can arise through mismatched synchronisations, e.g. `|<System2,System3>` (from figure 1) contains a timelock at

time 2, which arises because **System2** must have performed (and thus, synchronised on) action **xxx** by the time **t** reaches 2 while **System3** does not start offering **xxx** until after **t** has past 2.

The interesting difference between these two varieties of timelock is that the first one locks time, but it is not action locked. However, the second reaches a state in which neither time passing or action transitions are possible.

A relevant property which appears in the literature is that of time reactivity. It ensures that a system is time action lock free.

Definition 1 *A system is said to be time reactive if it can never reach a state in which neither time or action transitions can be performed.*

Action Locks. Timelocks are much more serious faults than action locks, since the latter generate *local deadlocks*, however, cannot prevent an independent process from evolving. A TA is action locked when it reaches a state from which, however long time is allowed to pass, an action will never be possible. The natural interpretation of action lock in the setting of timed systems is as follows.

Definition 2 *A state $[l, v]$ of a TA A is an action lock, denoted $AL([l, v])$, if and only if, $\forall t \in \mathbb{R}^{+0}\ ([l, v+t] \in [\![A]\!].1 \implies [l, v+t] \stackrel{e}{\not\Rightarrow})$, where $[l, v+t] \in [\![A]\!].1$ implies $[l, v+t]$ is reachable from $[l, v]$ by the definition of $[\![\,]\!]$. A TA A contains an action lock iff $\exists s \in [\![A]\!].1 \,.\, AL(s)$.*

3. TIMELOCKS

Zeno Timelocks. Using an approach of Tripakis [9] we introduce a static construction which ensures zeno timelock freeness. The idea is to ensure that for each loop in an automaton, time must pass by at least $\epsilon \in \mathbb{R}^+$ on every iteration (this is similar to imposing time guardedness in timed process algebra). First a definition.

Definition 3 *For $A \in TA$ we define a structural loop to be a sequence of distinct transitions, $l_0 \xrightarrow{e_1, g_1, r_1} l_1 \xrightarrow{e_2, g_2, r_2} \ldots. \xrightarrow{e_n, g_n, r_n} l_n$, s.t. $l_0 = l_n$.*

A is called strongly non-zeno if, for every such structural loop, there exists a clock $c \in A.5$, $\epsilon \in \mathbb{R}^+$ and $0 \leq i, j \leq n$ s.t., (1) $c \in r_i$; and (2) c is bounded from below in step j, i.e. $(c < \epsilon) \cap g_j = false$.

Clearly, **System1** of figure 1 fails to be strongly non-zeno since a suitable $\epsilon \in \mathbb{R}^+$ does not exist. The following result was presented in [9].

Proposition 1 *If $A \in TA$ is strongly non-zeno then $Tr(A)$ does not contain a path that is both infinite and yields a timelock.*

In addition, strong non-zenoness is well behaved through parallel composition. Specifically, the following result was presented in [9]. It ensures that we cannot generate new zeno timelocks through parallel composition.

Proposition 2 *If $A_1,...,A_n \in TA$ are strongly non-zeno then $|\langle A_1, ..., A_n\rangle$ is also strongly non-zeno.*

Also although we have no empirical evidence, in accordance with [9], we believe that in practice, specifications will invariably be strongly non-zeno.

The Nature of Synchronisation. Perhaps the most counter-intuitive aspect of the timelock story is the manner in which timelocks can arise from mis-matched synchronisations, such as the composition in figure 1. If we consider how this problem arises we can see that it is caused by the particular interpretation of urgent interaction employed in TA.

It is without doubt true that facilities to express urgency are required. However, it is our perspective that while urgency is needed, currently it is given an excessively strong formulation. We illustrate the issue with the following example.

Example. Consider the specification of the Dying Dining Philosophers problem. The scenario is basically the same as the Dining Philosophers except here we have extra constraints which state that philosophers *die* if they do not eat within certain time periods.

For example, if at a particular state, Aristotle must eat within 10 time units to avoid death, in TA his situation could be represented as state `l0` of timed automaton **`Aris1`** in figure 2. In addition, if say the fork he requires is being used by another philosopher, the relevant global behaviour of the rest of the system might correspond to **`Rest1`** in state `m0` (see figure 2 again).

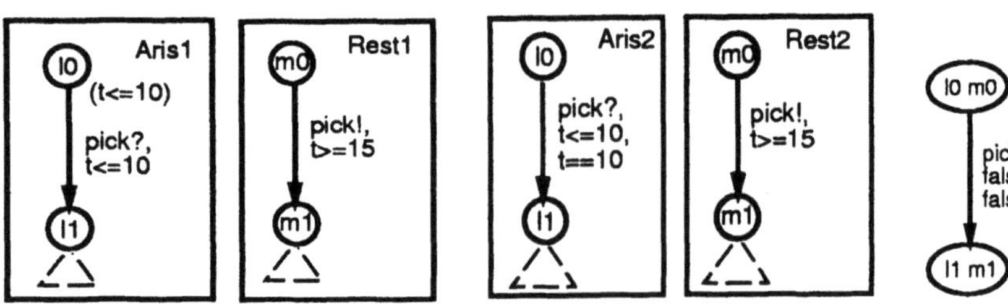

Figure 2. Dying Dining Philosophers Automata

However, the formulation **`|<Aris1,Rest1>`** will timelock when `t` reaches 10. This seems counter-intuitive. Aristotle knows he must pick-up his fork by a certain time otherwise drastic consequences will result for him (this is why he "registers" his `pick` request as urgent). However, if he *locally* fails to have his requirement satisfied, he cannot *globally* prevent the rest of the world from progressing, rather a *local* deadlock should result. As a consequence Aristotle might be dead, but as we all know, "the world will go on!"

Conceptually what is happening is that Aristotle is enforcing that his `pick` action must be taken *even if it is not possible*, i.e. it is not enabled. However, we would argue that urgency can only be forced if an action is possible. In other words, it should only be possible to make an action urgent if it is enabled, i.e.

must requires may or, in other terms, you can only force what is possible.

One way in which such an interpretation of urgency has previously been obtained is through only allowing urgency to be applied to internal actions. This is the so called *as soon as possible* (asap) principle [8], much discussed in the timed process algebra community. This property indeed prevents the occurrence of timelocks due to synchronisation mismatches, but unfortunately, it is not a suitable solution for timed automata. This is because TA do not have a hiding operator. In timed process algebra with asap the hiding operator, which turns observable into internal actions, has an important role since (implicitly) it makes actions urgent. The absence of hiding in TA means that we cannot (selectively) take an observable action that results from synchronising half actions and turn it into an (urgent) internal action.

Thus, now we consider an alternative framework for TA specification - *Timed Automata with Deadlines* (TADs) which was initially devised by Bornot and Sifakis [2, 3] and with which we can obtain the synchronisation interpretation we desire. Our presentation follows that in [4], with some refinements.

TADs Basics. For a full introduction to TADs, we refer the interested reader to [2, 3]; here we highlight the main principles.

Deadlines on Transitions. Rather than placing invariants on states, deadlines are associated with transitions. In order to do this, transitions are annotated with 4-tuples, (e, g, d, r), where e is the transition label; g is the guard; d is the deadline; and r is the reset set. Conceptually, deadlines state when transitions *must* be taken and taken immediately. Since we have deadlines on transitions there is no need for invariants on states. It is also assumed that the constraint, $d \Rightarrow g$ holds, which ensures that if a transition is forced to happen it is also able to happen. As a result of this constraint, *TADs are time reactive.*

(Timewise) Priorities. By restricting guards and deadlines in choice contexts, prioritised choice can be expressed, e.g. if we have two transitions, $\mathtt{b1} = (\mathtt{e1}, \mathtt{g1}, \mathtt{d1}, \mathtt{r1})$ and $\mathtt{b2} = (\mathtt{e2}, \mathtt{g2}, \mathtt{d2}, \mathtt{r2})$, and we are at a state with a choice between them, then we can give b2 priority over b1 by restricting the guards and deadlines of b1 to $(\mathtt{e1}, \mathtt{g1'}, \mathtt{d1'}, \mathtt{r1})$. The form of priority we use is to enforce the following restricted guard and deadline, $\mathtt{g1'} = \mathtt{g1} \wedge \Box\neg\mathtt{g2}$ and $\mathtt{d1'} = \mathtt{d1} \wedge \mathtt{g1'}$, where $\Box$ is the temporal operator henceforth. This ensures that b1 is only enabled if g1 holds and there is *no point in the future* at which g2 will hold.

Parallel Composition with Escape Transitions. The TADs framework employs a different parallel composition operator to that arising in standard timed automata. The key idea is that of an *escape transition.* These are the local transitions of automaton components that are combined when generating a synchronisation transition. Thus, not only are synchronisations included, but component transitions of the synchronisation are as well. The timewise priority mechanism is then used to give the synchronisation transition highest priority. Intuitively, the escape transitions can only happen if the synchronisation transition will *never* be enabled.

In fact, in addition to ensuring time reactivity, the TADs framework limits the occurrence of action locks. Specifically, the escape transitions allow the components of a parallel composition to escape a potential action lock by evolving locally.

We briefly review the definition of TADs. An arbitrary *TAD*, has the form, $(L, l_0, \rightarrow, C)$, where L is a finite set of locations; l_0 is the *start location*; and C is the set of clocks.

$\rightarrow \subseteq L \times \mathbb{A} \times CC_C \times CC_C \times \mathbb{P}(C) \times L$ is a transition relation. A typical element of which is, (l_1, e, g, d, r, l_2), where $l_1, l_2 \in L$; e labels the transition; g is a guard; d is a deadline; and r is a reset set. $(l_1, e, g, d, r, l_2) \in \rightarrow$ is typically written, $l_1 \xrightarrow{e,g,d,r} l_2$.

As was the case with TA, TADs are semantically interpreted as transition systems. The following two inference rules are used for this,

$$(S1)\ \frac{l \xrightarrow{e,g,d,r} l' \quad g(v)}{[l,v] \overset{e}{\Longrightarrow} [l', r(v)]} \qquad (S2)\ \frac{\forall l' \,.\, l \xrightarrow{e,g,d,r} l' \implies \forall t' < t \,.\, \neg d(v+t')}{[l,v] \overset{t}{\Longrightarrow} [l, v+t]}$$

Now we define the semantic map $[\![\,]\!]$ from TADs to transition systems as follows, $[\![(L, l_0, \rightarrow, C)]\!] = (S, s_0, \Rightarrow)$ where, $S = \{\, s' \in L \times \mathbb{V}_C \mid \exists s \in S, y \in Lab \,.\, s \overset{y}{\Longrightarrow} s' \,\} \cup \{\, [l_0, v_0] \,\}$; $s_0 = [l_0, v_0]$; and $\Rightarrow \subseteq (L \times \mathbb{V}_C) \times Lab \times (L \times \mathbb{V}_C)$ satisfies $(S1)$ and $(S2)$. Notice that, once again, S only contains reachable states.

In addition, we will use the function:

$$\theta_B(l) = \{ (e,g,d,r) \mid \exists l' . l \xrightarrow{e,g,d,r} l' \wedge e \in B \}$$

In [4] we considered three different TADs parallel composition rules - standard TADs, Sparse TADs and TADs with Minimal Priority Escape Transitions. The first of these coincides with Bornot and Sifakis's definition [2, 3], while the latter two were developed by us. We argued in favour of the latter two, since standard TADs composition generated too many escape transitions. Here we re-iterate the Sparse TADs and TADs with Minimal Priority Escape Transitions definitions.

Sparse TADs. This is a minimal TADs approach, in which we do not generate *any* escape transitions. The following parallel composition (denoted $||^s$) rules are used:

$$\frac{u[i] \xrightarrow{x?,g_i,d_i,r_i} u[i]' \quad u[j] \xrightarrow{x!,g_j,d_j,r_j} u[j]'}{||^s u \xrightarrow{x,g',d',r_i \cup r_j} ||^s u[i'/i, j'/j]} \qquad \frac{u[i] \xrightarrow{x,g,d,r} u[i]' \quad x \in CA}{||^s u \xrightarrow{x,g,d,r} ||^s u[i'/i]}$$

where $1 \leq i \neq j \leq |u|$, $g' = g_i \wedge g_j$ and $d' = g' \wedge (d_i \vee d_j)$.

These rules prevent uncompleted actions from arising in the composite behaviour; they only arise in the generation of completed actions, while (already) completed actions offered by components of the parallel composition can be performed independently. This definition has the same spirit as the normal TA rules of parallel composition. The difference being that here we have deadlines which we constrain during composition to preserve the property $d \Rightarrow g$, and hence to ensure time-reactivity.

Furthermore as a consequence of these characteristics of sparse TADs we have revised the interpretation of synchronisation in the manner we proposed. For example, if we consider again the Dying Dining Philosophers illustration, the obvious TADs formulation of `Aris1` and `Rest1`, are `Aris2` and `Rest2` shown in figure 2. Now sparse TADs composition of the two TADs yields the behaviour shown on the right of figure 2, which is action locked. This is the outcome that we were seeking. Since the `pick` synchronisation is not enabled, urgency cannot be enforced. This is reflected in both the guard and deadline in figure 2 being *false*.

TADs with Minimal Priority Escape Transitions. The idea here is the same as standard TADs, but rather than just giving escape transitions lower priority than their corresponding synchronisation, we also give them lower priority than other completed transitions. Thus, a component can only perform an escape transition if the component will never be able to perform a completed transition. This seems appropriate as our view of escape transitions is that they should only be performed as a very last resort - when the choice is between performing them or reaching an "error" state.

Letting, $1 \leq j \neq i \leq |u|$, the parallel composition (denoted $||^m$) rules are:

$$(R1) \quad \frac{u[i] \xrightarrow{x?,g_i,d_i,r_i} u[i]' \quad u[j] \xrightarrow{x!,g_j,d_j,r_j} u[j]'}{||^m u \xrightarrow{x,g',d',r_i \cup r_j} ||^m u[i'/i, j'/j]}$$

$$(R2) \quad \frac{u[i] \xrightarrow{x,g,d,r} u[i]' \quad x \in CA}{||^m u \xrightarrow{x,g,d,r} ||^m u[i'/i]} \qquad (R3) \quad \frac{u[i] \xrightarrow{a,g,d,r} u[i]' \quad a \in HA}{||^m u \xrightarrow{a,g'',d'',r} ||^m u[i'/i]}$$

where, $g' = g_i \wedge g_j$, $d' = g' \wedge (d_i \vee d_j)$ and $1 \leq k \neq i \leq |u|$ in,

$$\begin{aligned} g'' &= g \wedge \bigwedge\{ \Box\neg q.2 \,|\, q \in \theta_{CA}(u[i]) \} \wedge \\ &\quad \bigwedge\{ \Box\neg(q.2 \wedge q'.2) \,|\, q \in \theta_{HA}(u[i]) \wedge q' \in \theta_{\{\overline{q.1}\}}(u[k]) \} \\ d'' &= d \wedge g'' \end{aligned}$$

(R1) is the normal synchronisation rule; *(R2)* defines interleaving of completed transitions; and *(R3)* defines interleaving of incomplete, i.e. escape, transitions. In this final rule, g'' holds when, (1) g holds; (2) it is not the case that an already completed transition from $u[i]$ could eventually become enabled; and (3) it is not the case that an incomplete transition (including a itself) offered at state $u[i]$ could eventually be completed. Furthermore, the definition ensures that $d \Rightarrow g$ and thus that time reactivity is preserved. In addition, we again obtain the "weaker" handling of urgency in synchronisation that we seek.

4. ACTION LOCKS

The last section and the TADs framework in general provide a means to compose automata together without generating timelocks. This then raises the issue of whether the same can be done for action locks.

It is clear that independent parallel composition (both in an untimed and a timed setting) preserves action lock freedom (see [5] for a formal justification). However, interaction free parallel composition is of limited value. Thus, here we consider how the same action lock compositionality property can be obtained but while allowing interaction between processes. Our definition builds upon TADs with Minimum Priority Escape Transitions.

Consider the parallel composition $||^a A$ where A is a vector of TADs in which the component automata have disjoint clock sets. This is necessary to avoid action locks arising as a result of component automata resetting the clocks used by other components (see [5] for further justification). Letting, $1 \leq i \neq j \leq |u|$, the product rules for $||^a A$ are,

$$(RCA) \quad \frac{u[i] \xrightarrow{x?,g_i,d_i,r_i} u[i]' \quad u[j] \xrightarrow{x!,g_j,d_j,r_j} u[j]'}{||^a u \xrightarrow{x,g',d',r_i \cup r_j} ||^a u[i'/i, j'/j]}$$

$$(RIA) \quad \frac{u[i] \xrightarrow{x,g,d,r} u[i]' \quad x \in CA}{||^a u \xrightarrow{x,g,d,r} ||^a u[i'/i]} \qquad (RHA) \quad \frac{u[i] \xrightarrow{a,g,d,r} u'[i] \quad a \in HA}{||^a u \xrightarrow{a,g'',d'',r} ||^a u[i'/i]}$$

where, $g' = g_i \wedge g_j$, $d' = g' \wedge (d_i \vee d_j)$ and $1 \leq k \neq i \leq |u|$ in,

$$\begin{aligned} g'' \quad &= \quad (g \wedge \bigwedge\{\Box\neg q.2 \,|\, q \in \theta_{CA}(u[i])\} \;\; \wedge \\ &\quad\;\; \bigwedge\{\Box\neg(q.2 \wedge q'.2) \,|\, q \in \theta_{HA}(u[i]) \;\wedge\; q' \in \theta_{\overline{\{q.1\}}}(u[k])\}) \vee d'' \\ d'' \quad &= \quad d \wedge \bigwedge\{\neg q.3 \,|\, q \in \theta_{CA}(u[i])\} \;\; \wedge \\ &\quad\;\; \bigwedge\{\neg(q.2 \wedge q'.2 \wedge (q.3 \vee q'.3)) \,|\, q \in \theta_{HA}(u[i]) \wedge q' \in \theta_{\overline{\{q.1\}}}(u[k])\} \end{aligned}$$

(RCA) is the (now) familiar "conjunctive" synchronisation rule, with the deadline constraint ensuring that $d \Rightarrow g$ and thus preserving time reactivity. **(RIA)** gives the also familiar interleaved modelling of independent parallelism. **(RHA)** generates escape transitions in order to avoid action locks, with the guard and deadline constructions controlling when the escape transitions can occur. We justify our guard and deadline definitions now.

The guard in (RHA). This is a disjunction between the guard construction for escape transitions presented in section 2 and the deadline (d''). We justify the guard based disjunct (i.e. the first) here. A later point justifies disjoining with the deadline.

The basic idea of this first disjunct, is to enable the product to escape action locks resulting from mismatched synchronisations. As a simple illustration of this consider `A0` and `A1` in figure 3. Both of these TADs are action lock free. However, if just rules *(RCA)* and *(RIA)* are used the composition of `A0` and `A1` will action lock immediately as neither synchronisation can be fulfilled. However, application of rule *(RHA)* in conjunction with *(RCA)* and *(RIA)* will allow the action lock to be escaped, as shown in composition (i) in figure 3.

The deadline in (RHA). The definition of d'' has a similar shape to the guard construction we just considered, however, the temporal operators are not included. The construction states that, the deadline (d'') of the escape transition holds if and only if,

1 the deadline of the corresponding component transition (d) holds;

2 no internal transition of the component is at its deadline; and

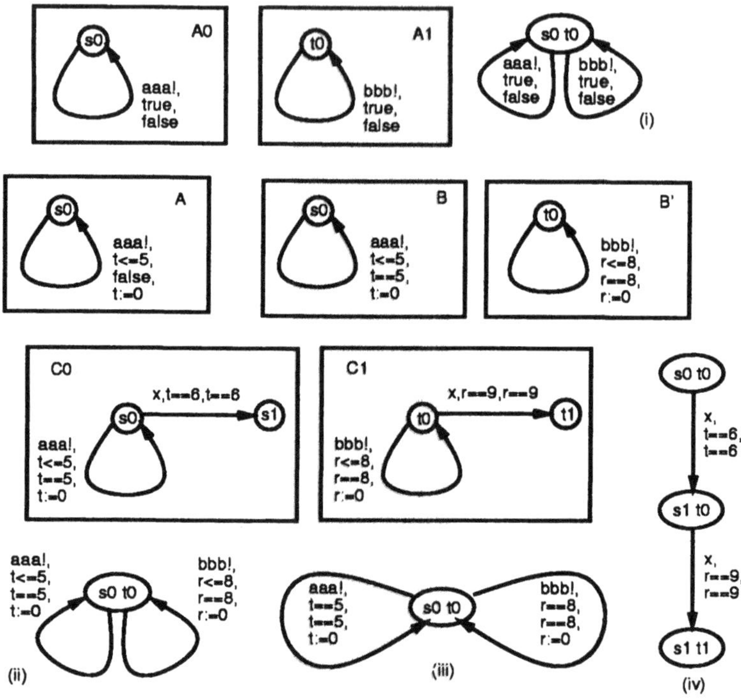

Figure 3. Assorted TADs

3 no synchronisation which includes a half action of the component is at its deadline.

The intuition behind the rule is that any (non competing) deadline that appears in the component but that does not arise in the product (because of a failed synchronisation) has its deadline preserved in an escape transition of the product. A deadline of a transition is *competing* at a state if the deadline of an alternative transition also holds there.

This deadline construction is motivated by the observation that in the majority of cases it is the deadline that ensures action lock freeness. For example, although the automaton `A` in figure 3 is strongly connected it is not action lock free. In particular, assuming `s0` is first entered with `t==0`, if it stays in state `s0` for longer than 5 time units, it will action lock. Furthermore, there is nothing constraining the length of time the automaton can idle in state `s0` as the deadline of the `aaa!` transition is `false`. However, (assuming `s0` is entered with `t<=5`) if the *false* deadline is replaced by, say, `t==5`, then it would be action lock free.

Now in order to obtain the action lock freeness property that we desire we need to guarantee that deadlines that ensure action lock freeness of component automata are preserved in the product (either through appearing as a result of rules *(RCA)* or *(RIA)* or by including relevant escape transitions). Our rule does this. Consider the two action lock free automata `B` and `B'` shown in figure 3. With just rules *(RCA)* and *(RIA)* the product of `B` and `B'` would be action locked. However, with *(RHA)* as well, the product automaton (ii) shown in figure 3 would result.

In fact, this product would have resulted from application of the rules presented in section 2 where the deadline is simply $d'' = d \wedge g''$. However, the example in figure 3 of two more action lock free TADs (`C0` and `C1`) shows that this is not sufficient in the general case. This is because according to the rules of section 2, the parallel composition of `C0` and `C1` would be as shown in figure 3 (iv) which will action lock at state `s1 t1`.

The problem is that the guards of the `aaa!` and `bbb!` escape transitions that the rules of section 2 generate, are false. This is because in both automata an internal action can eventually be taken and this internal action will take priority.

However, if we apply the rules *(RCA)*, *(RIA)* and *(RHA)* of $||^a$ then a product "behaviourally equivalent" to figure 3 (iii) results. This is because the deadline prevents clock `t` passing `5` and clock `r` passing `8`. Notice that the guard has been pruned to match the deadline. This ensures that the enabling of `aaa!` and `bbb!` is minimised to only what is required to preserve the desired action lock freeness property.

Also notice that this example illustrates why the priority enforced in the deadline has to be immediate and including temporal operators is inappropriate. Specifically, if a deadline d ensures action lock freedom then even if later transitions are possible the deadline must be preserved exactly in the product in order to prevent later transitions from being enabled which allow an action lock to be reached, e.g. the internal transitions in `C0` and `C1` above. This may not be the most refined solution since we might add an escape transition even though a later transition may prevent the action lock. However, it is not currently clear how to improve upon the approach.

Finally, we need to disjoin the deadline in g'' in order to ensure that $d \Rightarrow g$ and thus to preserve time reactivity. For example, without such a disjunct, the product of `C0` and `C1` would timelock when `t` reaches `5` as the guard on the `aaa!` transition from `s0 t0` would be *false*.

The central result of this section is given in the next theorem, it states that $||^a$ preserves action lock freeness. The proof of this result is not straightforward and due to space considerations it is not possible to include it here. However, the necessary theory, the full proof and illustration of use of the operator can be found in [5], which is available on the WWW.

Theorem 1
$\exists i(1 \leq i \leq |A|) \,.\, A[i]$ *is action lock free* $\Longrightarrow ||^a A$ *is action lock free.*

5. CONCLUSIONS

This paper builds from [4] by refining its timelock results and extending the results to action lock freeness. Although related to the work of Bornot and Sifakis [2, 3], our work is different. In particular, giving escape transtions lower priority than completed actions of a particular component, is unique to our work. Furthermore, such an interpretation is important since as we have argued, real-time structures such as timeouts are inappropriately expressed with the standard TADs parallel composition operator. In addition, in obtaining our compositionality results, the only constraint we impose on component automata is that they are time and / or action lock free. In contrast, Bornot and Sifakis require a number of well behavedness criteria to hold. This limits the generality of their approach when compared with ours.

Acknowledgements. The author has benefited greatly from discussions with Sebastian Bornot, Joseph Sifakis and Stavros Tripakis.

REFERENCES

[1] Johan Bengtsson, Kim G. Larsen, Fredrik Larsson, and Paul Pettersson amd Wang Yi. Uppaal - a tool suite for automatic verification of real-time system. In *Proceedings of the 4th DIMACS Workshop on Verification and Control of Hybrid Systems*, 1995.

[2] S. Bornot and J. Sifakis. On the composition of hybrid systems. In *Hybrid Systems: Computation and Control*, LNCS 1386, pages 49–63, 1998.

[3] S. Bornot, J. Sifakis, and S. Tripakis. Modeling urgency in timed systems. In *Compositionality, COMPOS'97*, LNCS 1536, 1997.

[4] H. Bowman. Modelling timeouts without timelocks. In *ARTS'99, Formal Methods for Real-Time and Probabilistic Systems, 5th International AMAST Workshop*, LNCS 1601, pages 335–353. Springer-Verlag, 1999.

[5] H. Bowman. On time and action lock free description of timed systems. Technical Report 16-99, Computing Laboratory, University of Kent at Canterbury, 1999. available at http://www.cs.ukc.ac.uk/people/staff/hb5/pubs.local.

[6] H. Bowman, G. Faconti, J-P. Katoen, D. Latella, and M. Massink. Automatic verification of a lip synchronisation algorithm using UPPAAL. *Formal Aspects of Computing*, 10(5-6):550–575, August 1998.

[7] C.Daws, A.Olivero, S.Tripakis, and S.Yovine. The tool KRONOS. In *Hybrid Systems III, Verification and Control*, LNCS 1066. Springer-Verlag, 1996.

[8] T. Regan. Multimedia in temporal LOTOS: A lip synchronisation algorithm. In *PSTV XIII, 13th Protocol Spec., Testing & Verification*. North-Holland, 1993.

[9] S. Tripakis. Verifying progress in timed systems. In *ARTS'99, Formal Methods for Real-Time and Probabilistic Systems, 5th International AMAST Workshop*, LNCS 1601. Springer-Verlag, 1999.

COMPILING REAL-TIME SCENARIOS INTO A TIMED AUTOMATON*

Aziz Salah, Rachida Dssouli, Guy Lapalme
Département d'Informatique et de Recherche Opérationnelle, Université de Montréal, H3T 1J4 Montreal, PQ, Canada
{salah,dssouli,lapalme}@iro.umontreal.ca

Abstract In this paper, we aim at synthesizing an executable specification for a real-time system by integrating real-time scenarios into a timed automaton. A scenario represents a partial description of a system behavior. A formal semantics is given for the model of a scenario and is used to compile a scenario into a timed automaton. The compilation algorithm is generalized to integrate several scenarios into a single timed automaton which simulates the behaviors specified by the scenarios. The results of the compilation algorithm are independent of the order in which the scenarios are added.

Keywords: real-time system, timed automata, formal specification, scenarios integration

1. INTRODUCTION

A real-time reactive system interacts with its environment under strict timing constraints. The formal specification of such systems is a difficult task that often designers prefer to avoid to move directly from informal specifications to implementation. The detection of design errors and thus the reliability of the system become difficult to evaluate increasing the development costs.

A scenario is a natural means to specify interactions of a system with its environment. The use of scenarios reduces the complexity of systems' specification since a scenario specifies a partial behavior of the system. Our main motivation is to ease the specification of real-time reactive systems by automating some activities in the specification process. Our approach consists of synthesizing an executable specification from a set

*Partially supported by France Telecom

of provided scenarios. The resulting system specification represents a prototype in the form of a timed automaton [AD94].

Most approaches based on scenarios [HSG+94, Gli95, SDLV00] concentrate on the requirements acquisition phase at the very beginning of the system development process. Scenario based approaches may be compared based on many criteria like, for instance, the input scenarios formalism, the method of integrating scenarios and its target formalism. The method of integration is the most important criterion since any approach is based on it. The integration of scenarios consists of merging many scenarios into a global model expressed in the target formalism. The main challenge is how to characterize the steps of a scenario so that occurrences of the same steps can be identified in other scenarios. Some approaches use a manual labeling to identify scenarios' steps while others use regular grammar[HSG+94]. Our approach uses the values of variables, which define the state of the system, to characterize scenarios' steps. We consider two kinds of variables, discrete variables describing system properties and continuous variables (clocks) measuring time.

A scenario is defined by sequences of interactions between the system and its environment. This work is limited to sequential systems in which only one interaction is allowed to be performed at a time. We consider two basic types of interactions namely the reception and the transmission of a message which are both viewed as observable actions. We define a model of scenarios and then we define its formal semantics which is used for compilation. To compile a single scenario into a timed automaton, the scenario is first transformed into an intermediate flat format which we will refer to by the set of *rule-actions* of the scenario. When several scenarios are integrated, the set of all rule-actions resulting from all scenarios is compiled into a single timed automaton. Our method of scenario integration is insensitive to the order of adding scenarios so the specification of a system can be built incrementally by adding scenarios to the current prototype.

The rest of the paper is organized as follows. In section 2, we present preliminaries and notations related to the definition of the timed automata model and discrete variables of a system. In section 3, we describe the syntax of a system specification, the formal semantics of a scenario and define the timed automaton of a scenario. Section 4 is dedicated to the construction algorithm of the scenario timed automaton. We illustrate, in section 5, the use of our algorithm to synthesize a timed automaton of a scenario. Section 6 generalizes this algorithm to compile several scenarios into one timed automaton and discusses the approach.

2. PRELIMINARIES

This section is dedicated to definitions and related notations. We first recall briefly the timed automata model and semantics, and then we present system discrete variables syntax.

2.1. Timed Automata

Timed Automata (TA) model uses dense model of time for modeling timed control. TA [AD94] is an extended labeled transition system with a finite number of real variables called clocks. Clocks values increase at the same rate. Let H be the set of clock variables and $\Phi(H)$ the set of clock constraints. Each $\phi \in \Phi(H)$ is defined by the grammar $\phi ::= x\#c \mid x - y\#c \mid \phi \wedge \phi \mid True$, where x,y are clocks in H and $\# \in \{\leq, <, =, \geq, >\}$ and c a constant in $\mathbb{N}$. A clock interpretation θ is a mapping from H to the set of non-negative reals. $\Theta(H)$ denotes the set of clock interpretations. For each $\theta \in \Theta(H)$ and a clock constraint $\phi \in \Phi(H)$, $\phi(\theta)$ is a boolean value describing whether θ satisfies ϕ or not. Given a non negative real d, $\theta + d$ denotes the clock interpretation which maps each clock x to the value $\theta(x) + d$.

For $\lambda \subset H$, $\theta[\lambda]$ is the clock interpretation which assigns the value 0 to each $x \in \lambda$ and agrees with θ over the other clocks. λ represents a clock assignment. Given a clock constraint $\phi \in \Phi(H)$, $\phi[\lambda]$ denotes the clock constraint satisfied by all interpretations $\theta[\lambda]$ such that θ satisfies ϕ.

Definition 1. *Each Timed Automaton (TA) is a tuple* $A = (L_A, L_A^o, M_A, T_A, H_A)$ *where,*

- L_A *is a finite set of locations*
- $L_A^o \subset L_A$ *is a set of initial locations*
- M_A *is a finite set of labels*
- H_A *is a finite set of clocks*
- $T_A \subset L_A \times M_A \times \Phi(H_A) \times 2^{H_A} \times L_A$ *is the set of transitions*
- *Inv is a mapping that assigns to each location in* L_A *a clock constraint from* $\Phi(H_A)$. *This constraint is called the invariant of the location.*

Definition 2. *A labeled transition system is a tuple* $\Sigma = (Q, Q_o, \rightarrow)$ *where* Q *is a set of states,* Q_o *is the set of initial states and the relation* $\rightarrow \subset Q \times Act \times Q$ *is the set of transitions, where Act denotes the set of labels. We write* $q \xrightarrow{a} q'$ *iff the transition* $(q, a, q') \in \rightarrow$.

To define the semantics of a TA A, a labeled transition system Σ_A is associated with it. Each state of Σ_A is a pair $(s, \theta) \in L_A \times \Theta(H_A)$ such that θ satisfies $Inv(s)$. Σ_A has two types of transitions:

- transitions modeling a time elapsing of a duration d: $(s, \theta) \xrightarrow{d} (s, \theta + d)$, such that for all $d' \leq d$, $\theta + d'$ satisfies $Inv(s)$, and
- transitions $(s, \theta) \xrightarrow{m} (s', \theta[\lambda])$ such that $(s, m, \phi, \lambda, s') \in T_A$, θ satisfies $\phi \wedge Inv(s)$ and $\theta[\lambda]$ satisfies $Inv(s')$.

In a discrete transition system, traces describe system runs. For a TA A, the notion of next state has no meaning, in fact, a transition $e \xrightarrow{d} e'$ of Σ_A models that the system passes through all states $e + d'$ such that $0 \leq d' \leq d$. We introduce now the notion of step as in [HNSY94, ACD93] to define a TA run.

Definition 3. *Let A be a TA, Σ_A its associated transition system and e, e' two states of Σ_A. e' is reachable in one step from e, denoted $e \overset{d,m}{\triangleright} e'$, if Σ_A allows the transitions $e \xrightarrow{d} e + d$ and $e + d \xrightarrow{m} e'$ for $d \geq 0$ and $m \in M_A \cup \{0\}$. A run σ is a (finite or infinite) sequence of steps $\sigma = e_0 \overset{d_1,m_1}{\triangleright} e_1 \overset{d_2,m_2}{\triangleright} e_2 \overset{d_3,m_3}{\triangleright} \ldots$*

Definition 4. *let $\sigma = e_0 \overset{d_1,m_1}{\triangleright} e_1 \overset{d_2,m_2}{\triangleright} e_2 \overset{d_3,m_3}{\triangleright} e_3 \ldots$ be a run. The timed trace of σ is defined as the sequence $((\theta_1, l_1) \ldots (\theta_j, l_j) \ldots)$ such that if $e \overset{d,m}{\triangleright} e'$ is the jth element of σ for which $e = (s, \theta)$and $m \in M_A$ then $\theta_j \overset{def}{=} \theta + d$ and $l_j \overset{def}{=} m$.*

2.2. Discrete Variable Constraints and Assignments

Let v be a system discrete variable in V. Let $Dom(v)$ denote a finite and discrete domain of v values. Let $\Psi(V)$ denote the set of variable constraints ψ defined by the grammar $\psi ::= v\#c \mid \psi \wedge \psi \mid \neg\psi \mid True$, where $v \in V$, $c \in Dom(v)$ and $\#$ is a binary relation of $Dom(v) \times Dom(v)$. Let $\Omega(V)$ be the set of variable interpretations. Each variable interpretation $\omega \in \Omega(V)$ maps a value in $dom(v)$ to each variable in V. $\psi(\omega)$ denotes a boolean value describing whether ω satisfies ψ or not.

Variable assignments allow variable modifications as a transformation functions. For example, given $v_1, v_2 \in V$ and an assignment $\delta = \{v_1 := v_2 + 3, v_2 := 1\}$and a variable interpretation $\omega \in \Omega(V)$, $\omega[\delta]$ denotes a new variable interpretation defined by $\omega[\delta](v_1) = \omega(v_2) + 3$, $\omega[\delta](v_2) = 1$ and $\omega[\delta](v) = \omega(v)$ for $v \in V - \{v_1, v_2\}$. There is no limitation on assignment statements expressions but the resulting value of each variable

```
V={A_status,A_signal,B_status,B_signal}
Dom(A_status) = {BUSY, IDLE}
Dom(A_signal) = {NONE,TONE,BUSY_TONE,DIALING(B),ECHO_RING(B),TALKING(B)}
Dom(B_status) = {BUSY , IDLE} ;
Dom(B_signal) = {NONE,TONE,BUSY_TONE,RING(A),TALKING(A)};
H={h}
```

Figure 1. An example of discrete variables set V and clocks set H

must remain in its domain. We write $\Delta(V)$ to denote the set of variable assignments. Given a constraint variable $\psi \in \Psi(V)$ and an assignment $\delta \in \Delta(V)$, the constraint $\psi[\delta] \in \Psi(V)$ designates the constraint satisfied by $\omega[\delta]$ such that the variable interpretation $\omega \in \Omega(V)$ satisfies ψ. Figure 1 shows the example of a description of a set V and a set H for a telephone switch controller system.

In the remaining part of this document we'll adopt the following notations: θ denotes a clock interpretation in $\Theta(H)$, ω represents a variable interpretation in $\Omega(V)$, φ and ϕ denote clock constraints in $\Phi(H)$, ψ represents a variable constraint in $\Psi(V)$, $\delta \in \Delta(V)$ denotes a variable assignment and $\lambda \subset H$ represents a clock assignment.

3. SPECIFYING A REAL-TIME REACTIVE SYSTEM BEHAVIOR

A real-time reactive system is a process characterized by a continuous interaction with its environment under strict timing constraints. This work is restricted to sequential processes. Process behaviors are specified by a set of scenarios. Each scenario is represented by sequences of actions. An action in a scenario describes the states of the process before and after the execution of the action.

Our main goal is to synthesize a TA from the provided scenarios. This TA represents a prototype of the desired system. Let P be the system process. P is a tuple (S, H, V, Act) where S is a set of scenarios, V a set of discrete system variables, H a set of clocks and Act a set of observable labels used during interaction with the environment. The syntax for clocks in H and discrete variable in V have been already defined in subsections 2.1 and 2.2 respectively. In the remainder of this section, we define scenarios syntax and their semantics.

3.1. Scenario Formalization

The scenarios in S describe possible behaviors of a process P. Each scenario action describes the changes in the process's state after the execution of the action or after a deadline. Each state of the process is a pair $e = (\omega, \theta) \in \Omega(V) \times \Theta(H)$ composed of a variable interpretation ω and a clock interpretation θ. Let us define some notations: $\psi(e) = \psi(\omega)$, $\varphi(e) = \varphi(\theta)$, $e[\delta] = \omega[\delta]$ and $e[\lambda] = \theta[\lambda]$.

Definition 5. *A scenario sc is a tree* $sc = (\mathcal{N}, \mapsto, \mathcal{A})$ *where:*

- $\mathcal{N} = \mathcal{N}_p \cup \mathcal{N}_s$ *such that* $\mathcal{N}_p$ *and* $\mathcal{N}_s$ *are respectively the sets of primary and secondary vertices of sc. We write* $\mathcal{N}_p = [N_1, N_2, \dots, N_n, N_{n+1}]$, $n+1 = |\mathcal{N}_p|$. *Secondary vertices are leaves of the tree.*
- $\mathcal{A}$ *is the set of actions that label the tree edges. Each* $a \in \mathcal{A}$ *is a tuple* $(\varphi_a, \psi_a, \phi_a, lab_a, \delta_a, \lambda_a)$ *where:*
  - $\varphi_a \in \Phi(H)$ *and* $\psi_a \in \Psi(V)$ *are constraints on the state of process.*
  - $lab_a \in Act$ *is a label of an observable action*
  - $\phi_a \in \Phi(H)$ *is an enabling constraint of the action a*
  - $\delta_a \in \Delta(V)$ *is an assignment updating the state of the process after the execution of the action a.*
  - λ_a *is a clock assignment that occurs after the execution of the action a.*
- $\mapsto \subset \mathcal{N} \times \mathcal{A} \times \mathcal{N}$ *is the set of edges.*
- *Each primary vertex* N_i *for* $1 \leq i \leq n$: $N_i = [a_{i1}, \dots, a_{ik_i}]$ *such that* $(N_i, a_{i1}, N_{i+1}) \in \mapsto$. a_{i1} *is called the primary action of the vertex* N_i.

For $1 < j \leq k_i$, the actions a_{ij} of the vertex N_i are alternatives of the primary action a_{i1}. These alternatives model timers expirations or any other interruption.

The semantics of a scenario action is formalized by associating with each primary vertex a pair of constraints $(\psi_{N_i}, \varphi_{N_i}) \in \Psi(V) \times \varphi_{N_i} \in \Phi(H)$ defining the context of the vertex. *To execute an action, the state of the process has to satisfy the context constraints of its vertex.* The context constraints of a vertex are defined by the following sequence:

$(\psi_{N_1}, \varphi_{N_1}) = (\psi_{a_{11}}, \bigwedge_{1 \leq k \leq k_1} \varphi_{a_{1k}})$, and for $2 \leq i \leq n$

$$(\psi_{N_i}, \varphi_{N_i}) = (\psi_{N_{i-1}}[\delta_{a_{(i-1)1}}] \wedge \psi_{a_{i1}}, \bigwedge_{1 \leq k \leq k_i} \varphi_{a_{ik}} \wedge (\varphi_{N_{i-1}} \wedge \phi_{a_{(i-1)1}})[\lambda_{a_{(i-1)1}}])$$

and $(\psi_{N_{n+1}}, \varphi_{N_{n+1}}) = (\psi_{N_n}[\delta_{a_{n1}}], (\varphi_{N_n} \wedge \phi_{a_{n1}})[\lambda_{a_{n1}}])$

Let a_{ij} be an action of N_i. While the state e of the process P satisfies the condition $\psi_{N_i}(e) \wedge \psi_{a_{ij}}(e) \wedge \varphi_{N_i}(e)$, the process P may either execute the action a_{ij} provided that $\phi_{a_{ij}}(e)$ is true, or let the time passes while the state of the process satisfies φ_{N_i}. When P executes a_{ij} from a state e, it moves to a new state $e' = (e[\delta_{a_{ij}}], e[\lambda_{a_{ij}}])$.

3.2. Timed Automaton of a Scenario

According to the expression of a context constraints of a primary vertex, the execution of an action depends on the context created by the preceding actions in the scenario. By merging scenario action and the context constraints of its primary vertex, we obtain a self contained rule-action which is independent from the execution context. A rule-action contains all the information about the states of the process and its condition of activity before and after the execution of the rule-action. Let $\mathcal{R}(sc)$ designate the set of rule-actions resulting from the scenario sc. Given an action a_{ij} of the vertex N_i, the resulting rule-action is a tuple $r = (\psi_r, \varphi_r, lab_r, \phi_r, \delta_r, \lambda_r, \varphi'_r)$ where:

- $\psi_r = \psi_{N_i} \wedge \psi_{a_{ij}}$
- $\varphi_r = \varphi_{N_i}$
- $lab_r = lab_{a_{ij}}$
- $\phi_r = \phi_{a_{ij}}$
- $\delta_r = \delta_{a_{ij}}$
- $\lambda_r = \lambda_{a_{ij}}$
- For φ'_r, two cases are to be distinguished:

$$\varphi'_r = \begin{cases} \varphi_{N_{i+1}} & \text{if } 1 \leq i \leq n \text{ and } j = 1 \\ (\varphi_{N_i} \wedge \phi_{a_{ij}})[\lambda_{aij}] & \text{if } 1 \leq i \leq n \text{ and } 2 \leq j \leq k_i \end{cases}$$

To define the semantics of a scenario sc and its set of rule-actions $\mathcal{R}(sc)$ we will associate to the scenario sc a labeled transition system Σ_{sc} which we refer to by the labeled transition system of the scenario sc, and is defined as follows:

Definition 6. *Each rule-action $r \in \mathcal{R}(sc)$, implies that Σ_{sc} will contain the two following sets of transitions:*

- *immediate transitions $e \overset{lab_r}{\rightarrow} e'$ such that $e = (\omega, \theta)$ and $e' = (\omega[\delta_r], \theta[\lambda_r])$ provided that $\psi_r(\omega) \wedge \varphi_r(\theta) \wedge \phi_r(\theta) \wedge \varphi'_r(\theta[\lambda_r])$ is true*
- *elapsing time transitions $e \overset{d}{\rightarrow} e'$ such that $e = (\omega, \theta)$ and $e' = (\omega, \theta + d)$ provided that $\psi_r(\omega) \wedge (\forall 0 \leq d' \leq d \,.\, \varphi_r(\theta + d'))$ is true*

Definition 7. *Let Σ_A be the labeled transition system of the TA A. An observational timed trace is a timed trace in Σ_A from which all invisible labels ε are removed. $OTT(e)$ denotes the set of observational timed traces that start from the state e of Σ_A.*

Definition 8. *Let Σ_A and $\Sigma_{A'}$ be the labeled transition system of the TA A and A' respectively. We assume that A and A' have the same* *observational labels.* Σ_A *and* $\Sigma_{A'}$ *are observational timed trace equivalent iff for each state* e *in* Σ_A*, there exists a state* e' *in* $\Sigma_{A'}$ *such that* $OTT(e) = OTT(e')$ *and for each state* e' *in* $\Sigma_{A'}$*, there exists a state* e *in* Σ_A *such that* $OTT(e) = OTT(e')$.

Theorem 1. *Given a scenario* $sc \in S$*, there exists a TA called TA of the scenario sc, noted* A_{sc}*, such that its labeled transition system* $\Sigma_{A_{sc}}$ *is observational timed trace equivalent to* Σ_{sc}.

The proof of this theorem will be given in the next section. A location of the TA A_{sc} is characterized by a pair (ω, φ) where φ represents the invariant of the location (ω, φ). Each location (ω, φ) includes all of the states (ω, θ) such that θ satisfies φ. The rule-actions of a scenario are not necessarily pairwise-disjoint (i.e for a given scenario sc, there may exist two rule-actions $r_1, r_2 \in \mathcal{R}(sc)$, ω and θ such that $\psi_{r_1}(\omega) \wedge \psi_{r_2}(\omega) = True$ and $\varphi_{r_1}(\theta) \wedge \varphi_{r_2}(\theta) = True$). Let us consider the set composed of locations (ω, φ_r) and $(\omega[\delta_r], \varphi'_r)$ for all $r \in \mathcal{R}(sc)$. Each element of this set must be split into new elements so that in the resulting set $L_{A_{sc}}$ no two elements will have a common state. The set $L_{A_{sc}}$ is the set of locations for the TA of the scenario. Let us now define its set of transitions $T_{A_{sc}}$ which is composed of two types of transitions:

- transitions resulting from a rule-action $r \in \mathcal{R}(sc)$ like $((\omega, \varphi), lab_r, \phi_r, \lambda_r, (\omega[\delta_r], \varphi'))$ provided that:
  - $(\omega, \varphi) \in L_{A_{sc}}$, $(\omega[\delta_r], \varphi') \in L_{A_{sc}}$ such that $\psi_r(\omega)$ is true, and
  - $\varphi \wedge \varphi_r = \varphi$ and $\varphi' \wedge \varphi'_r = \varphi'$
- transitions labeled by ε which we considere as the invisible label or the empty word. These transitions guarantee the time progress.

4. CONSTRUCTION OF THE TIMED AUTOMATON OF A SCENARIO

Given a scenario sc, the construction of its TA A_{sc} takes two phases. First, the locations resulting from rule-actions of $\mathcal{R}(sc)$ are split to produce $L_{A_{sc}}$, and then transitions are added according to the rule-actions of $\mathcal{R}(sc)$ to build the final $T_{A_{sc}}$.

4.1. Phase 1: Construction of the Locations Set of A_{sc}

To split the locations resulting from $\mathcal{R}(sc)$, we'll use an efficient algorithm based on the generic algorithm proposed by Bouajjani and al. [BFH+92] to obtain a minimal split. The generic algorithm generates a minimal transition system directly from an implicit description (like a program). This algorithm is useful in our context since our rule-actions are implicit description of the labeled transition system of a scenario. Let us briefly describe this algorithm and adapt it in order to construct the locations set $L_{A_{sc}}$.

Let $(Q, Q_o, \rightarrow)$ be a labeled transition system. For a partition ρ of Q and a state $q \in Q$, $post_\rho(q)$ denotes the set of ρ elements that are immediately reachable from q. The function $post_\rho$ is extended to the subsets of Q by $post_\rho(C) = \bigcup_{q \in C} post_\rho(q)$ where $C \subset Q$.

Given a class $C \in \rho$, C is said to be stable with respect to the partition ρ of Q iff $\forall C' \in \rho$:

$$((\exists q_1 \in C, \exists q'_1 \in C' \,.\, q_1 \xrightarrow{a} q'_1) \text{ implies } (\forall q_2 \in C\ \exists q'_2 \in C' \,.\, q_2 \xrightarrow{a} q'_2)$$

The minimization algorithm uses $split(C, \rho)$ operator which splits the class C into stable classes with respect to ρ such that:

$$split(C,\rho) = \{C_1, \ldots, C_k \mid (\forall 1 \le i,j \le k \,.\, C_i \cap C_j = \emptyset) \text{ and } (C_1 \cup \cdots \cup C_k = C) \text{ and } (\forall i \,.\, C_i \text{ stable}) \text{ and } (\forall i,j \,.\, C_i \cup C_j \text{non stable})\}$$

The minimization algorithm (figure 2)progressively refines the partition ρ starting from the initial partition ρ_o. *Acc* denotes the set of reachable states from the initial states. *Stb* contains the stable classes with respect to ρ. The algorithm stops when all of the reachable classes are stable with respect to ρ.

```
procedure Minimization
  ρ := ρo ; Acc := {C ∈ ρ|Q0 ∩ C ≠ ∅}; Stb := ∅;
  while Acc ≠ Stb do
    choose C ∈ Acc − Stb
    D := split(C, ρ)
    if D = {C} then
      Stb := Stb ∪ {C} ; Acc := Acc ∪ post_ρ(C)
    else
      Acc := (Acc − {C}) ∪ {C' ∈ D|C' ∩ Qo ≠ ∅}
      Stb := Stb − {C' ∈ Stb|C ∈ post_ρ(C')}
      ρ := (ρ − {C}) ∪ D
```

Figure 2. Minimization algorithm

To apply the minimization algorithm let us consider the TA A defined by:

- $H_A = H$
- $M_A = \{lab_r \,.\, r \in \mathcal{R}(sc)\}$
- $L_A = \{(\omega, \varphi_r) \in \Omega(V) \times \Phi(H) \mid r \in \mathcal{R}(sc) \text{ and } \psi_r(\omega)\} \cup \{(\omega[\delta_r], \varphi'_r) \in \Omega(V) \times \Phi(H) \mid r \in \mathcal{R}(sc) \text{ and } \psi_r(\omega)\}$

- $L_A^o = \{(\omega, \varphi_{N_1}) \in \Omega(V) \times \Phi(H) \mid \psi_{N_1}(\omega)\}$
- $T_A = \{((\omega, \varphi_r), \phi_r, lab_r, \lambda_r, (\omega[\delta_r], \varphi_r')) \mid r \in \mathcal{R}(sc) \text{ and } \psi_r(\omega)\}$
- $\forall(\omega, \varphi) \in L_A, Inv(\omega, \varphi) = \varphi$

The locations of L_A are the resulting locations from the rule-actions of $\mathcal{R}(sc)$. The TA A contains all the immediate transitions of A_{sc} but not all transitions modeling time elapsing. The TA A is not minimal because the rule-actions of $\mathcal{R}(sc)$ are not necessarily pairwise-disjoint. The locations of L_A will be split by the minimization algorithm in order to build a minimal TA. A minimal TA means that for a given pair (ω, θ), if there exists a location (ω, φ) in this TA such that $\psi(\omega) \wedge \varphi(\theta)$ is true, then this location is unique.

The minimization algorithm is applied to the labeled transition system of the TA A and with the following initial ρ_o:

$$\rho_o = \{(\psi_r, \varphi_r).r \in \mathcal{R}(sc)\} \cup \{(\psi_r[\delta_r], \varphi_r').r \in \mathcal{R}(sc)\} \quad (1)$$

Elements of ρ_o are classes. $[\varphi]$ and $[\psi]$ denote the set of clock interpretations and variable interpretations respectively which satisfy φ and ψ respectively. For a class $C \in \rho$ denoted (ψ_C, φ_C), C is the set $[\psi_C] \times [\varphi_C]$.

For a constraint $\varphi \in \Phi(H)$, $[\varphi]$ is a convex polyhedron of $\mathbb{R}^{|H|}$. Let us define the operators '$\cap$' and '$-$' on classes of ρ:

$$C \cap C' = (\psi_C \wedge \psi_{C'}, \varphi_C \wedge \varphi_{C'}) \qquad C - C' = (\psi_C \wedge \neg\psi_{C'}, \varphi_C \wedge \neg\varphi_{C'})$$

The intersection of convex polyhedrons is convex but their union is not necessarily convex. So $[\varphi_C \wedge \neg\varphi_{C'}]$ may not be convex. For a class $C = (\psi_C, \varphi_C)$, $[\varphi_C]$ must be convex because φ_C will be the invariant of a location of the TA under construction. Thus, let us define the operator $convex(\varphi)$ which transforms $[\varphi]$ into a partition of convex polyhedrons.The $convex$ operator is extended to classes by the expression $convex(\psi, \varphi) \stackrel{def}{=} \{(\psi, \varphi') | \varphi' \in convex(\varphi)\}$.

The algorithm of $split(\rho, C)$ (figure 3) uses the function $pre_r(C)$ denoting the class of elements from which C can be reached by transitions resulting from r.

The minimization algorithm assumes that ρ_o is a partition but in our case, initial ρ_o defined by the formula (1), is not necessarily a partition since the rule-actions of $\mathcal{R}(sc)$ may be not disjoint. The classes of ρ_o are reachable because they are resulting from $\mathcal{R}(sc)$, and therefore there exists an iteration in the minimization procedure where ρ becomes a partition and remains so until the end.

In the remainder of the paper, ρ designates the resulting partition from the minimization procedure (ie all of ρ's classes are stable). The

```
function split(ρ, C)
  Z := {C}
  for each C' ∈ ρ
    for each r ∈ R
      P := ∅
      U := pre_r(C')
      for each X ∈ Z do
        W := U ∩ X
        if (W = ∅ or W = X) then P = P ∪ {X}
        else P := P ∪ {W} ∪ convex(X − W)
      Z := P
  return (Z)
```

Figure 3. Split function

set of locations $L_{A_{sc}}$ may now be explicitly described by the following expression:

$$L_{A_{sc}} = \{(\omega, \varphi) \in \Omega(V) \times \Phi(H) \mid \exists C \in \rho \,.\, \psi_C(\omega) \text{ and } \varphi = \varphi_C\}$$

4.2. Phase 2: Construction of the Transitions Set of A_{sc}

As mentioned earlier, the set of transitions $T_{A_{sc}}$ contains two types of transitions:

- transitions deducted from $\mathcal{R}(sc)$ like $((\omega, \varphi), lab_r, \phi_r, \lambda_r, (\omega[\delta_r], \varphi'))$ such that $r \in \mathcal{R}(sc)$, $\psi_r(\omega)$, $[\varphi] \subset [\varphi_r]$ and $[\varphi'] \subset [\varphi'_r]$ and
- ε-transitions modeling time elapsing and labeled by the invisible action ε. Timing constraints of ε-transitions will be defined further in the paper.

For example, given two locations $(\omega, h < 3)$ and $(\omega, h \geq 3)$ of $L_{A_{sc}}$ where h is a clock of H. To model continuous time elapsing, we need to add the ε-transition $t = ((\omega, h < 3), \varepsilon, h = 3, \emptyset, (\omega, h \geq 3))$. Since the invariant of the location $(\omega, h < 3)$ is $(h < 3)$, the ε-transition t will never be fired as the constraint $((h < 3) \wedge (h = 3))$ can't never be satisfied. We also know that as soon as time leaves the zone $[h < 3]$, it enters to the zone $[h \geq 3]$. To avoid this topological problem, the invariant of the location $(\omega, h < 3)$ is relaxed to $(\omega, h \leq 3)$. For preserving the same transition system, the previous constraint $h < 3$ is added to the guarding constraint of the ingoing and the outgoing transitions of the relaxed location. The complete description of adding ε-transitions and

function *add_ε-transition*

  let $T_{A_{sc}} := \{((\omega,\varphi), lab_r, \phi_r, \lambda_r, (\omega[\delta_r], \varphi')) \mid$

    $r \in \mathcal{R}(sc),\ \psi_r(\omega),\ [\varphi] \subset [\varphi_r]$ and $[\varphi'] \subset [\varphi'_r]\}$

  for each $(\omega,\varphi_1),(\omega,\varphi_2) \in L_{A_{sc}}$ **such that** $\varphi_1 \prec \varphi_2$

    if $\overline{\varphi_1} = \varphi_1$

    then $s := (\omega,\varphi_2)$; $t_\varepsilon := ((\omega,\varphi_1), \varepsilon, \overline{\varphi_1 \wedge \varphi_2}, \emptyset, (\omega, \overline{\varphi_2}))$

    else $s := (\omega,\varphi_1)$; $t_\varepsilon := ((\omega,\overline{\varphi_1}), \varepsilon, \overline{\varphi_1 \wedge \varphi_2}, \emptyset, (\omega,\varphi_1))$

    $T_{A_{sc}} := Relax\_Inv(T_{A_{sc}}, s)$; $T_{A_{sc}} := T_{A_{sc}} \cup \{t_\varepsilon\}$

  return $T_{A_{sc}}$

function $Relax\_Inv(\mathcal{T}, (\omega,\varphi))$

  for each $t \in \mathcal{T}$ **such that** $t = ((\omega,\varphi), lab, \phi, \lambda, s')$ and $lab \neq \varepsilon$

    $\mathcal{T} := \mathcal{T} - \{t\} \cup \{((\omega,\overline{\varphi}), lab, \phi \wedge \varphi, \lambda, s')\}$

  for each $t \in \mathcal{T}$ **such that** $t = (s, lab, \phi, \lambda, (\omega,\varphi))$ and $lab \neq \varepsilon$

    $\mathcal{T} := \mathcal{T} - \{t\} \cup \{(s, lab, \phi \wedge \varphi[\lambda], \lambda, (\omega,\overline{\varphi}))\}$

  $L_{A_{sc}} := L_{A_{sc}} - \{(\omega,\varphi)\} \cup \{(\omega,\overline{\varphi})\}$

  return $\mathcal{T}$

Figure 4. Adding ε-transitions and locations relaxing functions

relaxing locations is given by functions *add_ε-transition* and *Relax_Inv* (figure 4). These algorithms use the following definitions:

Definition 9. *Given $\varphi \in \Phi(H)$, the relaxed constraint of φ denoted $\overline{\varphi}$, is the clock constraint in which the relation $<$ and $>$ are respectively replaced by $\leq$ and $\geq$ in the expression of φ.*

Definition 10. *We write $\varphi_1 \prec \varphi_2$ iff $[\varphi_1] \cap [\varphi_2] = \emptyset$ and $([\varphi_1] \cap [\overline{\varphi_2}] \neq \emptyset$ or $[\overline{\varphi_1}] \cap [\varphi_2] \neq \emptyset)$*

The resulting sets $T_{A_{sc}}$ and $L_{A_{sc}}$ after the *add_ε-transition* function call represent respectively the set of transitions and the set of locations of the TA A_{sc}. The set of labels is $M_{A_{sc}} = M_A \cup \{\varepsilon\}$.

4.3. Observational Timed Trace Equivalence

Σ_{sc} and $\Sigma_{A_{sc}}$ are the labeled transition systems of the scenario sc and its TA A_{sc} respectively. A_{sc} may contain some ε-transitions while Σ_{sc} does not. $\Sigma_{A_{sc}}$ and Σ_{sc} must be equivalent according to theorem 1. The objective of this equivalence is to allow us to ignore ε labels in the timed traces of A_{sc} to be able to compare them with the timed traces of Σ_{sc} since ε-transitions are not observable by the environment. Let us first define what we mean by runs and timed traces for Σ_{sc}. If we consider that Σ_{sc} is a labeled transition system of a fictive TA, the runs and timed traces of Σ_{sc} may be defined exactly as we did previously in definitions

3 and 4 for a TA. The next proposition defines an equivalence between Σ_{sc} and $\Sigma_{A_{sc}}$:

Proposition 1. *$\Sigma_{A_{sc}}$ and Σ_{sc} are observational timed trace equivalent.*

Proof. (sketch) Given a state e of Σ_{sc} there exists $r \in \mathcal{R}(sc), \omega$ and θ such that $e = (\omega, \theta)$ and at least one of the two following cases is true : 1)$\psi_r(\omega)$ and $\varphi_r(\theta)$, or 2) $\omega = \omega'[\delta_r]$ and $\psi_r(\omega')$ and $\varphi'_r(\theta)$. The states of $\Sigma_{A_{sc}}$ are in the form of $((\omega, \varphi), \theta)$ since (ω, φ) may represent a location in A_{sc}. There exists a state $e' = ((\omega, \varphi), \theta)$ of $\Sigma_{A_{sc}}$ resulting from the split of either the class (ψ_r, φ_r) or $(\psi_r[\delta_r], \varphi'_r)$. One can then proof $OTT(e) = OTT(e')$. □

According to the previous proposition, Σ_{sc} et $\Sigma_{A_{sc}}$ are equivalent from the point of view of observational timed trace equivalence. The construction of A_{sc} and the proposition 1 provide a proof of theorem 1.

The resulting partition ρ from the construction of A_{sc} allows to reduce the combinatorial explosion of the locations in A_{sc}. ρ abstracts A_{sc} into a compact TA A_{sc}/ρ such that:

- $L_{A_{sc}/\rho} = \rho$
- $L^o_{A_{sc}/\rho} = \{C \in \rho | C \cap L^o_A \neq \emptyset\}$
- $T_{A_{sc}/\rho} = \{(C, m, \phi, \lambda, C') | \ (s, m, \phi, \lambda, s') \in T_{A_{sc}}, s \in C \text{ and } s' \in C'\}$

A location of the TA A_{sc}/ρ groups together all the location in A_{sc} which are imperceptible if we do not take into account the values of the discrete variable. Such representation is more suitable as an input to verification tools since it is more compact than the initial TA A_{sc}.

5. APPLICATION: TELEPHONE SWITCH CONTROLLER

This section illustrates the construction of a TA from a simple scenario. The process P models a telephone switch controller. Discrete variables set V and clocks set H are those described in figure 1. In this example, we will construct the TA of the following call establishment scenario:

> "When the user A is idle, if the controller receives the *pickup*(A) message, it sends the tone to user A. If user A calls user B before 30 time units then the controller sends the ring to B provided that he's idle. But if user A does nothing during 30 time units, the controller sends him the *busy_tone*(A) message. While the user B terminal is ringing, if user B picks up before 60 time units, the controller establishes the call, else the later is canceled and the controller sends *busy_tone*(A)".

The representation of the tree of this scenario was removed from this document, for lack of space, but it can be restored from the first column of table 1 according to definition 5. Table 1 describes the set of rule-actions resulting from the call establishment scenario. The construction of the TA scenario uses as input the rule-actions of table 1 and its initial distribution classes ρ_o. The abstraction of the scenario TA defined by the resulting partition ρ is drawn in figure 5.

a_{ij}	Rule-action r							ρ_o
	φ_r	Lab_r	ψ_r	ϕ_r	δ_r	λ_r	φ'_r	
a_{11}	$True$	$pickup(A)$	$A\_stat{=}IDLE$ $\wedge A\_sig{=}NONE$	$True$	$A\_stat{:=}BUSY$	$\{h\}$	$h{=}0$	C_1 C_2
a_{21}	$h{=}0$	$send\_tone(A)$	$A\_stat{=}BUSY$ $\wedge A\_sig{=}NONE$	$h{=}0$	$A\_sig{:=}TONE$	$\{h\}$	$h{\leq}30$	C_2 C_3
a_{31}	$h{\leq}30$	$dialing(B)$	$A\_stat{=}BUSY$ $\wedge A\_sig{=}TONE$	$h{<}30$	$A\_sig{:=}$ $DIALING(B)$	$\{h\}$	$h{=}0$	C_3 C_4
a_{32}	$h{\leq}30$	$busy\_tone(A)$	$A\_stat{=}BUSY$ $\wedge A\_sig{=}TONE$	$h{=}30$	$A\_sig{:=}$ $BUSY\_TONE$	$\{h\}$	$True$	C_3 C_8
a_{41}	$h{=}0$	$ring(A,B)$	$A\_sig{=}$ $DIALING(B)$ $\wedge A\_stat{=}BUSY$ $\wedge B\_stat{=}IDLE$ $\wedge B\_sig{=}NONE$	$h{=}0$	$A\_sig{:=}$ $ECHO\_RING(B)$, $B\_sig{:=}RING(A)$, $B\_stat{:=}BUSY$	$\{h\}$	$h{\leq}60$	C_5 C_6
a_{51}	$h{\leq}60$	$pickup(B)$	$A\_sig{=}$ $ECHO\_RING(B)$ $\wedge B\_sig{=}RING(A)$ $\wedge A\_stat{=}BUSY$ $\wedge B\_stat{=}BUSY$	$h{<}60$	$A\_sig{:=}$ $TALKING$, $B\_sig{:=}$ $TALKING$	$\emptyset$	$True$	C_6 C_7
a_{52}	$h{\leq}60$	$busy\_tone(A)$	$A\_sig{=}$ $ECHO\_RING(B)$ $\wedge B\_sig{=}RING(A)$ $\wedge A\_stat{=}BUSY$ $\wedge B\_stat{=}BUSY$	$h{=}60$	$A\_sig{:=}$ $BUSY\_TONE$, $B\_sig{:=}NONE$, $B\_stat{:=}IDLE$	$\{h\}$	$True$	C_6 C_9

Table 1. Description of the rule-action set $R(sc)$ for the call establishment scenario. Each row in the table, contains the original scenario action, its corresponding rule-action and its two resulting classes in the initial ρ_o

6. METHOD OF SCENARIOS INTEGRATION

The objective of the integration of scenarios is to merge many of them into a single TA. The construction of the scenario TA algorithm may be applied on any rule-actions set. Let A_R be the TA synthesized from the set of rule-actions R. Based on this notation, the TA of a scenario sc may also be denoted $A_{\mathcal{R}(sc)}$. Particularly, for the scenarios sc_1 and sc_2, their respective TA A_{sc_1} and A_{sc_2} are merged into one TA by the operator $\oplus$ defined as follows:

$$A_{sc_1} \oplus A_{sc_2} = A_{\mathcal{R}(sc_1)} \oplus A_{\mathcal{R}(sc_2)} \stackrel{def}{=} A_{\mathcal{R}(sc_1) \cup \mathcal{R}(sc_2)}$$

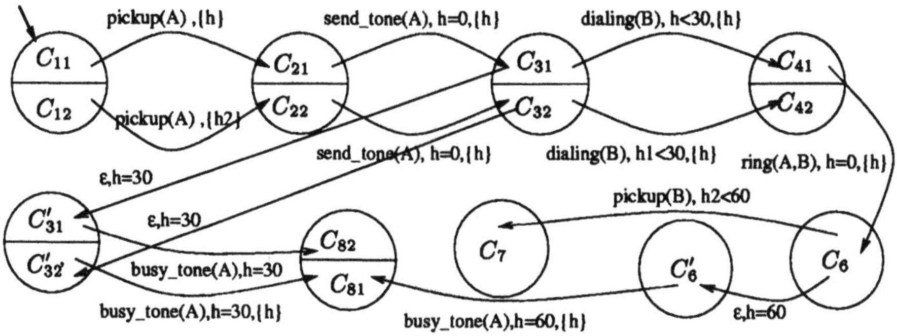

Figure 5. Abstract Timed automata of the call establishment scenario. The classes, for example, C_{11} and C_{12} classes result from the split of the class C_1 and so on.

Proposition 2. *For each state e in Σ_{sc} ($sc \in \{sc_1, sc_2\}$) there exists a state e' in $\Sigma_{A_{sc_1} \oplus A_{sc_2}}$ such that $OTT(e) \subset OTT(e')$. The reverse is false because of the possible overlapping of the scenarios.*

The proposition 2 allows to construct one TA $\mathcal{A}$ which models the process $P = (S, H, V, Act)$ and to integrate all of its scenarios. The TA $\mathcal{A}$ results from the compilation of the rule-actions set $\bigcup_{sc \in S} \mathcal{R}(sc)$. Given $S = \{sc_1, \ldots, sc_n\}$, the following formula describes the above stated compilation:

$$\mathcal{A} = A_{sc_1} \oplus A_{sc_2} \oplus \cdots \oplus A_{sc_n} = A_{\mathcal{R}(sc_1) \cup \mathcal{R}(sc_2) \cup \cdots \cup \mathcal{R}(sc_n)} \tag{2}$$

According to proposition 2, $\mathcal{A}$ includes all the behaviors allowed by each integrated scenario. $\mathcal{A}$ also includes some extra behaviors which may result from some overlapping scenarios. It is possible to identify these extra behaviors for an eventual validation process. The salient features of the integration algorithm may be inferred from the formula (2) as follows :

- The specification of a system is incremental. Assuming that the current specification of an existing system results from the compilation of a set of scenarios S, the system extension to support new services consists of adding new scenarios to the current specification. Assume S' is the set of those new scenarios. The whole system prototype is now the TA $A_R \oplus A_{R'}$ where $R = \bigcup_{sc \in S} \mathcal{R}(sc)$ and $R' = \bigcup_{sc \in S'} \mathcal{R}(sc)$. Each intermediate prototype of the system may be checked for the detection of possible features interaction.

- The operator $\oplus$ is commutative and associative as the union $\cup$. So, the order in which scenarios are added to construct the specification doesn't matter.

- Previous compilation results are reused when a new scenario sc is added. In the formula $A_{\bigcup_{sc \in S} \mathcal{R}(sc)} \oplus A_{sc'}$, the first operand is not recomputed but just reused to get the new prototype of the system.

7. CONCLUSION

In this paper, our main contribution is the development of an algorithm for compiling many real-time scenarios into a single timed automaton which represents an executable prototype of the system. This algorithm is insensitive to the order in which scenarios are integrated.

First we have proposed a syntax and a formal semantics for scenarios. Then we defined a flat format of a scenario as a set of self contained rule-actions. The compilation algorithm synthesizes a timed automaton from a set of rule-actions. The set of rule-actions may come from more than one scenarios and therefore the resulting timed automaton represents the integration of these scenarios. The compact representation format of the timed automaton produced by our integration algorithm is compatible to use with model checking tools like KRONOS and UPPAAL.

As future work, we wish to further define operators for explicit integration of the scenarios. These operators are used as directives of compilation.

REFERENCES

[ACD93] Rajeev Alur, Costas Courcoubetis, and David Dill. Model-checking in dense real-time. *Information and Computation*, 104(1):2–34, may 1993.

[AD94] R. Alur and D. Dill. A Theory of Timed Automata. *Theoretical Computer Science*, 126:183–235, 1994.

[BFH+92] A. Bouajjani, J.-C. Fernandez, N. Halbwachs, P. Raymond, and C. Ratel. Minimal state graph generation. *Science of Computer Programming*, 18, 1992.

[Gli95] M. Glinz. An Integrated Formal Model of Scenarios Based on State-charts. In *Software Engineering - ESEC'95*, pages 254–271. Springler LNCS 989, 1995.

[HNSY94] Thomas A. Henzinger, Xavier Nicollin, Joseph Sifakis, and Sergio Yovine. Symbolic model checking for real-time systems. *Information and Computation*, 111(2), jun 1994.

[HSG+94] P. Hsia, J. Samuel, J. Gao, D. Kung, Y. Toyoshima, and C. Chen. Formal approach to scenario analysis. *IEEE Software*, 11:33–41, march 1994.

[SDLV00] A. Salah, R. Dssouli, G. Lapalme, and D. Vincent. Vers un environnement de création de services fondé sur les scénarios enrichis. In *Proceedings of CFIP*, 2000.

DERIVING PARAMETER CONDITIONS FOR PERIODIC TIMED AUTOMATA SATISFYING REAL-TIME TEMPORAL LOGIC FORMULAS*

Akio Nakata and Teruo Higashino
Dept. of Informatics and Mathematical Science, Osaka University
Toyonaka, Osaka 560-8531, Japan
{nakata,higashino}@ics.es.osaka-u.ac.jp

Abstract A symbolic model checking method for parametric periodic timed automata is proposed. The method derives symbolically the weakest condition for parameters such that the specified control state of a periodic timed automaton satisfies some temporal properties. Unlike several existing parametric symbolic model checking methods, the proposed method is 'on-the-fly' — it does not unnecessarily check all the states. Instead, it traverses some necessary part of the computation tree to derive the weakest condition. We show that if we constrain a timed automaton to be *periodic*, i.e. if we force a timed automaton to return to its initial state periodically at the specified constant time, we have only to traverse at most the first 3 periods in the infinite computation tree. In the proposed method, we can avoid a costly (and generally undecidable) fixpoint-calculation for dense-time-domain state sets, and derive the weakest condition for parameters of a timed automaton to satisfy given temporal properties written in a real-time temporal logic formula.

Keywords: symbolic model checking, real-time periodic system, temporal logic

1. INTRODUCTION

Model checking[1] have been recognized as one of very useful and effective methods for designing reliable hardware/software systems. Especially, in recent years, real-time systems have been developed for the areas that high reliability is required, such as aircraft/train/car controlling, nuclear reactors, medical devices and other real-time systems which may be produced a lot and

*This work is partially supported by Semiconductor Technology Academic Research Center (STARC), Japan.

hard to modify in later (e.g. hardware chips or embedded systems). Model checking techniques may be very useful for developing such reliable real-time systems to ensure that the system's design written in some formal model satisfies the required properties such as safety, liveness, and fairness.

The classical model checking method is not *parametric*, that is, to check whether a behavioral specification written in some state model, satisfies some requirement specification (property) written in temporal logic, all the parameters in the specification must be fixed to some concrete values. In recent years, several *symbolic* model checking methods are proposed [2, 3, 4, 5]. Symbolic model checking is a method to construct a set of states (state-sets) which satisfy given temporal properties by representing infinite or finitely huge state-sets symbolically and using a symbolic calculation. The crucial part of symbolic calculation is a fixpoint calculation. There are some symbolic representations of state-sets, such as BDDs[2], which enable us to compute fixpoints efficiently. Although BDDs themselves are aimed for compression of finitely-huge state space, if we adopt some symbolic representations such as Presburger Arithmetic[6], we can extend the method to possibly infinite state space. Such an approach is especially useful for *parametric analyses*, that is, we can obtain the representation of the set parameter variables in order that the behavioral specification satisfies the requirement specification, instead of repeatedly guessing concrete parameter values and checking satisfiability.

More recently, [3] proposed a semi-decision procedure to derive a symbolic representation of parametric states of a hybrid automaton (an extension of a timed automaton) in order to reach some given state-sets. They adopt first-order theory with addition on real-numbers[6] as symbolic representation of state-sets. Although the satisfiability of the first-order theory with addition on real-numbers is decidable, fixpoint calculation is very costly and generally undecidable. [4] proposed some approximation techniques to cope with such undecidability of fixpoint calculations using Presburger Arithmetic as symbolic representation of state-sets, but the domain of the variables is restricted to integers. On the other hand, [5] proposed an algorithm to obtain the condition of parameters in order that the given *non-parametric* state model on dense time domain satisfies the given parametric temporal logic formula. However, they only allow to write parameters to temporal logic formulas, not in a timed automaton. In a realistic system design process, we usually want to choose parameter values of models (implementations) rather than in temporal logic (specifications).

Thus, we propose a decision algorithm to derive a set of parameters of a subclass of a timed automaton model which may contain parameters (*parametric timed automata*[7]) and satisfies a formula of a real-time extension of CTL[1]. In our method, parameters are allowed in both a model and a temporal logic formula. We adopt formulas of first-order theory with addition on real-numbers as symbolic representation of sets of parameter values, as similar to [3]. In order to reduce the size of the intermediate symbolic representation, we decompose the given problem on-the-fly to several subproblems, and recursively solve the subproblems to construct the entire condition for parameters. Unlike [3], in this approach, we need not encode symbolically the entire state space (it tends to be very long) and only the necessary part of the tree is traversed. Specifically, we compute the weakest condition $WPC(s, f)$ of parameters in order that state s of the given model should satisfy the given temporal property f. First, we define $WPC(s, f)$ as a recursive function such as

$$WPC(s, f) \stackrel{\text{def}}{=} F(WPC(s_1, f_1), \cdots, WPC(s_k, f_k)),$$

where $F()$ is a functional on first-order formulas, each s_i is either s or some *next* state of s, and each f_i is either a *proper* subformula of f or some *derived* formula of f (not necessarily a subformula of f). Basically, we can compute the weakest condition $WPC(s, f)$ if the application of the recursive definition of $WPC(s, f)$ is ensured to terminate. However, that is not the case in general. If the model contains some loops, such a recursive application does not terminate. To cope with the problem, we find some subclass of the timed automata such that we need not to explore the infinite computation tree. When the model is a *periodic* timed automaton, that is, after some fixed time period it returns to its initial state and repeats its behavior periodically, we have only to check some finite part of the infinite computation tree and can output the result.

In our method, $WPC(s, f)$ is ensured to be obtained after a fixed steps of recursive computations, thus we can avoid costly fixpoint calculations of first order theory on real-numbers. Moreover, in our on-the-fly approach, intermediate results need to compute $WPC(s, f)$ are kept small compared to the state-space construction approaches. Using dynamic programming, we can also avoid duplicate computation of $WPC(s_i, f_i)$'s when the pair of s_i and f_i is the same. Moreover, each subcondition $WPC(s_i, f_i)$ can be computed in parallel, so we can easily parallelize our method using parallel processors to improve the efficiency.

The rest of this paper is organized as follows. In Section 2, we introduce our model, periodic timed automata. In Section 3, we give a definition of the logic, real-time CTL. In Section 4, we explain our method to obtain $WPC(s, f)$ in detail. Section 6 concludes this paper.

Related Works Several parametric model checking method have been proposed for real-time models[7, 8, 9] and value-passing I/O models[10]. For given two parametric state models, Refs.[7, 8, 10] have proposed the methods to derive the parameter conditions to make one model a correct implementation of another model. [7] has adopted language inclusion as an implementation relation, while [8, 10] has adopted bisimulation equivalence. [8] is an extension of [10] to a timed model. The proposed method in [9] takes a non-parametric state model on discrete time domain and a temporal logic formula which may have some parameters bounded by quantifiers, and checks whether the model satisfies the logic formula. Although some very interesting properties may be expressed using parameters, it simply checks whether or not the given model satisfies the given property. In addition, they consider that the time domain is integers.

2. PARAMETRIC AND PERIODIC TIMED AUTOMATA

In this section, we formally define our model, periodic timed automata. In addition to the traditional theory of Timed Automata[11], we introduce parameters on any (discrete or dense) domain. Then, we define a (non-periodic) parametric timed automaton model. Our definition of timed automata is essentially the same but slightly different from the parametric timed automata in [7], since we generally allow timing constraints to be first-order formulas with addition on real-numbers.

Let Act, Var, $Pred(Var)$ denote the set of all actions, the set of all variables, and the set of all formulas of first order theory with addition on real-numbers over Var, respectively. We also denote the set of real-numbers by $\mathbf{R}$ and the set of non-negative real-numbers by $\mathbf{R}^+$

Definition 2.1 *A* parametric timed automaton *is a tuple* $\langle S, C, PVar, E, Inv(), s_{init}\rangle$, *where* S *is a finite set of control states,* $C \subseteq Var$ *is a finite set of* clocks, $PVar \subseteq Var$ *is a finite set of* parameters, $E \subseteq S \times Act \times Pred(Var) \times 2^C \times S$ *is a transition relation,* $Inv() : S \mapsto Pred(Var)$ *is an* invariant condition *for each state,* s_{init} *is the initial state. We write* $s_i \xrightarrow{a,P,r} s_j$ *if* $(s_i, a, P, r, s_j) \in E$. □

Informally, a transition $s_i \xrightarrow{a,P,r} s_j$ means that action a can be executed from s_i when the values of both clocks and parameters satisfy the formula P (called a *guard condition*), and after executed, the state moves into s_j and clocks in the set r are reset to zero. In any state s, values of all clocks increase continuously at the same speed, representing the time passage. Note that the values of clocks (and parameters if any) can never violate the invariant condition $Inv(s)$. Intuitively, $Inv(s)$ represents the range of values (e.g. minimum and maximum values) allowed for clocks (and parameters). Thus, time passage at state s will stop when the value of some clock will exceed the maximum value specified by $Inv(s)$. When time passage stops, some executable action is forced to execute, representing *urgency*[12] of the action. Also, any incoming transition of the state s' violating $Inv(s')$ is not allowed.

Example 2.1 *Fig. 1 is a simple example of a parametric timed automaton. In Fig. 1, a set of parameters is* $\{x, y, z\}$, *a set of clocks is* $\{c, c'\}$, *the initial state*

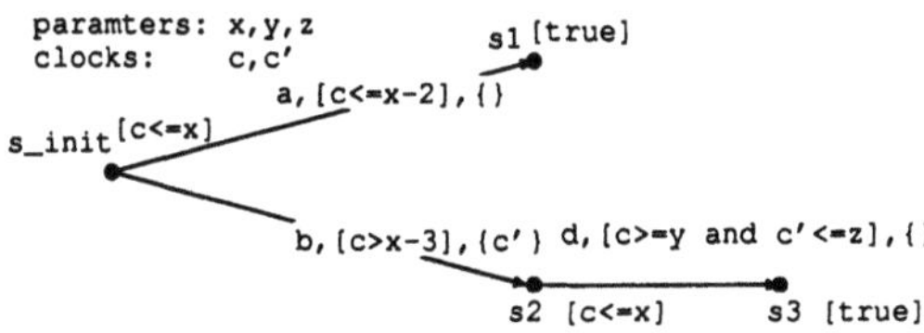

Figure 1. Example of Parametric Timed (DAG-)Automata

is s, and the transition $s[c \le x] \overset{a,[c \le x-2],\{\}}{\longrightarrow} s_1[true]$ *means that at state s, for a given value of the parameter x, some time may be elapsed (i.e. the clock c increases) while c satisfies the invariant* $[c \le x]$*, and when* $c \le x - 2$ *holds, action a can be executed, no clocks are reset to zero, and the state changes to* s_1 *(the invariant of* s_1 *is* $[true]$*, which means that any clock and parameter values are allowed at* s_1*). Similarly, the transition* $s[c \le x] \overset{b,[c>x-3],\{c'\}}{\longrightarrow} s_2[c \le x]$ *means that some time may be elapsed while c satisfies the invariant* $[c \le x]$*, and when* $c > x - 3$ *holds, action b can be executed, clock* c' *is reset to zero, and the state changes to* s_2*. In the transition* $s_2[c \le x] \overset{d,[c \ge y \wedge c' \le z],\{\}}{\longrightarrow} s_3[true]$*, a guard condition for both clocks c and* c' *are specified using parameters y and z.* □

Formal semantics of timed automata is defined as follows. The values of clocks and parameters are given by a function $\sigma : (C \cup PVar) \mapsto \mathbf{R}$. We refer to such a function as a *value-assignment.* We represent a set of all value-assignments by Val. We write $\sigma \models P$ if a formula $P \in Pred(Var)$ is true under a value-assignment $\sigma \in Val$. The semantic behavior of a parametric timed automaton is given as a semantic transition system on *concrete states.* A concrete state is represented by (s, σ), where s is a control state and σ is a value-assignment. Let $CS \stackrel{\text{def}}{=} \{(s, \sigma) | s \in S, \sigma \in Val\}$ be a set of concrete states. The semantic transition system consists of *delay-transitions* and *action-transitions.* A delay transition represents a time passage within the same control state $s \in S$, whereas an action transition represents an execution of an action which changes the control state to the next one s'. Formally, the semantic transition system is defined as follows.

Definition 2.2 *For any value-assignment* σ*,* $t \in \mathbf{R}^+$*, and* $r \subseteq C$*, let* $\sigma + t$ *and* $\sigma[r \to 0]$ *be the value-assignments such that*

$$(\sigma + t)(x) \stackrel{\text{def}}{=} \begin{cases} \sigma(x) + t & \text{if } x \in C, \\ \sigma(x) & \text{otherwise.} \end{cases}$$

$$(\sigma[r \to 0])(x) \stackrel{\text{def}}{=} \begin{cases} 0 & \text{if } x \in r, \\ \sigma(x) & \text{otherwise.} \end{cases}$$

A semantic transition system *for a parametric timed automaton* $\langle S, C, PVar, E, Inv(), s_{init}\rangle$ *is a labelled transition system on concrete states CS, where the transition relation is defined by the following rules:*

- $(s, \sigma) \xrightarrow{t} (s, \sigma + t)$ *if* $t \in \mathbf{R}^+$ *and* $(\sigma + t) \models Inv(s)$,
- $(s, \sigma) \xrightarrow{a} (s', \sigma[r \to 0])$ *if* $s \xrightarrow{a,P,r} s'$, $\sigma \models P$ *and* $\sigma[r \to 0] \models Inv(s')$.

□

The major difference of periodic timed automata from normal parametric timed automata is that it checks the elapsed time since it is started, and if it is equal to the specified period T, then it resets to its initial state. Moreover, it is assumed that only finitely bounded actions can be performed before returning to the initial state. To ensure the above properties, we define a periodic timed automaton as one obtained by adding reset transitions to a parametric timed automaton with no loops (we refer to such a parametric timed automaton as a *parametric timed DAG-automaton*). Formally it is defined as follows.

Definition 2.3 *A* parametric timed DAG-automaton *is a parametric timed automaton whose transition graph has no directed cycles, (i.e. it is a Directed Acyclic Graph(DAG)).* □

The parametric timed automaton in Example 2.1 is a parametric timed DAG-automaton since its transition graph is a tree (so it is also a DAG).

Definition 2.4 *A* periodic timed automaton *is a parametric timed automaton which is obtained from a parametric timed DAG-automaton by adding the special reset transition (called a* return transition*)* $s \xrightarrow{i,[c_p=T],C} s_{init}$ *for state s, where* s_{init} *is the initial state, C is a set of all clocks,* $c_p \in C$ *is a special clock which keeps the elapsed time from the initial state* s_{init} *(no other transition can reset this clock),* $T \in \mathbf{R}^+$ *is a* period*,* $i \in Act$ *is a special reset action.* □

Example 2.2 *Fig. 2 is an example of a periodic timed automaton. This example is a modified version of Example 2.1 where a special clock* c_p *and some return transitions such as* $s_1 \xrightarrow{i,[c_p=T],\{c,c',c_p\}} s$ *are added. Note that we allow periodic timed automata to terminate instead of returning to the initial state, such as the state* s_3 *in Fig. 2.* □

3. REAL-TIME CTL

In this section, we define RPCTL, a Real-time and Parametric extension of Computation Tree Logic(CTL)[1] including some operators in ACTL[13][1].

[1]Unlike [1] and many timed extensions of CTL such as TCTL[3], the next operator of RPCTL is attributed by an action name (similar to Hennessy-Milner Logic[14], or ACTL[13]) so that we can verify the properties

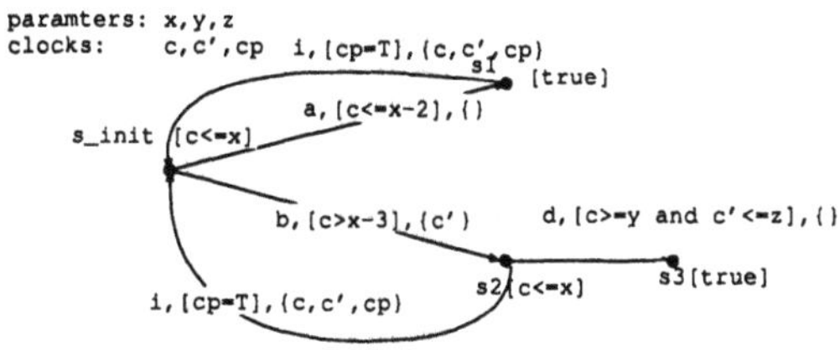

Figure 2. Example of Periodic Timed Automata

f ::=	$true$	(universally valid)
|	$false$	(universally invalid)
|	$\neg f$	(negation)
|	$f \wedge f$	(conjunction)
|	$f \vee f$	(disjunction)
|	$f \Rightarrow f$	(implication)
|	$\langle a \rangle_{\sim p} f$	(existential 'next' operator)
|	$[a]_{\sim p} f$	(universal 'next' operator)
|	$f EU_{\sim p} f$	(existential 'until' operator)
|	$f AU_{\sim p} f$	(universal 'until' operator)
|	$EF_{\sim p} f$	(existential 'eventually' operator)
|	$AF_{\sim p} f$	(universal 'eventually' operator)
|	$EG_{\sim p} f$	(existential 'always' operator)
|	$AG_{\sim p} f$	(universal 'always' operator)

Figure 3. Syntax of RPCTL

In compared to TCTL[3], we do not adopt the *freeze quantifer* as a primitive operator of RPCTL.

Definition 3.1 *The syntax of* RPCTL *formula is defined by the BNF in Fig. 3, where $a \in Act$ is an action name, p is a linear expression which may contain parameter variables, and $\sim \in \{<, \leq, >, \geq, =\}$ is a comparison operator. We may omit '$\sim p$' specifier, and in that case '≥ 0' is assumed.* □

RPCTL is a logic to specify a temporal property at the state of of a parametric timed automaton for its succeeding behavior using temporal operator with timing constraints which may contain parameters. Intuitive meaning of basic constructs of RPCTL is as follows. '$true$' holds at any concrete state. '$\neg f$' holds at a concrete state (s, σ) if and only if f does not hold at (s, σ). '$false$' never holds at any concrete state, which is equivalent to $\neg true$. '$f_1 \wedge f_2$' holds if and only if both f_1 and f_2 hold. '$f_1 \vee f_2$' and '$f_1 \Rightarrow f_2$' are also defined similarly to classic propositional logic. '$\langle a \rangle_{\leq p} f$' holds at (s, σ) if and only if there exists *some* transition from (s, σ) performing a within p units of time, such that f holds at the next (control) state. Since we can define similarly if

when we view the model as a mealy-machine, or (timed-extension of) process algebra with observation semantics (i.e. processes are identified by observing actions, not states) such as CCS, CSP or LOTOS.

$(s,\sigma) \models true.$

$(s,\sigma) \models \neg f \stackrel{\text{def}}{=} (s,\sigma) \not\models f.$

$(s,\sigma) \models f_1 \wedge f_2 \stackrel{\text{def}}{=} (s,\sigma) \models f_1$ and $(s,\sigma) \models f_2.$

$(s,\sigma) \models \langle a \rangle_{\sim p} f \stackrel{\text{def}}{=}$

there exists some transition sequence $(s,\sigma) \stackrel{t}{\longrightarrow} (s,\sigma + t) \stackrel{a}{\longrightarrow} (s',\sigma')$ such that $t \sim p$ and $(s',\sigma') \models f.$

$(s,\sigma) \models f_1 EU_{\sim p} f_2 \stackrel{\text{def}}{=}$

there exists some transition sequence

$(s,\sigma) = (s_1,\sigma_1) \stackrel{t_1}{\longrightarrow} (s_1,\sigma_1 + t_1) \stackrel{a_1}{\longrightarrow} \cdots \stackrel{t_{k-1}}{\longrightarrow} (s_{k-1},\sigma_{k-1} + t_{k-1}) \stackrel{a_{k-1}}{\longrightarrow} (s_k,\sigma_k)$

and some non-negative real-number t_k, s.t. $(s_k,\sigma_k + t_k) \models f_2$ and $t_1 + \cdots + t_k \sim p$

and for any $i(1 \leq i \leq k)$and for any $t_i'(0 \leq t_i' < t_i)$, $(s_i,\sigma_i + t_i') \models f_1$

$(s,\sigma) \models f_1 AU_{\sim p} f_2 \stackrel{\text{def}}{=}$

for any transition sequence such that

$(s,\sigma) = (s_1,\sigma_1) \stackrel{t_1}{\longrightarrow} (s_1,\sigma_1 + t_1) \stackrel{a_1}{\longrightarrow} \cdots \stackrel{t_{k-1}}{\longrightarrow} (s_{k-1},\sigma_{k-1} + t_{k-1}) \stackrel{a_{k-1}}{\longrightarrow} (s_k,\sigma_k)$

and for any nonnegative real-number t_k, $(s_k,\sigma_k + t_k) \models f_2$ and $t_1 + \cdots + t_k \sim p$

and for any $i(1 \leq i \leq k)$ and for any $t_i'(0 \leq t_i' < t_i)$, $(s_i,\sigma_i + t_i') \models f_1.$

Figure 4. Semantics of RPCTL

$\sim$ is other than $\leq$ (case of $\geq, <, >, =$), we only mention the case of $\leq$ in the following explanation. '$[a]_{\leq p} f$' holds if and only if for *any* transition from the state performing a within p units of time, f holds at the next state, which is the same as $\neg\langle a \rangle_{\leq p} \neg f$. '$f_1 EU_{\leq p} f_2$' holds if and only if there exists *some* transition sequence such that f_2 eventually holds within c units of time and until then, f_1 always holds. '$f_1 AU_{\leq p} f_2$' holds if and only if for *any* transition sequence, f_2 eventually holds within p units of time and until then, f_1 always holds. We can also use '$EF_{\leq p} f$', '$AG_{\leq p} f$', '$AF_{\leq p} f$' and '$EG_{\leq p} f$' as abbreviations for $true\ EU_{\leq p} f$, $\neg EF_{\leq p} \neg f$, $true\ AU_{\leq p} f$ and $\neg AF_{\leq p} \neg f$, respectively.

In general, we write $M,(s,\sigma) \models f$ to mean that an RPCTL formula f is satisfied by a concrete state (s,σ) of a parametric timed automaton M. If there are no confusions, we omit M and just write $(s,\sigma) \models f$. The formal definition of the relation $\models$ is as follows. We only give the definitions for six primitive constructs, *true*, $\neg f$, $f_1 \wedge f_2$, $\langle a \rangle_{\sim p} f$, $f_1 EU_{\sim p} f_2$ and, $f_1 AU_{\sim p} f_2$. The rest of the constructs can be specified similarly to the above constructs.

Definition 3.2 *The relation $(s,\sigma) \models f$ is formally defined in Fig. 4.* □

Example 3.1 *The RPCTL formula*

$$[a_1]_{<q_1}((\langle a_2 \rangle true) EU_{\geq q_2}(\langle a_3 \rangle true))$$

means that for every state reachable after executed action a_1 within q_1 units of time, there exists an execution path such that action a_2 is always executable until a_3 becomes executable after q_2 units of time elapsed. Note that q_1 and q_2 are parameter variables. □

4. DERIVATION OF THE WEAKEST CONDITION OF PARAMETERS

Now we describe our method to derive symbolically the weakest condition of parameters $WPC(s, f)$ in order that state s of the periodic timed automaton satisfies the RPCTL property f. To begin with, we give a precise description of our problem.

Definition 4.1 *Let M be a parametric timed automaton, s be a state of M, and f be an RPCTL formula. The* parameter condition derivation problem *is to derive a first-order formula $WPC(s, f)$ such that*

$$\sigma \models WPC(s, f) \text{ iff } (s, \sigma) \models f. \qquad \square$$

At first, we give an algorithm to solve the parameter condition derivation problem for parametric timed DAG-automata, and then we extend it to periodic timed automata.

4.1. Case of DAG Models

As mentioned in Section 1, we define $WPC(s, f)$ as a recursive function such that

$$WPC(s, f) \stackrel{\text{def}}{=} F(WPC(s_1, f_1), \ldots, WPC(s_k, f_k)).$$

Here we give a concrete definition of function $WPC(s, f)$ for each construct of RPCTL formula f.

Definition 4.2 *Let s and f be a state of a parametric timed automaton and an RPCTL formula f, respectively. Then, function $WPC(s, f)$ is defined in Fig. 5. In Fig. 5, C is a set of all clocks, r, r_i, etc. denote subsets of clocks, $P[C + t/C]$ ($P[0/r]$) represents a first order formula P whose every free occurrence of each variable $x \in C$ ($x \in r$) is replaced with $x + t$ (0, respectively).* □

The meaning of the definition of $WPC(s, f)$ is as follows. If f is one of *true* or $f_1 \wedge f_2$, the definition of $WPC(s, f)$ is straightforward. The case of $f = \neg f'$ is less obvious, but since we have defined $WPC(s, f)$ as the weakest condition, $\sigma \not\models WPC(s, f)$ immediately implies $\sigma \models WPC(s, \neg f)$, and vice versa. Hence we have $WPC(s, \neg f) = \neg WPC(s, f)$.

Consider the case of $f = \langle a \rangle_{\sim p} f'$. Suppose σ is a value-assignment such that $(s, \sigma) \models \langle a \rangle_{\sim p} f'$. From Definition 3.2, there must exist a concrete transition

$$
\begin{aligned}
WPC(s, true) &\stackrel{\text{def}}{=} true \\
WPC(s, \neg f) &\stackrel{\text{def}}{=} \neg WPC(s, f) \\
WPC(s, f_1 \wedge f_2) &\stackrel{\text{def}}{=} WPC(s, f_1) \wedge WPC(s, f_2) \\
WPC(s, \langle a \rangle_{\sim p} f) &\stackrel{\text{def}}{=} \exists t (0 \leq t \wedge t \sim p \wedge (Inv(s) \wedge \\
& \bigvee_{i \in I(s,a)} \{P_i \wedge (Inv(s_i) \wedge WPC(s_i, f))[0/r_i]\})[C + t/C]) \\
& \text{where } I(s,a) = \{i | s \stackrel{a,[P_i],r_i}{\longrightarrow} s_i\}, \\
WPC(s, f_1 EU_{\sim p} f_2) &\stackrel{\text{def}}{=} \exists t (0 \leq t \wedge \\
& \forall t'((0 \leq t' \wedge t' \leq t) \Rightarrow WPC(s, f_1)[C + t'/C]) \wedge \\
& (Inv(s) \wedge (t \sim p \wedge WPC(s, f_2) \vee \bigvee_{i \in I(s)} \{P_i \wedge WPC(s_i, f_1 EU_{\sim(p-t)} f_2)\}))[C + t/C]) \\
& \text{where } I(s) = \{i | s \stackrel{a_i,[P_i],r_i}{\longrightarrow} s_i\}, \\
WPC(s, f_1 AU_{\sim p} f_2) &\stackrel{\text{def}}{=} \forall t (0 \leq t \Rightarrow \\
& \forall t'((0 \leq t' \wedge t' \leq t) \Rightarrow WPC(s, f_1)[C + t'/C]) \wedge \\
& (Inv(s) \Rightarrow (t \sim p \wedge WPC(s, f_2) \wedge \bigwedge_{i \in I(s)} \{P_i \Rightarrow WPC(s_i, f_1 AU_{\sim(p-t)} f_2)\}))[C + t/C])
\end{aligned}
$$

Figure 5. Function $WPC(s, f)$

sequence $(s, \sigma) \stackrel{t}{\longrightarrow} (s, \sigma + t) \stackrel{a}{\longrightarrow} (s', \sigma')$ such that $t' \sim p$ and $(s', \sigma') \models f'$. Thus, the following conditions must also hold:

- some timed automaton transition $s \stackrel{a,P,r}{\longrightarrow} s'$ must exists.
- $\sigma + t$ must satisfy both $Inv(s)$ and P (Definition 2.2).
- σ' is a value-assignment $\sigma + t$ whose values of the clocks in r are reset to zero, i.e. $\sigma' = (\sigma + t)[r \to 0]$ (recall that $\sigma + t$ is the same value-assignment as σ except all clock values are increased by t, and $\sigma[r \to 0]$ is the same as σ except all the clocks in r are reset to zero, as defined in Definition 2.2) and it satisfies $Inv(s')$.
- $(s', \sigma') \models f'$, i.e. $\sigma' \models WPC(s', f')$.

Hence, we obtain a necessary condition

> "there exists some non-negative real-number t and some transition $s \stackrel{a,P,r}{\longrightarrow} s'$, such that $\sigma \models (t \sim p)$, $\sigma + t \models Inv(s) \wedge P$ and $(\sigma + t)[r \to 0] \models Inv(s') \wedge WPC'(s', f')$"

for σ to make state s satisfy f. We can rewrite '$\sigma + t \models Inv(s) \wedge P$' to the condition of σ, such as $\sigma \models (Inv(s) \wedge P)[C + t/C]$. By the same way, we can also rewrite $(\sigma + t)[r \to 0] \models Inv(s') \wedge WPC'(s', f')$ as $\sigma \models (Inv(s') \wedge$

$WPC(s', f'))[C + t/C, 0/r]$. Therefore, the following condition holds:

$$\sigma \models (0 \leq t \wedge t \sim p \wedge Inv(s) \wedge P \wedge (Inv(s') \wedge WPC(s', f'))[0/r])[C + t/C].$$

Since it is sufficient that some non-negative real-number t and some transition $s \xrightarrow{a,P,r} s'$ exist, we can weaken the above condition as:

$$\sigma \models \exists t(0 \leq t \wedge t \sim p \wedge Inv(s) \wedge \bigvee_{i \in I(s,a)} \{P_i \wedge (Inv(s_i) \wedge WPC(s_i, f'))[0/r]\})[C + t/C].$$

where $I(s, a) \stackrel{\text{def}}{=} \{i | s \xrightarrow{a,P_i,r_i} s_i\}$ is a set of indices of transitions whose source node is s and action name is a. We can easily prove that this is the weakest condition of σ such that $(s, \sigma) \models f$, and t is some fresh variable which does not appear in either $Inv(s)$, P_i, $Inv(s_i)$ or $WPC(s_i, f')$.

Consider the case of $f = f_1 EU_{\sim p} f_2$. Similar to above, suppose σ is a value-assignment such that $(s, \sigma) \models f_1 EU_{\sim p} f_2$. From Definition 3.2, there must exist some transition sequence $(s, \sigma) = (s_1, \sigma_1) \xrightarrow{t_1} (s_1, \sigma_1 + t_1) \xrightarrow{a_1} \cdots \xrightarrow{t_{k-1}} (s_{k-1}, \sigma_{k-1} + t_{k-1}) \xrightarrow{a_{k-1}} (s_k, \sigma_k) \xrightarrow{t_k} (s_k, \sigma_k + t_k)$, such that $(s_k, \sigma_k + t_k) \models f_2$, $t_1 + \cdots + t_k \sim p$, and for any j $(1 \leq j \leq k)$ and for any t'_j $(0 \leq t'_j < t_j)$, $(s_j, \sigma_j + t'_j) \models f_1$ holds. To obtain a recursive definition of $WPC(s, f)$, we divide the premise of the above statement into two cases, $k = 1$ (f_2 holds at the current state s_1) and $k \geq 2$ (f_2 holds at some future state s_k).

[Case $k = 1$]: If $k = 1$, then there must exist a transition sequence $(s, \sigma) \xrightarrow{t} (s, \sigma + t)$ such that $(s, \sigma + t) \models f_2$, $t \sim p$ and for any t' $(0 \leq t' < t)$, $(s, \sigma + t) \models f_1$ holds. Similar to the case of $f = \langle a \rangle_{\sim p} f'$, the weakest condition of σ is obtained as follows:

$$\sigma \models \exists t(0 \leq t \wedge t \sim p \wedge (Inv(s) \wedge WPC(s, f_2))[C + t/C] \wedge \forall t'((0 \leq t' \wedge t' \leq t) \Rightarrow WPC(s, f_1)[C + t'/C])) \quad (1)$$

[Case $k \geq 2$]: If we assume $k \geq 2$, then there must exist a transition sequence $(s, \sigma) = (s_1, \sigma_1) \xrightarrow{t_1} (s_1, \sigma_1 + t_1) \xrightarrow{a_1} (s_2, \sigma_2)$ such that $(s_2, \sigma_2) \models f_1 EU_{\sim(p-t_1)} f_2$ holds, and for any t'_1 $(0 \leq t'_1 \leq t_1)$, $(s_1, \sigma_1 + t'_1) \models f_1$ holds. Considering that there may exist multiple transitions $s \xrightarrow{a_i,P_i,r_i} s_i$, the weakest condition of σ is obtained as follows:

$$\sigma \models \exists t[0 \leq t \wedge (Inv(s) \wedge \bigvee_{i \in I(s)} \{P_i \wedge WPC(s_i, f_1 EU_{\sim(p-t)} f_2)\})[C + t/C] \wedge \forall t'[(0 \leq t' \wedge t' \leq t) \Rightarrow WPC(s, f_1)[C + t'/C]]] \quad (2)$$

where $I(s) \stackrel{\text{def}}{=} \{i | s \stackrel{a_i, P_i, r_i}{\longrightarrow} s_i\}$ is a set of indices of transitions whose source node is s.

Therefore, the general case is (1) or (2), that is,

$$\begin{aligned} \sigma \models\ & \exists t[0 \le t \wedge \forall t'[(0 \le t' \wedge t' \le t) \Rightarrow WPC(s, f_1)[C + t'/C]]] \\ & \wedge(Inv(s) \wedge (t \sim p \wedge WPC(s, f_2) \vee \\ & \bigvee_{i \in I(s)} \{P_i \wedge WPC(s_i, f_1 EU_{\sim(p-t)} f_2)\}))[C + t/C] \end{aligned}$$

The case of $f = f_1 AU_{\sim p} f_2$ is similar, and we omit the detailed description due to space limitation.

If the transition graph contains no loops, there are no cases that $WPC(s, f)$ is recursively called during the computation of $WPC(s, f)$ itself. Thus, the function call $WPC(s, f)$ is ensured to terminate. Hence, a recursive function $WPC(s, f)$ is an algorithm to obtain the parameter condition for DAG-formed models (i.e. parametric timed DAG-automata).

Theorem 4.1 *For every state s of a parametric timed DAG-automaton M and every RPCTL formula f, a recursive function $WPC(s, f)$ always terminates and returns a correct solution of a parameter condition derivation problem, i.e.:*

$$\forall \sigma.[\sigma \models WPC(s, f) \text{ iff } (s, \sigma) \models f]. \qquad \square$$

4.2. Case of Periodic Models

If parametric timed automata have some loops, the algorithm $WPC(s, f)$ in Theorem 4.1 may not terminate. In this section, we prove that if models are periodic timed automata, we have only to check a finite fragment of the computation tree to derive the weakest condition of parameters.

At first, we introduce the notion of unfolding. Replace all returning transitions of a periodic timed automaton (Fig. 6-(a)) with transitions to the special terminating state. We obtain the corresponding parametric timed DAG-automata (Fig. 6-(b)). Then, attach the copy of the corresponding parametric timed DAG-automaton to the special terminating state of itself. Finally, we have the parametric timed DAG-automaton which represents the first 2 periodic behavior (Fig. 6-(c)). We refer to such a model as a 2-*unfolding* of the periodic timed automaton. Similarly, we can also define a k-unfolding of a periodic timed automaton as the parametric timed DAG-automaton which represents the first k periodic behavior.

For any RPCTL operators *except* 'until' operators EU and AU (and all the operators derived from EU and AU such as AG, AF, EF and EG), the weakest condition of parameters are obtained by just checking the current state (and the next state if it is 'next' operators) and checking for the *proper* subformulas recursively. Since the size of each subformula is strictly decreasing, the recursion eventually terminates and we can obtain the result. However, it is not the case for the 'until' operators EU and AU. The definition of $WPC(s, f_1 EU_{\sim p} f_2)$ contains not only recursive calls for its proper subformulas f_1 and f_2, but also the recursive call for the formula $f_1 EU_{\sim(p-t)} f_2$, which is not a proper subformula. This means that it may execute the function call of the form $WPC(s_i, f_1 EU_{\sim(p-t_1-\ldots-t_k)} f_2)$ forever if the model contains some loops. Our result is that without loss of generality, we have only to check 3-unfolding of periodic timed automata for each subformula including 'until' operators (EU and AU).

Formally, it is proved by the following lemma:

Lemma 4.1 *For any concrete state (s, σ) of periodic timed automata, the following condition holds:*

$$(s, \sigma) \models f_1 EU_{\sim p} f_2 \textit{ if and only if } (s, \sigma) \models f_1 EU_{\sim p'} f_2$$

where T is the period of the periodic timed automata, $p' \stackrel{\text{def}}{=} p - m \times T$ and m is the minimum nonnegative integer s.t. $p - m \times T < 3T$. The same condition also holds for $f_1 AU_{\sim p} f_2$.

(proof) Let α, α', etc. denote finite execution paths such as $(s_1, \sigma_1) \xrightarrow{t_1} (s_1, \sigma_1 + t_1) \xrightarrow{a_1} (s_2, \sigma_2) \xrightarrow{t_2} \ldots \xrightarrow{a_{k-1}} (s_k, \sigma_k)$. Let $ET(\alpha)$ denote the execution time of the path α, i.e., $ET(\alpha) = t_1 + \cdots + t_{k-1}$. Consider there exists a path α which begins with (s, σ) such that $(s, \sigma) \models f_1 EU_{\sim p} f_2$ (which implies $ET(\alpha) \sim p$). The case of $0 \le ET(\alpha) < 3T$ is trivial, since if $(s, \sigma) \models f_1 EU_{\sim p} f_2$ holds for α, p must be less than $3T$ for any case of $\sim \in \{<, \le, >, \ge, =\}$, which implies $p = p'$ from the definition of p', and thus obviously $(s, \sigma) \models f_1 EU_{\sim p'} f_2$ holds, and vice versa. Consider the case of $ET(\alpha) \ge 3T$. As illustrated in Fig. 7, there must also exist a path α' containing at least one cycle, such that $0 \le ET(\alpha') < 3T$ and $(s, \sigma) \models f_1 EU_{\sim p} f_2$ holds. α' is obtained by removing cycles beginning with the initial state from α until $ET(\alpha) < 3T$ holds, while leaving at least one cycle.[2] Obviously, the relation between $ET(\alpha)$ and $ET(\alpha')$ is $ET(\alpha') = ET(\alpha) - m \times T$, where m is the number of the removed cycles. Therefore, $ET(\alpha) \sim p$ implies $ET(\alpha') \sim p'$ where $p' = p - m \times T$, and immediately we have $(s, \sigma) \models f_1 EU_{\sim p'} f_2$. Conversely, if there exists a path α' containing at least one cycle, such that $0 \le ET(\alpha') < 3T$ and $(s, \sigma) \models f_1 EU_{\sim p'} f_2$,

[2] Note that in a periodic timed automaton, all clocks are initially zero and reset to zero when returned to the initial state, whereas any other variables (parameters) are never modified during execution. So, the possible behaviour from the initial state is always the same, no matter how it is reached. Thus, after removing cycles beginning with the initial state from the execution path α, it is still an executable path. Moreover, f_2 holds at the last state of the path and along with the path, f_1 always holds until f_2 holds. Therefore, if $ET(\alpha) \sim p$ holds, then $f_1 EU_{\sim p} f_2$ still holds on the shortened path.

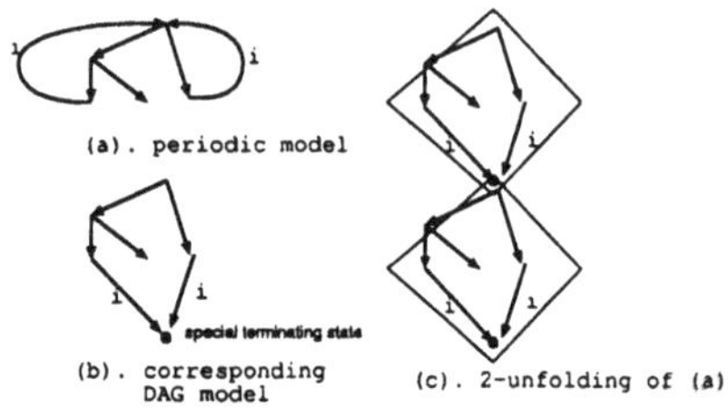

Figure 6. Unfolding of Periodic Timed Automata

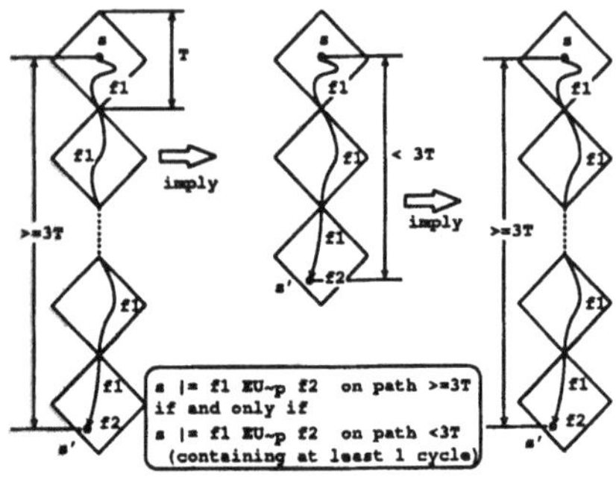

Figure 7. Illustration of Lemma 4.1

there must also exist a path α such that $ET(\alpha) \geq 3T$ and $(s, \sigma) \models f_1 EU_{\sim p} f_2$, as also illustrated in Fig. 7. In this case, α is obtained by duplicating intermediate cycles of α' (α' should have at least one cycle by the assumption) repeatedly until $ET(\alpha) \geq 3T$ and $ET(\alpha) \sim p$ hold. The case of $f_1 AU_{\sim p} f_2$ is similar (we consider a path which *violates* $f_1 AU_{\sim p} f_2$, instead), and we omit the detailed proof due to the space limitation. □

We define another algorithm $WPC3(s, f)$ instead of $WPC(s, f)$ for periodic timed automata. $WPC3(s, f)$ is almost the same as $WPC(s, f)$, except that $WPC3(s, f)$ derives the weakest parameter condition such that s satisfies f within 3 periods.

By Lemma 4.1, it is sufficient to consider finite paths whose execution time is at most $3T$ in order to derive the weakest condition of parameters. Therefore, we obtain the following main theorem:

Theorem 4.2 *For every state s of a periodic timed automaton M and every RPCTL formula f, a recursive function $WPC3(s, f)$ always terminates and returns a correct solution of a parameter condition derivation problem, i.e.:*

$$\forall \sigma.[\sigma \models WPC3(s, f) \textit{ iff } (s, \sigma) \models f]$$ □

5. COMPUTATIONAL COMPLEXITY

In this section, we evaluate the computational complexity of the functions $WPC(s, f)$ and $WPC3(s, f)$. Direct implementation of the recursive function $WPC(s, f)$ (and $WPC3(s, f)$) may perform very inefficiently because of repetitive function calls of the same tuple of arguments (s, f), which is redundant. Thus, we evaluate the complexity when we use a cache to avoid unnecessary computation. We assume that the function $WPC(s, f)$ has its own cache table, which has the entry (s, f, P) if $WPC(s, f)$ has already been computed and the result is P. If such a function is called, the cache table is checked first, and if its value has already been computed, the cached result is returned. Otherwise, the computation is performed, the result is registered into the cache, and the result is returned.

The complexity of the above algorithm is evaluated as follows. For simplicity, we here ignore the length of the resulting formulas (i.e. we ignore the cost to simplify and/or check satisfiability of the formulas). Under the assumption, time and space complexity coincides, that is, the time necessary to compute $WPC(s, f)$ is equal to the size of the cache table, which is equal to the maximum number of the different tuples of arguments with which $WPC(s, f)$ will be called. From the definition of $WPC(s, f)$, the number of the different tuples of arguments is bounded by $n \times (m_1 + m_2)$, where n is the number of control states reachable from s, and m_1 (m_2) is the number of the different subformulas of f (the number of the different derived formulas of the form $f_1 op_{\sim(p-t_1-\cdots-t_k)} f_2$ ($op \in \{EU, AU\}$), respectively). m_1 is equal to the number of nodes of the syntax tree of f. Thus, m_1 is $O(m \log m)$, where m is the length of f. m_2 is equal to $m_1 \times l$, where l is the depth of the DAG, i.e. the length of the longest directed paths from the root to leaf nodes of the DAG (here we refer to a leaf node of the DAG as a node which has no outgoing edges). Obviously, l does not exceed the number n of nodes. Overall, the number of different tuples of arguments (= the size of cache table) is $O(n \times m \log m \times n) = O(n^2 m \log m)$. Since we have only to perform at least one computation for each tuple of arguments, the number of all computation is also $O(n^2 m \log m)$. Therefore, the time and space complexity of $WPC(s, f)$ is $O(n^2 m \log m)$.

The complexity of $WPC3(s, f)$ is equal to the $WPC(s, f)$ for 3-unfolding DAG model. Let n denote the number of nodes of a periodic model. Then, from the definition of unfolding, the depth of the corresponding 3-unfolding is $3n$. Therefore, by replacing n with $3n$, we also conclude that the time and space complexity of $WPC3(s, f)$ is $O(n^2 m \log m)$.

Theorem 5.1 *The time and space complexity of $WPC(s, f)$ and $WPC3(s, f)$ for a DAG/periodic model M is $O(n^2 m \log m)$, where n is the number of control states in M, and m is the length of RPCTL formula f.* □

6. CONCLUDING REMARKS

In this paper, we propose a method to derive a parameter condition for a periodic timed automaton which satisfies a property written in a temporal logic RPCTL formula.

Although our method applies to only the restricted class of timed automata, many real-time applications such as audio/video streaming, time sharing task schedulers can be specified as periodic timed automata.

The future works are to develop and improve the efficiency of the implementation of the algorithm, to extend our algorithm to simplify the parameter condition and to handle multiple periodic timed automata which run concurrently. We are also considering to extend our method to handle some internal state variables of finite domain to improve expressiveness of the model.

REFERENCES

[1] E. M. Clarke, E. A. Emerson, and A. P. Sistla, "Automatic verification of finite state concurrent systems using temporal logic specifications," *ACM Trans. on Program Languages and Semantics*, vol. 8, no. 2, pp. 244–263, 1986.

[2] J. R. Burch, E. M. Clarke, K. L. McMilan, D. L. Dill, and L. J. Hwang, "Symbolic model-checking: 10^{20} states and beyond," *Information and Computation*, vol. 98, no. 2, pp. 142–170, 1992.

[3] R. Alur, T. A. Henzinger, and P. Ho, "Automatic symbolic verification of embedded systems," *IEEE Trans. on Software Engineering*, vol. 22, no. 3, pp. 181–201, 1996.

[4] T. Bultan, R. Gerber, and W. Pugh, "Symbolic model checking of infinite state systems using presburger arithmetic," in *Proc. of Int. Conf. on Computer Aided Verification (CAV'97)*, vol. 1254 of *Lecture Notes in Computer Science*, pp. 400–411, Springer-Verlag, 1997.

[5] F. Wang, "Parametric timing analysis for real-time systems," *Information and Computation*, vol. 130, no. 2, pp. 131–150, 1996.

[6] J. E. Hopcroft and J. D. Ullman, *"Introduction to Automata Theory, Languages and Computation"*. Addison-Wesley, 1979.

[7] R. Alur, T. A. Henzinger, and M. Y. Vardi, "Parametric real-time reasoning," in *Proc. 25th ACM Annual Symp. on the Theory of Computing (STOC'93)*, pp. 592–601, 1993.

[8] A. Nakata, T. Higashino, and K. Taniguchi, "Time-action alternating model for timed LOTOS and its symbolic verification of bisimulation equivalence," in *Proc. of Joint Int'l Conf. on Formal Description Techniques for Distributed Systems and Communication Protocols, and Protocol Specification, Testing, and Verification (FORTE/PSTV'96)* (R. Gotzhein and J. Bredereke, eds.), pp. 279–294, IFIP, Chapman & Hall, 1996.

[9] E. A. Emerson and R. J. Trefler, "Parametric quantitative temporal reasoning," in *Proc. IEEE Int'l Conf. on Logic in Computer Science (LICS'99)*, pp. 336–343, 1999.

[10] M. Hennessy and H. Lin, "Symbolic bisimulations," *Theoretical Computer Science*, vol. 138, pp. 353–389, 1995.

[11] R. Alur and D. Dill, "A theory of timed automata," *Theoretical Computer Science*, vol. 126, pp. 183–235, 1994.

[12] T. Bolognesi and F. Lucidi, "LOTOS-like process algebras with urgent or timed interactions," in *Formal Description Techniques, IV* (K. R. Parker and G. A. Rose, eds.), pp. 249–264, IFIP, Elsevier Science Publishers B.V.(North-Holland), 1992.

[13] R. De Nicola and F. W. Vaandrager, "Action versus state based logics for transition systems," in *Proc. of Ecole de Printemps on Semantics of Concurrency* (I. Guessarian, ed.), vol. 469 of *Lecture Notes in Computer Science*, pp. 407–419, Springer-Verlag, 1990.

[14] M. Hennessy and R. Milner, "Algebraic laws for nondeterminism and concurrency," *J. ACM*, vol. 32, pp. 137–161, Jan. 1985.

Part Four

Process Algebra

PAMR: A PROCESS ALGEBRA FOR THE MANAGEMENT OF RESOURCES IN CONCURRENT SYSTEMS*

Manuel Núñez and Ismael Rodríguez
Dept. Sistemas Informáticos y Programación
Universidad Complutense de Madrid, E-28040 Madrid. Spain.
{mn,ir}@sip.ucm.es

Abstract In this paper we present a *process algebra for the management of resources* in concurrent systems. Our aim is to define a formal framework that can help in the task of specifying systems that depend, for their execution, on a set of *resources* that they use. Usually,sy stems consist in a set of processes. In order to improve their performance, these processes will be able to exchange resources among them. In our language, processes will consist in a *behavior* (formalized as a LOTOS process) and in information about the resources that they own. Systems will be defined as the parallel composition of a set of processes. We will study some examples applying the features of PAMR. These examples will try to show the usefulness of our language for specifying and analyzing concurrent systems where resources play an important role.

Keywords: Extension of FDT's; Semantical Foundations; Process Algebras.

1. INTRODUCTION

During the last two decades Process Algebras have been used to specify and verify different kinds of systems. Nevertheless, most process algebraic models lack the ability to appropriately express the relation between the behavior of processes and the resources that they use. Actually,r esources are usually modeled as *processes*. In this paper we will present a process algebra to deal with systems where resources must be taken into consideration. In order to illustrate the kind of systems that we will deal with, let us introduce the following simple running example. Let us consider a system consisting in the parallel execution of n subsystems ($P_1 \ldots P_n$) and m different kinds of resources that these subsystems may use (let us suppose that the total quantity of the resource i is

*Research supported in part by the CICYT project TIC 2000-0701-C02-01.

equal to x_i). The performance of these subsystems depends on these resources (for example, the portion of memory used by each subsystem, time quantum of CPU, time quantum of access to the bus, etc). Each subsystem P_j has an initial distribution of resources $x_1^j \ldots x_m^j$, that is, in the beginning, subsystem j owns x_i^j units of the resource i. Given the fact that the quantity of resources that subsystems own cannot be bigger than the total amount, we work under the constraint: $\forall\, 1 \leq i \leq m : \sum_j x_i^j \leq x_i$. Finally, let us suppose that subsystems have a *preference* on how they *like* resources. For example, suppose that a subsystem P_{j_1} runs at the same speed if we replace one unit of the resource i_1 by four units of the resource i_2,while another subsystem P_{j_2} runs at the same speed if one unit of the resource i_1 is replaced by two units of the resource i_2. In particular, P_{j_1} will perform better if we replace three of its units of i_2 by one additional unit of i_1.

In order to describe this kind of systems, we will introduce a language where the usual description of the behavior of processes is extended with additional information. In our language, a process will not only take into account its activity (that is, which actions can be performed at a given point of time) but it will also consider which resources can be used during its execution. Besides, subsystems may *exchange* resources with other subsystems. For instance, in the example given above, if P_{j_1} gives to P_{j_2} three units of i_2 and receives from P_{j_2} one unit of i_1, then both subsystems run faster. So, our language will provide mechanisms to, starting with an initial distribution of resources, accomplish a better performance of the studied system.

Example 1 Consider a system where two programs are running. Using operating systems terminology suppose that one of the programs is an I/O-bound process (e.g. a process printing a paper) while the other one is a CPU-bound process (e.g. a process compiling the LaTeX source of another paper). Even if the initial distribution would assign the same resources to both processes (in terms of CPU use and access to I/O), it is clear that both processes would perform better if they exchange their resources in the adequate way. That is, even though the I/O-bound process will keep some time quantum of CPU, it is willing to reduce it (with respect to the original distribution) in order to get better access to I/O, and the other way around for the CPU-bound process. □

We will define our language in two steps. First, we consider a *base language* where we can specify the usual behavior of *processes*. For this purpose, we will use a LOTOS like language. As we already said, processes will also contain information about the available resources: The resources that they own, the resources that they need to perform actions, etc. In particular, there will be a function indicating their preferences on resources: A *utility function*. This function will compute values according to the actions that the process is able to perform immediately. For example, consider a program that initially behaves as a CPU-bound process and after a while its behavior is as an I/O-bound process. This must be reflected in its utility function. In the beginning, it will prefer to keep a bigger time quantum of CPU, but in the end, its utility function will report bigger values if it keeps better access to I/O.

The second step consists in the combination of processes to build *systems*. A system will be the parallel composition of a set of processes. Processes will be able to communicate with each other by two kinds of operations. First, we will have the usual mechanism for synchronization in parallel operators, that is, there will be some actions that processes may perform asynchronously and some actions that they have to perform synchronously. More important, in order to improve their performances, processes will be able to exchange resources. Besides, *harmful* exchanges will be forbidden. For instance, it will not be allowed that P_{j_2} exchanges three units of i_2 by one unit of i_1 with P_{j_1} because both subsystems get worse, and thus the whole system deteriorates. We will consider a parallel operator where communications are not restricted to be exactly between two processes. For instance, we will allow that three processes synchronize in an action a or that four processes exchange, in one step, resources among them.

Regarding exchange of resources, we will consider two *policies*. The first one, that we call *preserving utility*, will allow exchanges of resources only in the case that at least one of the processes improves its situation, and no process gets worse after the exchange. For example, this policy would not allow P_{j_1} to *trade* with P_{j_2} one unit of i_1 by one unit of i_2, because even though P_{j_1} improves its situation we have that P_{j_2} gets worse. The second policy,that we call *maximizing utility*, will not be based on the particular situations of processes but in the situation of the system. In this case, exchanges will be allowed only if they improve the overall performance of the system. For example, this policy would allow the previous exchange only if the *profit* for P_{j_1} is bigger than the *loss* for P_{j_2}. In fact, this policy is very similar to consider that processes are not the owner of resources. This is so because part of the resources of a process can be *expropriated* (as far as the whole system improves).

In terms of related work, our language is based on a previous proposal [12]. In this paper we have extended that language to get a more appropriate framework for the specification of the systems that we are dealing with. Nevertheless, the theoretical framework introduced in [12], where some concepts of microeconomic theory were used, can be inherited by adding some slight modifications. There are models to specify systems sharing resources (e.g. [5]), but in this case resources are just accessed, not traded; this access induces some delays in the behavior of processes. If we consider the maximizing utility policy, our proposal is somehow related to the *ODP trading function* [9]. Nevertheless, under a preserving utility policy, it is the case that a process only uses (and nobody else can use them) the resources that it owns. Besides, under this policy, trade permanently transfers the *ownership* of the traded resources. Finally, management of resources appears in fields like operating systems or concurrent programming. Resources are usually owned by a *mediator* which allows the processes to use them. So, given the fact that we separate the behavior of a process and how it manages its resources, we think that our language can be successfully used for the specification of this kind of systems.

The rest of the paper is organized as follows. In Section 2 we define our language in terms of processes and systems. In Section 3 we present two examples of the use of our language and we study the absence of deadlocks for the defined systems. Finally, in Section 4 we present our conclusions and some directions for further research. An extended version of this paper, where a bisimulation semantics is defined and more complex examples are presented, can be found in [13].

2. DEFINITION OF THE LANGUAGE

In this section we present PAMR as well as its operational semantics. First, we will introduce the notion of *process*. Then, we will say that a system is the parallel composition of several *communicating* processes. We will define the operational semantics of systems by means of three rules. The first rule will describe how processes exchange resources among them. Processes will exchange resources until no more *useful* exchanges are possible. Then, the last two rules define how systems perform actions (possibly by synchronizing among the corresponding processes). We will finish this section by presenting some useful results for verifying some properties in Section 3.

Before we present the formal definition of our language, we introduce some mathematical notation which will be used in the rest of the paper.

Definition 1W e consider $\mathbf{R}_+ = \{x \in \mathbf{R} \mid x \geq 0\}$. By abuse of notation, we will consider $\frac{r}{0} = \infty$. Let $r \in \mathbf{R}_+$. Then, $\texttt{trunc}(r)$ denotes the natural number resulting from discarding the decimal part of r (e.g. $\texttt{trunc}(1.2) = 1$).

We will usually denote *vectors* in $\mathbf{R}^n$ (for $n \geq 2$) by $\bar{x}, \bar{y}, \ldots$. Given $\bar{x} \in \mathbf{R}^n$, x_i denotes its *i-th* component. We extend some usual arithmetic operations to vectors. Let $\bar{x}, \bar{y} \in \mathbf{R}^m$. We define $\bar{x} + \bar{y} = (x_1 + y_1, \ldots, x_n + y_n)$. We write $\bar{x} \leq \bar{y}$ if for any $1 \leq i \leq n$ we have $x_i \leq y_i$.

We will usually denote *matrices* in A^{n*m} (for $n, m \geq 2$, and a set A)by calligraphic letters $\mathcal{E}, \mathcal{E}_1 \ldots$. Let $\mathcal{A} \in A^{n*m}$, and i, j be such that $1 \leq i \leq n$ and $1 \leq j \leq m$. We will denote the $(n * (i-1) + j)th$ component of $\mathcal{A}$ by $\mathcal{A}_{ij}$, that is, if we consider $\mathcal{A}$ as a matrix, this component corresponds to the element located at file i and column j.

Given a set A, $\mathcal{P}(A)$ denotes the set containing all the subsets of A. □

$$\text{(PRE)}\frac{}{\mu;B\xrightarrow{\mu}B} \qquad \text{(PAR1)}\frac{B_1\xrightarrow{\mu}B_1' \wedge \mu\notin A}{B_1\|_A B_2\xrightarrow{\mu}B_1'\|_A B_2}$$

$$\text{(CHO1)}\frac{B_1\xrightarrow{\mu}B_1'}{B_1+B_2\xrightarrow{\mu}B_1'} \qquad \text{(PAR2)}\frac{B_1\xrightarrow{a}B_1' \wedge B_2\xrightarrow{a}B_2' \wedge a\in A}{B_1\|_A B_2\xrightarrow{a}B_1'\|_A B_2'}$$

$$\text{(CHO2)}\frac{B_2\xrightarrow{\mu}B_2'}{B_1+B_2\xrightarrow{\mu}B_2'} \qquad \text{(PAR3)}\frac{B_2\xrightarrow{\mu}B_2' \wedge \mu\notin A}{B_1\|_A B_2\xrightarrow{\mu}B_1\|_A B_2'}$$

$$\text{(HID1)}\frac{B_1\xrightarrow{\mu}B_1' \wedge \mu\notin A}{\texttt{hide } A \texttt{ in } B_1\xrightarrow{\mu}\texttt{hide } A \texttt{ in } B_1'} \qquad \text{(HID2)}\frac{B_1\xrightarrow{a}B_1' \wedge a\in A}{\texttt{hide } A \texttt{ in } B_1\xrightarrow{\tau_a}\texttt{hide } A \texttt{ in } B_1'}$$

$$\text{(REC)}\frac{B\{X:=B/X\}\xrightarrow{\mu}B'}{X:=B\xrightarrow{\mu}B'}$$

Figure 1. Operational Semantics for the Base Language.

The behaviors of processes will be defined by means of a usual process algebra. In this case, we will consider a simple LOTOS like *base language*. First, we give an auxiliary definition introducing the sorts for actions.

Definition 2 Let $\texttt{Act}$ be the set of *visible* actions ($a, b\ldots$ range over $\texttt{Act}$). Let $\texttt{Act}_\tau$ be the set of *internal* actions ($\mu, \mu'\ldots$ range over $\texttt{Act} \cup \texttt{Act}_\tau$). We suppose that $\tau \in \texttt{Act}_\tau$, that there exists a bijection $f : \texttt{Act} \longrightarrow (\texttt{Act}_\tau - \{\tau\})$, and $\texttt{Act} \cap \texttt{Act}_\tau = \emptyset$. Given a visible action $a \in \texttt{Act}$,we will denote $f(a)$ by τ_a. We denote by $\texttt{ACT}$ the *set of actions*, that is, $\texttt{ACT} = \texttt{Act} \cup \texttt{Act}_\tau$. Finally, let *Id* be a set of (basic) process identifiers. □

Let us note that we will have not only a unique internal action, but a whole set of them. In addition to the usual τ action, we consider an internal action for each of the visible actions. For an *external* observer, all the internal actions will be equal. The difference among them comes from the fact that they will need different resources to be performed. So, if a process needs a set of resources $\bar{x}$ to perform a visible action a, and this action is *hidden*, the resulting action, that is τ_a, will need the same amount of resources $\bar{x}$ to be performed. Sets of internal actions appear in other models for concurrent processes (for example, for I/O automata [10]).

Definition 3 The set of *basic processes*, denoted by,$\mathcal{B}$ is given by the BNF-expression $B ::= \text{stop} \mid X \mid \mu\,; B \mid B + B \mid B \|_A B \mid \text{hide } A \text{ in } B \mid X := B$, where $\mu \in \texttt{ACT}$, $A \subseteq \texttt{Act}$, and $X \in Id$. □

The operational semantics for the base language is given in Figure 1, and it is standard. Let us remind that $B\{B'/X\}$ represents the replacement of the free occurrences of the variable X in B by the term B'. Let us also remark that, in rule (HID2), the result of hiding a visible action a is not τ but τ_a. The following definition will be used later on. It computes the actions that a basic process may perform immediately.

Definition 4 Let $B \in \mathcal{B}$. We define its *set of immediate actions*, denoted by $\mathtt{Imm}(B)$, as the set $\mathtt{Imm}(B) = \{\mu \in \mathtt{ACT} \mid \exists\, B' \in \mathcal{B} : B \xrightarrow{\mu} B'\}$. □

As we sketched in the introduction, a process will not only consist in a behavior. On the contrary, a process will keep track of the resources assigned to it and of some information relating resources and actions. Specifically, a process P will be defined as a tuple $(B, \bar{x}, u, n, c)$ where the intuitive meaning of the components is:

- $B \in \mathcal{B}$ is a *basic process* indicating the behavior of P.
- $\bar{x} \in \mathbf{R}_+^m$ indicates that P owns x_i units of the i-*th* resource, for any $1 \leq i \leq m$.
- u is a function indicating *preferences* between *baskets* of resources. This function takes as parameters a set of actions (the actions that B may perform immediately) and a set of resources, and returns a real number. For example, $u(A, \bar{x}) < u(A, \bar{y})$ means that if $\mathtt{Imm}(B) = A$ then P would prefer to own the basket $\bar{y}$ to $\bar{x}$. So, processes will try to increase the value of u, given A, by exchanging resources with other processes.
- n is a function relating *resources* and speed of execution. This will be modeled by a function taking as parameters an action and a set of resources: $n(a, \bar{x}) = \infty$ means that a needs more than $\bar{x}$ to be performed; $n(a, \bar{x}) = r \in \mathbf{R}_+$ means that a takes r units of time to be performed if the process has the use of the resources $\bar{x}$. A necessary condition for P to perform a is $n(a, \bar{x}) < \infty$. Let us remark that in some situations there will be a strong relation between the functions u and n: Bigger utilities will be produced by bigger amounts of (or better) resources, and this will imply faster performance of actions. So, we may have relations like $n(a, \bar{x}) = \frac{K}{u(\{a\}, \bar{x})}$, for a given constant $K \in \mathbf{R}_+$. Also note that this is not always the situation. For instance, there can be actions that do not need any resources to be performed, actions that need only a strict subset of them, etc. That is why we have preferred to keep both functions.
- c is a function indicating the *consumed* resources after performing an action. In some situations, the execution of an action needs to consume some resources (for example, writing in a bounded buffer); in other situations, resources are created after performing actions (for example, a pop operation from a bounded stack). These situations will be modeled by a function that takes as inputs an action and a set of resources, and returns a set of resources. That is, $c(a, \bar{x}) = \bar{y}$ means that, after performing

a, the set of resources of the process varies from $\bar{x}$ to $\bar{y}$. For example, if $\bar{z} = c(a, \bar{x}) - \bar{x}$ then $z_i > 0$ indicates that, after performing a, the process has created z_i units of the *i-th* resource. A necessary condition for a process to perform an action a is $c(a, \bar{x}) \geq \bar{0}$. We do not allow *debts* because they could generate inconsistencies. For example, such debts could produce that two processes use simultaneously a printer.

Definition 5 Let us consider that there exists a number $m > 0$ of different resources. We say that the tuple $P = (B, \bar{x}, u, n, c)$ is a *process* if $B \in \mathcal{B}$ (the *basic process* defining the behavior of the process), $\bar{x} = (x_1, \ldots x_m) \in \mathbb{R}_+^m$ (the *amounts* of resources owned by P), $u : \mathcal{P}(\mathtt{ACT}) \times \mathbb{R}_+^m \longrightarrow \mathbb{R}$ (the *utility function*), $n : \mathtt{ACT} \times \mathbb{R}_+^m \longrightarrow \mathbb{R}_+ \cup \{\infty\}$ (the *necessity* function), and finally $c : \mathtt{ACT} \times \mathbb{R}_+^m \longrightarrow \mathbb{R}^m$ (*consumption* of resources function).

Given a process P_i, we will usually consider $P_i = (B_i, \overline{x_i}, u_i, n_i, c_i)$, that is, indices will denote the process to which $B, \bar{x}, \ldots$ are related. □

The utility and necessity functions must fulfill some simple *rational* properties. We suppose that utility does not decrease if resources increase, that is, for any $A \in \mathcal{P}(\mathtt{ACT})$, and any $1 \leq i \leq m$, we have $\frac{\Delta u(A, \bar{x})}{\Delta x_i} \geq 0$. Given the fact that utility functions will be always applied to the initial actions of a behavior B, we suppose $u(\emptyset, \bar{x}) = 0$. This property means that a *deadlocked* process does not need any resources, so they will be shared by the rest of the processes (because those resources do not add any utility to the owner). In microeconomic theory,ther e are other restrictions that are usually imposed (strict monotonicity, convexity, etc) but they are not needed in our current framework (see [11, 12] for more details). Regarding the necessity function, we will impose the condition that the function n is non-increasing with the number of resources, that is, for any $1 \leq i \leq m$ and $\mu \in \mathtt{ACT}$, we have $\frac{\Delta n(\mu, \bar{x})}{\Delta x_i} \leq 0$. Moreover, we will suppose that $n(a, \bar{x}) = n(\tau_a, \bar{x})$, that is, a visible action a needs the same resources that its internal counterpart τ_a. Finally, let us comment on a *trick* that we will intensively use in the examples of Section 3. There are many situations where resources cannot be split. For example, it makes no sense to consider that a process *owns* half of a printer. We will indicate that half of a printer is so *useful* as no printer at all by using utility functions like $u(A, (x_1, \ldots, x_i, \ldots, x_n)) = f(A, x_1, \ldots, \mathtt{trunc}(x_i), \ldots, x_n))$.

We will extend the operational transitions for behaviors (as defined in Figure 1) to processes. The idea is that a process may perform a transition $\xrightarrow{\mu}$ if its corresponding behavior so does, and the process has enough resources. Formally, if $P = (B, \bar{x}, u, n, c)$ then we have the following rule:

$$(\text{PRO})\frac{B \xrightarrow{\mu} B' \ \wedge \ n(\mu, \bar{x}) < \infty \ \wedge \ c(\mu, \bar{x}) \geq \bar{0}}{P \xrightarrow{\mu} P'}$$

where $P' = (B', c(\mu, \bar{x}), u, n, c)$. Let us note that we have overloaded the symbol $\longrightarrow$: It is used both to denote transitions of behaviors and transitions of processes.

A *system* will be the parallel composition of a (finite, non-empty) collection of processes. Processes will communicate in two ways: Either by synchronizing in the execution of actions or by exchanging resources. Given the nature of the systems that we would like to describe, we have decided to use a relatively complex parallel operator. Specifically, our parallel operator is a simplification of [7] where m among n cooperation is not considered (similar proposals appear in [3, 6]). The idea is that synchronizations are not restricted to be binary. The parallel operator will have a tuple of subsets of $\mathtt{Act}$ as parameter: $(A_1, \ldots, A_n)$. So, if we have the parallel composition of $P_1, \ldots, P_n$, a process P_i will perform the actions in $\mathtt{ACT} - A_i$ asynchronously; if $a \in A_i$ then any P_j such that $a \in A_j$ must synchronize with P_i in order to perform a. Similarly, exchanges of resources may involve more than two processes.

Definition 6 Le t $A_1, \ldots A_n \subseteq \mathtt{Act}$. A *system* S consists in the parallel composition of n processes $P_1, \ldots P_n$ synchronizing, respectively, in the set A_i. We denote the system S by $\|_n^{A_i} P_i$. We denote the set of systems by $\mathcal{S}$. □

From now on we will assume that systems are initially *compatible* with the available resources. That is, if we want to specify a system having the use of a vector of resources $\bar{r}$ by $S = \|_n^{A_i} P_i$ then the sum of the total quantity of resources initially assigned to $P_1, \ldots, P_n$ must be equal to $\bar{r}$, that is, $\sum_i \overline{x_i} = \bar{r}$. In the following we introduce the operational rules for systems. The first rule uses two predicates that will be defined in forthcoming definitions.

Definition 7(*The Exchange Rule*). Let $S = \|_n^{A_i} P_i$ be a system, where for any $1 \leq i \leq n$ we have $P_i = (B_i, \overline{x_i}, u_i, n_i, c_i)$. The operational transitions denoting exchange of resources that S may perform are given by the rule:

$$(\text{Par1})\frac{\mathtt{valid}(S,\mathcal{E}) \wedge \mathtt{allowed}(S,\mathcal{E})}{S \overset{\mathcal{E}}{\rightsquigarrow} \|_n^{A_i} P_i' \left[\forall\, 1 \leq i \leq n:\; P_i' = (B_i, \overline{x_i} - \sum_j \mathcal{E}_{ij} + \sum_j \mathcal{E}_{ji}, u_i, n_i, c_i)\right]}$$

where $\mathcal{E} \in (\mathbf{R}_+^m)^{n*n}$. We denote by $\rightsquigarrow^*$ the reflexive and transitive closure of $\rightsquigarrow$. We say that S is a *local equilibrium*, denoted by $S \not\rightsquigarrow$, if there do not exist S' and $\mathcal{E}$ such that $S \overset{\mathcal{E}}{\rightsquigarrow} S'$. □

Intuitively, a transition $\|_n^{A_i} P_i \overset{\mathcal{E}}{\rightsquigarrow} \|_n^{A_i} P_i'$ indicates that P_i gives to P_j the quantities of resources indicated by $\mathcal{E}_{ij}$ and receives from it the quantities given by $\mathcal{E}_{ji}$. The total amount of resources that P_i owns is equal to its original resources minus the resources that it gives plus the resources that it receives. As we will see later, processes will not be allowed to perform actions until the system reaches a local equilibrium. The idea is that an equilibrium represents a *good* distribution of resources (because no more *useful* exchanges can be made) and so processes must be *delayed* until a local equilibrium is reached. If we allow a process P_i to perform transitions before the system is an equilibrium (for example, as soon as $n_i(a, \overline{x_i}) < \infty$ and $c(a, \bar{x}) \geq \bar{0}$ hold) then the system will not be working at its *best* possible performance(s).

Now we will define the predicates $\texttt{valid}(S, \mathcal{E})$ and $\texttt{allowed}(S, \mathcal{E})$. The first predicate holds if the processes do not give more resources that the ones that they initially own, and the diagonal of the matrix is filled with $\bar{0}$.

Definition 8 Let $S = \|_n^{A_i} P_i$ be a system, where for any $1 \leq i \leq n$ we have $P_i = (B_i, \overline{x_i}, u_i, n_i, c_i)$. We say that $\mathcal{E} \in (\mathbf{R}_+^m)^{n*n}$ is a *valid exchange matrix for S*, denoted by $\texttt{valid}(S, \mathcal{E})$,if for any $1 \leq i \leq n$ we have $\sum_j \mathcal{E}_{ij} \leq \overline{x_i}$ and $\mathcal{E}_{ii} = \bar{0}$. □

Regarding the predicate $\texttt{allowed}(S, \mathcal{E})$ we have several possibilities. This choice would depend on the kind of policy[1] that we want to implement. As we said in the introduction we consider two possible policies. The first policy, the *preserving utility* policy, will allow exchange of resources only if, after the exchange, at least one process improves and no process gets worse. Intuitively, processes are the owners of the resources and they will not give up them if they do not receive a *compensation.* The second policy, the *maximizing utility* policy, will allow exchanges if the overall situation of the system improves. In order to measure the improvement of the system, we consider the sum of the utilities of all the processes.[2] As we commented in the introduction of this paper, such a policy could be interpreted by thinking that processes are not the *owners* of resources because they can be *expropriated* without any compensation. Nevertheless, this policy is more efficient than a totally centralized system of resources delivery because it avoids the return and redelivery of the same resources to a process. In economic terms, the first of these policies can be seen as a kind of *market economy* while the second one could be interpreted as a kind of *planned economy.*

Definition 9 Let $S = \|_n^{A_i} P_i$ be a system, where for any $1 \leq i \leq n$ we have $P_i = (B_i, \overline{x_i}, u_i, n_i, c_i)$,and let $\mathcal{E}$ be a matrix such that $\texttt{valid}(S, \mathcal{E})$.

The $\texttt{allowed}(S, \mathcal{E})$ predicate under a *Preserving Utility Policy* is defined as $\forall\, 1 \leq i \leq n$ we have $u_i(\texttt{Imm}(B_i), \overline{x_i}) \leq u_i(\texttt{Imm}(B_i), \overline{x_i} - \sum_j \mathcal{E}_{ij} + \sum_j \mathcal{E}_{ji})$ and $\exists\, 1 \leq k \leq n:\ u_k(\texttt{Imm}(B_k), \overline{x_k}) < u_k(\texttt{Imm}(B_k), \overline{x_k} - \sum_j \mathcal{E}_{kj} + \sum_j \mathcal{E}_{jk})$.

The $\texttt{allowed}(S, \mathcal{E})$ predicate under a *Maximizing Utility Policy* is defined as $\sum_i u_i(\texttt{Imm}(B_i), \overline{x_i}) < \sum_i u_i(\texttt{Imm}(B_i), \overline{x_i} - \sum_j \mathcal{E}_{ij} + \sum_j \mathcal{E}_{ji})$. □

[1]The choice of a *good* policy is not a trivial task. Actually, it is impossible to choose a *perfect* policy. This problem is related with the social welfare aggregator problem. Arrow's *impossibility theorem* shows that there does not exist such an aggregator fulfilling a certain set of *desirable* properties (see [1] for more details).
[2]Using microeconomics terminology, this is a *Benthan* social welfare aggregator.

From now on, if the results depend on the chosen policy, we will indicate under which one we are working. Once we have finished the definition of rule (Par1), let us remark that exchange of resources produces a lot of nondeterminism. For example, consider the processes P_{j_1} and P_{j_2} presented in the introduction. For any $a > 0$, and $2 \le x \le 4$, exchanges where P_{j_1} receives a units of the resource i_1 from P_{j_2} and P_{j_1} gives $x \cdot a$ units of the resource i_2 from P_{j_2} will be allowed.

Definition 10 (*The Synchronization Rules*). Let $S = \|_n^{A_i} P_i$ be a system. The operational transitions denoting the actions that S may perform are given by the rules:

$$\text{(Par2)}\frac{S\not\rightsquigarrow \wedge P_j \xrightarrow{\mu} P'_j \wedge \mu \notin A_j}{S \xrightarrow{\mu} \|_n^{A_i} P''_i} \left[\forall\, 1\le i\le n\colon P''_i = \begin{cases} P'_j & i = j \\ P_i & i \ne j \end{cases}\right]$$

$$\text{(Par3)}\frac{S\not\rightsquigarrow \wedge P_j \xrightarrow{a} P'_j \wedge a\in A_j \wedge \forall\, 1\le k\le n\colon (a\in A_k) \Longrightarrow (P_k \xrightarrow{a} P'_k)}{S \xrightarrow{a} \|_n^{A_i} P''_i} \left[\forall\, 1\le i\le n\colon P''_i = \begin{cases} P'_i & \text{if } a \in A_i \\ P_i & \text{if } a \notin A_i \end{cases}\right]$$

□

Let us note that we have overloaded the symbol denoting transitions. The condition $S \not\rightsquigarrow$ indicates that the system has reached a local equilibrium. If S is not a local equilibrium, more useful exchanges can be made (i.e. the system may improve) and so actions should not be performed. Rule (Par2) is applied for interleaving actions. In this case, the only process that changes is the one involved in the transition. Rule (Par3) says that if a process P_j may perform an action a belonging to its synchronization set A_j, then all the processes having a in their synchronization sets must also perform a; all these processes will perform this action synchronously.

Finally,let us remark that our process algebra is a conservative extension of LOTOS: It is enough to consider that the functions u and n are always equal to 0 and that $c(\mu, \bar{x}) = \bar{x}$, for any $\mu \in \texttt{ACT}$.

2.1. Some Properties of the Language

In this section we study some properties that local equilibria fulfill according to the functions defining the involved processes. These properties will be used in the next section of this paper when analyzing the defined specifications. The first result says that any system may evolve into a local equilibrium.

Lemma 1F or any system S there exists S' such that $S \rightsquigarrow^* S' \not\rightsquigarrow$. □

The proof of this result is immediate. Let us remark that if S is already a local equilibrium then $S' = S$ (that is, there is a $\rightsquigarrow^*$ derivation with length equal to zero). The following result states that once a local equilibrium has been reached, no process will keep resources that do not add utility to it if there exists another process that uses the resource.

Lemma 2 Let $S = \|_n^{A_i} P_i$. Let us suppose that there exist two processes P_i, P_j, $\epsilon > 0$,and $1 \leq r \leq m$ such that for any $\overline{x}$ we have

- $u_i(\mathtt{Imm}(B_i), \overline{x}) = u_i(\mathtt{Imm}(B_i), (x_1, \ldots, x_r - \epsilon, \ldots, x_n))$,and
- $u_j(\mathtt{Imm}(B_j), \overline{x}) < u_j(\mathtt{Imm}(B_j), (x_1, \ldots, x_r + \epsilon, \ldots, x_n))$.

Under the previous conditions, for any S' such that $S \rightsquigarrow^* S' \not\rightsquigarrow$ we have that $x'_{ir} \leq x_{ir} - \epsilon$, where x_{ir} denotes the amount of the resource r owned by P_i in S (similar for x'_{ir} and S').

Proof Sketch: By contradiction. Suppose $x'_{ir} > x_{ir} - \epsilon$. We have that, under both policies, there exists an exchange where P_i gives part of the resource r to a process P_k (there is at least a process increasing its utility by increasing the amount of r). So, S' is not a local equilibrium. □

An immediate corollary of the previous result is that if the condition holds for any ϵ such that $x_{ir} \geq \epsilon > 0$ then $x'_{ir} = 0$. Note that in the previous result is essential to suppose that another process increase its utility by increasing its amount of the resource r. The next result states that under a maximizing utility policy, all the local equilibria that can be reached from a system S have the same *global* utility.

Definition 11 Let $S = \|_n^{A_i} P_i$ be a system, where for any $1 \leq i \leq n$ we have $P_i = (B_i, \overline{x_i}, u_i, n_i, c_i)$. The *total* utility of S, denoted by $\mathtt{total}(S)$, is defined as $\sum_i u_i(\mathtt{Imm}(B_i), \overline{x_i})$. □

Theorem 1 Let $S = \|_n^{A_i} P_i$ be a system, where for any $1 \leq i \leq n$ we have $P_i = (B_i, \overline{x_i}, u_i, n_i, c_i)$. Let S_1, S_2 be systems such that $S \rightsquigarrow^* S_1 \not\rightsquigarrow$, and $S \rightsquigarrow^* S_2 \not\rightsquigarrow$. If the *maximizing utility* policy is being used then we have $\mathtt{total}(S_1) = \mathtt{total}(S_2)$.

Proof: Let $S_1 = \|_n^{A_i} P_{1i}$ and $S_2 = \|_n^{A_i} P_{2i}$, where for any $1 \leq i \leq n$ we consider $P_{1i} = (B_i, \overline{x_{1i}}, u_i, n_i, c_i)$ and $P_{2i} = (B_i, \overline{x_{2i}}, u_i, n_i, c_i)$. Suppose that $\mathtt{total}(S_1) > \mathtt{total}(S_2)$. Let $\mathcal{E}$ be an exchange matrix such that for any $1 \leq i \leq n$ we have $\sum_j \mathcal{E}_{ji} - \mathcal{E}_{ij} = x_{1i} - x_{2i}$ and $\mathtt{valid}(S_2, \mathcal{E})$. Given the fact that $\sum_i x_{1i} = \sum_i x_{2i}$, such a matrix fulfilling $\mathtt{allowed}(S_2, \mathcal{E})$ always exists. Therefore, there exists S' such that $S_2 \overset{\mathcal{E}}{\rightsquigarrow} S'$, which represents a contradiction. □

Note that different local equilibria may have different assignment of resources among processes; the previous result only assures that the sum of the utilities is the same, not that the composition of the baskets are equal. Note that this result does not hold under a preserving utility policy: Different equilibria do not have necessarily the same total utility.

3. EXAMPLES

In this section we will show how **PAMR** can be used to specify and analyze concurrent systems where resources play an important role. We will present two classical examples: The dining philosophers and consumers/producers. Besides, we will study the absence of deadlock in these systems.

In the following we assume that an undefined value of a function is set to an arbitrary value. Actually, these cases will not be possible because they will correspond with a set of actions (resp. with an action) not reachable by the corresponding processes.

3.1. The Dining Philosophers

In this classical problem, five (male) philosophers stay by a dining table, which contains five forks. We will consider that these are the *resources* of the system: $fork_1, \ldots, fork_5$. These philosophers have only two tasks: To think and to eat. When a philosopher wants to eat, he must take both forks staying besides his dish (we suppose that philosopher i must take forks i and $(i \bmod 5) + 1$). The behavior of the philosophers can be described as:

$$Philosopher_i := think_i \,;\, eat_i \,;\, Philosopher_i$$

We suppose that the initial distribution of forks gives to philosopher i the i-th fork, that is, $\overline{x_i} = (\delta_{i1}, \ldots, \delta_{i5})$ where $\delta_{ij} = 1$ if $i = j$,and $\delta_{ij} = 0$ otherwise. Utility functions are defined as:

$$u_i(\{think_i\}, \bar{x}) = 0 \qquad u_i(\{eat_i\}, \bar{x}) = \mathtt{trunc}(fork_i) \cdot \mathtt{trunc}(fork_{(i \bmod 5)+1})$$

That is, if a philosopher wants to eat, he gets utility only if he owns both forks; otherwise, it will be the same to have one or none. Besides, a philosopher gets no utility by holding *useless* forks. We set $u_i(\{think_i\}, \bar{x})$ to an arbitrary value because no resources are needed to think. The necessity functions are defined as follows:

$$n_i(think_i, \bar{x}) = K_i \qquad n_i(eat_i, \bar{x}) = \frac{E_i}{u_i(\{eat_i\}, \bar{x})}$$

In this case, the time that philosophers spend thinking may be different, but does not depend on the resources. Besides, we also assume that they eat at different speeds. Finally, no resources are consumed by performing actions, so the consumption function is defined as $c_i(a, \bar{x}) = \bar{x}$. In conclusion, the system can be defined as

$$Dining\_Philosophers = \|_5^{A_i} P_i$$

where for any $1 \leq i \leq 5$, $A_i = \emptyset$ and $P_i = (Philosopher_i, \overline{x_i}, u_i, n_i, c_i)$.

The following result shows that this system cannot get deadlocked. As in the next example, the definition of the utility functions plays an important role in the absence of deadlocks.

Lemma 3 (*Absence of Deadlocks for Dining_Philosophers*). Consider a system S such that

$$Dining\_Philosophers \rightsquigarrow^* S_1 \xrightarrow{a_1} S_1' \rightsquigarrow^* S_2 \cdots \xrightarrow{a_n} S_n' \rightsquigarrow^* S$$

If S is not a local equilibrium then there exist $S', \mathcal{E}$ such that $S \overset{\mathcal{E}}{\rightsquigarrow} S'$; otherwise, there exist S', a such that $S \xrightarrow{a} S'$.

Proof: The first case is trivial from Lemma 1. If a philosopher is willing to think, this action may be performed (no resources are needed). Suppose that all the philosophers desire to eat, and none of them can. This implies that all of them have zero utility. Regardless of the chosen policy, S is not a local equilibrium, because any exchange where some philosopher takes his two forks is allowed (note that fractions of forks will not be exchanged because they do not increase utility). □

3.2. The Bounded-Buffer Producers/Consumers Problem

Consider n producers and m consumers. The former have access to a buffer where they can place their products; these products will be taken by the latter. The buffer is bounded: A consumer cannot get out a product if the buffer is empty and a producer cannot put in a product if the buffer is full. In order to avoid any damage in the structure, the buffer must be accessed preserving mutual exclusion. The behaviors of the involved processes are given by:

$$Producer := produce \; ; \; enqueue_1 \; ; \; enqueue_2 \; ; \; Producer$$

$$Consumer := dequeue_1 \; ; \; dequeue_2 \; ; \; consume \; ; \; Consumer$$

In order to visualize how mutual exclusion is implemented, the *enqueue* and *dequeue* operations are split into two different steps. The initial resources of the system are: a unit of *mutual exclusion,* zero *produced* units, and p free *places*. We assume that these *resources* are randomly distributed among the processes in such a way that $\sum_{i=1}^{n+m} \overline{x_i} = (1, 0, p)$. We will define a unique utility function for all consumers and producers:

$$u(\{enqueue_1\},\bar{x}) = \begin{cases} 1 & \text{if } \mathtt{trunc}(x_1)\cdot\mathtt{trunc}(x_3) \geq 1 \\ 0 & \text{otherwise} \end{cases}$$

$$u(\{enqueue_2\},\bar{x}) = \begin{cases} 2 & \text{if } x_1 = 1 \\ 0 & \text{otherwise} \end{cases}$$

$$u(\{dequeue_1\},\bar{x}) = \begin{cases} 1 & \text{if } \mathtt{trunc}(x_1)\cdot\mathtt{trunc}(x_2) \geq 1 \\ 0 & \text{otherwise} \end{cases}$$

$$u(\{dequeue_2\},\bar{x}) = \begin{cases} 2 & \text{if } x_1 = 1 \\ 0 & \text{otherwise} \end{cases}$$

$$u(\{produce\},\bar{x}) = K_1 \qquad u(\{consume\},\bar{x}) = K_2$$

Let us note that the order structure of the buffer is not fully represented: The distinct products or distinct free places are not distinguished. Regarding the necessity function, it can be defined from the utility function as:

$$n(a,\bar{x}) = \frac{K_3}{u(\{a\},\bar{x})}$$

Let us remark that if a process does not own the necessary resources, its utility function will return 0, and so, the necessity function will be equal to infinite (let us remember that we consider $\frac{r}{0} = \infty$). In this example, the last two resources are created and consumed, so we have to define the value of the consumption function in the appropriate way:

$$\begin{array}{lcllcl} c(enqueue_1,\bar{x}) & = & (x_1,x_2,x_3-1) & c(enqueue_2,\bar{x}) & = & (x_1,x_2+1,x_3) \\ c(dequeue_1,\bar{x}) & = & (x_1,x_2-1,x_3) & c(dequeue_2,\bar{x}) & = & (x_1,x_2,x_3+1) \\ c(produce,\bar{x}) & = & \bar{x} & c(consume,\bar{x}) & = & \bar{x} \end{array}$$

Finally, the system may be specified as:

$$Bounded\_Buffer = \|_{n+m}^{A_i} P_i$$

where, for any $1 \leq i \leq n$ we have $P_i = (Producer, \overline{x_i}, u, n, c)$, and for any $n+1 \leq j \leq n+m$ we have $P_j = (Consumer, \overline{x_j}, u, n, c)$. Besides, for any $1 \leq k \leq n+m$ we have $A_k = \emptyset$.

One important property that we must have is *mutual exclusion*. The next result states that accesses to the buffer are done by preserving this property: Once a process performs an action belonging to $\{enqueue_1, dequeue_1\}$, it is assured that this process will perform the corresponding *second part* of the performed action ($enqueue_2$ and $dequeue_2$ respectively) before any other process performs an action belonging to $\{enqueue_1, dequeue_1\}$.

Lemma 4 Le t S be a system such that

$$Bounded\_Buffer \leadsto^* S_1 \xrightarrow{a_1} S_1' \leadsto^* S_2 \xrightarrow{a_2} \cdots \leadsto^* S \not\to$$

Let us suppose $S \xrightarrow{a} S' \rightsquigarrow^* S'' \not\rightarrow$, where $a \in \{enqueue_1, dequeue_1\}$ and a has been performed by P_j (that is, $P_j \xrightarrow{a} P_j'$). For any transition $S'' \xrightarrow{b} S'''$ such that $b \in \{enqueue_1, dequeue_1, enqueue_2, dequeue_2\}$ we have that if $a = enqueue_1$ then $b = enqueue_2$ (resp. if $a = dequeue_1$ then $b = dequeue_2$) and this transition has been performed by the evolution of P_j' from S' to S''.

Proof Sketch: Let us consider the producers case (for consumers is similar). Under the *maximizing utility policy*, when a producer performs its $enqueue_1$ action, for any reachable local equilibrium, the unit of mutual exclusion resource is given again to that producer because no other process would have more utility than this one by owning it (note that this producer will have utility 2). Under the *preserving utility policy*, when a producer performs an $enqueue_1$ action, the mutual exclusion resource cannot be taken by another process because the utility of the owner would decrease. □

Lemma 5 (*Absence of Deadlocks for Bounded_Buffer*). Let us consider a system S such that $Bounded\_Buffer \rightsquigarrow^* S_1 \xrightarrow{a_1} S_1' \rightsquigarrow^* S_2 \cdots \xrightarrow{a_n} S_n' \rightsquigarrow^* S$. If S is not a local equilibrium then there exist $S', \mathcal{E}$ such that $S \overset{\mathcal{E}}{\rightsquigarrow} S'$; otherwise, there exist S', a such that $S \xrightarrow{a} S'$.

Proof Sketch: If there are no products then any producer will be able to produce. If there are no free places, any consumer will be able to consume. In both policies, the mutual exclusion resource will be taken (and owned) by a process only if it makes that process to perform an action. Therefore, at least one process will get the resources it needs. □

4. CONCLUSIONS AND FUTURE WORK

In this paper we have presented a formalism to specify systems where resources can be exchanged among subsystems. These exchanges will improve the performance of the system. We have studied some (theoretical) properties of the language. Finally, we have specified and studied two examples where the characteristics of our language have been shown.

The referees of this paper have suggested new interesting directions of research. For example, necessity functions play no relevant role in the current theory. We would like to define a real-time extension of our language where these functions play a fundamental role: They will induce delays in the behavior of processes. These delays can be either deterministic or can be defined by means of a random variable. In the latter case, we will consider some of the current models of stochastic process algebras (e.g. [2, 4, 8]). Besides, we will consider that the exchange of resources takes time. So, sequences of $\rightsquigarrow$ transitions should be grouped. Another interesting point is that utility functions (as well as necessity and consumption functions) are *static*, that is, they do not change along the performance of the system. Once time is taken into account, it is straightforward to extend the previous framework to consider *dynamic* functions. Finally, the relation between our framework and management of software projects should be explored. Indeed, the tasks of a project can be seen as processes while the members of the project can be seen as the resources *used* by the tasks.

Acknowledgments. We would like to thank the referees of this paper for the careful reading and the very interesting suggestions to improve the current framework. We would also like to thank the referees of [12] for pointing out some interesting directions which have been reflected in this paper. Finally, we would like to thank David de Frutos, Natalia López, Javier Rodríguez, and Fernando Rubio for helpful discussions on the topic of this paper.

REFERENCES

[1] K.J. Arrow. *Social Choice and Individual Values.* Wiley,2nd edition, 1963.

[2] M. Bernardo and R. Gorrieri. A tutorial on EMPA: A theory of concurrent processes with nondeterminism, priorities, probabilities and time. *Theoretical Computer Science*, 202:1–54, 1998.

[3] T. Bolognesi. A graphical composition theorem for networks of LOTOS processes. In *10th International Conference on Distributed Computing Systems*, pages 88–95. IEEE-CS Press, 1990.

[4] M. Bravetti and R. Gorrieri. The theory of interactive generalized semi-markov processes. To appear in Theoretical Computer Science, 2001.

[5] P. Brémond-Grégoire and I. Lee. A process algebra of communicating shared resources with dense time and priorities. *Theoretical Computer Science*,189(1-2):179–219, 1997.

[6] J. Davies and S. Schneider. A brief history of timed CSP. *Theoretical Computer Science*, 138:243–271, 1995.

[7] H. Garavel and M. Sighireanu. A graphical parallel composition operator for process algebras. In *Formal Description Techniques for Distributed Systems and Communication Protocols (XII), and Protocol Specification, Testing, and Verification (XIX)*, pages 185–202. Kluwer Academic Publishers, 1999.

[8] H. Hermanns, U. Herzog, and J.-P.Katoe n. Process algebra for performance evaluation. To appear in Theoretical Computer Science, 2001.

[9] ISO/IEC. ODP Trading Function. Draft International Standard 13235, ISO - Information Processing Systems, 1995.

[10] N.A. Lynch and M.R. Tuttle. Hierarchical correctness proofs for distributed algorithms. In *6th ACM Symp. on Principles of Distributed Computing*, pages 137–151, 1987.

[11] A. Mas-Colell, M.D. Whinston, and J.R. Green. *Microeconomic Theory.* Oxford University Press, 1995.

[12] M. Núñez. Including microeconomic theory into FDTs: A first approach. Available at: `http://dalila.sip.ucm.es/~manolo/papers/exchange.ps.gz`, 2000.

[13] M. Núñez and I. Rodríguez. PAMR: A process algebra for the management of resources in concurrent system. Available at: `http://dalila.sip.ucm.es/~manolo/papers/pamr.ps.gz`, 2001.

A SYMBOLIC SEMANTICS AND BISIMULATION FOR FULL LOTOS

Muffy Calder
Department of Computing Science,
University of Glasgow, Glasgow G12 8QQ
muffy@dcs.gla.ac.uk

Carron Shankland
Department of Computing Science and Mathematics,
University of Stirling, Stirling FK9 4LA
carron@cs.stir.ac.uk

Abstract
A *symbolic* semantics for Full LOTOS in terms of *symbolic transition systems* is defined; the semantics extends the (infinitely branching) standard semantics by giving meaning to data parameterised behaviours, and provides a finitely branching representation for behaviours. Symbolic bisimulation is defined.
This extends our previous work [14], making the definitions more amenable to automated reasoning and processes with recursion.

Keywords: LOTOS, symbolic transition systems, symbolic bisimulation.

1. INTRODUCTION

LOTOS [11] is a message passing process algebra which combines two orthogonal languages: a process language, known as Basic LOTOS, with features from both CSP and CCS, and the equational abstract data type language ACT ONE. LOTOS is an ISO standard [11] whose given semantics is in terms of structured labelled transition systems. In this semantics (referred to here as the *standard semantics*), each data variable in a process is instantiated by every possible value of its corresponding type, resulting in infinite transition systems (both in breadth and in depth).

This approach has several drawbacks. First, it is impossible to use standard (finite state) model-checking techniques over infinite transition systems; the usual solution is to restrict the underlying datatypes. Indeed, any kind of automated reasoning becomes difficult, if not impossible. Second, because the data values are embedded in the transitions, any uniformities in the actions of the processes are lost. For example, the process description may make it clear that a particular action happens when the value of some variable lies between 3 and 42 (say), but that information is much harder to extract from the labelled transition system directly, especially if there are an infinite number of branches. Finally, as a consequence of this approach it is not possible to reason about partial, or data parameterised, behaviour expressions. Our experiences with LOTOS applications (e.g. [15, 16]) indicate that this is highly desirable.

The advantage of the standard approach is that it easily accommodates multi-way synchronisation, i.e. associative synchronisation between two or more processes. Multi-way synchronisation has led to a particular constraint-oriented style of specification in the LOTOS community which is particularly useful when building a system by layering behaviour on a simple building block.

The problem we address here and in [14] is how to reason over potentially infinite LOTOS processes while retaining multi-way synchronisation and without restricting the datatypes. Since the addition of data to the language is the reason for the problem, some sort of separation of the concerns of data and processes seems appropriate. There are three kinds of solution to this problem. The first is to get rid of data altogether in a brute force manner [1] which changes the behaviour of the process. The second is to construct a process representation of the data type [2, 9]. This approach only converts data operations into process operations; the data values are still present therefore the process remains infinite branching. The third solution is to adopt a symbolic approach such as the symbolic semantics for message passing CCS in [10]. We drew on this approach for our earlier work [14] in defining a symbolic semantics and bisimulation for LOTOS. That definition did not deal effectively with recursive processes, and could not be easily implemented (since it required an infinite stock of new names). These problems are addressed here and the definitions considerably revised. The symbolic approach is also taken by Eertink [8]. While Eertink's semantics achieves a separation of the concerns of data and process, without losing information, it is rather operational, concentrating on using the semantics in a simulation tool. There are no equivalences or preorder relations associated with the semantics.

Our motivation is to define an entire symbolic framework for reasoning about LOTOS specifications including a symbolic semantics, appropriate logics, relations and tools. It is important that this new framework preserves the distinguishing features of LOTOS (presented in Section 2). In this paper we present the foundation of the framework: symbolic transition systems (STSs) and the axioms and rules for generating an STS from a (possibly open) LOTOS behaviour expression (Section 3). In order to define relations over STSs we must first define the notion of substitution over STSs (Section 4). We follow this with the definition of symbolic bisimulation (Section 5). Symbolic bisimulation should not lead to processes being distinguished which are not in the standard semantics; similarly processes which are distinguished under the standard semantics are not identified in the symbolic semantics. We have proven this to hold for unparameterised processes, but the proof is the subject of a different paper. Broadly, we follow the approach taken in [10] for symbolic transition graphs and message passing CCS, but our approach differs in several significant ways to accommodate the particular features of LOTOS. Finally, we draw our conclusions and discuss future directions.

2. DISTINGUISHING FEATURES OF LOTOS

We assume the reader is familiar with the standard LOTOS syntax and semantics [11]. An accessible tutorial to LOTOS is [12]. Here we present three related features distinguishing LOTOS from most of the standard process algebras, particularly message passing CCS: *multi-way (broadcast) synchronisation*, *value negotiation*, and *selection predicates*. These make it non-trivial to directly apply the notion of symbolic transition and bisimulation of [10].

Multi-way synchronisation means that when two actions synchronise, with possibly some data exchange taking place, the resulting action may be involved in further synchronisation. As a result of multi-way synchronisation, it makes less sense in LOTOS to refer to"input" events and "output" events. In LOTOS an event *offers* a single value or a set of values drawn from a particular sort; these two cases are distinguished by the use of ! or ? respectively.

Multiway synchronisation is achieved in the underlying transition system by *encoding* data into transitions in both *!* and *?* events. This can be seen clearly by referring to the rules from the standard [11] for generating a transition system from action prefix events. The rule for *!* events is straightforward:

$$g!E;\ P \xrightarrow{gv} P$$

where E is a data expression (with no variables in the case of the standard LOTOS semantics) and $v = [E]$ (i.e. the equivalence class of E). Perhaps less obvious is the rule for ? events:

$$g?x\!:\!S;\ P \xrightarrow{gv} P[v/x]$$

where v is a ground term of sort S, and $P[v/x]$ denotes the substitution of v for x in P. This rule gives us an axiom *schema* for each $?x$ event. For example, see Fig. 1; each possible value of Nat results in a transition.

Thus, ? event offers correspond to a (possibly infinite) choice over all values of the data type. The binding of x is defined at this point, i.e. the semantics is *early*. A *late* semantics (where binding of variables to values is delayed as long as possible) has no counterpart in the (standard) concrete semantics. This is in constrast to value-passing CCS where both kinds of semantics are possible.

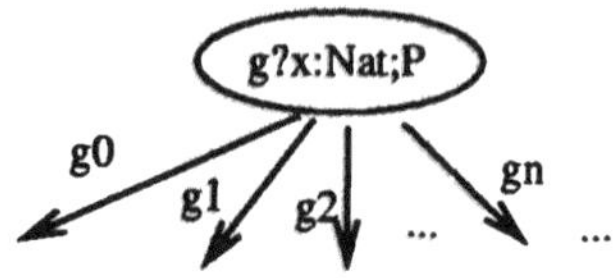

Figure 1. Standard semantics of *g?x:Nat* event offer

Encoding of values in the transitions affects the rules for synchronised parallelism. Consider a rule for CCS style two-way synchronisation. We use LOTOS syntax here for comparison:

$$\frac{g!E;\ P_1 \xrightarrow{gv} P_1 \qquad g?x:S;\ P_2 \xrightarrow{gx} P_2}{g!E;\ P_1|[g]|\,g?x:S;\ P_2 \xrightarrow{\mathbf{i}} P_1|[g]|\,P_2[v/x]}$$

There is a single transition associated with the ? offer (labelled with x) and a clear indication that value passing is occurring: x gets bound to the value v. However, this approach is clearly limited to two-way synchronisation because the transition label becomes the unobservable **i** action which may not synchronise with any other action.

Contrast this with the LOTOS approach to synchronisation:

$$\frac{g!E;\ P_1 \xrightarrow{gv} P_1 \qquad g?x:S;\ P_2 \xrightarrow{gv} P_2}{g!E;\ P_1|[g]|\,g?x:S;\ P_2 \xrightarrow{gv} P_1\,|[g]|\,P_2}$$

Here, one of the transitions generated by the axiom schema for ? offers is chosen to match the transition generated by the ! offer (where $v \equiv [E]$). That is, there may be lots of other potential transitions from the state $g?x : S;\ P_2$ labelled with g and some value, but only one which is labelled with the exact value v. In fact, this is only one case of the rule for parallelism. In particular, the processes in the premise may be a further parallel combination, yielding multi-way synchronisation. Different combinations of offers are also permitted, i.e. ! and ? offers can synchronise in any combination. The most unusual case is *value negotiation*, which arises when $g?x : S$ and $g?y : S$ synchronise, the result being that for every possible value, x and y are bound to the same value thereafter. Both values must have the same type.

Selection predicates may restrict data further. For example, transitions where $x \geq 42$ are denied by $g?x : S[x < 42]$. If the selection predicate evaluates to *false*, the event itself is prevented from occurring. In the case of value negotiation if both ? offers are qualified by selection predicates, then the value must satisfy both predicates. An important distinction from other process algebras is that selection predicates can refer to data in the current event (typically the guards found in other process algebras refer to data from previous events).

So, to preserve multi-way synchronisation and selection predicates in LOTOS, we cannot simply employ the CCS approach to ? offers. We therefore define a new semantics based on *symbolic* transition systems.

3. SYMBOLIC TRANSITION SYSTEMS FOR LOTOS

3.1 Preliminaries

We assume a countable set of *variables*, *Var*, ranged over by *x*, *y*, etc., and a (possibly infinite) set of *values*, *Val*, ranged over by v. We also assume a set of *data expressions*, *Exp*, which includes *Var* and *Val* and is ranged over by E, and a set of *Boolean expressions*, *BoolExp*, ranged over by *b*, including the constants tt (true) and ff (false). We also assume that we have a set of gates, G, ranged over by g. The set of events, denoted Act, ranged over by α, comprises *SimpleEv* and *StructEv*. The set of simple events, *SimpleEv*, ranged over by a, is defined as $G \cup \{\mathbf{i}, \delta\}$. (Recall that in LOTOS **i** is the internal event and δ is the exit event.) The set of structured events, *StructEv*, contains all gate-expression combinations gE, as well as all combinations δE.

We assume the existence of the *flattening* function of the standard semantics [11]. The flattening function ensures that the given specification adheres to the LOTOS syntax, but also removes all hierarchical structure, ensures uniqueness of variable names, and that all names and types used are previously defined. The resulting object is called a *canonical LOTOS specification.*

We follow several naming conventions of the standard, including the use of a variable name to stand for a more complex structure including information not just about the name of a variable, but also its type and scope.

Variables and Substitutions. Variables and substitutions are over *data*, and *typed*, although we do not make this explicit, as noted above.

We assume a set **new-var** of fresh variable names. Strictly speaking, any reference to this set requires a context, i.e. the variable names occuring so far. For simplicity, we will assume that this context can be inferred, as required.

A *substitution* is a partial function from *Var* to *Var* ∪ *Val*, written as $[z/x]$ where z is substituted for x. A substitution is denoted σ, and the composition of two substitutions σ_1 and σ_2 is denoted $\sigma_1\sigma_2$, where σ_2 has precedence.

Structured Events. *Multiple* data offers at a gate, e.g. $g!x!y?n : \mathrm{Nat};\ P$, are allowed by LOTOS syntax. For simplicity in the following we will assume that only one event offer can occur at a gate. The obvious generalisation to lists of event offers can be easily made. The function **name()** : Act $\cup\ \{\delta, \mathbf{i}\} \rightarrow G \cup \{\delta, \mathbf{i}\}$ extracts the gate name from a structured event and is defined in [11].

Free and Bound Variables. The variables occurring in a data expression E are given by *vars*(E). A behaviour expression may contain *free* and *bound* data variables. The free variables of behaviour expressions, denoted *fv*(P), are defined in Definition 5 of the Appendix.

Informally, free variables arise through usage in an expression where the variable name has been previously bound in one of several ways: as a formal process parameter, by a ? event, a let clause, or an enable (>>) with *accept* clause. For example, in $g?x;\ g!x;\ exit$, all occurrences of x are bound, but in $g!x;\ exit$, x is free.

3.2 Symbolic Transition Systems

Following [10], *symbolic transition systems* (STS) are transition systems separating data from process behaviour by making the data symbolic. We define an STS to be a labelled transition system with variables, both in states and transitions, and conditions which determine the validity of a transition.

Definition 1 *(Symbolic Transition Systems)*
A symbolic transition system consists of:

- *A (nonempty) set of states.*
  Each state T is associated with a set of free variables, denoted $fv(T)$.
- *A distinguished initial state,* T_0.
- *A set of transitions* $T \xrightarrow{b\ \ \alpha} T'$ *such that* $fv(T') \subseteq fv(T) \cup fv(\alpha)$ *and* $fv(b) \subseteq fv(T) \cup fv(\alpha)$ *and* $\#(fv(\alpha) - fv(T)) \leq 1$.

Following convention, we shall often identify an STS with its initial state. Since one possible interpretation of states is to view them as labelled by behaviour expressions, the set of free variables of an STS T, $fv(T)$, can be defined as $fv(P)$, where P is the behaviour expression labelling T.

3.3 Intuition

We give a symbolic semantics for LOTOS by associating a symbolic transition system with each LOTOS behaviour expression P, written STS(P). Before giving the axioms and rules for the symbolic semantics, we give an example illustrating the concrete (Fig. 2) and symbolic (Fig. 3) semantics for the behaviour expression

$$g?x\!:\!\text{Nat}[x < 10];\ h?y\!:\!\text{Nat};\ h!x;\ stop$$

In the standard semantics, query offers are instantiated by explicit data offers. Therefore, in Fig. 2, the ? offers correspond to either many or an infinite number of transitions, each labelled by a concrete offer.

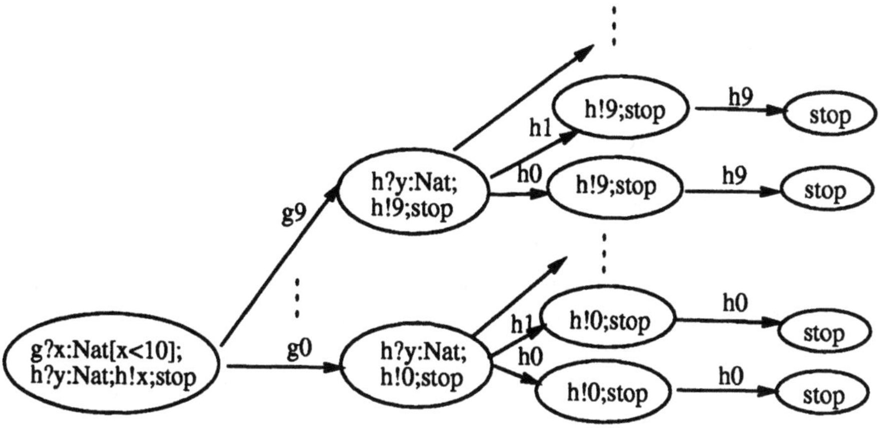

Figure 2. Standard "Concrete" Transition System

In the symbolic semantics, *open* behaviour expressions label states (e.g. $h!x;$ ***stop***), and transitions offer variables. The range of permissible values for the variables is determined by the Boolean conditions. Whereas the system in Fig. 2 has infinite branching, the system in Fig. 3 has only finite branching.

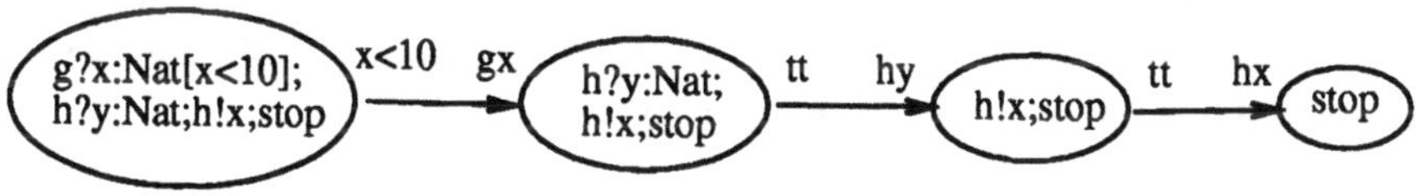

Figure 3. Symbolic Transition System

3.4 Key Features

Many of the rules given in the following section are very similar to those in the LOTOS standard; only a simple notational change to make the transition a symbolic one is required. We note here the main departures from the standard semantics, and differences from [10] and [14]).

- The syntactic distinction between the two kinds of data offer, i.e. between ? and !, has been lost. That is, both are represented by a transition labelled by a gate, an expression (possibly a simple variable) and a Boolean condition. Each offer is a *set* of values constrained in some way – expressed both by the form of the expression and by the condition of the transition. The type of offer can be determined by examining the free variables of the associated states, i.e. if $\mathit{fv}(E) \subseteq \mathit{fv}(T)$ then this is a free variable (in a ! offer, introduced previously), and if $\mathit{fv}(E) \not\subseteq \mathit{fv}(T)$ then this is a new variable (in a ? offer, introduced in this transition).

- Guarding, prefix and parallelism are the only rules which alter transition conditions, but unlike the LOTOS standard we do not evaluate those conditions while constructing the STS.
- Transitions associated with ? events may introduce new variables, in order to avoid variable name capture. For example, the query variable in $g?x : S;\ P\ |||\ g!x;\ Q$ needs to be renamed in order avoid capturing the free variable in the right hand side. New variables may be necessary even when every ? variable in a specification is unique. For example, when a process is invoked more than once, e.g. $P[g]\ \ |||\ \ P[g]$ where $P[g] = g?x : S;\ P'$, then one of the x variables must be assigned a unique name to avoid confusion. We assume that we may perform alpha conversion (renaming of free variables) whenever necessary.

  This is a major difference from our previous work [14], in which new variables were introduced at every data offer and synchronisation. This style of semantics gave a very clear distinction between information about a value (in the Boolean condition) and the value itself (denoted by a variable name), but would be difficult to implement since so many new names are required. Hence our revised version presented here.
- Transitions may have conditions which are not satisfiable, and may arise through unsatifiable guards in the LOTOS description, or through unsatisfiable combinations obtained through synchronisation.

3.5 Axioms and Rules of Transition

In this section we give the rules to generate STS(P), the symbolic semantics of behaviour P, from a given canonical LOTOS behaviour expression P. Following the LOTOS standard the rules are grouped according to syntactic structures (shown in the boxes). Axioms show the transition and resulting state from a given portion of LOTOS syntax (on the left of the transition). Rules show the same, but with some preconditions given above the line. Note that relabelling is not part of standard LOTOS syntax, but is introduced in order to define process instantiation (as in the LOTOS standard).

We give here a complete set of rules for deriving a symbolic semantics for a LOTOS expression; however, the most interesting rules are those for prefix, exit, guard, and parallelism.

prefix axioms

$$a;\ P \xrightarrow{\mathtt{tt}\quad a} P$$

$$g\ d;\ P \xrightarrow{\mathtt{tt}\quad gE'} P$$

$$E' = \begin{cases} E & \text{if} \quad d = !E \\ x & \text{if} \quad d = ?x : S \end{cases}$$

$$g\ d[SP];\ P \xrightarrow{SP \qquad gE'} P$$

$$E' = \begin{cases} E & \text{if} \quad d = !E \\ x & \text{if} \quad d = ?x:S \end{cases}$$

exit axioms

$$\text{exit} \xrightarrow{tt \quad \delta} \text{stop}$$

$$\text{exit(ep)} \xrightarrow{tt \quad \delta E'} \text{stop}$$

$$E' = \begin{cases} E & \text{if ep} = E \\ z & \text{if } E = \textbf{any } S \quad \text{where } z \in \textbf{new-var}. \end{cases}$$

let rule

$$\frac{P[E/x] \xrightarrow{b \quad \alpha} P'}{let\ x : S = E\ in\ P \xrightarrow{b \quad \alpha} P'}$$

choice range rules

$$\frac{P[g_i/g] \xrightarrow{b \quad \alpha} P'}{choice\ g\ in\ [g_1, \ldots, g_n]\ [\,]\ P \xrightarrow{b \quad \alpha} P'}$$

for each $g_i \in \{g_1, \ldots, g_n\}$

$$\frac{P \xrightarrow{b \quad \alpha} P'}{choice\ x : S\ [\,]\ P \xrightarrow{b \quad \alpha} P'}$$

par rule

$$\frac{P[g_1/g]\ op\ \ldots\ op\ P[g_n/g] \xrightarrow{b \quad \alpha} P'}{par\ g\ in\ [g_1, \ldots, g_n]\ op\ P \xrightarrow{b \quad \alpha} P'}$$

where *op* is one of the parallel operators, $||$, $|||$, or $|[h_1, \ldots, h_m]|$, for some gate names $h_1, \ldots, h_m$.

hide rules

$$\frac{P \xrightarrow{b \quad \alpha} P'}{hide\ g_1, \ldots, g_n\ in\ P \xrightarrow{b \quad i} hide\ g_1, \ldots, g_n\ in\ P'}$$

if $\textbf{name}(\alpha) \in \{g_1, \ldots, g_n\}$

$$\frac{P \xrightarrow{b \quad \alpha} P'}{hide\ g_1, \ldots, g_n\ in\ P \xrightarrow{b \quad \alpha} hide\ g_1, \ldots, g_n\ in\ P'}$$

if $\textbf{name}(\alpha) \notin \{g_1, \ldots, g_n\}$

accept rules

$$\frac{P_1 \xrightarrow{b \quad \alpha} P_1'}{P_1 \gg accept\ x:S\ in\ P_2 \xrightarrow{b \quad \alpha} P_1' \gg accept\ x:S\ in\ P_2}$$

if **name**$(\alpha) \neq \delta$

$$\frac{P_1 \xrightarrow{b \quad \delta E} P_1'}{P_1 \gg accept\ x:S\ in\ P_2 \xrightarrow{b \quad i} P_2[E/x]}$$

Similarly for $\gg$ with no data.

disable rules

$$\frac{P_1 \xrightarrow{b \quad \alpha} P_1'}{P_1 [> P_2 \xrightarrow{b \quad \alpha} P_1' [> P_2}$$

if **name**$(\alpha) \neq \delta$

$$\frac{P_1 \xrightarrow{b \quad \alpha} P_1'}{P_1 [> P_2 \xrightarrow{b \quad \alpha} P_1'}$$

if **name**$(\alpha) = \delta$

$$\frac{P_2 \xrightarrow{b \quad \alpha} P_2'}{P_1 [> P_2 \xrightarrow{b \quad \alpha} P_2'}$$

general parallelism rules (synchronising)

$$\frac{P_1 \xrightarrow{b_1 \quad g} P_1' \qquad P_2 \xrightarrow{b_2 \quad g} P_2'}{P_1 |[g_1, \ldots, g_n]| P_2 \xrightarrow{b_1 \wedge b_2 \quad g} P_1' |[g_1, \ldots, g_n]| P_2'}$$

where $g \in \{g_1, \ldots, g_n, \delta\}$

$$\frac{P_1 \xrightarrow{b_1 \quad gE_1} P_1' \qquad P_2 \xrightarrow{b_2 \quad gE_2} P_2'}{P_1 |[g_1, \ldots, g_n]| P_2 \xrightarrow{b_1 \wedge b_2 \wedge E_1 = E_2 \quad gE_1} P_1' |[g_1, \ldots, g_n]| P_2'}$$

when $vars(b_1 \cup E_1) \cap vars(b_2 \cup E_2) = \emptyset$.

general parallelism rules (not synchronising)

$$\frac{P_1 \xrightarrow{b \quad \alpha} P_1'}{P_1 |[g_1, \ldots, g_n]| P_2 \xrightarrow{b\sigma \quad \alpha\sigma} P_1'\sigma |[g_1, \ldots, g_n]| P_2}$$

name$(\alpha) \notin \{g_1, \ldots, g_n, \delta\}$

$$\sigma = \begin{cases} [z/x] & \text{if } \alpha = gx \text{ and } x \in vars(P_2) \quad \text{where } z \in \textbf{new-var}. \\ [\,] & \text{otherwise} \end{cases}$$

Similarly for P_2.

choice rules

$$\frac{P_1 \xrightarrow{b \quad \alpha} P_1'}{P_1 \; [\,] \; P_2 \xrightarrow{b \quad \alpha} P_1'}$$

$$\frac{P_2 \xrightarrow{b \quad \alpha} P_2'}{P_1 \; [\,] \; P_2 \xrightarrow{b \quad \alpha} P_2'}$$

guard rule

$$\frac{P \xrightarrow{b \quad \alpha} P'}{([SP] \text{ -> } P) \xrightarrow{b \wedge SP \quad \alpha} P'}$$

stop rule

stop generates no rules.

instantiation rule

$$\frac{P[g_1/h_1, \ldots, g_n/h_n][E_1/x_1, \ldots, E_m/x_m] \xrightarrow{b \quad \alpha} P'}{p[g_1, \ldots, g_n](E_1, \ldots, E_m) \xrightarrow{b \quad \alpha} P'}$$

where $p[h_1, \ldots, h_n](x_1, \ldots, x_m) := P$ is a process definition

parenthesis rule

$$\frac{P \xrightarrow{b \quad \alpha} P'}{(P) \xrightarrow{b \quad \alpha} P'}$$

relabel rule

$$\frac{P \xrightarrow{b \quad \alpha} P'}{(P)[g_1/h_1, \ldots, g_n/h_n] \xrightarrow{b \quad \alpha'} (P')[g_1/h_1, \ldots, g_n/h_n]}$$

$$\alpha' = \begin{cases} h_i & \text{if } \alpha = g_i \text{ and } g_i \in \{g_1, \ldots, g_n\} \\ h_i E & \text{if } \alpha = g_i E \text{ and } g_i \in \{g_1, \ldots, g_n\} \\ \alpha & \textit{otherwise} \end{cases}$$

4. STATE EQUIVALENCE AND SUBSTITUTION

When constructing STSs from behaviour expressions, straightforward syntactic substitution on behaviour expressions was employed. Thus, state equivalence is defined purely by syntactic equivalence of LOTOS behaviours. However, in order to define equivalence relations, preorders, or logics over STSs and to ensure that cycles (such as might arise from recursive processes) are handled correctly, we must define *substitution* on STSs. This is new from [14].

In [10], this problem is solved by introducing the concept of a "term": a node in a symbolic transition system paired with a substitution. The same solution can be adapted for LOTOS.

Formally, a *term* consists of an STS, T, paired with a substitution, σ such that $domain(\sigma) \subseteq fv(T)$. We write this as T_σ to indicate that the substitution is not applied directly to T, and use t and u to range over terms.

Definition 2 gives the rules for transitions on terms, given the corresponding transitions on STSs. Only three cases are required: dataless transitions, transitions arising from ! offers, and transitions arising from ? offers. In all cases, $\sigma' = fv(T') \lhd \sigma$, that is, the restriction of σ to include only domain elements in the set $fv(T')$. The definition of free variables is extended to terms in the obvious way. Terms, rather than STSs, are used as the basis for defining the bisimulation in the next section, and the logic in [5, 6].

Definition 2 *(Transitions between terms)*

$$
\begin{array}{llll}
T \xrightarrow{b \quad a} T' & \textit{implies} & T_\sigma \xrightarrow{b\sigma \quad a} T'_{\sigma'} & \\
T \xrightarrow{b \quad gE} T' & \textit{implies} & T_\sigma \xrightarrow{b\sigma \qquad gE\sigma} T'_{\sigma'} & \textit{where } fv(E) \subseteq fv(T) \\
T \xrightarrow{b \quad gx} T' & \textit{implies} & T_\sigma \xrightarrow{b\sigma[z/x] \quad gz} T'_{\sigma'[z/x]} & \textit{where } x \notin fv(T), z \notin fv(T_\sigma)
\end{array}
$$

We note that while the vast majority of LOTOS specifications give rise to finite STSs, infinite (depth) STSs are still possible, for (data) parameterised processes. For example, *P(x) = g!x; P(x+1)* has an infinite (depth) STS. We do not consider the possibility of infinite depth a major drawback for our approach, but are beginning to investigate ways of dealing with it.

5. SYMBOLIC BISIMULATION

Standard bisimulation on transitions without data requires only the gate names to be the same. The obvious extension to transitions with data requires both gate and value to be matched exactly. In the symbolic world this is not appropriate. Consider, a symbolic transition stands for a set of concrete transitions. There are many ways to split this set into subsets; each characterisation yields a different symbolic representation. Therefore, in symbolic bisimulation, all gate names have to match exactly, but there need not be a one to one correspondence between values. A single transition in one process can be matched by one or more transitions in the other.

For example, consider the processes in Figure 4. The processes are clearly bisimilar, subject to the bisimilarity of T' with $U1$ and $U2$, but one transition in T, $x \in A \cup \bar{A}$, matches two transitions in U, $x \in A$ or $x \notin A$.

This idea of partitioning data into different sets, according to some Boolean expressions, is the crux of the following definition of symbolic bisimulation. The use of the partition means that each bisimulation is a parameterised family of relations, where the parameters are the predicates. Whereas Hennessy and Lin [10] define both an *early* and a *late* bisimulation equivalence, only an *early* bisimulation is meaningful in our context (as explained in Section 2).

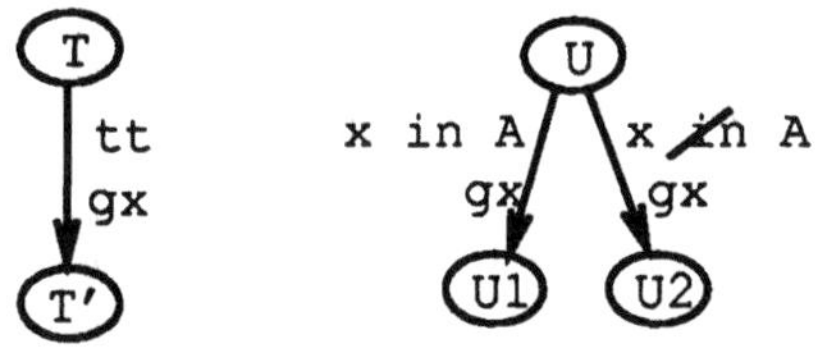

Figure 4. Bisimulation example

We give a definition of "layered" symbolic bisimulation, written $\sim_i^b$, where i is the depth of bisimulation and b is the initial context (usually simply tt).

We assume a function $new(t, u)$ which, given two terms t and u, returns a variable which is not among the free variables of either t or u.

Definition 3 *(Symbolic Bisimulation on terms)*
For all b, a Boolean expression, t and u, terms:

1 $t \sim_0^b u$.

2 For all $n > 0$, $t \sim_{n+1}^b u$ *iff*

(a) **(dataless case)**
if t has a transition $t \xrightarrow{b_t\ \alpha} t'$ then there is a finite set of Boolean conditions B over $fv(t)$ such that $(b \wedge b_t) \Rightarrow \bigvee B$ and for each $b' \in B$ there is a transition $u \xrightarrow{b_u\ \alpha} u'$ such that $b' \Rightarrow b_u$ and $t' \sim_n^{b'} u'$.

(b) **(data case, no new variable)**
if t has a transition $t \xrightarrow{b_t\ gE_t} t'$, where $fv(E_t) \subseteq fv(t)$, then there is a finite set of Boolean conditions B over $fv(t) \cup \{z\}$ such that $(b \wedge b_t \wedge z = E_t) \Rightarrow \bigvee B$, where $z = new(t, u)$, and for each $b' \in B$ either

there is a transition $u \xrightarrow{b_u\ gE_u} u'$, where $fv(E_u) \subseteq fv(u)$, and $b' \Rightarrow b_u$ and $b' \Rightarrow E_t = E_u$ and $t' \sim_n^{b'} u'$

or

there is a transition $u \xrightarrow{b_u\ gz} u'$ such that $b' \Rightarrow b_u$ and $t' \sim_n^{b'} u'$

(c) **(data case, new variable)**
if t has a transition $t \xrightarrow{b_t\ gz} t'$, where $z = new(t, u)$, then there is a finite set of Boolean conditions B over $fv(t) \cup \{z\}$ such that $(b \wedge b_t) \Rightarrow \bigvee B$ and for each $b' \in B$ either

there is a transition $u \xrightarrow{b_u\ gE_u} u'$, where $fv(E_u) \subseteq fv(u)$, and $b' \Rightarrow b_u$ and $b' \Rightarrow z = E_u$ and $t' \sim_n^{b'} u'$

or

there is a transition $u \xrightarrow{b_u\ gz} u'$ and $b' \Rightarrow b_u$ and $t' \sim_n^{b'} u'$

(d), (e), (f) Symmetrically, the transitions of u must be matched by t.

We may be relating processes that are parameterised. Therefore, the free variable (parameter) must be matched accordingly.

Definition 4 *($\sim^b$ for parameterised processes)*
If $fv(t) = \{x\}$, $fv(u) = \{y\}$, and $z = new(t, u)$ then

$$t \sim^b u \quad \textit{iff} \quad \forall z.t_{[z/x]} \sim^{b[z/x,z/y]} u_{[z/y]},$$

We use $\sim^b$ to denote the largest symbolic bisimulation, for a given b.

The role of the partition is to provide a step between the Boolean conditions of term t and those of term u. This can be clearly seen in case *(a)*. The intuition for cases *(b)* and *(c)* of Definition 3 is as follows. For case *(b)* we assume that the data of the t transition is a value (expression). The role of the new variable z is to provide a common language for matching transitions. In the context of b', a member of the partition, the expression E_t can either be matched by a u transition with an equivalent value E_u, or a u transition with a new variable z. The rules for transitions between terms (Definition 2) allow the new variable z to be used without explicit renaming at this point. The conditions for matching vary depending on what sort of u transition is matched. Essentially, if a new variable is matched then conditions relating to the data are captured exactly by the condition b_u. If a data expression is matched then information about data is given both by b_u and the expression E_u. Case *(c)* considers the situation in which the data of the t transition is a new variable z.

The resulting bisimulation is a Boolean condition-indexed relation. So, in most cases, when $t \sim^b u$, and t evolves to t', and u evolves to u', and $t' \sim^{b'} u'$, then b' is a different condition to b, i.e. different states in the symbolic transition systems will be related by different members of the family of relations. This is because, in a typical symbolic transition system, restrictions on data increase with depth.

We have no space here to include an example; see [7].

6. CONCLUSIONS AND FURTHER WORK

We have defined a symbolic semantics for LOTOS in terms of symbolic transition systems, and symbolic bisimulation over those transition systems. Broadly speaking, we have adopted the approach of [10]; however, the features of LOTOS, especially the need to accomodate multi-way synchronisation and the resulting model of value passing, mean that this is not a straightforward adaptation of the theory presented in [10].

Our symbolic approach eliminates infinite branching which has been a major source of difficulty in reasoning about LOTOS specifications. The inference rules for the semantics are relatively simple and intuitive, and the meaning of unparameterised processes is the same as in the standard semantics (although we did not show that here).

We have only considered strong bisimulation here, though other relations (e.g. weak bisimulation) can be defined. While we have a means of checking whether a given relation is a symbolic bisimulation we have not given here an effective method of constructing that relation. However, it is fairly easy to see that the partition has to be derived from (the cross product of) the conditions in each transition system; an algorithm has been developed and implemented in Haskell. Worst case complexity is exponential, but early recognition of zeros (in the Boolean expressions) can lessen the pain somewhat.

Related work includes the definition of a modal logic FULL [5] which is adequate with respect to a version of the bisimulation defined here [6]. That is, it distinguishes and identifies exactly the same processes as that bisimulation. Tools to support model checking of FULL with respect to LOTOS specifications (or STSs) are also being developed based on several different implementation paradigms [3, 4, 13]. Further work is to add a notion of state (in the imperative sense) to deal with infinite (depth) transition systems.

Acknowledgments

The authors would like to thank Savi Maharaj for useful input on the definition of bisimulation, and Ed Brinksma for many fruitful discussions on reasoning about LOTOS. Carron Shankland thanks the British Council, the Nuffield Foundation and the Engineering and Physical Sciences Research Council (under the project "Developing Implementation and Extending Theory: A Symbolic Approach to Reasoning about LOTOS") for their support.

REFERENCES

[1] T. Bolognesi, editor. Catalogue of LOTOS Correctness Preserving Transformations. Technical Report Lo/WP1/T1.2/N0045, The LOTOSPHERE Esprit Project, 1992.

[2] E. Brinksma. From Data Structure to Process Structure. In K.G. Larsen and A. Skou, editors, *Proceedings of CAV 91*, LNCS 575, pages 244–254, 1992.

[3] J. Bryans and C. Shankland. Implementing a modal logic over data and processes using XTL. In this volume. Kluwer, 2001.

[4] J. Bryans, A. Verdejo, and C. Shankland. Using Rewriting Logic to implement the modal logic FULL. In D. Nowak, editor, *AVoCS'01: Workshop on Automated Verification of Critical Systems*, 2001. Oxford University Computing Laboratory technical report PRG-RR-01-07.

[5] M. Calder, S. Maharaj, and C. Shankland. A Modal Logic for Full LOTOS based on Symbolic Transition Systems. *The Computer Journal*, 2001. In press.

[6] M. Calder, S. Maharaj, and C. Shankland. An Adequate Logic for Full LOTOS. In J. Oliveira and P. Zave, editors, *Formal Methods Europe'01*, LNCS 2021, pages 384–395. Springer-Verlag, 2001.

[7] M. Calder and C. Shankland. A Symbolic Semantics and Bisimulation for Full LOTOS. Technical Report TR-2001-77, University of Glasgow, 2001. Extended version.

[8] H. Eertink. *Simulation Techniques for the Validation of LOTOS Specifications*. PhD thesis, University of Twente, 1994.

[9] R. Gotzhein. Specifying Abstract Data Types with LOTOS. In B. Sarikaya and G.V. Bochmann, editors, *Protocol Specification, Testing, and Verification, VI*, pages 15–26. Elsevier Science Publishers B.V. (North-Holland), 1987.

[10] M. Hennessy and H. Lin. Symbolic Bisimulations. *Theoretical Computer Science*, 138:353–389, 1995.

[11] International Organisation for Standardisation. *Information Processing Systems — Open Systems Interconnection — LOTOS — A Formal Description Technique Based on the Temporal Ordering of Observational Behaviour*, 1988.

[12] L. Logrippo, M. Faci, and M. Haj-Hussein. An Introduction to LOTOS: Learning by Examples. *Computer Networks and ISDN Systems*, 23:325–342, 1992.

[13] P. Robinson and C. Shankland. Implementing the modal logic FULL using Ergo. In D. Nowak, editor, *AVoCS'01: Workshop on Automated Verification of Critical Systems*, 2001. Oxford University Computing Laboratory technical report PRG-RR-01-07.

[14] C. Shankland and M. Thomas. Symbolic Bisimulation for Full LOTOS. In M. Johnson, editor, *Algebraic Methodology and Software Technology, Proceedings of AMAST 97*, LNCS 1349, pages 479–493. Springer-Verlag, 1997.

[15] M. Thomas. The Story of the Therac-25 in LOTOS. *High Integrity Systems Journal*, 1(1):3–15, 1994.

[16] M. Thomas. Modelling and Analysing User Views of Telecommunications Services. In *Feature Interactions in Telecommunications Systems*, pages 168–183. IOS Press, 1997.

Appendix: Auxiliary Definitions

Definition 5 *(Free Variables)*
The free variables occurring in an expression is $fv(E) = vars(E)$, where $vars(E)$ denotes the variables occurring in expression E. The free variables of behaviour expression P, $fv(P)$, is defined below. Similarly for the set of free variables of an action α, $fv(\alpha)$.

$$
\begin{array}{lcl}
fv(\mathbf{stop}) & = & \{\} \\
fv(\mathbf{exit}) & = & \{\} \\
fv(\mathbf{exit}(x)) & = & \{x\} \\
fv(P[g]) & = & \{\} \\
fv(P[g](x_1,\ldots,x_n)) & = & \{x_1,\ldots,x_n\} \\
fv(g;\ P) & = & fv(P) \\
fv(g?x:S[SP];\ P) & = & (vars(SP) \cup fv(P)) \setminus \{x\} \\
fv(g!x[SP];\ P) & = & \{x\} \cup vars(SP) \cup fv(P) \\
fv([SP]\text{->}\ P) & = & vars(SP) \cup fv(P) \\
fv(\mathbf{let}\ x = E\ \mathbf{in}\ P) & = & vars(E) \cup (fv(P) \setminus \{x\}) \\
fv(\mathbf{hide}\ g\ \mathbf{in}\ P) & = & fv(P) \\
fv(P_1 * P_2) & = & fv(P_1) \cup fv(P_2), \\
 & & \text{where } * = [\,],\ [>,\ \gg,\ |[g_1,\ldots,g_n]|,\ ||,\ ||| \\
fv(P_1 \gg \mathbf{accept}\ x : S\ \mathbf{in}\ P_2) & = & fv(P_1) \cup (fv(P_2) \setminus \{x\}) \\
fv(\mathbf{choice}\ g\ \mathbf{in}\ [g_1,\ldots,g_n]\ [\,]\ P) & = & fv(P) \\
fv(\mathbf{choice}\ x : T\ [\,]\ P) & = & fv(P) \setminus \{x\} \\
fv(\mathbf{par}\ g\ \mathbf{in}\ [g_1,\ldots,g_n]\ op\ P) & = & fv(P) \text{ where } op \text{ is one of the parallel operators}
\end{array}
$$

IMPLEMENTING A MODAL LOGIC OVER DATA AND PROCESSES USING XTL

Jeremy Bryans
Department of Computing Science and Mathematics,
University of Stirling, UK
jwb@cs.stir.ac.uk

Carron Shankland
Department of Computing Science and Mathematics,
University of Stirling, UK
ces@cs.stir.ac.uk

Abstract The modal logic FULL is designed to capture strong bisimulation over early symbolic transition systems (STSs) for full LOTOS. It provides a compact way of expressing and verifying properties involving both data and transitions. In this paper we present a restricted prototype implementation of a model checker for LOTOS for queries written using the FULL logic. The model checker is developed within the CADP package using XTL.

Keywords: Modal Logic, Symbolic LOTOS, CADP, XTL

1. INTRODUCTION

Model checking (11) has proved to be a valuable verification method in recent years, and the size of systems which can realistically be verified is constantly increasing. However, there is still much room for improvement, and many techniques have been proposed to increase the state space of verifiable systems, including symmetry (13), abstraction (10) and symbolic methods (6). A particular problem is the inclusion of data in systems, since this can often lead to infinite state space. Typically this is dealt with by restricting the size of the data type. An alternative approach is to deal with data symbolically (16; 9). The state space is reduced by grouping transitions according to the kind of data passed, or the properties of that data. Thus, instead of investigating a single transition for every data value, groups of transitions are formed and verification is performed at the level of the groups.

Previous work (7; 8) has established a formal framework for the symbolic interpretation of LOTOS behaviours. The formal description technique LOTOS (18) was chosen due to its popularity and applicability to a wide range of applications (e.g. protocols and services (1), distributed systems (22), and as a semantics for higher level languages such as feature descriptions (26) and use-case maps (2)). Further, there are well established tools such as CADP (14) for reasoning about LOTOS behaviours. LOTOS is particularly amenable to the symbolic approach because it allows the description of process flow of control as well as allowing data to be passed between processes. Such data can affect the process flow of control and therefore cannot simply be ignored in verification.

On top of the symbolic interpretation of LOTOS an equivalence relation (9) and a modal logic called FULL (7) have been defined. The logic provides a way of expressing properties involving both data and transitions. Currently tools are being developed to support reasoning in all parts of the framework, including model checkers for the logic (5; 24). As an interim step, a prototype of the model checker has been implemented within CADP using XTL (21), a functional-type programming language allowing description of computation over graphs. This approach has limitations. The underlying semantics of CADP is not symbolic therefore the prototype is necessarily limited. The logic remains unchanged syntactically, but its expressive power is reduced since only properties over finite data types can be expressed. So, here we do not exploit the main advantage of our symbolic framework, which is to represent infinitely branching systems by finitely branching ones. However, the advantage of the approach is integration with a range of verification tasks already implemented in CADP, and the ability to experiment with the logic at an early stage.

The purpose of this paper is to present the XTL implementation of FULL, illustrate the sorts of symbolic properties (albeit over processes which have a concrete representation in CADP) which can be expressed and verified automatically, and discuss the ongoing plans for symbolic reasoning about LOTOS behaviours. We begin by introducing the CADP toolkit, summarising the main capabilities of interest to us. In Section 3 we present an overview of the logic FULL. We assume the reader is familiar with Full LOTOS, or at least process algebra. A more detailed introduction may be found in (7), where we discuss further the effects on the logic of the restrictions imposed by CADP. The main section of the paper is devoted to the implementation of modal operators with data using XTL. Finally we evaluate the success of this experiment, and present directions of ongoing and future work.

2. CADP

The CAESAR/ALDEBARAN development package (CADP) is a versatile multi-functional tool offering many different formal verification techniques, from interactive simulation through to compositional verification based on refinement. It is based around a common format for explicit Labeled Transition Systems (LTSs), known as Binary Coded Graphs, or BCGs. In particular, it accepts Full LOTOS as an input formalism, and offers a model-checking algorithm for expressions written in eXecutable Temporal Logic, or XTL. The BCG basis of CADP is finite and therefore all processes handled by CADP must use finite datatypes of up to 256 values.

XTL is a functional-type programming language designed to allow an easy, compact representation of common temporal logic operators. They are evaluated over LTSs encoded as BCGs. XTL provides low level operators which access the states, labels, and transitions of the LTS, and also operators to determine the successors and predecessors of states and transitions. A number of modal logics (eg. HML (17) and CTL (11)) have already been successfully encoded within XTL. CADP is therefore an ideal tool within which to build a prototype model checker for FULL, although the finite basis of BCGs means that we will not be able to exploit the full expressive power of the logic. This is discussed in more detail in the next section.

3. THE LOGIC FULL

The FULL logic was designed as part of an ongoing research project (12) to develop a framework for reasoning about Full LOTOS, i.e. the processes and the data. In this section we present an informal introduction to the symbolic logic FULL.

The syntax of FULL is based on a variant of HML (25), with quantifiers over data added. It is made up of two parts. The first set of formulae, ranged over by Φ, applies to closed terms. The second set, ranged over by Λ, applies to terms with a single free variable, as would arise from a LOTOS process with a single parameter. (The extension to multiple free variables is straightforward.)

Definition 1 *Syntax of FULL*

$$\begin{aligned} \Phi \; ::= \;& b \mid \Phi_1 \wedge \Phi_2 \mid \Phi_1 \vee \Phi_2 \mid [a]\Phi \mid \langle a\rangle\Phi \\ & \mid \langle \exists y\, g\rangle\Phi \mid [\exists y\, g]\Phi \mid \langle \forall y\, g\rangle\Phi \mid [\forall y\, g]\Phi \\ \Lambda \; ::= \;& \exists y.\Phi \mid \forall y.\Phi \end{aligned}$$

where b is some Boolean expression, $a \in G \cup \{\mathbf{i}, \delta\}$, $g \in G \cup \{\delta\}$, G is the set of gate names, $\mathbf{i}$ is the internal event, δ is the exit event and y denotes a variable name.

The simple event operators are taken from HML (25) and are well known.

$\langle a \rangle \Phi$: it is possible to do an a event (and then satisfy Φ), and

$[a]\Phi$: after every a event, Φ is satisfied. (May hold vacuously.)

In this logic, the novel operators are the four *quantified modal* operators. They are formed as combinations of the modes $\langle\ \rangle$ and $[\ ]$ and the quantifiers $\exists$ and $\forall$: $\langle \exists y\, g \rangle$, $[\exists y\, g]$, $\langle \forall y\, g \rangle$ and $[\forall y\, g]$. For each of these combination operators the meaning is determined using the quantifier to range over the value (y) being passed, and the mode to refer to the gate (g) at which it is being passed. For simplicity, we assume that a single data value is passed at a gate. The extension to multiple data offers at a gate is straightforward.

To introduce and illustrate these operators, we consider the example of the process P in Figure 1, which is taken from the introductory paper (7). This illustrates a number of the capabilities of the FULL logic.

When encoding the process P within CADP, we used the library NUM10.lib, which gives the natural numbers from 1 to 10 and offers the operators GT, LT and EQ ($>$, $<$ and $=$). This allows the finite BCG representation of P given in Figure 2. As noted above, the data type may have up to 256 values, but 10 makes the example easier to present.

Informally, the formula $\langle \exists y\, G \rangle \Phi$ is satisfied by a process if it can perform some transition with data y at a G gate, and the subsequent process satisfies the formula Φ. For example, the formula $\langle \exists y\, G \rangle \langle K \rangle tt$ is satisfied by the process P, because there are values of y (in particular: 4, 5 and 9) which can lead to a state satisfying $\langle K \rangle tt$.

The formula $\langle \forall y\, G \rangle \Phi$ is satisfied by a process if all the values that y can take have at least one G transition that leads to a process that satisfies Φ. For example, the formula $\langle \forall y\, G \rangle \langle H \rangle tt$ is satisfied by the process P, because every y value has a possible G transition leading to a state satisfying $\langle H \rangle tt$.

The formula $[\exists y\, G]\Phi$ is satisfied by a process if there is some value that y can take, and every time the process performs a G transition with that value, the subsequent state satisfies Φ. For example, the formula $[\exists y\, G]\langle H \rangle tt$ is satisfied by the process P, because if $y = 1, 2, 3, 5, 6, 7, 8$ or 10 then it is always possible to perform an H action afterwards.

The formula $[\forall y\, G]\Phi$ is satisfied by a process if for every possible value of y, every G transition leads to a state satisfying Φ. For example, the formula $[\forall y\, G](\langle H \rangle tt \vee \langle K \rangle tt)$ is satisfied by the process P, because every possible transition leads to a state which can either perform an H or a K action.

The logic is presented more completely in (7), where it is defined formally in terms of symbolic transition systems as derived from LOTOS (9). In order to determine the meaning of the logic (normally defined over symbolic transition systems) over Binary Coded Graphs, an equivalent interpretation of the logic over labelled transition systems is used. (See Appendix.) There is a direct

```
process P [G,H,K] : exit :=
      G?x:Num10 [x LT 5]; H; exit
   [] G!4; K; exit
   [] G?x:Num10 [x EQ 5]; (H; exit [] K; exit)
   [] G!5; H; exit
   [] G?x:Num10 [x GT 5]; H; exit
   [] G!9; K; exit
endproc
```

Figure 1. process P

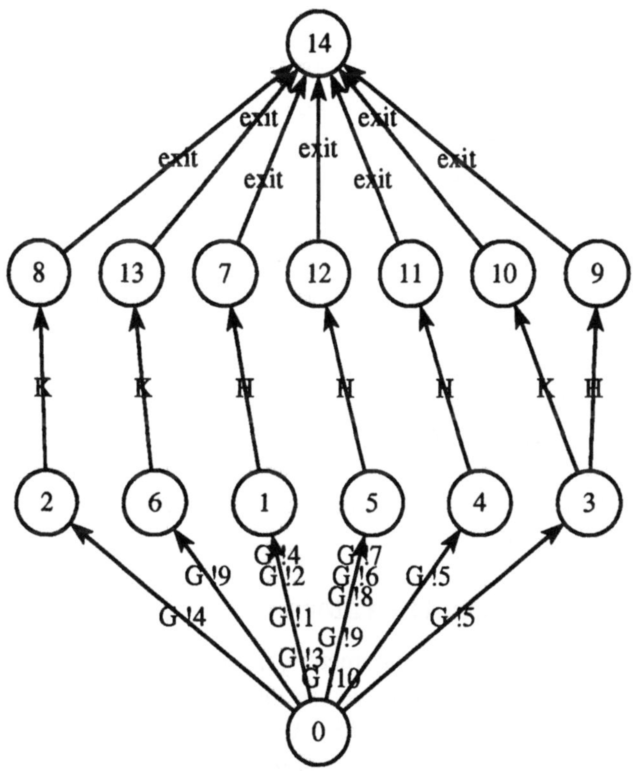

Figure 2. Binary Coded Graph for process P

mapping between BCGs and LTSs. The equivalence of the two definitions of the logic is shown in (4).

The expressivity of the logic may be considered to be parameterised by the number of values of the data types used. For this prototype, CADP restricts us to using data types with only 256 values, therefore the expressivity of the logic is clearly restricted, since when interpreted over a symbolic transition system an infinite number of data types may be considered.

In practice, this restriction may not affect the verification process. It may be that only 256 distinct values are required to distinguish processes and properties. If this where not a possibility, then CADP would be almost useless as a verification tool.

As soon as more than 256 values may be used in one branching point the logic of the prototype can no longer be used to distinguished processes which can be distinguished by FULL.

Another consideration is efficiency. A symbolic transition system allows many (possibility infinitely many) transitions to be represented finitely (possibility by just one transition). If a data variable can have 256 values, CADP will construct a BCG with 256 corresponding transitions, and all of these transitions must be explored when verifying properties. Therefore more work has to be done than in the symbolic case, in which the same transitions may be represented by just one transition.

4. IMPLEMENTING FULL WITHIN XTL

We show in this section how to implement the simple modal operators ($\langle g\rangle$ and $[g]$) and the quantified modal operators ($[\exists y\, g]$, $\langle \exists y\, g\rangle$, $[\forall y\, g]$ and $\langle \forall y\, g\rangle$) within the XTL language. The boolean operators $\wedge$, $\vee$ and $\neg$ have direct translations within XTL and are not shown here.

In XTL, a formula is associated with the largest set of states which satisfy it. Since the BCGs are finite these sets of states are necessarily finite.

To illustrate some useful XTL constructs, we begin with the simple modal operators, involving no data.

4.1 Simple Modal Operators

The $\langle a\rangle\Phi$ operator has two parameters, the simple gate name a and the formula Φ. To represent the gate a we use the labelset A, and the formula Φ is represented by the set of states Phi. Any state S will satisfy this operator, provided at least one of the outgoing edges of S is labelled a, (i.e. `exists T : edge among out (S) in (label (T) among A)`), and the resultant state of that edge satisfies the formula Φ (i.e. `(target (T) among Phi)`). We define $\langle a\rangle\Phi$ as `DIA(A,Phi)` within XTL by

```
def DIA (A : labelset, Phi : stateset) : stateset =
```

```
    { S : state where
    exists T : edge among out (S) in
        (label (T) among A) and (target (T) among Phi)
    end_exists }
end_def
```

Note here that any labelset A and any stateset Phi will be acceptable to `DIA`, but that to remain true to the FULL definition, A should be a single label representing the simple event a, and the stateset Phi should represent the FULL formula Φ. This stateset can be supplied as the result of a nested function. For example, the formula $\langle a \rangle \langle b \rangle tt$ would be implemented as `DIA(a, DIA(b, tt))`, where `tt` represents all states that satisfy *true*, i.e. all states.

The $[a]\Phi$ operator has the same two parameters, the gate name a and the formula Φ. A state S will satisfy this operator, provided for any appropriately labelled outgoing edges of S, the resultant state of that edge satisifies the formula Φ. This may be vacuously true if no edges are appropriately labelled. The differences from the implementation of $\langle a \rangle \Phi$ are therefore the use of `forall` instead of `exists`, and `implies` instead of `and`. Again, although BOX takes a set of labels A and a set of states Phi, A should be a single label representing the simple event a, and the stateset Phi should represent the FULL formula Φ.

We define $[a]\Phi$ as `BOX(A,Phi)` within XTL by

```
def BOX (A : labelset, Phi : stateset) : stateset =
    { S : state where
          forall T : edge among out (S) in
              ((label (T) among A) implies
                  (target (T) among Phi))
          end_forall }
end_def
```

4.2 Quantified Modal Operators

The quantified modal operators are the ones which allow us to refer to the data part of Full LOTOS.

To understand the implementation of a particular quantified modal operator, we begin by considering a general formula $OP(y, g)\Phi$, where OP is one of the four quantified modal operators, y is a variable of type *ytype* and g is a gate name, of type *ytype*. We deconstruct Φ into a form $\Phi_y \wedge \Phi_1$, where Φ_y contains all the restrictions on the variable y, and Φ_1 contains none of them.

The interpretation of each operator over a BCG can be understood as matching a particular pattern in the matrix of reachable states, as shown below.

The matrix in Figure 3 is formed from the BCG (Figure 2) of process `P`, where the headers of the columns are the labels of the outgoing transitions

$\Phi_1 = \langle H \rangle tt$	Outgoing transitions									
	G!1	G!2	G!3	G!4	G!5	G!6	G!7	G!8	G!9	G!10
P1	T	T	T	T	o	o	o	o	o	o
P2	o	o	o	F	o	o	o	o	o	o
P3	o	o	o	o	T	o	o	o	o	o
P4	o	o	o	o	T	o	o	o	o	o
P5	o	o	o	o	o	T	T	T	T	T
P6	o	o	o	o	o	o	o	o	o	F

Figure 3. Matrix representing the process P and $\langle H \rangle tt$

from the initial state, and the labels on the rows are all the reachable states from the initial state.

For a particular label and state, the relevant position in the matrix can have one of three labels:

- T: a transition marked with this label can reach this state, and Φ_1 holds,
- F: a transition marked with this label can reach this state, but Φ_1 does not hold, or
- o: a transition marked with this label cannot reach this state.

The pattern on the matrix is then determined solely by Φ_1, i.e. the part of the formula which is not concerned with the variable y. Similar matrices can be built for other process states and other FULL formulae Φ.

The formula Φ_y, which contains all the restrictions on y, determines which columns need to be considered when evaluating the complete formula $\Phi_y \wedge \Phi_1$. The validity of different FULL operators can be related to different patterns within the matrix.

As an example, let us consider the formula $(y = 4) \wedge \langle H \rangle tt$, so Φ_y is $(y = 4)$ and Φ_1 is $\langle H \rangle tt$. We will consider prefixing this by each of the four modal operators in turn.

- $\langle \exists y\, G \rangle \Phi$ needs just *one place* in the matrix to be labelled T. But since our example is $(y = 4) \wedge \langle H \rangle tt$ then this place must be in the $G!4$ column, because in every other column the formula $(y = 4)$ is not true.
- $[\exists y\, G]\Phi$ needs *one column* in the matrix which contains only the letters T and o in order to hold. But our specific example restricts this to the $G!4$ column, and it therefore does not hold. If instead we had Φ defined as $(y = 5) \wedge \langle H \rangle tt$ the formula $[\exists y\, G]\Phi$ would be true.
- $\langle \forall y\, G \rangle \Phi$ needs *at least one T in every column* in order to hold. Allowing Φ to restrict y, although syntactically possible, makes the whole formula

immediately false because the quantifier ranges over all possible values of y. Clearly, formulae such as $\langle \forall y\, G \rangle (y = 4)$ must be false, since y can have values other than 4. However, the formula $\langle \forall y\, G \rangle \Phi_1$ holds.

- $[\forall y\, G]\Phi$ needs *no Fs anywhere* in the table in order to hold. Again, restricting y in Φ immediately forms a false formula. In this case, since the table contains Fs, no formula $[\forall y\, G]\Phi$ holds. But if we constructed the table $\Phi_1 = \langle H \rangle tt \vee \langle K \rangle tt$ then $[\forall y\, G]\Phi_1$ would hold.

IMPLEMENTATION In order to interpret $OP(y, G)\Phi$ within XTL, we will break Φ into $\Phi_y \wedge \Phi_1$ as described above. There will be no other variables in Φ because any variables will have been bound in a previous step.

Each of the implementations of the quantified modal operators will receive two parameters: a labelset A which (loosely) corresponds to the combination of the gate g and the formula Φ_y, and a stateset Phi which corresponds to the formula Φ_1. The implementation will return the set of all states which satisfy the operator. A formula is satisfied by a process if the initial state is in this returned set.

Whereas for the simple modal operators we assumed the labelset was in fact a singleton (to conform with FULL syntax), for the quantified modal operators the labelset will typically not be a singleton. Although in FULL we write a symbolic expression such as $[\exists y\, G]\Phi$ this has to be expanded by XTL into a concrete set of labels. We use the (built in) XTL macro `EVAL_A(A)` to generate the labelset of the actions which satisfy the predicate `A`. For example, all y actions possible at a `G` gate are therefore evaluated as `EVAL_A(G?y:ytype)`, yielding the set `{"G!1", "G!2", "G!3", ..., "G!10"}`. This labelset is then supplied to the implementation of a quantified modal formula.

This is the point at which the restrictions placed on our implementation of the FULL model checker by the BCG implementation become evident. If we were adhering strictly to the semantics as expressed in the Appendix, then the labelset could be infinite (ranging across all values of the data type). Using the BCG representation we quantify over finite types.

The labelset may also encode Φ_y. To restrict y to values which satisfy Φ_y we use `EVAL_A(G?y:ytype where` Φ_y`)`. The type `ytype` can be a simple XTL type, or one of the types created for the LOTOS process. If we wish to use a created type we need to be careful to provide C implementations for the constructors and operators within the type definition, and to reiterate these within the `.xtl` file. In our example the CAESAR compiler identifies the NUM10 values as integers, and we can use `integer` in the XTL queries. But if we used the strings `one`, `two`, ... etc. instead, then the compiler would implement these as character strings, and we would be forced to provide a C function to convert these back to NUM10 values within the `.xtl` file.

$\langle \exists y\, G \rangle \Phi$ A state S will satisfy this operator, provided there is some outgoing edge in A which hits the target set *Phi* (i.e. `exists T : edge among out (S)`). In matrix terms, finding one T anywhere would be sufficient, provided A made no restrictions on y. Otherwise the T would have to be in one of the columns to which y was restricted. This additional constraint is implemented wholly by the use of the `EVAL_A(G?y:ytype where` Φ_y`)` expression in generating a restricted labelset. See Section 5.1 for an example of this sort of formula. We can define $\langle \exists y\, G \rangle \Phi$ within XTL as

```
def DIAEVAL (A : labelset, Phi : stateset) : stateset =
          { S : state where
              exists T : edge among out (S) in
            (if label (T) among A
              then (target(T) among (Phi))
              else false
             end_if)
              end_exists }
end_def
```

The XTL definition above is a reasonably direct translation of the semantics of $\langle \exists y\, G \rangle \Phi$ as given in the Appendix. The `if` statement corresponds to finding one T in the matrix (with the right sort of label).

$[\exists y\, G]\Phi$ A state S will satisfy this operator, provided there is some label L in A (`exists L : label among A`) such that every outgoing edge labelled with L hits the target set *Phi* (the `forall` expression with nested `if`). That is, there is a whole column in the matrix containing only the letters T and o. The `else true` implements the fact that this operator can be vacuously true if there is a label L in A which is not assigned to any outgoing edge, i.e. the corresponding column contains only the letter o. We define $[\exists y\, G]\Phi$ within XTL as

```
def BOXEVAL (A : labelset, Phi : stateset) : stateset =
          { S : state where
              exists L : label among A in
                  (forall T : edge among  out (S) in
                    if   label (T) = L
                     then target (T) among (Phi)
                     else true
                    end_if
                   end_forall)
              end_exists }
end_def
```

It is harder to see a correspondence between the semantics of this operator as expressed in the Appendix and the XTL implementation above. The finite BCG semantics means that we effectively translate an expression of the form $[\exists y\, G]\Phi$ into $[g1]\Phi \;\vee\; [g2]\Phi \;\vee\; \ldots \;\vee [g10]\Phi$, where each value of the data type yields a corresponding hardwired [ ] operator.

$\langle\forall y\, G\rangle\Phi$ To implement the universally quantified operators we rely on the duality of $\exists$ and $\forall$, and $\langle\;\rangle$ and [ ], as pointed out in (7). In XTL, `not (Phi)` is the complement of the set `Phi`. We define $\langle\forall y\, G\rangle$ as

```
def DIAAVAL (A : labelset, Phi : stateset) : stateset =
  not (BOXEVAL (A, not (Phi)))
end_def
```

taking advantage of the fact that $neg(\langle\forall y\, G\rangle\Phi)$ is $[\exists y\, G]neg(\Phi)$.

$[\forall y\, G]\Phi$ We define $[\forall y\, G]\Phi$ as

```
def BOXAVAL (A : labelset, Phi : stateset) : stateset =
 not (DIAEVAL (A, not (Phi)))
end_def
```

taking advantage of the fact that $neg([\forall y\, G]\Phi)$ is $\langle\exists y\, G\rangle neg(\Phi)$.

5. THE MODEL CHECKER

To model check a LOTOS process within CADP using the FULL logic, we use the macro SAT, defined as

```
macro SAT (Phi) =
   PRINT_FORM( if Phi includes init then (TRUE)
                                     else (FALSE)
               end_if)
end_macro
```

This accepts a formula *Phi* and prints TRUE if the initial state of the process is included in the stateset representing *Phi*, and FALSE otherwise.

We can then model check a process using

```
$xtl test.xtl process.bcg
```

The file `process.bcg` is the output from the `caesar.adt` and `caesar` components of the CADP tool. These take a LOTOS specification and produces a Binary Coded Graph(BCG). The file `test.xtl` contains a series of SAT commands. For convenience, SAT commands can also take a text parameter, which is returned with the model checker output. This is optional, but such comments can serve as a reminder of the formula being tested.

5.1 Examples

Here we use again the process P from Section 3.

1 The query *is it possible for process* P *to perform a G!4 action?* which is captured as $P \models \langle \exists y\, G\rangle(y = 4)$ using FULL, is written

```
SAT( "<E y:G>(y=4)tt: ",
   ( DIAEVAL(EVAL_A(G?y:integer where y=4),TRUE)))
```

and because (EVAL_A(G?y:integer where y=4)) evaluates to "G !4" the formula becomes

```
if DIAEVAL({"G !4"}, TRUE) includes INIT then TRUE
     else FALSE
end_if
```

and for DIAEVAL({"G !4"}, TRUE) to be true for process P, the initial state simply needs one outgoing *G!4* action. This is clearly true (see Figures 2 and 3), so the output is

```
<E y:G>(y=4)tt:  TRUE
```

2 The query *for all values y, is it possible to do a G!y action followed by an H action?* which is written as $P \models \langle \forall y\, G\rangle\langle H\rangle tt$ using FULL, is expressed as

```
SAT( "<V y:G><H>tt: ",
   (DIAAVAL(EVAL_A(G?y:integer),
         (DIA(EVAL_A(H),TRUE)))))
```

The parameter (EVAL_A(G?y:integer) evaluates to the set

```
{"G !1", "G !2", "G !3", "G !4", "G !5",
 "G !6", "G !7", "G !8", "G !9", "G !10"}
```

the restriction to ten elements coming from the process P, which is limited to the library NUM10.lib. This means every transition possible for P must be considered. The operator $\langle \forall y\, G\rangle$ requires that each label is on a transition leading to at least one state where the formula Φ is true. This is true here (see Figure 3, one T in every column), and so the output is

```
<V y:G><H>tt: TRUE
```

3 The query *for all values y, after any G!y action is it possible to do an H action?* is written within FULL as $P \models [\forall y\, G]\langle H\rangle tt$ and is expressed as

```
SAT( "[V y:G]<H>tt:  ",
   ( BOXAVAL(EVAL_A(G?y:integer),
           (DIA(EVAL_A(H),TRUE)))))
```

The parameter `EVAL_A(G?y:integer)` generates the full list of possible transitions again, but this time the query fails, because, for example, it is possible to take a transition labelled $G!4$ to a state which only permits a K action. It doesn't matter that there is a successful *G!4* transition which can subsequently perform an H action.

The reasonably direct translation from FULL to XTL can be seen from these simple examples. The only tricky part is the need to provide requirements on a value Φ_y to the point at which that value is introduced.

6. CONCLUSIONS AND FURTHER WORK

We have successfully implemented a model checker for the FULL logic over LOTOS (with data), and we are planning to extend this work in a number of directions.

We have modelled the Remote Procedure Call case study (3; 15) using LOTOS within CADP, and performed some simple queries on it using FULL. We have yet to fully explore this example.

The FULL logic cannot as yet express mu-calculus type queries (20) with infinitely repeating patterns, but we are currently working on extending the logic in this direction. Further, for ease of presentation, the FULL logic is currently limited to single data values being passed at gates, and for practical use it would be preferable to allow multiple data values. For example, if `G?y1:bool?y2:integer` was a gate, we would like to be able to identify all labels that matched `G!true?y2:integer`. Implementing the mu-calculus operators in XTL should be straightforward, but currently expressing pattern matching queries within XTL is not feasible.

The CADP toolkit provides a very accessible way of building a prototype model checker, and we are very pleased with the results obtained. However the BCG graphs are only ever finitely branching, because the LOTOS processes are restricted by the datatypes used. In order to reason about symbolic aspects of FULL allowing infinite data types we are considering two other approaches: theorem proving (using the Ergo tool) (24) and rewriting logic (5). Much interesting work on combining theorem proving and model checking is available (23; 19), and rewriting logic, while a comparatively recent technique, has already been successfully used to implement a number of simple model checkers (27). These techniques also have the advantage of allowing integration of reasoning about data and reasoning about processes.

Acknowledgments

Thanks to Radu Mateescu for the help provided for our work with CADP. The authors would like to thank the Engineering and Physical Sciences Research Council for supporting this research under the project award "Developing Implementation and Extending Theory: A Symbolic Approach to Reasoning about LOTOS".

REFERENCES

[1] I. Ajubi, G. Scollo, and M. van Sinderen. Formal Description of the OSI Session Layer. In P.H.J. van Eijk, C.A. Vissers, and M. Diaz, editors, *The Formal Description Technique LOTOS*, pages 89–210. Elsevier Science Publishers B.V. (North-Holland), 1989.

[2] D. Amyot, L. Charfi, N. Gorse, T. Gray, L. Logrippo, J. Sincennes, B. Stepien, and T. Ware. Feature Description and Feature Interaction Analysis with Use Case Maps and LOTOS. In M. Calder and E. Magill, editors, *Feature Interactions in Telecommunications and Software Systems VI.* IOS Press, May 2000.

[3] M. Broy, S. Merz, and K. Spies, editors. *Formal Systems Specification*, volume 1169 of *Lecture Notes in Computer Science.* Springer-Verlag, 1996.

[4] J. Bryans, M. Calder, S. Maharaj, B. Ross, and C. Shankland. Report on Developing a Symbolic Framework for Reasoning about Full LOTOS. Unpublished, 2001.

[5] J. Bryans, A. Verdejo, and C. Shankland. Using Rewriting Logic to implement the modal logic FULL. In D. Nowak, editor, *AVoCS'01: Workshop on Automated Verification of Critical Systems*, 2001. Oxford University Computing Laboratory technical report PRG-RR-01-07.

[6] J. R. Burch, E. M. Clarke, D. E. Long, K. McMillan, and D. L. Dill. Symbolic model checking for sequential circuit verification. *IEEE Transactions on Computer Aided Design of Integrated Circuits*, 13(4):401–424, 1994.

[7] M. Calder, S. Maharaj, and C. Shankland. A Modal Logic for Full LOTOS based on Symbolic Transition Systems. *The Computer Journal*, 2001. In press.

[8] M. Calder, S. Maharaj, and C. Shankland. An Adequate Logic for Full LOTOS. In J. Oliveira and P. Zave, editors, *Formal Methods Europe'01*, LNCS 2021, pages 384–395, 2001.

[9] M. Calder and C. Shankland. A Symbolic Semantics and Bisimulation for Full LOTOS. Technical Report TR-2001-77, University of Glasgow, 2001.

[10] E. M. Clarke, O. Grumberg, and D. E. Long. Model checking and abstraction. *ACM Transactions on Programming Languages and Systems*, 16(5):1512–1542, 1994.

[11] E. M. Clarke, O. Grumberg, and D. A. Peled. *Model Checking.* The MIT Press, 1999.

[12] Developing Implementation and Extending Theory: A Symbolic Approach to Reasoning about LOTOS. `http://www.cs.stir.ac.uk/diet/diet.html`, 2000.

[13] E.A. Emerson and A. P. Sistla. Symmetry and model checking. In C. Courcoubetis, editor, *Proceedings of the 5th Workshop on Computer Aided Verification*, pages 463–478, 1993.

[14] J.-C. Fernandez, H. Garavel, A. Kerbrat, R. Mateescu, R. Mounier, and M. Sighireanu. CADP(CAESAR/ALDEBARAN Development Package): A Protocol Validation and Verification Toolbox. In R. Alur and T. Henzinger, editors, *CAV'96*, volume 1102 of *Lecture Notes in Computer Science*, pages 437–440. Springer Verlag, 1996.

[15] R. A. Hardy. RPC Specification in LOTOS. Unpublished manuscript, 2000.

[16] M. Hennessy and H. Lin. Symbolic bisimulations. *Theoretical Computer Science*, 138:353–389, 1995.

[17] M. Hennessy and R. Milner. Algebraic laws for nondeterminism and concurrency. *Journal of the Association for Computing Machinery*, 32(1):137–161, 1985.

[18] ISO:8807. Information Processing Systems — Open Systems Interconnection — LOTOS — A Formal Description Technique Based on the Temporal Ordering of Observational Behaviour. International Organisation for Standardisation, 1988.

[19] J. J. Joyce and C-J. H. Seger. Linking BDD-based symbolic evaluation to interactive theorem proving. In *Proceedings of the 30th Design Automation Conference*. Association for Computing Machinery, 1993.

[20] D. Kozen. Results on the propositional mu-calculus. *Theoretical Computer Science*, 27:333–354, 1983.

[21] R. Mateescu and H. Garavel. XTL: A Meta-Language and Tool for Temporal Logic Model-Checking. In *Proceedings of the International Workshop on Software Tools for Technology Transfer*, July 1998.

[22] C. Pecheur. Using LOTOS for specifying the CHORUS distributed operating system kernel. *Computer Communications*, 15(2):93–102, March 1992.

[23] S. Rajan, N. Shankar, and M. K. Srivas. An integration of model checking with automated theorem proving. In P. Wolper, editor, *Proceedings of the 1995 Workshop on Computer-Aided Verification*, volume 939 of *LNCS*, pages 84–97. Springer-Verlag, 1995.

[24] P. Robinson and C. Shankland. Implementing the modal logic FULL using Ergo. In D. Nowak, editor, *AVoCS'01: Workshop on Automated Verification of Critical Systems*, 2001. Oxford University Computing Laboratory technical report PRG-RR-01-07.

[25] C. Stirling. Temporal Logics for CCS. In J. de Bakker, W.-P. de Roever, and G Rozenberg, editors, *Linear Time, Branching Time and Partial Order in Logics and Models for Concurrency*, volume 354 of *LNCS*, pages 66–672, REX School/Workshop, Noordwijkerhout, The Netherlands, May/June 1989. Springer-Verlag.

[26] Kenneth J. Turner. An architectural description of intelligent network features and their interactions. *Computer Networks*, 30(15):1389–1419, September 1998.

[27] Alberto Verdejo and Narciso Marti-Oliet. Implementing CCS in Maude. In *Proceedings of FORTE/PSTV 2000*, pages 351–366, 2000.

Appendix: The FULL logic definition

We give here the semantics of the logic, defined inductively over concrete labelled transition systems. It is assumed that $vars(\Phi) \cap vars(t) = \{\}$, i.e. the variables of the transition system and the variables of the logical formula are disjoint.

First we define $t \models \Phi$, denoting that a closed term t satisfies a closed modal formula Φ. The relation is defined inductively over the syntax of the logic by the equations of Definition 2.

The last two rules of the definition where t is an open term relate the (single) free variable of the parameterised transition system to the (single) free variable of a quantified logical property. The remaining part of the property is evaluated using the earlier $t \models \Phi$ definitions.

Definition 2 *Semantics of FULL over a labelled transition system*
Given any closed term t, the semantics of $t \models \Phi$ is given by:

$$
\begin{array}{lcl}
t \models b & = & b \equiv tt \\
t \models \Phi_1 \wedge \Phi_2 & = & t \models \Phi_1 \textit{ and } t \models \Phi_2 \\
t \models \Phi_1 \vee \Phi_2 & = & t \models \Phi_1 \textit{ or } t \models \Phi_2 \\
t \models \langle a \rangle \Phi & = & \textit{there is a } t' \textit{ s.t. } t \xrightarrow{a} t' \textit{ and } t' \models \Phi \\
t \models [a]\Phi & = & \textit{whenever } t \xrightarrow{a} t' \textit{ then } t' \models \Phi \\
t \models \langle \exists x\, g \rangle \Phi & = & \textit{for some value } v, \textit{ for some } t', \ t \xrightarrow{gv} t' \textit{ and } t' \models \Phi[v/x] \\
t \models \langle \forall x\, g \rangle \Phi & = & \textit{for all values } v, \textit{ for some } t', \ t \xrightarrow{gv} t' \textit{ and } t' \models \Phi[v/x] \\
t \models [\exists x\, g]\Phi & = & \textit{for some value } v, \textit{ whenever } t \xrightarrow{gv} t' \textit{ then } t' \models \Phi[v/x] \\
t \models [\forall x\, g]\Phi & = & \textit{for all values } v, \textit{ whenever } t \xrightarrow{gv} t' \textit{ then } t' \models \Phi[v/x]
\end{array}
$$

Given any term t with one free variable z the semantics of an open formula, $t \models \Lambda$, is given by:

$$
\begin{array}{lcl}
t \models \exists x.\Phi & = & \textit{for some value } v, \ t_{[v/z]} \models \Phi[v/x] \\
t \models \forall x.\Phi & = & \textit{for all values } v, \ t_{[v/z]} \models \Phi[v/x]
\end{array}
$$

For a more complete explanation of the logic, consult [7].

Part Five

Applications of Verification

FORMAL VERIFICATION OF PEEPHOLE OPTIMIZATIONS IN ASYNCHRONOUS CIRCUITS

Xiaohua Kong, Radu Negulescu
Montreal, Quebec, Canada H3A 2A7
Department of Electrical and Computer Engineering, McGill University
E-mail: {kong,radu}@macs.ece.mcgill.ca

Abstract This paper proposes and applies novel techniques for formal verification of peephole optimizations in asynchronous circuits. We verify whether locally optimized modules can replace parts of an existing circuit under assumptions regarding the operation of the optimized modules in context. A verification rule related to assume-guarantee and hierarchical verification is presented, using relative timing constraints as optimization assumptions. We present the verification of speed-optimizations in an asynchronous arbiter as a case study.

Keywords: Peephole optimization, formal verification, peephole rule, relative timing, process spaces

1. INTRODUCTION

The dramatic increase of integrated circuit complexity raises the need for guarantees of correctness early in the design flow, instead of leaving such guarantees to testing after the design is completed. Presently, simulation is widely used to provide guarantees of correctness of designs. Unfortunately, simulation typically covers only a small subset of system behaviors, especially for asynchronous circuits, where several interleavings of signal transitions must be taken into account. In contrast, formal verification checks the correctness of a system under all possible signal interleavings. In this paper, we verify whether an implementation meets the specification by checking a refinement partial order on processes; we refer to this verification approach as *refinement checking*.

Typical verification problems for multi-component systems are PSPACE-complete (see e.g. [17]). Several *structured verification* approaches can be used to alleviate state explosion in refinement checking by breaking the verification of a large system into several sub-tasks, and establishing refinement individually for each subtask. One structured verification approach is *hierarchical verification*, where circuit descriptions are provided at several different levels of abstraction. At each level, the circuit is treated as an interconnection of modules. Refinement is checked only between successive levels: the higher level is treated as specification and the lower level is treated as implementation. The description of circuit behavior at a level between the top and bottom levels is called an *intermediate specification*. Hierarchical verification can reduce computational costs if the verification tasks at successive levels involve fewer components than the overall verification task. Another structured verification approach is *assume-guarantee* reasoning, which is essentially an induction-like argument that breaks the circularity of mutual reliance of components in a system.

The application of hierarchical verification depends on finding suitable abstractions for components, which is difficult for systems that have poor modularity. For example, peephole optimizations in digital circuits often introduce additional dependencies between the optimized submodules and the rest of the system, making it difficult to obtain suitable abstractions for the application of hierarchical verification.

Correspondingly, we propose verification techniques that adapt assume-guarantee and hierarchical verification rules to the verification of peephole optimizations. Using relative timing constraints as optimization assumptions, we can verify a circuit after peephole optimizations by: 1) verifying the corresponding circuit before optimization, while it has high modularity; 2) verifying the local replacement between circuit before optimization and circuit after optimization. If proper constraints are given, even optimizations that change the interface of a submodule can be verified quickly based on the verification of the circuit before optimization.

The framework for formal verification used in this paper is process spaces [19, 20]. Our hierarchical verification is related to the methods in [5] and [6]. Assume-guarantee rules are addressed in [5], [4], [15], [10], [11]. We extend the application of such rules by introducing optimization assumptions in verification. A distinctive point of our approach is that arbitrary processes can be used as optimization assumptions, regardless of connectivity; the choice of such processes affects the result mainly through computational costs, as explained in Section 3. In particular, our optimization assumptions can be processes representing relative timing constraints. Also, our proposed verification technique does not rely on induction properties; the proof of Theorem 1 is given without reference to the structure of executions.

2. PRELIMINARIES

2.1 Process Spaces

Process spaces [19, 20] are a general theory of concurrency, parameterized by the execution type. Systems are represented in terms of their possible executions, which can be taken to be sequences of events, functions of time, etc., depending on the level of detail desired in the analysis. In this paper, executions are taken to be traces (sequences of events); for trace executions, close relationships exist to several previous treatments of concurrency, such as [6] and [9].

Let E be the set of all possible executions. A *process* p is a pair (X, Y) of subsets of E such that $X \ll Y = E$. A process represents a contract between a device and its environment, from the device viewpoint. Executions in X? Y, called *goals*, denoted by **g** p, are legal for both the device and the environment. Executions from outside X, called *escapes*, denoted by **e** p, represent bad behavior on the part of the device. Finally, executions from outside Y, called *rejects*, denoted by **r** p, represent bad behavior on the part of the environment. We also use **as** p (accessible) and **at** p (acceptable) to denote X and Y respectively.

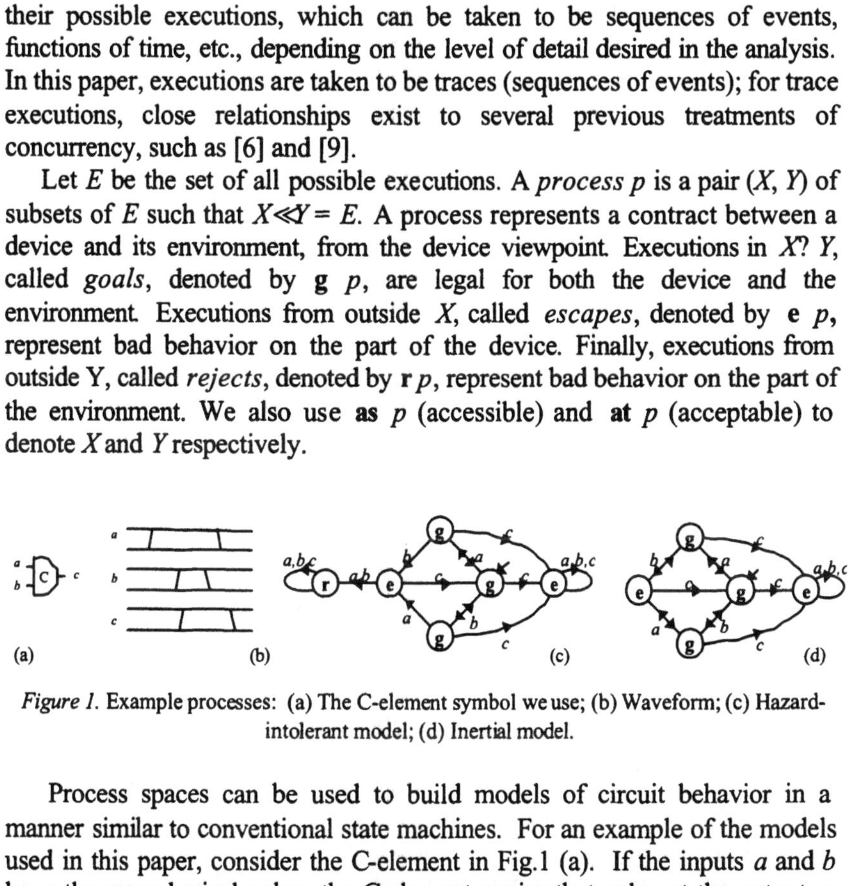

Figure 1. Example processes: (a) The C-element symbol we use; (b) Waveform; (c) Hazard-intolerant model; (d) Inertial model.

Process spaces can be used to build models of circuit behavior in a manner similar to conventional state machines. For an example of the models used in this paper, consider the C-element in Fig.1 (a). If the inputs a and b have the same logical value, the C-element copies that value at the output c; otherwise, the output value remains unchanged. Waveforms are represented by finite sequences of actions corresponding to signal transitions, such as *abcbac* for the waveform in Fig. 1 (b). In this paper, we use the term *trace* to refer to such a sequence of actions. We sometimes indicate that a certain action represents a rising or falling transition, as in $a+$ $b+$ $c+$ $b-$ $a-$ $c-$.

If all signals start low, the C-element can be represented by the process in Fig. 1 (c), where **r**, **g**, and **e** stand for reject, goal, and escape. Illegal output events lead to an escape state with self loops on all subsequent events, call it a *permanent escape*, and illegal input events lead to a reject state that cannot be left either, call it a *permanent reject.* The state where *ab* leads is also marked **e**, making it illegal for the device to complete its operation by stopping there.

The model in Fig. 1 (c) is a *hazard-intolerant* model. There are variations of the CMOS cell models, because, in the presence of hazards, the behavior of a CMOS cell is not fully standardized. A hazard is a situation where an output transition is enabled and then disabled without being completed. For example, execution *abb* is a hazard for the C-element in Fig. 1. Hazard-intolerant models simply require the environment to avoid hazards, by stating that each execution that includes a hazard will lead to a permanent reject. The model in Fig. 1 (d) is an *inertial* model. Inertial models ignore hazards by filtering out input pulses that are shorter than the delay of the gate.

In process spaces, processes can be used to model not only gates or cells, but also relative timing assumptions of the following form:

$$D\,(b_1\, b_2\, ...\, b_n) \; > \; D\,(a_1\, a_2\, ...\, a_m)$$

where a_1, ..., a_m, b_1, ..., b_n are events such that a_1 is the same as b_1, and the *D*s are the durations of the chains of events. Such a constraint, called a *chain constraint* [17], enforces that the *b* chain of events will not be completed before the *a* chain (unless one of the *a* or *b* actions involved occurs out of order).

Treating constraints as processes rather than linear inequalities permits us to deal with cases of deadlock and non-determinism, where the inequalities might not apply. Chain constraints can be implemented by transistor sizing. The absence of numerical information in the process models for chain constraints leads to more efficient verification using existing tools for non-timed analysis. Metric-free verification under relative timing constraints was presented in [19] and [17].

In this paper, we only use the following operations and conditions on processes:

?? *Refinement* is a binary relation, written $p \lll q$, meaning "process *q* is a satisfactory substitute for process *p*".

?? *Product* is a binary operation, written *p*?*q*, yielding a process for a system of two devices operating "jointly". Product is defined by **as** (*p*?*q*) = **as** *p* ? **as** *q* and **g** (*p*?*q*) = **g** *p* ? **g** *q*.

?? *Robustness* is a class of processes, written R_E, so that process *p* is in R_E if and only if **r** *p* = ? . Robustness represents a notion of absolute correctness: the device is "fool-proof" and can operate in any environment.

?? *Reflection*, written -*p* , defined by **as** (- *p*) = **at** *p* and **at** (- *p*) = **as** *p*, represents a swap of roles between environment and device.

In process spaces, refinement is reflexive, transitive, and antisymmetric; product is commutative, associative, and idempotent. Furthermore, for processes p, q, and r,

$$p \underline{\ll} q \quad ? \quad p \; ? \; r \underline{\ll} q \; ? \; r.$$

These properties suffice to break a verification problem into several layers of partial specifications, and each layer into several modules, and to verify only one module at a time instead of the overall problem in one piece.

Product admits an identity element, which is the process ? defined by g ? = E. Note that a process is robust if and only if that process refines ? .

Manipulations of finite-word processes are implemented by a BDD-based tool called FIREMAPS [17] (for finitary and regular manipulation of processes and systems). FIREMAPS implements the process space operations and conditions mentioned above, and has built-in constructors for hazard-intolerant and inertial models, and for chain constraints. In addition, if refinement does not hold, FIREMAPS can produce a *witness execution* that pinpoints the failure. Such witness executions are used for fault diagnosis.

2.2 Peephole Optimizations in Asynchronous Circuits

By *peephole optimizations* we mean local changes in circuit sub-modules that do not affect the rest of the circuit or the operation of the circuit as a whole. This sometimes involves changes in the interface of the respective sub-module to take advantage of signals available in other modules.

Peephole optimizations are often performed after high-level synthesis with significant gains. In [7], a peephole optimization step after synthesis by the methods in [1] and [3] is shown to produce gate count improvements up to a factor of five, and speed (cycle time) improvements up to a factor of two.

Due to their heuristic nature and limited scope, peephole optimizations can greatly benefit from automated verification. As unconventional designs, it is important that such optimizations be proven correct. Because of the small number of components involved, state explosion can be alleviated by localizing the verification.

On the other hand, flat verification of peephole optimizations against module specifications poses special problems. First, such verifications must usually take into account intricacies of switch-level and relative-time behaviors. Second, changes in the module interfaces need special techniques for modeling and incorporating the assumptions that justified the optimisations.

3. A STRUCTURED APPROACH FOR PEEPHOLE VERIFICATION

3.1 Assume-Guarantee Verification

Using hierarchical verification, a verification task p_1 ? $p_2 \preccurlyeq q$ can be split into three separate tasks that might have lower computational costs overall:

$$q_1 \text{ ? } q_2 \preccurlyeq q;\ p_1 \preccurlyeq q_1;\ p_2 \preccurlyeq q_2. \quad (3\text{-}1)$$

Here, p_1 and p_2 represent two submodules in the implementation, q_1 and q_2 represent the intermediate specifications of the respective submodules, and q represents the overall specification implemented by p_1 and p_2.

A frequently encountered difficulty, however, is loss of context information if operating assumptions are not modeled. Modules p_1 and p_2 may only "work correctly" in the presence of each other, in which case the refinement relations $p_1 \preccurlyeq q_1$ and $p_2 \preccurlyeq q_2$ only hold under certain assumptions. In Fig. 2, the OR gate does not directly refine the XOR gate specification. Nevertheless, if the environment guarantees that the inputs are mutually exclusive, a refinement check in context holds, verifying that the OR gate safely implements the XOR; see Example 1 in Subsection 3.2.

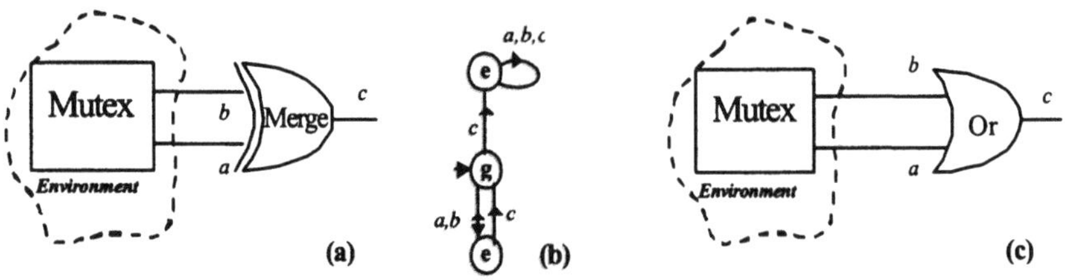

Figure 2. Replacement under assumptions: (a) XOR in a certain environment; (b) Inertial XOR; (c) OR in a certain environment.

In verification methods based on assume-guarantee rules (such as [5], [15], [10], and [11]), the correctness of the implementation relies on assumptions from the environment. According to assume-guarantee rules, a verification of the form p_1? $p_2 \preccurlyeq q$ can be decomposed as follows:

$$p_1 \text{ ? } q_2 \preccurlyeq q \text{ ? } q_1; \quad (3\text{-}2)$$
$$p_2 \text{ ? } q_1 \preccurlyeq q \text{ ? } q_2. \quad (3\text{-}3)$$

More precisely, it suffices to verify (3-2) and (3-3) for some q_1 and q_2 of a certain restricted form in order to establish p_1 ? $p_2 \preccurlyeq q$. However, note that q_1 and q_2 cannot be arbitrary. For instance, empty accessible sets of q_1 and q_2 trivially satisfy the decomposed verification tasks. Special conditions must be satisfied by q_1 and q_2 to break the circularity of reasoning and establish validity of p_1 ? $p_2 \lesssim q$. In previous methods, q_1 and q_2 are selected to have non-empty prefix-closed accessible sets, which justify a structural induction argument. In our method, we extend the rule so that q_1 and q_2 are arbitrary.

3.2 The Peephole Rule

Suppose that, in the original verification task, the refinement relation

$$p_1? \;\ldots r_i \;?\; \ldots \;?\; p_n \lesssim q$$

is checked. Here p_i (i=1..n) and q are arbitrary processes, representing the implementation components and specification, respectively; r_i is an arbitrary process, call it *peephole replacement,* which replaces p_i in a peephole optimization. Now, let d be an arbitrary process, call it *optimization assumption*, which formalizes the designer's hypothesis that made the replacement possible; and, let M be an arbitrary set of processes, call it *support model,* representing modules in the closed system, consisting of the implementation and the environment, that will guarantee the validity of the optimization assumption.

Theorem 1. *For processes p_1, ..., p_n,q, r_i, d and process set M:*

if $p_1? \;\ldots p_i \;?\; \ldots ?\; p_n \lesssim q$
and $(?\; d \;?\; S_E,\; M \;?\; \{p_j \,|\, j?i\} \ll \{-q\}$:
$\qquad\qquad r_i \;?\; d \lesssim p_i \;?\; (?_{m?M}\, m) \lesssim (?_{m?M}\, m) \;?\; d\;)$
then $p_1 \;?\; \ldots r_i \;?\; \ldots \;?\; p_n \lesssim q$.

We can phrase Theorem 1 informally as follows: if r_i refines p_i under constraint d, and d imposes no additional confinement over the system, then r_i can replace p_i in a refinement relation.

Proof:

$r_i \;?\; d \lesssim p_i$
? ? (by monotonicity w.r.t. product by $(?_{j?i}\, p_j) \;?\; (-q)$)
$(?_{j?i}\, p_j) \;?\; (-q) \;?\; r_i \;?\; d \lesssim (?_{j?i}\, p_j) \;?\; (-q) \;?\; p_i$
? ? (by commutativity of product)
$(?_{j?i}\, p_j) \;?\; (-q) \;?\; r_i \;?\; d \lesssim (?_k\, p_k) \;?\; (-q)$ (1)

$(?_{m?M}\, m) \lesssim (?_{m?M}\, m) \;?\; d$ and $M \;?\; \{p_j \,|\, j?i\} \ll \{-q\}$
? (by montonicity of product w.r.t. components from outside M)
$(?_{j?i}\, p_j) \;?\; (-q) \lesssim (?_{j?i}\, p_j) \;?\; (-q) \;?\; d$
? (by montonicity of product w.r.t. r_i)
$(?_{j?i}\, p_j) \;?\; (-q) \;?\; r_i \lesssim (?_{j?i}\, p_j) \;?\; (-q) \;?\; d \;?\; r_i$ (2)

(1) and (2)
? (by transitivity, since $(?_k p_k) \lll q$? $(?_k p_k)$? $(-q)$? $p_i \lll ?$)
$(?_{j?i} p_j)$? $(-q)$? $r_i \lll ?$
?
p_1? ... r_i ? ... ? $p_n \lll q$

?

Since Theorem 1 is specifically developed for verifying peephole optimizations, we refer to Theorem 1 as the *peephole rule*.

Example 1: The example in Fig. 2 illustrates a peephole optimization that replaces an XOR gate by an OR gate in a mutual exclusion environment. In this example, the peephole implementation is an inertial XOR gate; the peephole replacement is an inertial OR gate; the optimization assumption is "Pulses on *a* and *b* never overlap"; and, the support model is a Mutex component existing in the system.

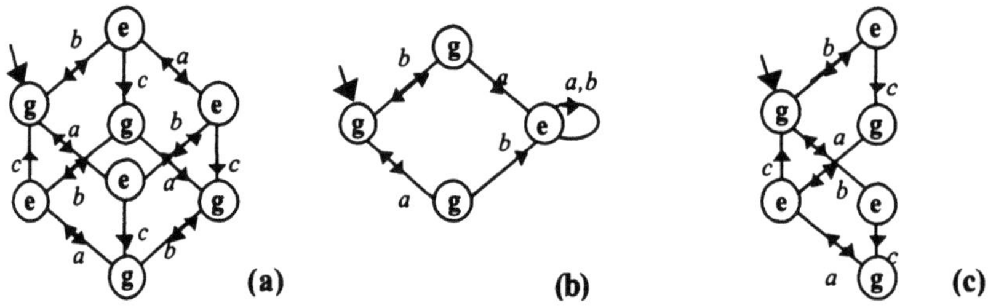

Figure 3. Usage of the peephole rule: (a) Inertial OR gate in open system; (b) Context constraint; (c) Process of OR gate under environment assumptions (context constraint). (Omitted edges on *a*, *b*, *c* lead to a permanent escape state.)

In the verification of Fig.2, assume refinement relation

$$p_{\mathrm{XOR}} ? \cdots p_{mutex} ? \cdots \lll q$$

holds, in which p_{XOR} represents the XOR gate and p_{mutex} represents an environment component which physically enforces mutual exclusion of signals *a* and *b*.

Let $M = \{p_{mutex}\}$, and let *d* be mutual exclusion constraint in Fig. 3 (b). Note that *d* is the product of the following relative timing constraints:

$D(a{+}\ a{-}) < D(a{+}\ b{+})$;
$D(b{+}\ b{-}) < D(b{+}\ a{+})$.

Because

$$p_{mutex} \preccurlyeq p_{mutex} ? d$$

and

$$p_{\mathrm{OR}} ? d \preccurlyeq p_{\mathrm{XOR}},$$

using the peephole rule, we have:

$$p_{\mathrm{OR}} ? \ldots p_{mutex} ? \ldots \preccurlyeq q.$$

Thus, the OR gate can safely replace the XOR gate in this system. ?

Theorem 1 relates to the assume-guarantee rule over a particular case. If p_1 from (3-2) is substituted by r_i from Theorem 1, q_1 from (3-2) is substituted by p_i from Theorem 1, and q_2 from (3-2) is substituted by the process $(?_{j?i}\, p_j)$, then the hypothesis of Theorem 1 implies inequation (3-2), which is part of the hypothesis of the assume-guarantee rule.

$$r_i ? d \preccurlyeq p_i \;?\; r_i ? (?_{j?i}\, p_j) ? d \preccurlyeq p_i ? (?_{j?i}\, p_j)$$
$$r_i ? (?_{j?i}\, p_j) \preccurlyeq p_i ? (?_{j?i}\, p_j) = p_i ? (?_{j?i}\, p_j) ? p_i \preccurlyeq q ? p_i$$

The remaining part of the hypothesis of the assume-guarantee rule, inequation (3-3), also follows from the hypothesis of Theorem 1 under the substitutions above, while p_2 from (3-2) can be substituted by any process p that refines $(?_{j?i}\, p_j)$.

Theorem 1 also relates to hierarchical verification over a particular case. If p_j is the intermediate specification for r_i, and M is the empty set, then $(?_{m?M}\, m) = ?$. Since ? is the most transparent process in a process space (Definition 2.13 in [17]) then $d = ?$. In this case, repeated application of Theorem 1 matches the hierarchical verification procedure as described by inequation (3-1).

3.3 Heuristics for Finding Verification Assumptions

Not all the optimization assumptions are guaranteed by the environment of the peephole alone. Since Theorem 1 has no restrictions on the connectivity of the processes involved, an optimization assumption may overlap both the peephole and the peephole environment. If the optimization assumption overlaps the peephole, we call the respective optimization assumption a *design assumption*; otherwise, we call the respective optimization assumption an *environment assumption.* Design assumptions are derived from properties of the circuit under verification which are known to the designer or verifier.

Notice that the peephole rule has no restrictions on the choice of the support model or the optimization constraint: system process subset M and process d in Theorem 1 are arbitrary. On the other hand, a poor selection of the support model will not lead to reductions of computational costs. For proper selection of the support model, one should consider the context of the circuit under verification. For example, in Fig. 3 (c), the process of "OR gate under environment assumption" has fewer accessible executions than the process in Fig. 3 (a) "OR gate in open system", because some of the possible behaviors are eliminated by an optimization assumption based on a support model which involves not just circuit elements, but also the environment of the circuit.

In our experiments, the costs of verification based on peephole rules are influenced mainly by a tradeoff between the complexity and the determinism of the assumptions used.

?? Keep the assumptions simple, as the complexity of verification increases with the number of assumption processes.

?? Make the assumptions efficient by eliminating as many as possible of the "don't-care behaviors" of the circuit under verification.

For instance, in the OR gate example above, a more effective assumption would only allow the pulses of *a* and *b* to alternate with pulses on *c*. If such an assumption can be guaranteed by the environment of the peephole, then not only we can perform stronger optimizations, but also we can verify them with less costs.

Presently, we mostly use relative timing constraints as optimization assumptions. By using relative timing constraints, verification does not need to start from a complete environment model, because a few hints from an incomplete environment model may suffice to guarantee the respective delay constraint as optimization assumption.

4. VERIFICATION OF THE ASP* ARBITER

A high-speed arbiter using the asP* protocol is reported in [8], with a non-optimized version and a speed-optimized version. The non-optimized implementation has good modularity, in the sense that the simple interfaces of the submodules achieve decoupling of the submodule designs. The optimized implementation achieves higher speed at some costs in modularity, by including more signals in the submodule interfaces. The high-level verification of the non-optimized version of this arbiter was reported in [13]. In this paper, we focus on the verification of the peephole optimisations.

4.1 Verification Strategy for the asP* Arbiter

The block diagram of the non-optimized version of the arbiter from [8] is shown in Fig. 4(a). The arbiter receives request events as input pulses and issues grant events as output pulses, after arbitration. Signals q_1 and q_2 are initially high, and the other signals are initially low. The rlatch component is a positive edge-triggered SR latch. When it receives input pulses (r_1 or g_1 for rlatch1), the rlatch converts them to output signal levels (y_1 for rlatch1). The dlatch in Fig. 4(a) converts pulses on the inputs to levels on the output. The two NOR gates generate grant pulses. The Mutex component ensures mutual exclusion between requests, and it is a four-phase component. For example, when request pulses from two channels arrive, rlatch1 and rlatch2 set y_1 and y_2 high. The Mutex arbitrates the input and gives grant to one, e.g. channel1, then q_1 is set to low and fires the rising edge of g_1. The rising edge of g_1 is propagated to dlatch and set f high, thereby resetting g_1; at the same time, feedback of g_1 pulse withdraws the request y_1. The pending request from channel2 gets a grant from Mutex and grant pulse g_2 to channel2 will be issued after an acknowledge pulse d_1 from channel1 is received. Notice that the grant pulse will be issued without waiting for the falling edge on the request pulse; only the rising edges of the r, g, and d signals are active.

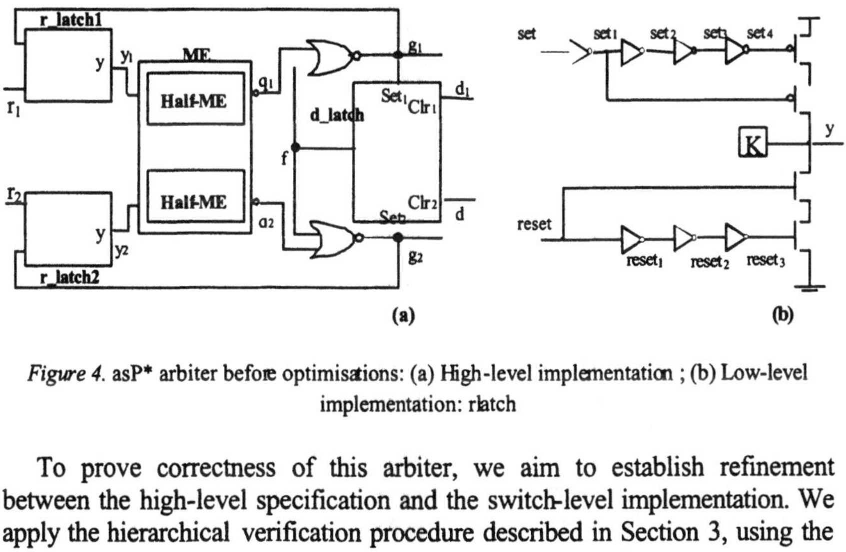

Figure 4. asP* arbiter before optimisations: (a) High-level implementation ; (b) Low-level implementation: rlatch

To prove correctness of this arbiter, we aim to establish refinement between the high-level specification and the switch-level implementation. We apply the hierarchical verification procedure described in Section 3, using the submodules in Fig. 4(a) as intermediate specifications. In the following, p_{Mutex}, $p_{rlatch1}$, $p_{rlatch2}$, p_{dlatch}, p_{NOR1} and p_{NOR2} denote processes for the switch-level implementation of submodules, q_{Mutex}, $q_{rlatch1}$, $q_{rlatch2}$, q_{dlatch}, q_{NOR1} and q_{NOR2} denote processes for the submodule interfaces, and $q_{arbiter}$ denotes the process for the overall transition-event specification of the arbiter. The verification proceeds on two levels, as follows:

On the first level, the refinement between the connection of intermediate specifications and overall specification is checked:

$$q_{Mutex} \,?\, q_{rlatch1} \,?\, q_{rlatch2} \,?\, q_{dlatch} \,?\, q_{\mathrm{NOR1}} \,?\, q_{\mathrm{NOR2}} \lesssim q_{arbiter}$$

Next, the refinements of submodule specifications against their switch level implementations are checked under certain environment assumptions:

$$p_{Mutex} \lll q_{Mutex}; \quad p_{rlatch1} \lll q_{rlatch1}; \; p_{rlatch2} \lll q_{rlatch2};$$
$$p_{dlatch} \lll q_{dlatch}; \quad p_{NOR1} \lll q_{NOR1}; \quad p_{NOR2} \lll q_{NOR2}.$$

In this paper, we call the verification on the first level high-level verification and the second level submodule verification. For the details of high-level verification and overall specification, we refer readers to [13].

4.2 Rlatch Verification

Submodule verification consists of refinement checks between switch level implementation of submodules and their intermediate specifications. We use rlatch as the example of refinement checking on this level. The representations of implementations at switch level follow [17].

The implementation of the rlatch from [8] is shown in the Fig. 4 (b). When a rising edge of signal *set* comes, the inverter chain of set (contains 4 inverters) generates a pulse upon the PMOS network, thus setting output *y* high. The same mechanism is used for reset. The box labeled *K* is a "keeper" circuit consisting of a weak inverter and a feedback inverter that form a loop, so that the value of the rlatch output can be kept when there are no set or reset paths enabled.

The rlatch is verified by checking the following refinement relationships:

$$p_{rlatch} = p_{set\_inv\_chain} \; ? \; p_{reset\_inv\_chain} \; ? \; p_{MOS} \lll q_{rlatch}$$

in which $p_{set\_inv\_chain}$ is the model of set inverter chain (4 inverters); $p_{reset\_inv\_chain}$ is the model of reset inverter chain (3 inverters); p_{MOS} is the model of the reset part of rlatch.

Some assumptions can be drawn from the environment of the rlatch in the asP* arbiter system. Interface signals *set*, *reset* and *y* of the rlatch are under constraints. For instance, in the channel1 of asP* arbiter, grant signal g_1 is used as reset signal of rlatch1. Notice that for rlatch1, g_1+ will not be triggered until y_1+ is propagated through Mutex and NOR gate. As a result, we can make an environment assumption for rlatch:

$$d_{env}\text{:} \;\; D\,(set{+}\; reset{+}) > D\,(set{+}\; y{+})$$

Correspondingly, the verification of the rlatch becomes:

$$d_{env} \; ? \; p_{rlatch} \lll q_{rlatch}$$

The verification result shows that the refinement relationship holds when the input pulse is wide enough to be caught by the inverter chain. This relative timing constraint is easily implemented by sizing, thus the rlatch implementation in Fig. 4(b) refines rlatch specification in typical environments for the asP* arbiter.

4.3 Peephole Optimizations of the asP* Arbiter

Fig. 5 (a) is the block diagram of asP* arbiter after optimization in [8]. Notice the change of submodules at their interfaces. (For example, the new interface of Mutex has 12 signals.) In Fig. 5 (a), signal *rr* of the Half_Mutex (Mutex/2 blocks in Fig. 5(a)) is connected to the internal nodes set_2 of rlatch (see fig. 4(b)); signal g_b is connected to the internal node set_1 of dlatch, which has the same setting-resetting structure as that for the rlatch [8].

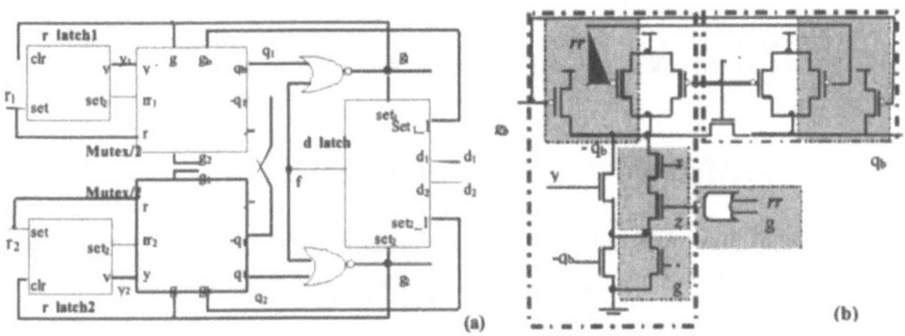

Figure 5. Peephole Optimization of asP* arbiter: (a) asP* arbiter block diagram after optimization; (b) Half-Mutex after optimisation.

Fig. 5 (b) shows the peephole optimizations over the half Mutex. Boxes in dark indicate modifications during the optimizations. Signals *rr* and g_b from the internal nodes of rlatch and dlatch are highlighted. For the details and motivation of these optimizations, we refer the reader to [8].

Notice that only Mutex's low-level structure is changed by optimization. For the other submodules, optimization only changes the interfaces by making some internal signals visible. As shown in Fig. 6(a), modules *rr*1, *rr*2, *gb*1 and *gb*2 are "abstracted" from rlatch1, rlatch2 and dlatch. Relations between models hold as follows:

$$p_{rlatch1} = p_{rr1} ? \ p_{rlatch1}; \quad p_{rlatch2} = p_{rr2} ? \ p_{rlatch2}; \quad p_{dlatch} = p_{gb1} ? \ p_{gb2} ? \ p_{dlatch}$$

By this module partition, we isolate unchanged modules (white blocks) in Fig. 4(a) from the models changed by optimization (dark blocks). The verification task after peephole optimization verification is:

$$p_{gb1} ? \ p_{gb2} ? \ p_{rr1} ? \ p_{rr2} ? \ p_{Mutex'} ? \ q_{rlatch1} ? \ q_{rlatch2} ? \ q_{dlatch} ? \ q_{\mathrm{NOR1}} ? \ q_{\mathrm{NOR2}} \leqq q_{arbiter}$$

In which $p_{Mutex'}$ represents the replacement of p_{Mutex} in optimization.

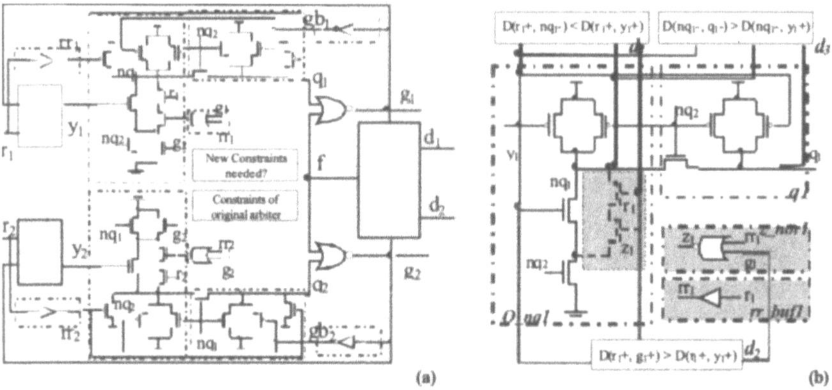

Figure 6. Peephole Optimization verification of asP* arbiter: (a) asP* arbiter after optimization; (b) Verification and result of optimization1 (half).

4.4 Verification of Peephole Optimizations

Applying the peephole rule, the verification of the peephole optimised Mutex amounts to verifying the following relation, for some constraint *d*:

$$p_{Mutex'} \text{ ? } d \lesssim q_{Mutex}$$

and to verifying that constraint *d* does not impose undue confinement on the system, by finding a suitable support model for *d*.

There are three optimizations addressed in [8]. The first optimization over half-Mutex is shown in Fig. 6(b). Changes involved in this optimization are highlighted. To apply the peephole rule, we take the following preparation steps:

?? First, a model of the optimization assumption is constructed. In Fig. 6(b), the highlighted part indicates the optimization assumption of optimization1, which "adds an additional 'bypass' to allow the rising edge of a request to be applied directly to the arbiter without incurring the delay of the rlatch" [8]. The relative timing constraint
d_1: $D(r_i{+}\, nq_i{-}) > D(r_i{+}\, y_i{+}) (i = 1, 2)$
in Fig. 6(b) comes from this statement.

?? Second, the refinement
p_{gb1} ? p_{gb2} ? p_{rr1} ? p_{rr2} ? q_{Mutex} ? $q_{rlatch1}$? $q_{rlatch2}$? q_{dlatch} ? q_{NOR1}
?? q_{NOR2} ? $d_1 \lesssim q_{arbiter}$
is checked and the relation holds.

?? In the last step, we choose $M = \{p_{gb1}, p_{gb2}, p_{rr1}, p_{rr2}, d_1\}$ as the initial support model of peephole verification, and optimization assumption $d = d_1$? p_{rr1}? p_{rr2}.

Two constraints are introduced in the procedure. One constraint:

$$d_2: \quad D(r_i+\, g_i-) > D(r_i+\, y_i+) \quad (i = 1, 2)$$

has as support model the components $p_{rlatch1}$ and $p_{rlatch2}$. The result of peephole verification is:

$$M = \{\, p_{gb1}, p_{gb2}, p_{rr1}, p_{rr2}, d_1, q_{rlatch1}, q_{rlatch2});$$
$$d = d_1 \; ? \; d_{rr1} ? \; d_{rr2} ? \; d_2;$$
$$p_{Mutex'} \; ? \; d \; ? \; d_3 \lll q_{Mutex}$$

in which d_3i s an extra delay constraint:

$$d_3: \quad D(nq_i\text{-}\, q_i-) > D(nq_i\text{-}\, q_i+)(i = 1, 2)$$

needed for the holding of the refinement.

The second and third optimizations are verified in the same way and the refinement only holds under certain additional constraints.

To examine the validity of the additional constraints detected in the peephole verification, we changed peephole in verification from Mutex to the whole arbiter. Applying peephole rules, we verify the implementation in Fig. 6 (a). The verification terminated and the result reported that no other relative timing constraints are needed for arbiter optimization:

$$M = \{d_{des1}, d_{des2}\};\; d = d_{des1} \; ? \; d_{des2}$$
$$p_{Mutex'} \; ? \; q_{rlatch1} \; ? \; q_{rlatch2} \; ? \; q_{dlatch} \; ? \; q_{NOR1} \; ? \; q_{NOR2} \; ? \; d_h \; ? \; d \lll q_{arbiter}$$
$$d_{des1}: \quad D(r_i+\, nq_i-) < D(r_i+\, y_i+) \quad \text{from optimization 1 } (i = 1, 2)$$
$$d_{des2}: \quad D(g_i+\, gb_i-) > D(g_i+\, y_i+) \quad \text{from optimization 2 } (i = 1, 2)$$

We conclude that the optimized arbiter in Fig. 5 (a) can substitute the arbiter in Fig. 4(a).

Acknowledgments

We thank Mark Greenstreet for numerous comments, and Mark Josephs for discussions that inspired Example 1. We gratefully acknowledge financial support from NSERC, FCAR, CFI, MEQ, and McGill University.

REFERENCES

[1] V. Akella and G. Gopalakrishnan. "SHILPA: A high-level synthesis system for self-timed circuits." In *Int. Conf. Computer-Aided Design (ICCAD)*, pp. 587–591, 1992.

[2] R. Alur, T. Henzinger, F. Mang, S. Qadeer, S. Rajamani, and S. Tasiran. "MOCHA: modularity in model checking." In *Computer-Aided Verification (CAV)*, pp. 521-525, 1998.

[3] E. Brunvand and R. F. Sproull. "Translating concurrent programs into delay-insensitive circuits." In *Int. Conf. Comput. Design*, pp. 262–265, 1989.

[4] E. M. Clarke, O. Grumberg, and D. E. Long. "Model checking and abstraction." In *Proc. Symp. on Principles of Programming Languages*, pp. 343-354, 1992.

[5] E. M. Clarke, D. E. Long, and K. L. McMillan. "Compositional Model Checking." In *Proc. Annual Symp. on Logic in Computer Science (LICS)*, pp. 353-362, 1989.

[6] D. L. Dill. *Trace Theory for Automatic Hierarchical Verification of Speed-Independent Circuits*. MIT press, 1989. (An ACM Distinguished Dissertation.)

[7] G. Gopalakrishnan. "Peephole Optimization of Asynchronous Macromodule networks." *IEEE Transitions on Very Large Scale Integration Systems*, (7)1: 1999.

[8] M.R. Greenstreet and T. Ono-Tesfaye. "A fast, asP* RGD arbiter." In *Proc. Int. Symp. on Advanced Research on Asynchronous Circuits and Systems*, pp. 173-85, 1999.

[9] C. A. R. Hoare. *Communicating Sequential Processes*. Prentice-Hall, 1985.

[10] T. Henzinger, S. Qadeer, and S. Rajamani. "You assume, We Guarantee: Methodology and Case Studies." In *Computer-aided Verification (CAV)*, pp. 440-451, 1998.

[11] T. Henzinger, S. Qadeer, S. Rajamani, and S. Tasiran. "An Assume-Guarantee Rule for Checking Simulation." In *Proc. Int. Conf. on Formal Methods in Computer-aided Design (FMCAD)*, pp. 421-432, 1998.

[12] S. Hauck. "Asynchronous design methodologies: an overview." *Proceedings of the IEEE*, Vol. 83, pp: 69-93, 1995.

[13] X. Kong, R. Negulescu. "Formal Verification of Pulse-Mode Asynchronous Circuits." In *Proc. Asia and South Pacific Design Automation Conference (ASP-DAC), Special Session on Asynchronous Circuits*, pp. 347–352, 2001.

[14] C. Loiseaux, S. Graf, J. Sifakis, A. Bouajjani, and S. Bensalem. "Property preserving abstractions for the verification of concurrent systems." *Formal Methods in System Design*, (6):1-35, 1995.

[15] K.L. McMillan. "A compositional rule for hardware design refinement." In *Computer-Aided Verification (CAV)*, pp 24-35, 1997.

[16] C. E. Molnar, I. W. Jones, B. Coates, and J. Lexau. "A FIFO ring oscillator performance experiment." *In Proc. International Symposium on Advanced Research in Asynchronous Circuits and Systems*, 1997.

[17] R. Negulescu. *Process Spaces and Formal Verification of Asynchronous Circuits*. Ph.D. thesis, University of Waterloo, Canada, 1998.

[18] R. Negulescu. "Process spaces." In *Proc. Int. Conf. on Concurrency Theory (CONCUR)*, pp. 196-210, 2000.

[19] R. Negulescu and A. Peeters. "Verification of speed-dependences in single-rail handshake circuits." In *Proc. Int. Symp. on Advanced Research in Asynchronous Circuits and Systems*, pp. 159-170, 1998.

[20] J. Sifakis. "Property preserving homomorphisms of transition systems." In *Proc. 4th Workshop on Logics of Programs*, June 1983.

SYMBOLIC VERIFICATION OF COMPLEX REAL-TIME SYSTEMS WITH CLOCK-RESTRICTION DIAGRAM

Farn Wang

Institute of Information Science, Academia Sinica

Taipei, Taiwan 115, ROC

farn@iis.sinica.edu.tw

Abstract Real-world real-time systems may involve many complex structures, which are difficult to verify. We experiment with the model-checking of an application-layer html-based web-camera which involves structures like event queues, high-layer communication channels, and time-outs. To contain the complexity, we implement our verification tool with a newly developed BDD-like data-structure, reduced CRD (Clock-Restriction Diagram), which has enhanced the verification performance through intensive data-sharing in a previous report. However, the representation of reduced CRD does not allow for quick test of zone containment. To this purpose, we thus have designed a new CRD-based representation, cascade CRD, which has given us enough performance enhancement to successfully verifying several implementations of the web-camera.

Keywords: Real-time systems, BDD, verification, Clock-Restriction Diagram

1. INTRODUCTION

Fully symbolic model-checking with BDD data-structure[8, 4] has achieved great success in hardware systems and become one of the standard technologies in Electronic Design Automation (EDA). Naturally, people are now looking forward to repeating the success in model-checking complex systems, which may involve structures like queues, stacks, clocks, communicaiton channels, ... Such structures usually cause great complexities to verify with BDD-like data-structures. In this work, we experiment with a real-world project, an html-based web-camera (`http://www.mmedia.com.tw`) implemented in application layer, and discuss how we have coped with the challenges in verification. Especially, we implemented the newly developed CRD (Clock-Restriction Diagram), which is a BDD-like data-structure for space-efficiency in timed automaton verification, in our model-checking tool and found that the representation of reduced CRD does not allows for efficient testing of zone containment. It was only with our new representation scheme of cascade CRD that we had been able to verify implementations of the web-cameras.

Most modern model-checkers for real-time systems are built around some symbolic manipulation procedures[12] of zones implemented in data-structures like DBM[10], NDD[1], CDD[7], RED[15], or CRD[16]. A zone means a behaviorally equivalent state subspace of a timed automaton and is symbolically represented by a set of difference constraints between clock pairs. It has been the general opinion that DBM (difference-bounded matrix) [10] is perhaps the best data-structure for the representation and manipulation of zones. For a long time, BDD-like structures[1, 7, 15, 16] have not performed as well as the popular DBM (difference-bounded matrix) [10] which is a 2-dimensional matrix and nothing BDD-like. But recently, there comes a new BDD-like data-structure, called CRD (Clock-Restriction Diagram) [17], which has shown efficiency in experiments (see appendix 6). CRD uses evaluation variables corresponding to the entries in DBM. Particularly, the default value of variables, say $x - x'$, is treated as $x - x' < \infty$ (i.e. no restriction). Since a CRD is actually a DBM set represented in decision diagram, it is neither a canonical representation of dense-domain state-spaces. The representation of CRD with zones with smallest set of inequalities[14], i.e. reduced CRD, was proposed in [17] to cope with space-complexity. But the drawback of reduced CRD is that zone containment cannot be determined efficiently. Thus many many zones, which are contained by others in the reachable state space representation, are calculated again and again. As a result, both space and time-complexities become unmanageable and our verification tool with reduced CRD could not finish verifying several implementations of the web-camera.

Note that the problem, with the zone containement relation, mentioned in the last paragraph cannot be simply handled by using the representation of CRD with all-pair shortest-path representation zones [10, 14] because as argued in [17], such representation is less space-efficient than reduced CRD. To overcome this obstacle, we here propose a new representation of CRD, the cascade CRD, which is not a representation of set of zones with minimal number of inequalities. For a zone with identical clock readings, cascade CRD may contain more inequalities than reduced CRD. But for a given state-space (i.e. a set of zones), cascade CRD may result in much less zones recorded than reduced CRD with efficient elimination of zones contained by others. It is with the cascade CRD that we have successfully verified several implementations of the web-camera system.

Here is our presentation plan. Section 2 briefly defines timed automata as our model for discussion. Section 3 basically restates part of [17] to give readers a background knowledge to the issues in verification research of timed automata and CRD. Section 4 presents cascade CRD and its computation. Section 5 reports our tool implementations and experiments with the web-camera. Appendix 6 shows how good our CRD-based verification tool is compared to two well-known verification tools and justifies our choice to continue experiments with CRD.

2. TIMED AUTOMATA

We use the widely accepted timed automata[2] as our model. A timed automaton is a finite-state automaton equipped with a finite set of clocks which can hold nonnegative real-values. It is structured as a directed graph whose nodes are modes (control locations) and whose arcs are transitions. The modes are labeled with invariance conditions while the transitions are labeled with triggering conditions and a set of clocks to be reset during the transitions. The invariance conditions and triggering conditions are Boolean combinations of inequalities comparing a clock with an integer. At any moment, the timed automaton can stay in only one mode. In its operation, one of the transitions can be triggered when the corresponding triggering condition is satisfied. Upon being triggered, the automaton instantaneously transits from one mode to another and resets clocks in the corresponding transition clock set label to zero. In between transitions, all clocks increase their readings at a uniform rate.

For convenience, given a set X of clocks, we use $B(X)$ as the set of all Boolean combinations of inequalities of the form $x - x' \sim c$ where $x, x' \in X \cup \{0\}$, "$\sim$" is one of $\leq, <, =, >, \geq$, and c is an integer constant.

Definition 1 <u>automata</u> A timed automaton A is given as a tuple $\langle X, Q, q_0, I, \mu, T, \gamma, \tau, \pi \rangle$ with the following restrictions. X is the set of clocks. Q is the set of modes. q_0 is the initial mode. $I \in B(X)$ is the initial condition on clocks. $\mu : Q \mapsto B(X)$ defines the invariance condition of each mode. T is a finite set of transitions. $\gamma : T \mapsto (Q \times Q)$ describes the source and destination modes of transitions. $\tau : T \mapsto B(X)$ and $\pi : T \mapsto 2^X$ respectively defines the triggering condition and the clock set to reset of each transition. ||

A valuation of a set is a mapping from the set to another set. Given an $\eta \in B(X)$ and a valuation ν of X, we say ν satisfies η, in symbols $\nu \models \eta$, iff when the variables in η is interpreted according to ν, η will be evaluated true.

Definition 2 <u>states</u> Given a timed automaton $A = \langle X, Q, q_0, I, \mu, T, \gamma, \tau, \pi \rangle$, A state ν of A is a valuation of $X \cup \{\texttt{mode}\}$ such that

- $\nu(\texttt{mode}) \in Q$ is the mode of A in ν with mode as a special auxiliary variable; and
- for each $x \in X$, $\nu(x) \in \mathcal{R}^+$ such that $\mathcal{R}^+$ is the set of nonnegative real numbers and $\nu \models \mu(\nu(\texttt{mode}))$. ||

For any $t \in \mathcal{R}^+$, $\nu + t$ is a state identical to ν except that for every clock $x \in X$, $\nu(x) + t = (\nu + t)(x)$. Given $\bar{X} \subseteq X$, $\nu\bar{X}$ is a new state identical to ν except that for every $x \in \bar{X}$, $\nu\bar{X}(x) = 0$.

Definition 3 runs Given a timed automaton $A = \langle X, Q, q_0, I, \mu, T, \gamma, \tau, \pi \rangle$, a ν-run is an infinite sequence of state-time pair $(\nu_0, t_0)(\nu_1, t_1) \ldots (\nu_k, t_k) \ldots\ldots$ such that $\nu = \nu_0$ and $t_0 t_1 \ldots t_k \ldots\ldots$ is a monotonically increasing real-number (time) divergent sequence, and for all $k \geq 0$,

- for all $t \in [0, t_{k+1} - t_k]$, $\nu_k + t \models \mu(\nu_k(\mathtt{mode}))$; and
- either $\nu_k(\mathtt{mode}) = \nu_{k+1}(\mathtt{mode})$ and $\nu_k + (t_{k+1} - t_k) = \nu_{k+1}$; or for some $w \in T$,
  - $\gamma(w) = (\nu_k(\mathtt{mode}), \nu_{k+1}(\mathtt{mode}))$; and
  - $\nu_k + (t_{k+1} - t_k) \models \tau(w)$; and
  - $(\nu_k + (t_{k+1} - t_k))\pi(w) = \nu_{k+1}$. ||

A safety requirement on timed automaton A can be written as a Boolean combination of clock constraints in $B(X)$ and mode restrictions in the form of $\mathtt{mode} = q$ meaning that A is currently in mode $q \in Q$. A run $\rho = (\nu_0, t_0)(\nu_1, t_1) \ldots (\nu_k, t_k) \ldots\ldots$ of A satisfies safety requirement η, in symbols $\rho \models \eta$, iff for all $k \geq 0$ and $t_k \leq t \leq t_{k+1}$, $\nu_k + t \models \eta$. We say $A \models \eta$ iff for all ν-runs ρ, $\nu \models I \wedge (\mathtt{mode} = q_0)$ implies $\rho \models \eta$. Our verification framework is safety analysis problem that when given A and η, asks whether $A \models \eta$.

3. BASICS

To better prepare readers for understanding of the materials, we restate some of the paragraphs form [17] in this section. Subsection 3.1 discusses the concept of zones. Subsection 3.2 formally defines CRD and its manipulations. Especially, subsection 3.2.2 illustrates how reduced CRD can be more space-efficient than the other technologies.

3.1 Zones, closure form, and reduced form

Most modern model-checkers are built around some symbolic manipulation procedures[12] of zones implemented in data-structures like DBM, NDD, CDD, RED, or CRD. A zone means a behaviorally equivalent state subspace of a timed automaton and is symbolically represented by a set of difference constraints between clock pairs. Previously, people still believed that DBM was the most efficient data-structure. DBM-technology generally handles the complexity of timing constant magnitude very well. But when the number of clocks increases, its performance also degrades rapidly.

Let $\mathcal{Z}$ be the set of integers. Given $c \geq 0$ and $c \in \mathcal{Z}$, let $\mathcal{I}_c$ be $\{\infty\} \cup \{d \mid d \in \mathcal{Z}; -c \leq d \leq c\}$. Also for any $d \in \mathcal{Z}$, $d + \infty = \infty + d = \infty$.

Given a safety analysis problem for a timed automaton A with biggest timing constant C_A used in A, a zone is a convex subspace of $\mathcal{R}^{|X|}$ constrained by half spaces represented by inequalities like $x - x' \sim d$, with $x, x' \in X \cup \{0\}$, $\sim \in \{``\leq", ``<"\}$, and $d \in \mathcal{I}_{C_A}$, such that when $d = \infty$, $\sim$ must be "<". For convenience, let $\mathcal{B}_c = \{(\sim, d) \mid \sim \in \{``\leq", ``<"\}; d \in \mathcal{I}_c; d = \infty \Rightarrow \sim = ``<"\}$. With respect to given X and C_A, the set of all zones is finite. Formally, a zone ζ can be defined as a mapping $(X \cup \{0\})^2 \mapsto \mathcal{B}_{C_A}$. Alternatively, we may also define a zone ζ as the set $\{x - x' \sim d \mid \zeta(x, x') = (\sim, d)\}$. In the following, we shall use the two equivalent definitions flexibly as we see fit.

There can be many zones representing the same convex subspace. A straightforward canonical representation of a zone-characterizable convex subspace is its zone in closure form (called shortest-path closure in [14]). A zone ζ is in closure form if and only if for any sequence of elements $x_1, \ldots, x_k \in X \cup \{0\}$, with $x_1 - x_k \sim d \in \zeta$ and $\forall 1 \leq i < k (x_i - x_{i+1} \sim_i d_i \in \zeta)$, either $d < \sum_{1 \leq i < k} d_i$ or $(d = \sum_{1 \leq i < k} d_i \wedge (\sim = ``\leq" \Rightarrow \bigwedge_{1 \leq i < k} \sim_i = ``\leq"))$. Intuitively, this means that every half space constraint has to be tight. We can artificially designate the closure form of each zone as our canonical representation of the corresponding state subspace characterized by the zone. For convenience, given a zone ζ, we let ζ^C be the notation for its closure form.

A few terms to define before we explain the second candidate for zone canonical representation. Two clocks $x, x' \in X \cup \{0\}$ are equivalent in a zone ζ, in symbols $x \equiv_\zeta x'$, iff $\exists d \in \mathcal{Z}(x - x' \leq -d \in \zeta^C \wedge x' - x \leq d \in \zeta^C)$. For convenience, assume that $X = \{x_1, \ldots, x_n\}$ and 0 is also named x_0. If $Y = \{x_{i_1}, x_{i_2}, \ldots, x_{i_k}\} \subseteq X \cup \{0\}$ is a maximal set of equivalent clocks in ζ such that $i_1 < i_2 < \ldots < i_k$, let $\min_{\equiv_\zeta}(x_{i_j}) = x_{i_1}$ for all $1 \leq j \leq k$. Given $x - x' \sim d \in \zeta^C$, (x, x') is redundant in ζ iff $x \not\equiv_\zeta x'$ and there is a $\bar{x} \in X \cup \{0\}$, with $x - \bar{x} \sim_1 d_1, \bar{x} - x' \sim_2 d_2 \in \zeta^C$, such that $x \not\equiv_\zeta \bar{x} \not\equiv_\zeta x'$ and $d = d_1 + d_2 \wedge (\sim = ``<" \Rightarrow (\sim_1 = ``<" \vee \sim_2 = ``<"))$.

Another candidate for the canonical representation of zones is the reduced form (called shortest-path reduction in [14]) which records only minimum number of constraints for each zone. We refer interested readers to [14, 17] for explanation how to convert a given zone ζ to its zone in reduced form, in symbols ζ^R. It is shown in [14] that $\zeta^C = (\zeta^R)^C$; and DBM with zones in reduced form can be used as a canonical representation of timed automaton convex states-spaces and can significantly save space in model-checking.

3.2 Clock Restriction Diagram

CRD is not a decision diagram for state space membership. Instead it is like a database for zones. We devise the new data-structure CRD exactly because CRD acts like a database (recording device) and is more suitable for comparison and manipulation of sets of clock diference constraints.

3.2.1 Previous data-structures. NDD[1] uses binary encoding for clock readings and its performance is very sensitive to timing-constant magnitude.

CDD[7] is a decision diagram for state-space membership. It has very similar structure to CRD. There are two major differences of CRD. First, a value d for evaluation variable, say $x - x'$, in CDD means $x - x' = d$ while a value like $(\sim, d)$ means $x - x' \sim d$. This is because CDD is designed to be decision diagram. Second, the default value of a variable in CRD is interpreted as $(<, \infty)$ while it is interpreted as $(-\infty, \infty)$ in CDD. To reduce information duplication in representations of state-spaces, CDD has to be transformed to closure form [10, 7] which records the shortest-path distances between all pairs of clocks and is very space-inefficient.

RED[15, 16] encodes the ordering of fractional parts of clock readings in the variable ordering and has achieved very high space-efficiency for systems with large number of clocks and small timing constants. RED is indeed a canonical representation of timed automaton state subspaces. But for large timing constants, RED's performance degrades rapidly.

At this moment, DBM is still the most popular and efficient data-structure. DBM-technology generally handles the complexity of timing constant magnitude very well. But when the number of clocks increases, its performance also degrades rapidly.

3.2.2 Definition . CRD is a directed acyclic graph for representation of sets of zones. It has similar structure as BDD without FALSE terminal. Each of the pairs $(x, x') \in (X \cup \{0\})^2$ is treated as an evaluation variable. By fixing an evaluation order, we can construct a CRD just as BDD, CDD, or RED. For example, given $C_A = 10$, the CRD for a set $\{\{0 - x_1 \leq -3, x_1 - x_3 < -4\}, \{0 - x_2 < -1, x_2 - x_1 < 6\}\}$ of two zones (constraints of the form $x - x' < \infty$ are omitted) is in figure 1(a). In CRD, a missing constraints on differences of clock pairs, say x, x', is interpreted as $x - x' < \infty$. Thus in the root vertex, even no constraint is on $0 - x_1$ in the latter zone, we still construct an arc with $0 - x_1 < \infty$ from the root vertex. This is one major difference of our CRD from decision diagrams like CDD which interprets a missing restriction on x, x' as $-\infty < x - x' < \infty$ with an implied lowerbound of $-\infty$ on $x - x'$.

An evaluation index $\omega : (X \cup \{0\})^2 \cup \{\text{true}\} \mapsto \{0, 1, \ldots, |(X \cup \{0\})^2|\}$ is a mapping such that $\omega(\text{true}) = |(X \cup \{0\})^2|$ and for every two $e, e' \in (X \cup \{0\})^2 \cup \{\text{true}\}, \omega(e) \neq \omega(e')$.

Definition 4 Clock Restriction Diagram (CRD) A CRD is a labeled directed acyclic graph $D = (V, \phi, E, \lambda)$, with single source and single sink, constructed under a given evaluation index ω such that

- V is the set of vertices;
- $\phi : V \mapsto (X \cup \{0\})^2 \cup \{\text{true}\}$ defines the evaluation variable at each vertex;

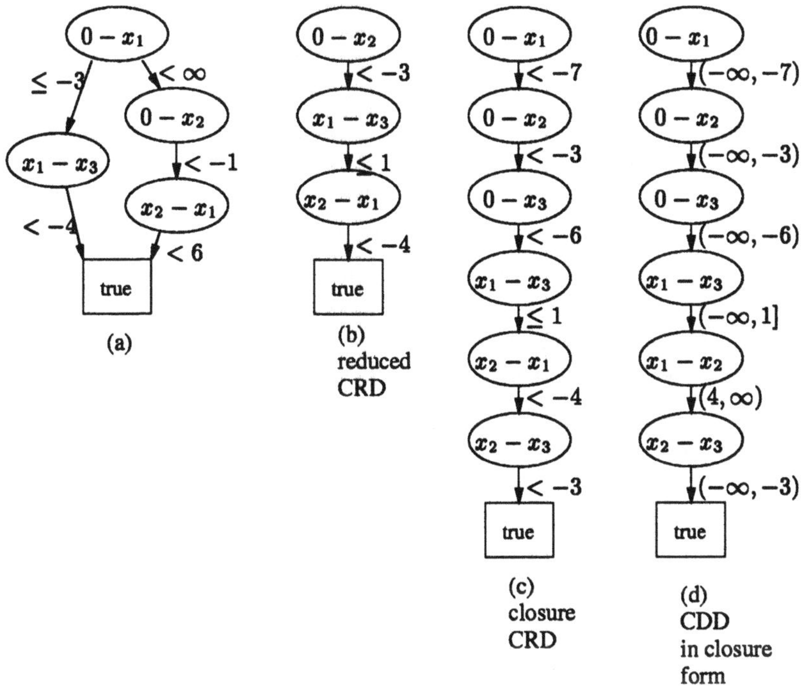

Figure 1. examples of CRD and comparison with CDD

- $E \subseteq V \times V$ is the set of arcs such that for every $(v, v') \in E$, $\omega(\phi(v)) < \omega(\phi(v'))$ (i.e., evaluation ordering must be respected); and
- $\lambda : E \mapsto \mathcal{I}_{C_A}$ such that for every $(v, v'), (v, v'') \in E$, $v' \neq v'' \Rightarrow \lambda(v, v') \neq \lambda(v, v'')$.

There is at most one $v \in V$ such that $\phi(v) = \text{true}$ and this v is the single sink of the CRD. ||

Since many zones can represent the same subspace, like DBM and CDD, neither is CRD a canonical representation of zone-characterized state spaces. To reduce information duplication in state-space representation, one solution is to convert all zones to their closure form[10, 7] (called shortest-path closure in [14]), which is the set of all pairwise clock difference constraints derived from the all-pair shortest-path distances, and only store their closure form. Such conversion is expensive and, as we shall illustrate in figure 1(b)-(d), incurs large space consumption with data-structures like CDD and CRD.

Also note that although zones in closure form can be used as canonical representation for convex state-spaces characterizable by zones, CRD with closure form zones is not a canonical representation of dense-time state-spaces. This is because CRD with closure form zones does not intrinsically have the capability to eliminate zones contained by the unions of other zones.

An alternative solution for space-efficient representation of state spaces is zones in their reduced form (called shortest-path reduction in [14]) which contains minimal number of clock difference constraints chosen by a policy. As shown in [14], DBM with zones in reduced form can be space-efficient. Wang has proposed to use CRD with zones in reduced form (or reduced CRD in short) as the representation for state-spaces to enhance verification efficiency [17]. Moreover, a space-efficient algorithm, with only four auxiliary variables, for computing reduced CRD has also been presented in [17].

Note again that reduced CRD is neither a canonical representation of dense-time state-space. Nevertheless, in average, CRD with zones in reduced form (reduced CRD in short) is much more space-efficient than previous data-structures [1, 7, 15, 16]. For example, given $X = \{x_1, x_2, x_3\}$ and $C_A = 10$, in figure 1(b),(c),(d), we have the representations of zone $\{0 - x_2 < -3, x_1 - x_3 \leq 1, x_2 - x1 < -4\}$ in reduced CRD (figure 1(b)), in CRD with zones in closure form (figure 1(c)), and in CDD with zones in closure form (called tightened form in [7]) without FALSE terminal vertex (figure 1(d)). It is easy to see that as the number of clocks increases, reduced CRD will perform better and better.

3.2.3 Set-oriented manipulations on CRD . For convenience of discussion, given a CRD, we may just represent it as the set of zones recorded in it. Set-union ($\cup$), set-intersection ($\cap$), and set-exclusion ($-$) of two zone sets respectively represented by two CRDs are straightforward. For example, given CRDs $D_1 : \{\zeta_1, \zeta_2\}$ and $D_2 : \{\zeta_2, \zeta_3\}$, $D_1 \cap D_2$ is the CRD for $\{\zeta_2\}$; $D_1 \cup D_2$ is the CRD for $\{\zeta_1, \zeta_2, \zeta_3\}$; and $D_1 - D_2$ is the CRD for $\{\zeta_1\}$. The complexities of the three manipulations are all $O(|D_1| \cdot |D_2|)$.

Set-extraction ($|$) selects zones satisfying certain features from a zone set. Suppose D is the CRD for $\{\zeta_1, \ldots, \zeta_k\}$ and D' is the CRD for $\{\zeta'_1, \ldots, \zeta'_h\}$, then $D|D' = \{\zeta_i \mid 1 \leq i \leq k; \exists 1 \leq j \leq h \forall x - x' \sim d \in \zeta'_j (d \neq \infty \Rightarrow x - x' \sim d \in \zeta_i)\}$. The complexity is also $O(|D_1| \cdot |D_2|)$.

Given two zones ζ_1 and ζ_2, $\zeta_1 * \zeta_2$ is a new zone representing the space-intersection of ζ_1 and ζ_2. Formally speaking, for every x, x' with $\zeta_1(x, x') = (\sim_1, d_1)$ and $\zeta_2(x, x') = (\sim_2, d_2)$, $\zeta_1 * \zeta_2(x, x') = (\sim_1, d_1)$ if $d_1 < d_2 \vee (d_1 = d_2 \wedge \sim_1 = \text{“}<\text{”})$; or $\zeta_1 * \zeta_2(x, x') = (\sim_2, d_2)$ otherwise. Space-intersection ($*$) of two CRDs D_1 and D_2, in symbols $D_1 * D_2$, is a new CRD for $\{\zeta_1 * \zeta_2 \mid \zeta_1 \in D_1; \zeta_2 \in D_2\}$. Our current implementation of the manipulation has complexity $O(|D_1|^2 \cdot |D_2|^2)$.

3.2.4 CRD and BDD. It is possible to combine CRD and BDD into one data-structure for fully symbolic manipulation. Since CRD only has one sink vertex: true, it is more compatible with BDD without FALSE terminal vertex which is more space-efficient than ordinary BDD. There are two things we need to take care of in this combination. The first is about the interpretation of default values of variables. In BDD, when we find a variable is missing during valuating variables along a path, the variable's value can be interpreted as either TRUE or FALSE. But in CRD, when we find a variable for constraint $x - x'$ is missing along a path, then the constraint is interpreted as $x - x' < \infty$.

The second is about the interpretation of CRD manipulations to BDD variables. Straightforwardly, "$\cup$" on Boolean variables is interpreted as "$\vee$" on Boolean variables. "$\cap$" and "$|$" on Boolean variables are both interpreted as "$\wedge$" on Boolean variables. $D_1 - D_2$ on Boolean variables is interpreted as $D_1 \wedge \neg D_2$ when the root variable of either D_1 or D_2 is Boolean. For $D_1 * D_2$, the manipulation acts as "$\wedge$" when either of the root variables are Boolean. Due to page-limit, we shall omit the proof for the soundness of such interpretation.

From now on, we shall call it CRD+BDD a combination structure of CRD and BDD.

3.2.5 Variable ordering in CRD. In our BDD+CRD, we found the following evaluation ordering quite efficient in our experiment.

- All discrete variables precede those clock inequality variables in the evaluation ordering.
- Let $0 \prec x_1 \prec \ldots \prec x_n$. For clocks $x, x', y, y' \in \{0, x_1, \ldots, x_n\}$, $x - x'$ precedes $y - y'$ iff either (1) $(x \prec y \wedge x \prec y') \vee (x' \prec y \wedge x' \prec y')$; or (2) $x \prec y \wedge x = y' \wedge x' = y$.

Especially, the condition (2) of item 2 puts variable like $x_2 - x_1$ immediately below $x_1 - x_2$ and allows us to efficiently check for some trivial negative cycles.

4. CASCADE CRD

Reduced CRD indeed can be very space-efficient, as reported in [17], for some applications. But it is more difficult to efficiently decide the containment relation between two zones with reduced CRD. For example, we may have the reduced CRD in figure 2(a) for the state-space represented by formulus (1) in the following.

$$\begin{array}{ll} & (x_1 - x_2 \leq -2 \wedge x_2 - x_4 \leq -3 \wedge x_4 - x_1 \leq 5 \wedge x_1 = x_3) \\ \vee & (x_1 - x_2 \leq -2 \wedge x_2 - x_4 \leq -3 \wedge x_4 - x_1 \leq 5) \end{array} \qquad (1)$$

As can be seen that the zone of the first conjunction is actually contained in that of the second. This containment relation can be easily computed with all-pair shortest-path difference relation between clock pairs. But when we transform a given CRD into its reduced CRD, there is not enough information left to efficiently derive such containment relation. In specific, we cannot use the simple reduce operations described in the last paragraph of subsection 3.2.3 to eliminate zones which are contained by others in the same CRD. Thus, many

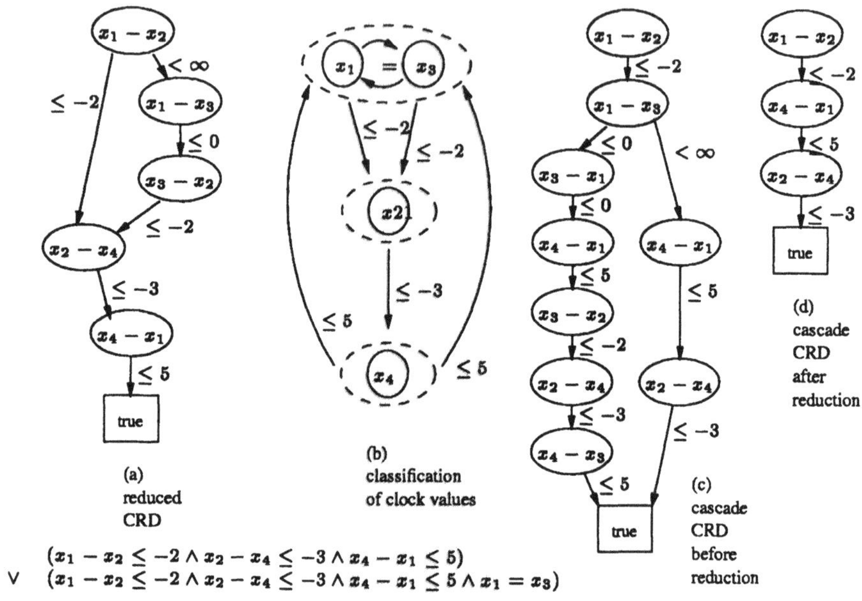

Figure 2. reduction in cascade CRD

many zones may be represented in CRD for the reachable zone set which are actually subsets of zones already generated before. This problem may result in huge waste in both space and computation time. Actually, in our experiment, none of the web-camera implementations can be verified with reduced CRD.

Here we propose another representation scheme of CRD, called cascade CRD, which allows us to efficiently control CRD complexity with the simple reduction operations mentioned at the end of subsection 3.2.3. For the convenience of defining cascade CRD, we first define the cascade form of zones. A cascade CRD is merely a CRD, all whose zones are represented in their cascade form. Just like closure form and reduced form of zones, the cascade form of a zone is also a canonical representation of zone-characterizable state-spaces. The idea is to add a few more inequalities to zones in reduced forms to facilitate the decision of zone containment relation. Cascade CRD is especially designed for those systems which can have states in which many clocks have identical values. This kind of states may happen because of clock-reset operations at a synchronization, between a sender and a receiver, which is very common in modelling tightly-coupled interactions in concurrent systems.

In figure 2(b), we illustrate how we compute the cascade zone from its non-cascade zone represented by the first conjunction of formulus (1), that is:

$$(x_1 - x_2 \leq -2 \wedge x_2 - x_4 \leq -3 \wedge x_4 - x_1 \leq 5 \wedge x_1 = x_3)$$

Given a zone ζ, we use ζ^S to denote the cascade form of ζ. The procedure to compute ζ^S from ζ can be presented as the following steps.

1 We classify the clocks in a state into equivalence classes, just as what has been done in [14]. Two clocks x, x' are in the same equivalence classes in a zone ζ iff $x - x' = d \in \zeta^C$ for some $d \in \mathcal{N}$. ($x - x' = d \in \zeta^C$ is a shorthand for $x - x' \leq d \in \zeta^C \wedge x' - x \leq -d \in \zeta^C$.) For example, in the first conjunction of formulus (1), all four clocks are in the same equivalence class; while in the second conjunction, only clock x_1, x_2, x_4 are in the same equivalence class.

   In the following, then we divide the task into two subtasks. The first task is for adding inequalities in ζ^S between those clocks in the same equivalence classes while the second task is for adding inequalities in ζ^S between those clocks in different classes.

2 **Case of clocks in the same equivalence class:** Suppose we are given an equivalence class of clocks in a zone ζ.

   (a) We first classify clocks in the given equivalence class into identity classes. Two clocks are in the same identity class if they have the same reading. For example, in the first conjunction of formulus (1), only clocks x_1, x_3 are in the same identity class and all other two clocks are not in the same identity class. But in the second conjunction, no two clocks are in the same identity class.

   (b) Then we arrange the identity classes in a linear sequence according to their less-than relation. For example, in the first conjunction of formulus (1), we have the following sequence.

   $$\{x_1, x_3\} \xrightarrow{-2} \{x_2\} \xrightarrow{-3} \{x_4\}$$

   Here the label of -2 is on the first "$\longrightarrow$" because $x_1 - x_2 \leq -2$ and $x_3 - x_2 \leq -2$. But in the second conjunction, the sequence for the equivalence class $\{x_1, x_2, x_4\}$ is

   $$\{x_1\} \xrightarrow{-2} \{x_2\} \xrightarrow{-3} \{x_4\}$$

   (c) For each two clocks x, x' in the same identity class, let $\zeta^S(x, x') = (\leq, 0)$.

   (d) For two clocks x, x' such that the identity class of x just precedes that of of x' in the linear sequence, let $\zeta^S(x, x') = \zeta^C(x, x')$.

   (e) For a clock x in the last identity class and a clock x' in the first identity class, let $\zeta^S(x, x') = \zeta^C(x, x')$.

3 **Case of clocks not in the same equivalence class:** Suppose we are given clock x, x' not in the same equivalence class. If x is in the last identity

class of its equivalence class and x' is in the first identity class of its equivalence class, then let $\zeta^S(x, x') = \zeta^C(x, x')$.

4 For any two inequivalent clocks x, x' such that (x, x') is redundant in ζ, we let $\zeta^S(x, x') = (<, \infty)$.

5 For all other clock difference relation not covered in the above-mentioned steps, we let $\zeta^S(x, x') = (<, \infty)$.

In the same style of the manipulation algorithm presented in [17], we can implement the above-mentioned procedure with symbolic CRD+BDD manipulation routines. But due to page-limit, we shall only present the above procedure. After CRD are stored with zones in cascade form (see for example in figure 2(c)), we can then use the reduction operation mentioned at the end of subsection 3.2.3 to efficiently eliminate many contained zones (see for example in figure 2(d)).

One good property of our cascade CRD is that for a zone ζ wihtout nontrivial identity classes (a class is trivial if it has only one element), then the corresponding zone in cascade form have the same number of inequalities as its counterpart in reduced form.

LEMMA 1 : Given a zone ζ such that no two clocks are in the same identity class in ζ, then ζ^S and ζ^R have the same number of inequalities of the form $x - x' \sim d$ with either "$\sim$"$\neq$"$<$" or $d \neq \infty$.

Proof : When there is no nontrivial identity class, we will skip step 2(c) in the our procedure. This skip makes our procedure essentially identical to the one for constructing zones in reduced form presented in [14], except for the following. For an equivalence class with clock x and without clock x', an inequality like $x - x' \sim d$ is called an incoming arc while one like $x' - x \sim d$ is called an outgoing arc. In the zones in reduced form constructed according to [14], only incoming and outgoing arcs to the clocks in the first trivial identity class in the sequence can be kept. But in our cascade zones, outgoing arcs are recorded for the clocks in the first trivial identity class in the sequence while incoming arcs are recorded for the clocks in the last trivial identity class in the sequence. With this way of accounting, it is easy that both cascade form and reduced form of the same zone have the same number of clock inequalities. ||

Lemma 1 shows that our cascade CRD only adds to complexity when it is making effect on its target zones.

5. IMPLEMENTATION AND EXPERIMENTS

We have implemented our CRD-technology in version 3.0 of our tool `red` which was previously announced in [15, 16] (version 1.0, 2.0) and supports the modelling and safety-analysis of real-time systems with multiprocesses, pointer data-structures, and synchronizations (synchronous send and receive) from one process to another. The new version, together with benchmarks, is now available at:

`http://www.iis.sinica.edu.tw/~farn/red`

Each process can use global and local variables of type clock, discrete, and pointer. Pointer variables either contain value NULL or the identifiers of processes. Thus in the models input to `red`, we allow complicate dynamic networks to be constructed with pointers.

At this moment, `red` supports backward reachability analysis. We have also implemented a reduction techniques in `red`. That is the reduction by elimination of inactive variables[13, 18] which is always executed. A variable is inactive in a state iff it is not read in any computation from the state before its content is overwritten. Contents of inactive variables can be omitted from state information without any effect on the computations.

We have tested our verifier with four implementations of the web-camera system, each with only one client. We have cooperated with Metamedia, a local company specializing in driver and peripheral software (`http://www.mmedia.com.tw`), to test our new CRD-technology. One of their current projects is web cameras with browser interface (HTML language). The product supports multi-user connections, with web browsers, to dynamically monitor remote activities through internet. The software operates in application layer with an event queue and software interrupt handling. We want to analyze how long the event queue buffer needs to be.

In the following, we shall first describe how we model the system behavior, and then present our analysis result in section 5. There are three parameters in the system: number of clients (in symbols $\#C$), maximal length of queue (L), and the communication delay between server and clients $[\beta, \alpha]$. We have modeled implementations of the system with various values of $\#C$, L, and $[\beta, \alpha]$. The total number of processes we need is $3 + 2\#C$. The processes are described in the following.

- The event queue which supports operations of dequeue and enqueue; and signalling of new event in the queue to the server. We need variables $e_0, e_1, \ldots, e_{L-1}$ to records the content of the queue; variable l to records the current length of the queue.
- The camera process is also modeled by the single-mode automaton which outputs a video image every 20 time units. The video-ready event will be caught by the queue process throught a synchronizer.
- $\#C$ timer processes for communication respectively with $\#C$ clients. A timer process signals the first timeouts when its corresponding client does not respond in 40 time units. Then the server will send a new frame to the client. If still no response is received in 30 time units, the channel is disconncted.

- $\#C$ client processes which acknowledges within time interval $[\beta, \alpha]$ the reception of a frame or dies.
- The server process which processes the events in the queue in the following way.
  - When the event is "VIDEO_READY" by camera, the server sends out a video-frame to each client. The processing time per client per video-frame is 5 time units. After processing the "VIDEO_READY" event, a timer is activated with timeout value 40 time units for each client.
  - When the event is "TIME_OUT$_j^1$" for client j at 40 time units, the server sends out a video-frame to client j again (in 5-time-unit processing time) and activates a timer with timeout value 30 time units.
  - When the event is "TIME_OUT$_j^2$" for client j at 30 time units, then the server thinks that the client is dead and disconnect the service to client i.
  - When the event is "ACK$_j$" (acknowledgment) from client i, then it disables the timer activated for client i.

Two variables, each with possible values, orthogonally generate the four implementations. The first variable is L, the event queue maximal length, with possible values of 1 and 2. The second variable is the response time interval from the client $[\beta, \alpha]$ with possible intervals of $[1, 1]$ and $[20, 40]$. The property to verify is whether the queue will overflow or not.

With reduced CRD, our tool cannot finish the verification task in 6 hours for any of the four implementations. But with cascade CRD, our tool is capable of coming up with the following performance data.

	$L = 1$	$L = 2$
$[\beta, \alpha] = [1, 1]$	no overflow/584s/812k	no overflow/2946s/3257k
$[\beta, \alpha] = [20, 40]$	overflow/2320s/1378k	no overflow/3072s/2829k

s: seconds; k: kilobytes of memory in data-structure;

The performance data is collected on a Pentium II 366MHz with 256MB memory running Linux.

6. CONCLUSION

CRD gains its power primarily from intensive data-sharing. But its manipulations are usually burdened with high overhead. At this moment, CRD-technology only shows its edge when number of clocks is large (see the appendix). But we believe that as a new data-structure for the verification of timed automata, there can be much room for improvement on CRD. Especially, we will try out various zone forms and verification techniques appropriate for CRD.

REFERENCES

[1] Asarin, Bozga, Kerbrat, Maler, Pnueli, Rasse. Data-Structures for the Verification of Timed Automata. Proceedings, HART'97, LNCS 1201.

[2] R. Alur, C. Courcoubetis, D.L. Dill. Model Checking for Real-Time Systems, IEEE LICS, 1990.

[3] F. Balarin. Approximate Reachability Analysis of Timed Automata. IEEE RTSS, 1996.

[4] J.R. Burch, E.M. Clarke, K.L. McMillan, D.L.Dill, L.J. Hwang. Symbolic Model Checking: 10^{20} States and Beyond, IEEE LICS, 1990.

[5] M. Bozga, C. Daws. O. Maler. Kronos: A model-checking tool for real-time systems. 10th CAV, June/July 1998, LNCS 1427, Springer-Verlag.

[6] J. Bengtsson, K. Larsen, F. Larsson, P. Pettersson, Wang Yi. UPPAAL - a Tool Suite for Automatic Verification of Real-Time Systems. Hybrid Control System Symposium, 1996, LNCS, Springer-Verlag.

[7] G. Behrmann, K.G. Larsen, J. Pearson, C. Weise, Wang Yi. Efficient Timed Reachability Analysis Using Clock Difference Diagrams. CAV'99, July, Trento, Italy, LNCS 1633, Springer-Verlag.

[8] R.E. Bryant. Graph-based Algorithms for Boolean Function Manipulation, IEEE Trans. Comput., C-35(8), 1986.

[9] E. Clarke, O. Grumberg, M. Minea, D. Peled. State-Space Reduction using Partial-Ordering Techniques, STTT 2(3), 1999, pp.279-287.

[10] D.L. Dill. Timing Assumptions and Verification of Finite-state Concurrent Systems. CAV'89, LNCS 407, Springer-Verlag.

[11] C. Daws, A. Olivero, S. Tripakis, S. Yovine. The tool KRONOS. The 3rd Hybrid Systems, 1996, LNCS 1066, Springer-Verlag.

[12] T.A. Henzinger, X. Nicollin, J. Sifakis, S. Yovine. Symbolic Model Checking for Real-Time Systems, IEEE LICS 1992.

[13] P.-A. Hsiung, F. Wang. User-Friendly Verification. Proceedings of 1999 FORTE/PSTV, October, 1999, Beijing. Formal Methods for Protocol Engineering and Distributed Systems, editors: J. Wu, S.T. Chanson, Q. Gao; Kluwer Academic Publishers.

[14] K.G. Larsen, F. Larsson, P. Pettersson, Y. Wang. Efficient Verification of Real-Time Systems: Compact Data-Structure and State-Space Reduction. IEEE RTSS, 1998.

[15] F. Wang. Efficient Data-Structure for Fully Symbolic Verification of Real-Time Software Systems. TACAS'2000, March, Berlin, Germany. in LNCS, Springer-Verlag.

[16] F. Wang. Region Encoding Diagram for Fully Symbolic Verification of Real-Time Systems. the 24th COMPSAC, Oct. 2000, Taipei, Taiwan, ROC, IEEE press.

[17] F. Wang. Clock Restriction Diagram: Yet Another Data-Structure for Fully Symbolic Verification of Timed Automata, Technical Report TR-IIS-01-002, IIS, Academia Sinica, Taiwan, ROC.

[18] F. Wang, P.-A. Hsiung. Automatic Verification on the Large. Proceedings of the 3rd IEEE HASE, November 1998.

[19] S. Yovine. Kronos: A Verification Tool for Real-Time Systems. International Journal of Software Tools for Technology Transfer, Vol. 1, Nr. 1/2, October 1997.

Appendix: Performance comparison between our CRD-based tool and Kronos and UPPAAL

The following table, from [17], shows performance data from Kronos, UPPAAL, and `red` w.r.t. implementations of benchmarks with various numbers of concurrent processes. Fischer's mutual exclusion protocol are modified from [3, 13, 15, 18]. CSMA/CD is extracted from [19] while FDDI is from [5, 11]. Information on the three benchmarks can be found in `http://www.iis.sinica.edu.tw/~farn/red`.

benchmarks	concurrency	Kronos	UPPAAL	`red` no CRD	`red` (w. CRD)
Fischer's mutual exclusion	3 processes	0.03s	0.014s	45.42s/1448k	1.12s/48k
	4 processes	0.14s	0.197s	871.2s/12703k	8.09s/162k
	5 processes	0.989s	5.94s	O/M	48.6s/471k
	6 processes	O/M	537.7s	O/M	278s/1271k
	7 processes	O/M	O/M	O/M	1593s/3208k
	8 processes	O/M	O/M	O/M	9352s/7788k
CSMA/CD	3 processes	0.032s	0.0046s	O/M	0.89s/60k
	4 processes	0.071s	0.028s	O/M	3.63s/121k
	5 processes	0.309s	0.216s	O/M	13.8s/278k
	6 processes	1.915s	3.45s	O/M	60.6s/707k
	7 processes	O/M	172s	O/M	227s/1781k
	8 processes	O/M	O/M	O/M	709s/4426k
	9 processes	O/M	O/M	O/M	3580s/10851k
FDDI token-ring passing	11 stations	399s	0.34s	N/A	1.98s/502k
	12 stations	O/M	0.487s	N/A	2.90s/591k
	20 stations	O/M	4.05s	N/A	27.28s/1676k
	30 stations	O/M	25.27s	N/A	151s/4165k
	40 stations	O/M	95.99s	N/A	465s/7939k
	50 stations	O/M	O/M	N/A	1107s/13513k
	60 stations	O/M	O/M	N/A	1828s/21014k

data collected on a Pentium III 800MHz with 256MB memory running LINUX;
s: seconds; k: kilobytes of memory in data-structure; O/M: Out of memory; N/A: not available;

For the Fischer's and CSMA/CD benchmarks, UPPAAL is invoked with options "`-aSWD`." The performance data shows that CRD-technology is still more space-efficient and scales better w.r.t. number of clocks. For the FDDI benchmark, UPPAAL is invoked with options "`-STDda`" where "`d`" is for depth-first-search and suits very well for highly synchronous algorithms like FDDI.

We remind the readers that compared to those well-established tools like Kronos and UPPAAL. For example, UPPAAL can pass most of the benchmarks in very short time if the option of "convex-hull over-approximation" is switched on. But since our experiment was focused to argue that "with similar and equivalent verification techniques, BDD-like data-structures can outperform DBM," we think it is better to just compare with Kronos and UPPAAL in the context of precise verification. Definitely, the success experience of Kronos and UPPAAL will be the guide for future enhancement of `red`. In the future, when such techniques are incorporated with CRD-based technology, we believe there will be much room for performance enhancement.

VERIFYING A SLIDING-WINDOW PROTOCOL USING PVS

Vlad Rusu
IRISA/INRIA, Campus Univ. de Beaulieu, 35042 Rennes Cedex, France
rusu@irisa.fr

Abstract We present the deductive verification of safety and liveness properties of a sliding-window protocol using the PVS theorem prover. The protocol is modeled in an operational style which is close to an actual program. It has parametric window sizes for both sender and receiver, and unbounded, lossy communication channels carrying unbounded data. The proofs are done using invariant-strengthening techniques, encoded as PVS automated strategies based on heuristics and decision procedures.

1. INTRODUCTION

In the rapidly growing field of formal verification, *deductive techniques* [13] and tools [4, 8, 14] are general and powerful ways to verify properties of hardware and software systems. However, many practitioners and researchers hesitate to use deductive methods, which are perceived to be difficult, time-consuming and requiring advanced theoretical knowledge. This is in contrast with *algorithmic techniques* [2, 5, 9] which are completely automatic once a suitable abstraction of the system has been found and entered into, *e.g.*, a model-checking tool.

The goal of this paper is to advocate the use of deductive techniques. Specifically, we show that although deductive verification is not fully automatic, it can be performed in a systematic way, using specific approaches, strategies, and automatic procedures on decidable subproblems that reduce the need for human intervention. This intervention remains necessary, but only as a general guideline to organize the verification process, i.e, to decompose the verification problem into manageable subproblems. Then, the subproblems can be solved in a highly automated manner.

As a case study we use an infinite-state sliding-window protocol for which we verify safety and liveness properties using the PVS theorem prover [14]. The protocol is a popular application for verification methods, both algorithmic and deductive. To our knowledge, our work is among the most general in terms of the properties verified and of the level of detail at which the protocol is modeled.

The rest of the paper is organized as follows. In Section 2 we describe the sliding-window protocol and related work on algorithmic and deductive approaches that have been applied to it. In Section 3 we briefly present the PVS theorem prover and describe, on a very simple example, the *invariant-strengthening* proof technique. In Section 4 we present in some detail our encoding of the sliding-window protocol in PVS, and in Section 5 we show how to prove safety and liveness properties by making extensive use of invariant strengthening. We conclude the presentation with ideas on how to further automate the deductive verification process. The specification and verification took three weeks to a moderately experienced PVS user and amounted to proving a total of eighteen theorems. PVS is available free of charge at `http://pvs.csl.sri.com`, and our sliding-window case study can be downloaded from `http://www.irisa.fr/pampa/perso/rusu/forte`.

2. RELATED WORK

We briefly present the sliding-window protocol and previous approaches that have been employed to verify it.

The Sliding-Window Protocol The sliding-window protocol (see, *e.g.*, [17]) is a member of the data link layer. Its main function is to deliver a sequence of messages between a sender and a receiver in the correct order, regardless of possible losses that can occur during the transmission on the communication channels. The general principle underlying the protocol is that of pipelining. Instead of waiting for an acknowledgment after each message, the sender will continue sending messages until it has in its buffer, or "window", a total of, say, `sw` unacknowledged messages.

Similarly, the receiver will not just discard a message that has not the next expected sequence number, but will store it in its window, provided it is at most `rw` many positions from the expected message. With this flexibility on both sides and using adequate values for the window sizes `sw` and `rw`, typically computed from other parameters, *e.g.*, the expected travel times of messages, the protocol can achieve very efficient transmission rates.

Verification of the Protocol The properties to be verified are divided into *safety* properties, which ensure that messages are delivered in the correct order, and *liveness* properties, which guarantee that every message sent is eventually delivered.

Algorithmic approaches. These approaches typically verify finite-state instances of the protocol. In [15] safety properties for small window and channel sizes are verified by brute-force model-checking. In [12] the authors use one specific equivalence relation that allows them to handle window sizes as high as seven and to prove both safety and liveness properties. More recently, the paper [6] shows how to verify both kinds of properties on a version of the protocol with channel size eleven and window sizes two for both sender and receiver, using a specialized data structure called a Queue Difference Diagram.

Deductive approaches. These approaches typically verify general infinite-state specifications of the protocol, *e.g.*, with unbounded window and channel sizes. An early deductive verification effort is presented in [1]. Only safety properties are verified, and a large number of invariants (over a thousand) were involved in the process. Another early work [3] presents the verification of safety and liveness properties of a specification of the protocol which includes real-time aspects. The level of automation of tools available at that time was low and the verification process included proving a large number (tens to hundreds) of trivial facts, *e.g.*, that addition is commutative.

A recent paper involving a much more automated approach is [16]. Here, the authors use the weak monadic second-order logic with one successor (WS1S), a decidable logic that is expressive enough to encode finite sets and sequences and limited integer operations. A higher-level language is used to describe both the protocol and safety properties, which are then translated to WS1S. Two versions of the protocol are verified. One version considers a sender with an unbounded window size, which can be nondeterministically modified on-the-fly. In the second version, the sender's window size is 256. The receiver's window size is, in both versions, one. A total of eight invariants are required to prove the main safety property. Liveness properties are not considered.

This work. We verify an operational description of the protocol which is very close to a programming language, with parametric window sizes for both sender and receiver and unbounded communication channels. As [16], we emphasize on generality, automation, and compactness (*i.e.*, there is a relatively small number of lemmas to prove), but treat the more general case of receiver's window size greater than one and verify both safety and liveness properties.

3. PVS AND EXTENDED AUTOMATA

In this section we briefly describe the PVS specification and verification system and the model of *extended automata* that can be used for modeling reactive programs such as communication protocols. We demonstrate on a simple example how to prove properties of extended automata using the invariant-strengthening technique and how this technique is implemented in PVS. To the best of our knowledge, invariant strengthening in PVS was first mentioned in [11].

PVS. The PVS system consists of an input language, a typechecker, and an interactive prover. The *input language* is typed higher-order logic with a rich type system including simple types such as booleans, enumerations, integers, and records, and more complex function types, subtypes, dependent types, and abstract datatypes. Having such an expressive language makes it easy to specify, *e.g.*, concurrent programs in a natural way, very close to a programming language. The drawback is that *typechecking* the input language is undecidable. However, PVS transforms this apparent weakness into an actual strength, because whenever the typechecker cannot decide whether an expression is type-correct, it generates a proof obligation. Most of the proof obligations can be discharged automatically, and obligations that cannot be proved often point to subtle errors in the specification.

A PVS *proof* is a tree, the root of which is the theorem being proved. The leaves of the tree are called *pending subgoals*. A proof proceeds as a sequence of commands, each of which transforms the proof tree by either proving a pending subgoal or by replacing a pending subgoal by a new set of pending subgoals. There are many proof commands, from propositional and first-order logic commands, to decision procedures and heuristic quantifier instantiation techniques, all of which can be combined into high-level, user-defined proof strategies.

Extended Automata. Extended automata are a computational model for reactive programs. An extended automaton A consists of a finite set of typed variables V, an initial condition Θ and a finite set of guarded transitions $\mathcal{T}$. The variables can be either control or data variables; the control variables are of a finite type *Location*. Each transition $\tau \in \mathcal{T}$ is labelled and consists of a guard and an assignment. For example, in the automaton illustrated in Fig. 1, the type *Location* consists of the two values l_1, l_2. There is one control variable pc and one integer data variable x. The initial condition is $\Theta: \quad pc = l_1 \wedge x = 0$. There are two transitions, $\tau_1 : \quad l_1 \overset{inc}{\to} l_2$ and $\tau_2 : \quad l_2 \overset{dec}{\to} l_1$. The guard of transition τ_1 is $pc = l_1$ and its assignment is $x := x + 1$.

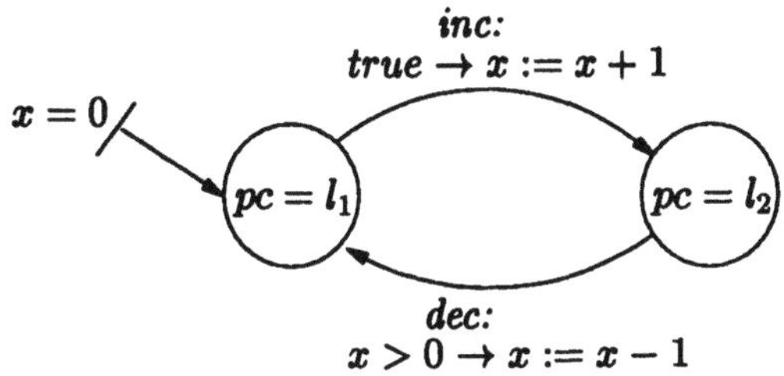

Figure 1. Example of Extended Automaton

A *state* is a type-consistent valuation of the variables. Each transition τ induces a *transition relation* ρ_τ relating the possible before and after states. The global transition relation of the system is $\rho_{\mathcal{T}} = \bigcup_{\tau \in \mathcal{T}} \rho_\tau$. A *run* of the automaton is an infinite sequence of states, in which the first state satisfies the initial condition and every two consecutive states are in the transition relation. For example, in the extended automaton illustrated in Fig. 1, there is only one initial state that has location l_1 and variable $x = 0$. From this state the program can take the transition *inc* (its guard is *true*), assigning 1 to x, moving to location l_2, etc.

An *invariant* is an assertion that holds at every state of every run. Invariants are the simplest form of safety properties. For example, it is not hard to show that $x \geq 0$ is an invariant of the automaton in Fig. 1.

Invariant strengthening. A state predicate is *inductive* if it holds initially and, if it is true at a given state s, then it is also true at all states s' that are successors of s through the transition relation $\rho_{\mathcal{T}}$. For example, in the automaton represented in Figure 1, it is not hard to check that the predicate $x \geq 0$ is inductive. Indeed, it is true initially, and from any state satisfying $x \geq 0$, any transition, *i.e.*, *inc* or *dec*, will lead to a state also satisfying the predicate.

Clearly, an inductive predicate is an invariant. The converse is not true: consider, for example, the predicate $pc = l_2 \supset x = 1$. It is not hard to see that it is an invariant, that is, on every run, whenever control is at location l_2, variable x equals 1. But the predicate is not inductive: by knowing only that it is true before transition *inc*, it cannot be inferred that it is still true after the transition is taken. This does not mean the predicate is not an invariant, it is just that we cannot prove this fact by using only the predicate itself as induction hypothesis. To succeed in the proof we need additional information, which can be obtained by *invariant strengthening*.

For example, consider again the predicate $\varphi: \quad pc = l_2 \supset x = 1$, a state s' of the automaton in Fig. 1 satisfying this predicate, and a state s from which s' is obtained by taking transition *inc*. Then, clearly, s satisfies $\psi: \quad pc = l_1 \supset x = 0$. The predicate ψ is a pre-condition for φ to hold. Now, it turns out that conjunction $\varphi \wedge \psi$ is inductive, thus, it is an invariant. In particular, we have just proved that φ is an invariant too.

Thus, invariant strengthening is the following process: to prove a predicate is an invariant, first try to prove that it is inductive. If this is the case, then the proof is done. Otherwise, compute the pre-condition of the predicate, and try to prove that the conjunction of the invariant with the pre-condition is inductive. The process can be iterated until an inductive invariant is obtained. This is not guaranteed to happen in general, because the problem of proving invariants of general extended automata is undecidable. However, by using this approach in an interactive theorem prover, the user can often detect infinite patterns of behavior and formulate a predicate which is not just the pre-condition of a predicate by one transition, but the fixpoint of an infinite sequence of such operations. We demonstrate this latter feature in the next section.

4. THE SLIDING-WINDOW PROTOCOL IN PVS

In this section we present the sliding-window protocol in more detail, then show how to encode it as an extended automaton in PVS.

Architecture of the Protocol The architecture of the protocol is represented in Figure 2. The sender obtains data from a FIFO stream of data called the `source`. Each data element is first saved into the sender's window, a FIFO buffer called `sndWindow`. The sender takes an element from its window, associates an index to it (a natural number keeping track of the order in which data has been obtained from the source), and sends the resulting record called a Message to the Message Channel `MsgChan`. The latter is a lossy FIFO, which may lose, but not reorder or create messages.

On the receiver side, messages are removed from `MsgChan` and, if a message's index is within the bounds of the receiver's window `rcvWindow`, it is stored there, otherwise, it is discarded. The receiver delivers contiguous sequences of messages to the external data `target`, and acknowledges the message with the highest index that it has delivered by sending the value of that index to the acknowledgment channel **AckChan** (another lossy queue). Finally, the sender reads acknowledgments from **AckChan**. An acknowledgment is considered valid if it is between the bounds defined by the bottom of the sender's window and the current top of that window, *i.e.*, up to the message number obtained from the data source. A valid acknowledgement **a** makes the bottom of the sender's window become **a+1**, meaning that all messages with value at most **a** are acknowledged.

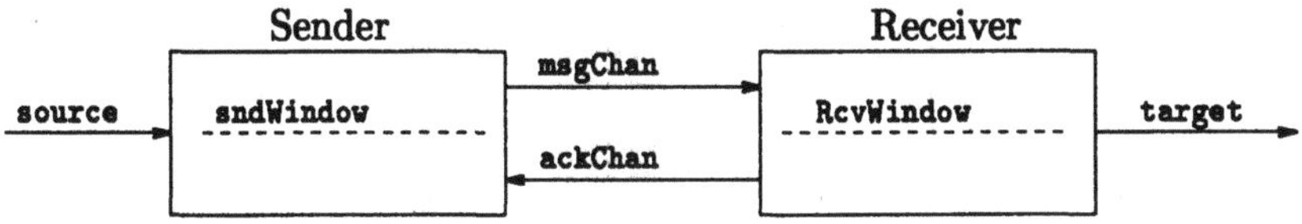

Figure 2. Architecture of the Sliding-Window Protocol

PVS encoding: declarations. First, the *labels* of the protocol are encoded into theory **Action_and_types** (Figure 3) that includes some type declarations and an **Action** abstract datatype to encode the externally visible actions of the protocol. The meaning of each action is explained in the comments. Note that actions have *parameters*, *e.g.*, **GET(d)** to express obtaining datum **d** from the data source. (Technically, in terms of abstract datatypes, **GET** is a constructor, **d** is an accessor, and **GET?** is a recognizer, but for better understanding it is preferable to see them as a declaration of actions with parameters.) Thus, the data being transferred between sender and receiver are natural numbers, and messages serve as recipients for the data together with an index (another natural number) that allows to reconstruct the order of messages, in case of losses during transmission.
The state of the automaton is encoded as the PVS record type **State** (cf. Fig. 4), with fields for variables of the sender, receiver, and channels:

Thus, the sender's window size **sw** and receiver's window size **rw** are strictly positive natural numbers. The receiver has three locations **receiving**, **flushing**, and **sendingAck**. Since there is only one location for the sender, it is not necessary to use a field for it. The meaning of the state variables is the following:

- variables for the data source: **source** is the infinite stream of data, modeled as a function from **nat** to **nat**. **sourceIndex** is the index of the next data element that will be passed to the sender,
- variables for the sender: **sndWindow** is the sender's window, modeled as a function from **nat** to **nat**. **sndLow** is the bottom of the sender's

```
Action_and_types : Theory
BEGIN
Msg: TYPE = [#data: nat, index: nat #]
Action : DATATYPE
BEGIN
  GET(d: nat): GET?                         %get datum d from data source
  SEND_MSG(m:Msg): SEND_MSG?           %send message m to message channel
  RCV_ACK(a:nat):RCV_ACK?                 % receive ack a from ack channel
  TIMEOUT: TIMEOUT?                            %sender receives a timeout
  RCV_MSG(m:Msg): RCV_MSG?         %receive message m from message channel
  PUT(d: int): PUT?                         %deliver datum d to data target
  SEND_ACK(a: nat): SEND_ACK?                    %send ack a to ack channel
  LOSE_MSG: LOSE_MSG?                          %environment loses message
  LOSE_ACK: LOSE_ACK?                   %environment loses acknowledgment
END Action
END Action_and_types
```

Figure 3. Sliding-Window Protocol: Types and Actions

```
sliding_window: THEORY
BEGIN
IMPORTING Action_and_types
sw, rw : posnat                         % sender and receiver window sizes
rcvControlType: TYPE={receiving,flushing,sendingAck} %receiver location
State: TYPE =                                        % global state type
 [# source: [nat-> nat], sourceIndex : nat,                % data source
    sndWindow: [nat-> nat],                             %sender's window
    sndLow,                                  %  bottom of sender's window
    sndWinIndex,                % counts data received from data source
    sndIndex: nat,                % counts data sent to message channel
    rcvControl : rcvControlType,     %control variable for the receiver
    rcvWindow : [nat-> int],                          %receiver's window
    rcvLow,                                % bottom of receiver's window
    headMsgChan,tailMsgChan: nat, msgChan: [nat->Msg],% message channel
    headAckChan,tailAckChan : nat, ackChan: [nat->nat],   % ack channel
    target :[nat->int],targetIndex : nat,                 % data target
    actual : boolean #] ...                       %rest of theory follows
```

Figure 4. Sliding-Window Protocol: State type

window, thus, at any moment in the execution, the current sender's window ranges from `sndWindow(sndLow)` to `sndWindow(sndLow+sw-1)`. The variable `sndWinIndex` is the position in the sender's window where the data source can place a data element, and variable `sndIndex` is the position of the next message that the sender will place of the message channel,

- variables of the receiver: `rcvControl` is the control variable. `rcvWindow` is the receiver's window, and `rcvLow` is the bottom of the receiver's window. Thus, at any moment in time, the receiver's window consists of elements from `rcvWindow(rcvLow)` to `rcvWindow(rcvLow+rw-1)`,

- variables of the channels: the message channel `msgChan` is a function from `nat` to `Msg`. It is a `fifo` whose active content at any moment in time ranges from `msgChan(tailMsgChan)` to `msgChan(headMsgChan-1)`. There are similar declarations for the acknowledgment channel `ackChan`.

Initial states and transition relation. The *initial states*, not shown here from lack of space, consist in setting every integer variable to zero, the receiver's control to `receiving`, and receiver's window to `Absent` (syntactic sugar for -1) to denote absence of any data in it: `FORALL(n: nat): s'rcvWindow(n) = Absent`.

The *transitions* of the model are encoded as a mapping `next` from `State` and `Action` to `State`. `Action` is the PVS abstract datatype depicted in Figure 3. There are nine transitions in all: four for the sender, three for the receiver, and two for the environment that may lose messages. The system works by nondeterministically firing an enabled transition, then choosing another transition and firing it, etc. A separate PVS theory `runs.pvs` (not shown here from lack of space) defines runs as sequences over a generic state type and the notions of invariants and inductive invariants.

The first transition of the sender specifies that a `GET` action with parameter `d` can fire if there is space in the window: `s'sndWinIndex < s'sndLow + sw`, and `d` is the next element of the data source. If this is the case, the sender's window is modified to hold the value `d` at position `sndWinIndex`, then, `sndWinIndex` and `sourceIndex` are both incremented:

```
GET(d) : IF s'sndWinIndex < s'sndLow + sw AND d = s'source(s'sourceIndex)
         THEN  s WITH [sndWindow:= s'sndWindow WITH[(s'sndWinIndex):= d],
                 sndWinIndex:=  s'sndWinIndex+1,
                 sourceIndex:= s'sourceIndex+1]
         ELSE s WITH [actual := false] ENDIF,
```

The second transition of the sender takes a message `m` whose data field is the element at position `sndIndex` of the sender's window, associates the value of that same index to it, places it on the message channel, and increments indices:

```
SEND_MSG(m):IF s'sndIndex<s'sndWinIndex AND m'data=s'sndWindow(s'sndIndex)
            AND m'index = s'sndIndex THEN
              s WITH [msgChan:= s'msgChan WITH [(s'headMsgChan):= m],
                      headMsgChan := s'headMsgChan+1,
                      sndIndex:= s'sndIndex+1]
```

```
            ELSE s WITH [actual := false] ENDIF,
```

A timer (not explicitly modeled in this specification) is started each time a SEND_MSG operation occurs. When a timeout occurs and there is no other action possible (*i.e.*, sender's window is full and all data from it has been sent to message channel and no acknowledgments are available), then prepare to start resending from bottom of sender's window:

```
TIMEOUT: IF s'sndWinIndex = s'sndLow + sw AND  s'sndIndex = s'sndWinIndex
         AND s'headAckChan = s'tailAckChan THEN
           s WITH [sndIndex := s'sndLow]
         ELSE s WITH [actual := false] ENDIF,
```

The last action of the sender consist in receiving an acknowledgement a (if available) from the acknowledgment channel. If a is in the sender's window, then we set the bottom of the sender's window sndLow to a+1, meaning that all messages with index at most a have been acknowledged and will never be sent again. The index sndIndex (of the next message to be sent) is set to the maximum between itself and a+1:

```
RCV_ACK(a): IF s'headAckChan>s'tailAckChan AND a=s'ackChan(s'tailAckChan)
            THEN IF a >= s'sndLow AND a < s'sndWinIndex THEN
                   s WITH[sndLow:= a+1,
                          sndIndex:= IF a>s'sndIndex THEN a+1
                                     ELSE s'sndIndex ENDIF,
                          tailAckChan := s'tailAckChan+1]
                 ELSE s WITH[tailAckChan := s'tailAckChan+1] ENDIF
            ELSE  s WITH [actual := false] ENDIF,
```

The first action of the receiver is to receive a message (if available) from the message channel. The control has to allow it (rcvControl = receiving). If the message's index is in the window of the receiver, it is stored at the correct position:

```
RCV_MSG(m): IF s'rcvControl = receiving AND s'headMsgChan > s'tailMsgChan
             AND  m = s'msgChan(s'tailMsgChan) THEN
               IF(m'index >= s'rcvLow AND m'index < s'rcvLow+rw) THEN
                  s WITH[rcvWindow:=s'rcvWindow WITH[(m'index):=m'data],
                         tailMsgChan:=s'tailMsgChan+1]
             ELSE s WITH [tailMsgChan:=s'tailMsgChan+1] ENDIF
            ELSE s WITH [actual:=false] ENDIF,
```

The second action of the receiver is to acknowledge data. The message with index just below the receiver window's bottom is acknowledged. The control has to allow it (rcvControl = sendingAck), then control goes back to receiving:

```
SEND_ACK(a): IF s'rcvControl =  sendingAck AND a = s'rcvLow-1 THEN
               s WITH[ackChan := s'ackChan WITH[(s'headAckChan) := a],
                      headAckChan := s'headAckChan+1,
                      rcvControl := receiving]
             ELSE s WITH [actual := false]  ENDIF,
```

The last action of the sender is also the most complex in the whole protocol. It consists in "flushing" the receiver window to the data target. This means sending to the data target the longest contiguous sequence (starting at the bottom of the sender's window) of data items that are not equal to `Absent`. The `rcvControl` field is used to control this process. At the beginning, `rcvControl = receiving`, and if the bottom of the receiver window contains `Absent`, then the state is left unchanged, because there is nothing to flush. Otherwise, the control switches to `flushing`, and stays there to copy data from receiver's window to data target until an `Absent` data item is met, in which case control goes to `sending_ack`:

```
PUT(d) : IF (s'rcvControl = receiving OR s'rcvControl = flushing) AND
         d = s'rcvWindow(s'rcvLow) THEN
           IF d /= Absent THEN
              s WITH[rcvControl := flushing,
                     target := s'target WITH[(s'targetIndex):= d],
                     targetIndex:= s'targetIndex+1,rcvLow := s'rcvLow+1]
           ELSIF s'rcvControl =flushing THEN
              s WITH [rcvControl := sendingAck]
           ELSE s ENDIF
        ELSE s WITH [actual := false] ENDIF,
```

Finally, there are two actions of the environment: losing a message and losing an acknowledgement (if there are any available). They consist in incrementing `tailMsgChan` and `tailAckChan`, respectively.

5. VERIFYING PROPERTIES

In this section we verify safety and liveness properties of the protocol.

Safety Properties

The main safety property required from the protocol is that the sequence of data delivered to the data target is a prefix of that obtained from the data source:

```
main_safety_property: THEOREM invariant(LAMBDA (s: State):
           FORALL (i: nat): (i < s'rcvLow => s'target(i) = s'source(i)))
```

The idea for proving Theorem `main_safety_property` is to prove the equality of sequences `source`, `target`, etc (cf. Fig. 2) from the "outside" to the "inside". We prove, that up to a given position, the data source equals the sender's window, the data target equals the receiver's window, and that the two windows are equal. Then, the equality of these sequences imply the `main_safety_property` theorem:

- $\forall i < sndWinIndex : sndWindow(i) = source(i)$
- $\forall i < rcvLow : rcvWindow(i) = target(i)$

- $rcvLow \leq sndWinIndex$
- $\forall i < rcvLow : rcvWindow(i) = sndWindow(i)$.

It should be noted that organizing the verification as above, *i.e.*, decomposing the problem into four subproblems, is a user-dependent choice. However, once the subproblems are formulated as lemmas, the process of proving them is completely systematic. It is an invariant-strengthening process: a PVS strategy for proving inductiveness of invariants is applied. If the property is inductive, the proof succeeds, otherwise, PVS returns a number of pending subgoals. All these subgoals correspond to transitions that do not preserve the property under proof. By examining the subgoals, we formulate auxiliary invariants that, if proved, would eliminate the pending subgoals and settle the original property.

1: proving $\forall i < sndWinIndex : sndWindow(i) = source(i)$. This is expressed in PVS as Lemma `source_equals_send` below. By applying our PVS strategy for proving inductive invariants, we obtain two pending subgoals, which suggest to prove the auxiliary Lemma `source_equals_send_1`:

```
source_equals_send: LEMMA invariant(LAMBDA (s: State): FORALL (i : nat):
                        i < s'sndWinIndex => s'source(i) = s'sndWindow(i))
source_equals_send_1: LEMMA invariant(LAMBDA(s: State):
                                               s'sndWinIndex=s'sourceIndex)
```

The conjunction of the predicates in the above lemmas is inductive, and the strategy for proving inductive invariants succeeds: the step is completed.

2: proving $\forall i < rcvLow : rcvWindow(i) = target(i)$. This is similar to Step 1 and is expressed in PVS as Lemma `target_equals_receive` below. The approach for proving it also takes one invariant-strengthening step.

```
target_equals_receive: LEMMA invariant(LAMBDA(s:State): FORALL (i: nat):
                            i < s'rcvLow => s'target(i) = s'rcvWindow(i))
```

3: proving $rcvLow \leq sndWinIndex$. This step and the following are more complex and we explain them in more detail. The property is expressed as:

```
rcvLow_leq_sndWinIndex : LEMMA invariant(LAMBDA(s: State) :
                                              s'rcvLow <= s'sndWinIndex)
```

The inductiveness strategy fails on the above lemma. The pending subgoal in the PVS proof identifies the PUT transition of the protocol (cf. Section 4) to be responsible for the failure. This is because the PUT transition increments variable `rcvLow`, thus, if `rcvLow <= sndWinIndex` holds before the transition is taken, it might not hold afterwards. The subgoal also suggests how to solve the problem: by showing (1): `rcvWindow(sndWinIndex) =`

`Absent`. This is because the PUT transition can increase `rcvLow` and invalidate Lemma `rcvLow_leq_sndWinIndex` only when both `rcvLow = sndWinIndex` and `rcvWindow(rcvLow) /= Absent` hold before the transition. By proving (1), we prove that the latter situation cannot happen.

Now, if we formulate (1) as a lemma and try to prove it by basic invariant strengthening we run into an infinite loop. Indeed, PVS requires to prove `rcvWindow(sndWinIndex+1) = Absent`, `rcvWindow(sndWinIndex+2) = Absent`, etc. Noticing this pattern we realize that the actual lemma we need to prove is the "fixpoint" of these properties:

```
rcvLow_leq_sndWinIndex_1: LEMMA invariant(LAMBDA(s: State):FORALL(i:nat):
                            i >= s'sndWinIndex => s'rcvWindow(i)= Absent)
```

We apply the inductiveness strategy to prove this lemma, and fail again. The pending subgoal now points to transition RCV_MSG of the protocol (cf. Section 4). Indeed, this transition modifies `rcvWindow`, which could invalidate the `rcvLow_leq_sndWinIndex_1` lemma. There is one pending subgoal, which suggests to prove the property (2) `msgChan(tailMsgChan)'index < sndWinIndex`. Indeed, in this case `rcvWindow` is modified at a position that is not referred to in the above lemma, thus, the lemma will not be invalidated if (2) holds. But attempting to prove (2) by simple invariant-strengthening actually points us to proving the more general lemma

```
rcvLow_leq_sndWinIndex_2: LEMMA invariant(LAMBDA(s: State):FORALL(i:nat):
             i < s'headMsgChan => s'msgChan(i)'index  < s'sndWinIndex)
```

The inductiveness strategy succeeds in proving this lemma, and Step 3 is done.

4: proving $\forall i < rcvLow : rcvWindow(i) = sndWindow(i)$. This property is expressed as the following PVS lemma:

```
send_equals_receive: LEMMA invariant(LAMBDA(s: State): FORALL (i : nat):
                         i < s'rcvLow => s'sndWindow(i) = s'rcvWindow(i))
```

The inductiveness strategy fails to prove the lemma, and leaves two pending subgoals. The first subgoal corresponds to the GET transition of the protocol (cf. Section 4). This is because that transition modifies `sndWindow`, which could invalidate the lemma. The second subgoal corresponds to the PUT transition, which modifies `rcvLow` and could also invalidate the lemma.

To prove the first subgoal it is enough to prove (3) `rcvLow<=sndWinIndex`. Indeed, the GET transition modifies `sndWindow` at position `sndWinIndex`, and (3) would guarantee that this modification has no influence on our lemma. We have already proved (3) in Lemma `rcvLow_leq_sndWinIndex`.

The second subgoal suggests to prove: if (4) `rcvWindow(rcvLow) /= Absent`, then (5) `rcvWindow(rcvLow) = sndWindow(rcvLow)`. This is because if (4) holds, then the PUT transition increments `rcvLow`, and what we have

to show is that the predicate in Lemma send_equals_receive holds when i = rcvLow, which is implied by (5).
Now, proving that (4) implies (5) by simple invariant strengthening would again lead us into an infinite loop, thus, the solution is to prove

```
send_equals_receive_1: LEMMA invariant(LAMBDA(s: State): FORALL(i: nat):
(i>=s'rcvLow AND s'rcvWindow(i)/=Absent)=> s'rcvWindow(i)=s'sndWindow(i))
```

This property is not inductive, and the strategy fails again, leaving two pending subgoals. The first subgoal is due to transition GET, which modifies sndWindow, and can be settled by proving that rcvWindow(sndWinIndex) = Absent always holds before this transition. This was already proved in Lemma rcvLow_leq_sndWinIndex_1. Finally, the second subgoal is due to transition RCV_MSG that modifies rcvWindow, and is settled by proving the following lemma, which is inductive and is proved automatically.

```
send_equals_receive_2:LEMMA invariant(LAMBDA(s: State): FORALL (i: nat):
   i<s'headMsgChan => s'msgChan(i)'data=s'sndWindow(s'msgChan(i)'index))
```

Liveness Properties

The main liveness property is that every data item from the data source is eventually delivered to the data target. If we call *fair* the runs that eventually obtain every data item from the source, and *live* the runs that eventually deliver every data item to the target, then we have to show that every fair run is live:

```
fair : PRED[(run)] = LAMBDA (r: (run)) : FORALL (m :nat): EXISTS (n:nat):
                                                   r(n)'sourceIndex > m
live : PRED[(run)] = LAMBDA (r: (run)) : FORALL (m :nat): EXISTS (n:nat):
                                                   r(n)'targetIndex > m
main_fairness_property: THEOREM FORALL(r:(run)) : fair(r) => live(r)
```

Here, (run) is the type of runs, defined in theory runs.pvs as the subtype of function from nat to State whose initial state satisfy the initial condition and such that any two consecutive states satisfy the transition relation. As for safety, the proof of Theorem main_fairness_property is divided into several steps and required to prove auxiliary invariants.

```
fair_aux1: PRED[(run)]=LAMBDA (r:(run)): FORALL(m:nat): EXISTS(n:nat):
                                                 r(n)'sndWinIndex > m
fairness_property_aux1: LEMMA FORALL(r:(run)): fair(r) => fair_aux1(r)
fair_aux2:PRED[(run)]=LAMBDA (r:(run)):FORALL (m:nat): EXISTS (n:nat):
                                                      r(n)'sndLow > m
fairness_property_aux2: LEMMA FORALL (r:(run)): fair_aux1(r) =>
                                                         fair_aux2(r)
safe_aux2: LEMMA invariant(LAMBDA(s:State):s'sndWinIndex<=s'sndLow+sw)
fair_aux3: PRED[(run)]=LAMBDA (r:(run)):FORALL m: nat):EXISTS (n:nat):
                                                      r(n)'rcvLow > m
safe_aux3bis : LEMMA invariant (LAMBDA (s:State) : FORALL (i:nat):
                        i < s'headAckChan => s'ackChan(i) < s'rcvLow)
safe_aux3 : LEMMA invariant (LAMBDA (s: State):  s'sndLow <= s'rcvLow)
```

```
fairness_property_aux3 : LEMMA FORALL (r:(run)) : fair_aux2(r) =>
                                                     fair_aux3(r)
fairness_property_aux4:LEMMA FORALL(r: (run)): fair_aux3(r) => live(r)
```

Lemmas `fairness_property_aux1` to `fairness_property_aux4` imply the main fairness theorem, and the verification is terminated.

6. CONCLUSION AND FUTURE WORK

In an attempt to demonstrate that deductive methods can be used in a systematic way to validate communication protocols, we present the verification of safety and liveness properties of a sliding-window protocol using the PVS specification and verification system. Some final remarks in favor of our thesis:

- The proofs of the safety properties are almost all the same. Indeed, most of the new proofs during the verification process were created by copy and paste, and only some quantifiers needed specific intervention to be properly instantiated. This was also true to some extent for the liveness properties,
- the strategy for proving inductive invariants suggests, in case of failure, new invariants to prove for settling the unproved subgoals. Thus, the method provides useful feedback for performing invariant strengthening,
- the new invariants did not need to be carefully analyzed and fully understood for proceeding in the proof. We have found the mechanized proof with PVS to be an automatic process, which does include the human in the loop, but does not excessively tax his patience or ingenuity.

We are currently working on techniques to further the automation of deductive verification. In particular, we are interested in a class of extended automata with integer variables and function symbols for which we study the automatic generation of auxiliary invariants and the automatic proof of inductiveness of invariants in the context of the automatically generated invariants. The class is expressive enough to cover the sliding-window protocol specification, and a large fraction of the properties presented in this paper.

Acknowledgment

Thanks to Duncan Clarke for carefully reading the manuscript and for useful comments and suggestions.

REFERENCES

[1] D. Brand and W. H. Joyner. Verification of HDLC. *IEEE Transactions on Communications* 30(5): 1136-1142, 1982.

[2] J.R. Burch, E.M. Clarke, K.L. McMillan, D.L. Dill and J. Hwang. Symbolic model checking: 10^{20} states and beyond. *Information and Computation*, 98(2):142–170, 1992.

[3] R. Cardell-Oliver. The specification and verification of sliding-window protocols in higher order logic. Technical Report No. 183, Computer Laboratory, University of Cambridge, 1989.

[4] C. Cornes, J. Courant, J.-C. Filliâtre, G. Huet, P. Manoury, C. Paulin-Mohring, C. Muñoz, C. Murthy, C. Parent, A. Saïbi, and B. Werner. The Coq Proof Assistant Reference Manual Version 6.1. Technical Report RT-0203, INRIA, July 1997.

[5] J-C. Fernandez, H. Garavel, A. Kerbrat, R. Mateescu, L. Mounier, and M. Sighireanu. CADP: A protocol validation and verification toolbox. *Computer-Aided Verification*, LNCS 1102, 1996.

[6] P. Godefroid and D.E. Long. Symbolic protocol verification using Queue BDDs. *Formal Methods in System Design*, 14(13):257-271, 1999.

[7] S. Graf and H. Saidi. Construction of abstract state graphs with PVS. In *Conference on Computer-Aided Verification*, LNCS 1254, 1997.

[8] M. Gordon and T.F. Melham. *Introduction to the HOL system.* Cambridge University press, 1994.

[9] G.J. Holzmann. The model checker Spin. IEEE Transactions on Software Engineering, 23(5): 279–295, 1997.

[10] N. Halbwachs, Y.E. Proy, and P. Roumanoff. Verification of real-time systems using linear relation analysis. *Formal Methods in System Design*, 11(2):157–185, 1997.

[11] K. Havelund and N. Shankar. Experiments in theorem proving and model checking for protocol verification. In *Formal Methods Europe*, LNCS 1051, 1996.

[12] R. Kaivola. Using compositional preorders in the verification of a sliding window protocol. *Computer-Aided Verification* LNCS 1633, 1997.

[13] Z. Manna and A. Pnueli. *Temporal verification of reactive systems.* Vol. 1: Specification, Vol. 2: Safety. Springer-Verlag, 1991 and 1995.

[14] S. Owre, J. Rushby, N. Shankar, and F. von Henke. Formal verification for fault-tolerant architectures: Prolegomena to the design of PVS. *IEEE Transactions on Software Engineering*, 21(2):107-125, 1995.

[15] J. L. Richier, C. Rodriguez, J. Sifakis, and J. Voiron. Verification in Xesar of the sliding-window protocol. In *Protocol Specification, Testing, and Verification, North-Holland, 1987.*

[16] M. Smith and N. Klarlund. Verification of a sliding-window protocol using IOA and MONA. In *Formal Description Techniques & Protocol Specification, Testing and Verification*, Kluwer Academic Publishing, 2000.

[17] A. Tanenbaum. *Computer Networks.* Prentice Hall, 1996.

Part Six

Test Sequence Derivation

TEST SEQUENCE SELECTION

David Lee and Ruibing Hao
Bell Labs Research China, Beijing, China

Abstract We study the test sequence selection problem. Given a large number of tests for a protocol system, we want to select a subset of tests so that the number of tests is reduced yet the test coverage is not sacrificed. We discuss the complexity of the test selection problem and propose a class of algorithms for different protocol system information requirements, test coverage criteria, and cost. We report experimental results on ISDN connection control protocol system Q.931.

Keywords: protocol system, test sequence, test coverage, extended finite state machine, reachability graph, NP-hard, algorithm, tree, sorting

1. INTRODUCTION

With advanced computer technology protocol systems are getting larger to fulfil complicated tasks. However, they are also becoming less reliable. Testing is an indispensable part of system design and implementations; yet it has proved to be a formidable task for complex systems. Because of its practical importance and theoretical interest, there have been a lot of activities on protocol system testing. There are conformance testing, interoperability testing, and performance testing. Conformance testing is to test conformance of system implementations to their specifications [Br, LY1, PBG]. Interoperability testing tends to uncover faults when different system components are interoperating or interfacing with each other [GHLS, KK]. Conformance and interoperability testing are designed to check the correctness of system behaviors whereas performance testing is related to the system performance, such as its transmission rate.

For conformance testing, test sequences are generated and then applied to the systems under test to conclude whether the implementations conform to their specifications. For the conducted tests to be conclusive, we have to generate test sequences with desired fault coverage. Furthermore, it is costly to execute tests, and we want to minimize the number of tests yet without sacrificing the fault coverage.

For certain complicated legacy protocol systems, such as 5ESS (AT&T/Lucent No. 5 Electronic Switching System), tests have been generated and applied to the systems over the years at different development stages and by different test engineers. There are thousands of test sequences in the available test set. To test such systems, it is impractical to generate tests from scratch, since it is often impossible to model the systems because they are too complex. Therefore, we want to use the available test set accumulated over years. However, we do not want to execute all of them, since there are too many and to run each test takes a substantial amount of time. A natural solution in practice is to select test sequences among the available test set.

The test selection problem was studied in [LPB] based on the valuations of the test sequences to be selected, and it was reduced to an optimisation problem over Boolean algebra. For testing in context, the problem was studied in [YCL]. An important issue of test selection is the possible loss of coverage. This problem was investigated in [LPB] for partially specified machines and also in [CV, VC, ZV] with an elegant metric of coverage. The selection criteria and their notations were studied in [Pa].

In this paper, we study the following test selection problem. We have a large set of test sequences, and we want to select a minimal subset of tests to execute without sacrificing the fault coverage. For a formal study, we use extended finite state machine to model protocol systems. It is only for convenience. Our results are easily adaptable to other equivalent mathematical models such as labeled transition systems. We discuss the complexity of the problem and propose efficient algorithms for the test sequence selection.

In Section 2 we describe an extended finite state machine model and its reachability graph. In Section 3 we formulate a problem of test sequence selection, discuss the coverage criteria, and study the problem complexity. In Section 4 we discuss efficient algorithms for the test selection. In Section 5 we discuss a more general problem when the reachability graph is not available. In Section 6 we report experimental results on ISDN Q.931 system testing. In Section 7 we conclude the paper with a comparison of all the algorithms studied.

2. BASICS

A finite state machine contains a finite number of states and produces outputs on state transitions after receiving inputs. It is often used to model control portions of protocol systems. However, data portions of protocols include variables and operations based on their values; ordinary finite state machines are not powerful enough to model in a succinct way the physical systems any more. Extended finite state machines (EFSM), which are finite state machines extended with variables, have emerged from the design and analysis of communication protocols [LY1]. For instance, IEEE 802.2 LLC [ANSI] is specified by 14 control states, a number of variables, and a set of transitions (pp. 75-117). For a formal definition of EFSM and the related concepts, see [LY1].

Each combination of a state and variable values is called a configuration. An EFSM usually has an initial state s_0 and all the variables have an initial value $\mathbf{x}_{init}$; we have the initial configuration $(s_0,\mathbf{x}_{init})$. A reachability graph consists of all the configurations and transitions, which are reachable from the initial configuration. It is a directed graph where the nodes and edges are the reachable configurations and transitions, respectively. Obviously, a control state may have multiple appearances in the nodes (along with different variable values) and each transition may appear many times as edges in the reachability graph.

For a path from the initial node (configuration) of the reachability graph, the input/output (I/O) labels on the transitions of the path provide an I/O sequence. Conversely, if an I/O sequence corresponds to a unique path from the initial node in the reachability graph, then the underlying EFSM is *deterministic*. Otherwise, it is *non-deterministic.* For clarity, in this paper we only consider systems, which are modelled by deterministic EFSM's. Our approaches can be modified to handle non-deterministic EFSM's, as we shall comment in the conclusion of the paper.

Given an input sequence, there is at most one path from the initial node in the reachability graph such that the transitions on the path are labelled with the input sequence, since the machine is deterministic. If such a path exists, then the given input sequence is *valid.* Otherwise, it is *invalid.*

Obviously, a valid input sequence corresponds to a unique I/O sequence, which are the labels on the corresponding unique path from the initial node.

We first study the test selection problem, assuming that the underlying system reachability graph is available. We then relax the problem with an assumption that only the EFSM specification of the underlying system is available. Finally, we present algorithms for a general case when no information of the underlying system is available for test selection.

3. TEST SEQUENCE SELECTION PROBLEM AND ITS COMPLEXITY

A test sequence (or a scenario) is valid input sequence of a protocol system that is modelled by an EFSM. Since the system is deterministic, there is a one-to-one correspondence between test sequences and paths in the reachability graph from the initial configuration. In practice, a test sequence usually consists of an I/O sequence; the input sequence is applied to the system under test, and the output sequence is to be observed from the system response. Therefore, a test sequence can be represented by a valid input sequence, a valid I/O sequence, or a path from the initial configuration in the reachability graph of the underlying protocol system. For convenience, we shall use these terms interchangeably.

Informally, the test sequence selection problem is: Given a set of test sequences S, select from S a subset of tests S^* with a desirable fault coverage. Fault coverage is essential for testing. However, it is often specified differently from different system models and practical needs.

For conformance testing of FSM's we may consider checking sequences [LY1, Mo], which guarantee that an implementation machine is structurally isomorphic to the specification machine. For FSM's there are polynomial time algorithms for constructing the checking sequences, however, even for medium size machines it is too long to be practical [YL1]. For EFSM's hundreds of thousands of states (configurations) are typical and it is in general impossible to construct checking sequences. For the test selection problem, if we are interested in checking sequences, the problem is even harder. Specifically, given a set of tests *S*, we want select a subset S^* from *S* such that S^* are checking sequences. The problem is NP-hard [Ya].

A commonly used criterion of fault coverage is to test each edge in the reachability graph at least once. It has been a well-accepted criterion in practice, and we first consider this criterion. Formally, given an EFSM **M**, let G be its reachability graph with an initial configuration v_0, which corresponds to the initial state of **M** and the initial variable values. A test sequence of **M** is a path in G from v_0. A covering test set is a set of test sequences such that each edge of G is covered by at least one of the test sequences. We want to select a subset of tests such that it is still a covering test set and contains a minimal number of tests:

Problem 1. Test selection with a reachability graph. Given the reachability graph of an EFSM and a covering test set S, select a subset of test sequences S^* from S, such that S^* is a covering test set with a minimal cardinality.

The problem is hard. We can reduce the set cover problem to this test selection problem. Since the set cover problem is NP-hard [GJ], we have:

Proposition 1. The test selection problem 1 is NP-hard.

Therefore, it is hard to obtain an optimal solution for test selection in general. We discuss heuristic methods next.

4. TEST SEQUENCE SELECTION WITH A REACHABILITY GRAPH

We consider the following greedy method. We first find a test sequence in S that covers a maximal number of uncovered edges of the reachability graph, and add this test to S^*. We then repeat the process until all the edges of the reachability graph are covered. Formally,

Algorithm 1.
input: reachability graph G and a covering test set S
output: a covering test set S, which is a subset of S*

1. mark all edges in graph G as uncovered;
2. $S^* = \Lambda$; /* an empty set */
3. **repeat**
4. find a test sequence s in S that covers a maximal number of uncovered edges of the graph G, and mark these edges as covered;
5. $S = S - \{s\}$, $S^* = S^* \cup \{s\}$;
6. **until** all edges in G are covered
7. **return** S^*

Figure 1. Algorithm 1

It takes polynomial time to find a test sequence s from S that covers a maximal number of uncovered edges. A straightforward approach is to examine each test sequence in turn, and it takes time proportional to the length of the test sequence examined. Suppose that there are N test sequences in S and the total length of all the test sequences in S is M. Then it takes time O(M) to select a test and time O(MN) in the worst case to determine the test set S^*.

Proposition 2. Algorithm 1 takes time O(MN) to select a covering test set where M and N are the total length and number of all the test sequences in the given covering test set.

Algorithm 1 provides a heuristic approach, that is, the selected test set is a covering set but may not be minimal. Instead of selecting one test at a time, we could select two tests at a time such that their combined coverage is maximal [C8]. It increases the cost but still runs in polynomial time. On the other hand, it may reduce the number of selected tests. In general, can we do better than Algorithm 1 in polynomial time? With a similar argument as in [LY2] it can be proved that we cannot do better than a logarithmic factor in polynomial time. That it, for any polynomial time algorithms, which select a covering set of cardinality N^*, there are cases such that $N^*=\Omega(N^{**}\log N^{**})$ where N^{**} is the cardinality of a minimal covering test set, unless of course P=NP.

It takes time O(MN) for Algorithm 1 to select a covering test set where M and N are usually large in practice. To reduce the cost we may want to avoid examining all the tests in the set. We can conduct the following preprocessing and variations of Algorithm 1.

Algorithm 1.1. We first sort the tests in S according to their lengths in a descending order, and then examine them in that order.

The rationale is: longer tests correspond to longer paths in the reachability graph and may cover more edges of the graph. Furthermore, if we have found a test, which covers k uncovered edges, then there is no need to examine tests of length k or less. Since there are N tests in the set, the added cost of sorting is O(NlogN), which is negligible since it is dominated by O(MN).

Algorithm 1.2. Select each test sequence to be examined from S uniformly at random.

As often observed in practice, randomization helps. See the experimental results in Section 6.

Algorithm 1.3. Sort the test sequences by their input symbols alphabetically rather than their lengths, remove all the tests which are a proper prefix of other tests, and then apply Algorithm 1.

Obviously, this preprocessing reduces redundant tests and we have fewer tests to process. We shall further discuss this variation in the next section.

5. TEST SEQUENCE SELECTION WITHOUT A REACHABILITY GRAPH

Often we only have a test set available but do not have the corresponding reachability graph of the underlying system or it is too costly to construct such a graph. The problem becomes harder since we do not know whether a selected test set covers all the edges in the reachability graph.

Suppose we have the underlying system specification EFSM yet without its reachability graph. We have:

Problem 2. Test selection with an EFSM. Given an EFSM and a covering test set S, select a subset of test sequences S^* from S, such that S^* is a covering test set with a minimal cardinality.

Similar to Problem 1, we have:

Proposition 3. The test selection problem 2 is NP-hard.

Therefore, it is hard to obtain an optimal solution for test selection in general. On the other hand, the remarks on the complexity of Problem 1 also apply to Problem 2.

We can use the following heuristic procedure. We examine all the test sequences in S, trace each sequence in the EFSM, and record all the edges covered in the corresponding reachability graph. We then select test sequences in S until all the recorded edges are covered. The selection procedure is similar to Algorithm 1, and the variations of Algorithm 1 also apply here. Note that we do not construct the reachability graph explicitly.

Algorithm 2.
*input: EFSM **M** and covering test set S*
output: a covering test set S, which is a subset of S*

1. $E = \Lambda$; /* identify all edges of reachability graph; E is an empty edge set */
2. **for** each test sequence s_i *in S*, $i = 1, \ldots, N$
3. $\quad u=u_0=< s_0\ x_{init} >$; /* u_0 is the initial configuration of **M** */
4. $\quad$ **for** $j = 1, \ldots, k_i$ /* k_i is the number of transitions in s_i */
5. $\quad\quad$ trace transition t_{ij} in **M** and determine next configuration v;
6. $\quad\quad$ **if** corresponding edge in reachability graph $(u,v) \notin E$
7. $\quad\quad\quad E = E \cup \{(u,v)\}$;
8. $\quad\quad u = v$;
9. set all edges in set E as uncovered; /* find a covering subset S^* */
10. $S^* = \Lambda$; /* an empty set */
11. **repeat**
12. $\quad$ find a test sequence s in S that covers a maximal number of uncovered edges in set E, and mark these edges as covered;
13. $\quad S = S - \{s\}$, $S^* = S^* \cup \{s\}$;
14. **until** all edges in E are covered
15. **return** S^*

Figure 2. Algorithm 2

Given the underlying EFSM and test set S, Algorithm 2 first collects all the edges of the corresponding reachability graph of the EFSM, and then selects a covering subset of tests. One could construct a reachability graph of the EFSM and then select tests using Algorithm 1. However, it is often a difficult task to explicitly construct a reachability graph due to the state explosion. Algorithm 2 avoids this process.

Algorithm 2 traces all the tests in Line 1-8 and records the edges in E. The total number of edges in the worst case can be proportional to M. Therefore, it takes time O(M) to check if an edge under consideration is in E, and the total time is $O(M^2)$. We can keep the recorded edges in set E in a sorted order (by their inputs, the start and end states and their associated variable values). It takes time proportional to O(logM) to check if an edge is in E, and the total time to execute Line 1-8 is O(MlogM). By a same argument as for Algorithm 1, the run time of Line 9-15 is O(MNlogM). In summary,

Proposition 4. Algorithm 2 takes time O(MNlogM) to select a covering test set where M and N are the total length and number of all the test sequences in the given covering test set.

Algorithm 2 is for a more general test selection Problem 2 where there is no reachability graph and also no need to construct one. However, there is an extra price to pay in terms of the run time, i.e., a factor of logM.

Often in practice the underlying system model may not be available, especially for those legacy systems, and we only have a set of tests to select. This is the most general case of the test selection problem:

Problem 3. Test selection without any information of the underlying protocol system. Given a covering test set S for a protocol system, for which there is no information available, select a subset of test sequences S^*, such that S^* is a covering test set.

Different from Problem 1 and 2, in general it is impossible to select a minimal covering test set since there is no information of the underlying protocol system.

For this problem we only consider the following redundancy criterion. Let s and s^* be two test sequences where s^* is a prefix of s. Let p and p^* be the corresponding paths from the initial configuration in the reachability graph, which we do not know. Since the corresponding EFSM is deterministic, p^* is a prefix of p as paths in the reachability graph, and we can remove s* from the test set without sacrificing the coverage. Consequently, if S is a covering test set, the test subset S^*, selected by discarding all the prefixes, is still a covering test set. As a matter of fact, in the worst case, we cannot reduce the tests anymore. Consider the following reachability graph. It consists of separate paths from the initial node. For test selection, we can only discard the paths (tests), which are a proper prefix of another path; any further reduction will lose the coverage.

The above observations lead to the following algorithm. Consider the alphabet set of all the input symbols. We sort all the input sequences in S alphabetically and then remove all the input sequences (test sequences), which is a prefix of another input sequence. To check whether a test sequence s_i is a prefix of another test, there is no need to check s_i again all the tests in S; since the test set is sorted, we only need to check s_i with the next test sequence s_{i+1} :

Lemma 1. A Test s_i in S is a prefix of another test if and only if it is a prefix of s_{i+1}.

Proof. We only need to prove that if s_i is a prefix of another test $s_i\ t$ then it is also a prefix of s_{i+1}. Since s_{i+1} is sorted in between s_i and $s_i\ t$, it must be of a form s_i r. Therefore, s_i is a prefix of $s_{i+1} = s_i$ r.

Algorithm 3.
input: test sequence set S and alphabet set of all the input symbols Σ
output: subset S of S with redundant tests removed*

```
sort all input sequences of S alphabetically;
S* = Λ;  /* an empty set */
for i = 1, …, N-1   /* N is the cardinality of S */
    if s_i is not a prefix of s_{i+1}
        S* = S* ∪ {s_i};
S* = S* ∪ {s_N};
return S*
```

Figure 3. Algorithm 3

Since we only remove the redundant tests, we have:

Proposition 5. Given a test set S, Algorithm 3 removes redundant tests and selects a subset of tests S^*. If S is a covering test set of the reachability graph of the underlying protocol system, then S^* remains a covering test set.

Note that Algorithm 3 may select test sequences that are redundant in terms of covering the reachability graph since we do not have its information.

Therefore, the resulting number of tests may be larger than that from Algorithm 1 and 2. See Section 6 for the experimental results.

We now discuss the algorithm complexity. For simplicity, we assume that each test sequence is of length k, i.e., it contains k input symbols. Suppose that there are N test sequences in the test set S. Since there are NlogN comparisons to sort S and each comparison takes time O(k), it takes time O(kNlogN) to sort all the tests in S. We then examine all the tests in the sorted order and remove all the prefixes; the total time is O(kN). Since the total lengths of all the tests in S is: M=kN, we have:

Proposition 6. It takes time O(MlogN) to select tests using Algorithm 3 where M and N are the total length and number of all the test sequences in the given test set.

The cost of the test selection of Algorithm 3 is dominated by sorting the test sequences. We now discuss another method, using trees. It is optimal in terms of run time; it takes time proportional to the total lengths of the test sequences to be selected.

We grow a tree as follows. Initially, we have a root node v_0. Each edge is labeled with an input symbol. For each input sequence (test), we walk down the tree as follows. At a node *u* of the tree with an input symbol *a* in the sequence, if there is an outgoing edge from *u* labeled with *a*, then we walk down along that edge, arrive at the end node *v*, and process the input symbol in the test sequence after *a*. However, if there are no outgoing edges labeled with *a*, then we add a tree edge (*u*,*v*) from *u*, label it with *a*, walk down along (*u*,*v*), and continue processing from *v* as before. Each test requires a tree walk. After all the tests have been processed, we can obtain the selected tests as follows. We only have to select the tests (or paths) from the root to a leaf node; all the other tests (paths) a prefix of these selected tests. Note that each step of the tree walk takes a constant time (if we have a bitmap at each node to keep track of the labels of all the outgoing edges). Therefore, it takes time proportional to the length of a test sequence to process. Hence the total cost is O(M).

Proposition 7. It takes time O(M) to select tests using Algorithm 4 where M is the total length of all the test sequences in the given test set.

Obviously, Algorithm 4 removes the prefixes and selects the same test set as Algorithm 3. It is more efficient but involves an on-line tree construction process. Again, if the given test set is a covering set then the selected subset is also a covering. But it may contain redundant tests in terms of covering the reachability graph, of which we have no information.

Algorithm 4.
input: test sequence set S and alphabet set of all the input symbols Σ
output: subset S of S with redundant tests removed*

```
V = {v0};  /* V is node set of tree T; each node keeps track of all
input symbols of its outgoing edges by a bitmap,  */
for i =1, ..., N   /* N is the number of test sequences in S */
   u = v0; /* a tree walk from initial node v0 */
      for j = 1, ..., k_i  /* k_i is number of input symbols in sequence s_i
*/
      let t_ij be transition of s_i under examination with input t_ij->a;
      if (u.bitmap(a)=0)   /* not recorded yet */
         initialize a new node v;
         construct a new tree edge (u, v);
         u.bitmap(a) =1;  /* record this input symbol at u */
         u =v; /* walk down tree to v along newly constructed edge */
      else /* input has been recorded and tree edge exits for a walk */
         u=v;
S* = Λ;  /* an empty set */
for each leaf node v of tree T;
      let s be test sequence corresponding to path from v0 to v;
      S*=S*∪{s};
return S*
```

Figure 4. Algorithm 4

6. EXPERIMENTS

We applied our test selection algorithms to a simplified version of Q.931 protocol. We now report experimental results. Due to the space limit, for the reachability graph of Q.931 and the test sequences, see [LH].

Q.931 is the connection control protocol for ISDN, comparable to TCP in the Internet protocol stack. It manages connection setup and teardown. However, it does not provide flow control nor perform retransmission, since the underlying layers are assumed to be reliable and the circuit-oriented nature of ISDN allocates bandwidth in fixed increments of 64 kbps. Similar to TCP, Q.931 documents both the protocol itself and a protocol state machine. For experiment purpose, we constructed an EFSM for Q.931 that covers most of the states and most of the message types. The corresponding reachability graph has 13 nodes and 19 transitions. We also have a set of 59 test sequences for Q.931 that were generated from an automated test generation tool.

We implemented Algorithm 1 and its three variations, and then applied to the 59 test sequence set with the reachability graph. We have:

Table 1. Comparison of Algorithm 1 and its variations

Algorithms	Number of selected test sequences
original test set	59
Algorithm 1	6
Algorithm 1.1 (sort-by-length)	9
Algorithm 1.2 (random-select)	8
Algorithm 1.3 (sort-by-alphabet)	6

From Table 1, all four algorithms reduce the number of final test sequences. According to the discussion in Section 4, Algorithms 1 and 1.3 tend to generate the least number of test sequences. Experimental results in Table 1 verify this observation; both algorithms have selected a set of 6 test sequences, which have the same edge coverage of the reachability graph as the original 59 sequences. Comparing with Algorithm 1 and 1.3, we observe that Algorithm 1.1 and 1.2 execute a little bit faster when processing the same set of input sequences. However, the number of test sequences selected by Algorithm 1.1 and 1.2 is larger than Algorithm 1 and 1.3 (9 and 8 compared to 6); efficiency is achieved with a cost of more redundancy in the tests selected.

We also compare Algorithm 3 and 4 with Algorithm 1 in the number of selected tests and run cost. They use the same set of test sequences as input. The following are the results:

Table 2. Comparison of Algorithm 3, 4 and 1

	Algorithm 3	Algorithm 4	Algorithm 1
Complexity	$O(MlogN)$	$O(M)$	$O(MN)$
number of selected tests	18	18	6

From Table 2, Algorithm 3 and 4 both have selected 18 sequences. The run time of Algorithm 4 is less than Algorithm 3. Comparing with Algorithm 1, both Algorithm 3 and 4 selected 12 more test sequences, since Algorithm 3 and 4 do not incorporate the information of the reachability graph of the underlying system. On the other hand, the cost of Algorithm 3 and 4 is much less than that of Algorithm 1.

7. CONCLUSION

In this paper, we have discussed the test sequence selection problem. Given a test sequence set, we want to select a subset of tests; without sacrificing the coverage we want to minimize the number of the selected tests to reduce the test execution time. We have proposed several algorithms with different required information and coverage, at different costs, and with different redundancy of the selected tests.

Algorithm 1 and its variations Algorithm 1.1-1.3 all require that a reachability graph of the underlying protocol system is available. They are designed for Problem 1 where the coverage is the reachability graph. Algorithm 2 relaxes the requirement; it needs an EFSM of the underlying system. It is designed for Problem 2 where the coverage is also the reachability graph yet only the EFSM of the underlying system is given. Algorithm 3 and 4 are for the most general case when no information is needed for the test selection except for the given test set. They are designed for Problem 3 where there is no information available of the underlying system, and the selected tests also cover the reachability graph if the given test set does.

Algorithm 1 is expensive; it tends to minimize the number of tests, i.e., to reduce the redundancy as much as possible. Algorithm 2 has a same coverage as Algorithm 1, consumes less space (no reachability graph is needed), but takes more time to run. It shows the advantage of using the compact representation EFSM rather than its reachability graph. However, the on-line construction and recording of its edges involve more complicated data structures and algorithms, and it takes longer to run the algorithm. Algorithm 3 costs less than Algorithm 1 and 2. Algorithm 4 is optimal, since its cost is proportional to the total length of tests to be selected; one has to at least read all of them to select.

Algorithm 1 and 2 require more information and is more expensive to run. But they select a smaller test set without sacrificing the coverage. Algorithm 3 and 4 require no information of the underlying system and cost less, but select more tests. Of course they also cover the reachability graph if the given test set does. However, they may contain redundant tests, which are not needed for the coverage of the reachability graph. But there is no way one can figure out since there is no information of the underlying protocol system at all.

In practice, often information of the underlying protocol system is not available or it is too expensive to obtain. In this case we resort to Algorithm 3 and 4. As demonstrated by the experiments, they can also reduce significantly the number of tests yet without sacrificing the coverage.

So far we assume that the underlying protocol system is deterministic, and in this case, a valid test sequence corresponds to a unique path in the reachability graph. If the system is non-deterministic, then a valid test sequence may correspond to multiple paths in the graph. Since the execution sequence is non-deterministic, the coverage of the edges in the reachability graph is probabilistic. While the union of all the possible paths, which are associated with a valid test sequence, is considered for coverage, we can only claim in a probabilistic sense.

ACKNOWLEDGEMENTS

Kun Du from Peking University implemented some of the algorithms. Gerard Holzmann and Margaret Smith from Bell Labs provided us with the set of Ubet scenarios used in our experiments. Our thanks go to the reviewers for their constructive comments and suggestions.

REFERENCES

[AHU] A.V. Aho, J. E. Hopcroft, and J. D. Ullman, *The Design and Analysis of Computer Algorithms*, Addison-Wesley, 1974.

[ANSI] International standard ISO 8802-2, ANSI/IEEE std 802.2, 1989.

[Br] E. Brinksma, *A Theory for the Derivation of Tests, Proc. PSTV*, pp. 63-74, 1988.

[BTV] E. Brinksma, J. Tretmans and L. Verhaard, *A Framework for Test Selection*, *Proc. PSTV*, pp. 233-248, 1991.

[C8] Anonymous Reviewer of PSTV-FORTE'2001.

[CV] J. A. Curgus and S. T. Vuong, *A Metric Based Theory of Test Selection and Coverage*, *Proc. PSTV*, 1993.

[GJ] M. R. Garey and D. S. Johnson, *Computers and Intractability: a Guide to the Theory of NP-Completeness*, W. H. Freeman, 1979.

[GHLS] N. Griffeth, R. Hao, D. Lee, and R. Sinha, *Integrated System Interoperability Testing with Applications to VoIP, Proc. FORTE/PSTV*, pp. 69-84, 2000.

[KK] S. Kang and M. Kim, *Interoperability Test Suite Derivation for Symmetric Communication Protocols, Proc. FORTE/PSTV*, 57-72, 1997.

[LH] D. Lee and R. Hao, *Test Sequence Selection*, Bell Labs Tech. Memo. 2001.

[LY1] D. Lee and M. Yannakakis, *Principles and Methods of Testing Finite State Machines - a Survey, The Proceedings of IEEE*, Vol. 84, No. 8, pp. 1089-1123, 1996.

[LY2] D. Lee and M. Yannakakis, *Optimization Problems from Feature Testing of Communication Protocols, Proc. ICNP*, pp. 66-75, 1996.

[LPB] G. Luo, A. Petrenko, and G. v. Bochmann, *Selecting Test Sequences for Partially Specified Nondeterministci Finite State Machines, Proc. IFIP 7th Int. Workshop on Protocol Test Systems*, 1994.

[Mo] E. F. Moore, *Gedanken-experiments on sequential machines, Automata Studies, Annals of Mathematics Studies*, Princeton University Press, no.34, pp.129-153, 1956.

[Pa] J. Pachl, *A Notation for Specifying Test Selection Criteria, Proc. PSTV*, pp. 71-84, 1990.

[PY] C. H. Papadimitriou, M. Yannakakis, *Optimization, Approximation and Complexity Classes, J. Comp. Sys. Sc.*, 43(3), 425-440, 1991.

[PBG] A. Petrenko, S. Boroday, R. Groz, *Confirming Configurations in EFSM, Proc. FORTE/PSTV*, 5-24, 1999.

[VC] S. T. Vuong and J. A. Curgus, *On Test Coverage Metrics for Communication Protocols, Proc. IWTCS*, 1991.

[Ya] M. Yannakakis, Private communication.

[YCL] N. Yevtushenko, A. Cavalli, L. Lima, *Test Suite Minimization for Testing in Context, Proc. IWTCS*, pp. 127-145, 1998.

[ZV] J. Zhu and S. T. Vuong, *Generalized Metric Based Test Selection and Coverage Measure for Communication Protocols, Proc. PSTV*, pp. 299-314, 1997.

EXECUTABLE TEST SEQUENCE FOR THE PROTOCOL DATA FLOW PROPERTY[1]

Wen-Huei (Paul) Chen
Department of Electronic Engineering, Fu Jen Catholic University, Taipei, Taiwan, R.O.C.
paultaipei@yahoo.com.tw

Abstract A new test sequence generation method is proposed for testing the conformance of a protocol implementation to its data portion modeled by an *Extended Finite State Machine* which is represented by a *Data Flow Digraph*. *All-Use* and *IO-df-chain* are two important criteria in selecting paths from the Data Flow Digraph in order to generate a test sequence which traces the data flow property, but it is a tedious process to select a path which satisfies the criteria while guaranteeing that the generated test sequence is *executable* (i.e., one that has feasible parameter values.) The proposed four-step method automatizes the selecting process as well as optimizing the test sequence length. First, the Data Flow Digraph is embedded with certain (but not all) parameter values so as to construct a *Behavior Machine Digraph* where executable test sequences can be directly generated. Second, executable test paths which trace every association defined by each criterion are generated from the Behavior Machine Digraph. Third, the Behavior Machine Digraph is embedded with these test paths so as to construct the *SelectUse* and *SelectIO* Digraphs. Finally, the *Selecting Chinese Postman Tours* of the two digraphs are used to generate the optimally executable test sequences that satisfy the All-Use and IO-df-chain criteria.

Keywords: Conformance testing, executable test sequence, data flow.

1. INTRODUCTION

A distributed system is composed of many *parties* (i.e., computers, instruments, etc.) remotely connected by communication *links* (i.e., cables,

---

[1] This work is supported by the National Science Council of Taiwan under Grant NSC892213E030025 and by Fu Jen University under a grant from the SVD section.

fibers, etc.) through which messages are transmitted. A *protocol* is the representation as well as the orderly exchange of these messages that must be agreed on by any party before using it, and a set of protocols is usually layered to establish a complex communicating behavior. In each party, a protocol is implemented in either software or hardware that has an upper (or lower) interface to the upper-layer (lower-layer) protocol(s) [9]. The objective of *protocol conformance testing* is to see if a protocol implementation conforms to the protocol specification defined as a standard. In a testing center, the protocol implementation is tested as a black box. Inputs are sent from an external tester to the implementation through the interfaces, and the outputs are checked to see if they are as expected. The sequence of input/output pairs is the *test sequence*, and the number of inputs is the test sequence *length* [5]. In general, such a test sequence is generated from the protocol specification.

The protocol specification contains a control and a data portion [2]. The control portion determines how messages are sent and received. It can be considered a *deterministic*[2] *Finite State Machine* (FSM) which contains *states* and *transitions* [5]. Initially, the FSM is in a specific state called the *initial state*. An input message (i.e., input or stimulus) will cause the FSM to generate output message(s) (i.e., outputs or responses) and to change from the current state to a new state; this process is a *transition*. The data portion specifies other functions (e.g., quality of service) that involve the parameter values associated with the messages. Informally, the data portion is described in words which are then formulated into a set of rules among parameter values [6]. Formally, the data portion is specified by an *Extended Finite State Machine (EFSM)* [2] which is extended from the FSM by introducing *variables* and *parameters*. Initially, the EFSM is in the initial state and all the variables are set to initial values. The EFSM can receive an input that has parameter values, which combine with certain variable values to define a logic function (i.e., *predicate*.) That input will cause the EFSM to change from the current state to a new state that depends on the truth value of the predicate, then the EFSM updates the variable values according to a computation function (i.e., *action*) while generating output(s) that have certain parameter values.

In a data portion described by a set of rules, the generated test sequence is *executable* if it has parameter values that do not violate any rule [1][6]. In [6], Kwast converts the FSM and rules into a *Behavior Machine Digraph*, in which paths can be used to generate the executable test sequence which verifies the rules. In [1], Chen converts the Behavior Machine Digraph into

---

[2] "deterministic" means that for each input there is at most one transition defined at each state.

a *Selecting Digraph* and proposes the *Selecting Chinese Postman Algorithm* to find a specific tour of the digraph which generates a minimum-length executable test sequence that verifies each rule at least once. In [2], Chen generalizes the Selecting Digraph into a *SelectO* digraph which is used to generate an executable test sequence that verifies both the control and data portions. However, the data portions of many protocols which are described in words cannot be formulated into rules. This paper is concerned with test sequence generation from the data portion specified by an EFSM.

The EFSM of a data portion can be represented by a Data Flow Digraph, where the inputs, outputs, predicates and actions of the EFSM are represented by a set of nodes [11]. A test sequence generated from a path of the digraph is *executable* (or *feasible*) if it has certain input parameter values which make each predicate along the path to remain true. Because it is infeasible to traverse all possible paths of the Data Flow Digraph, a test path is usually selected according to a criterion which involves a data flow property of the EFSM. In [12], a criterion is defined for observing the data flow abnormality of the EFSM due to a fault model. However, it is difficult for a test designer to construct a fault model that covers all possible faulty implementations of the EFSM. In [10][11], the *All-Use* and *IO-df-chain* criteria are defined for observing the data flow of variable values from where they are defined (or input) to where they are used (output.) They have become important criteria in EFSM testing.

A test path of the Data Flow Digraph is said to satisfy the All-Use (or IO-df-chain) criterion if it can trace all the variable value associations defined by that criterion. In order to trace[3] an association, the test path starts from the first node (i.e., the initial state) and proceeds to a specific node where a property of that association is exhibited, then returns to the first node to result in a *complete path* [11]. At the same time, we must ensure that the complete path allows a feasible assignment of input parameter values which makes all predicates along the path remain true. New complete paths must be obtained until that all associations defined by the criterion are traced. Unfortunately, this process is tedious and introduces a lot of overhead sequences in taking the EFSM to/from the initial state. In this paper, we are going to propose an automatic method to generate the executable test sequences that satisfy the All-Use and IO-df-chain criteria as well as optimizing the test sequence length.

We convert the Data Flow Digraph (the EFSM) into a *Behavior Machine Digraph*, the paths of which can be used to generate executable test sequences. Unlike the *Global FSM* which represents the complete behavior of the EFSM by enumerating all possible parameter and variable values as the inputs/outputs and states, the Behavior Machine only represents a partial

[3] such a trace is different from the "trace" used in verification for proving the property.

behavior which will be used for testing. That is, only certain parameter (or variable) values are embedded into the inputs/outputs (states) for constructing the Behavior Machine where all predicates are removed [4]. For instance, the Simple Connection Protocol of 4 states (see next section) is converted to a Global FSM of 768 states but is converted to a Behavior Machine of 6 states. Multiple paths of the Behavior Machine Digraph trace the same variable value association defined by the criterion, but only one needs to be included into the final test sequence. Thus, we use the Selecting Chinese Postman Algorithm to optimally select either one to construct a minimum-length executable test sequence which traces each association at least once.

In Section 2, the two criteria are reviewed. In Section 3, the method based on the first criterion is proposed. In Section 4, the method based on the second criterion is proposed. In Section 5, our conclusions are presented.

2. THE TWO DATA FLOW TESTING CRITERIA

Consider the Simple Connection Protocol (SCN). The control portion specifies how to establish/release a connection. To establish a connection, an input "CONreq" (i.e., connection request) from the upper interface causes the SCN to output "connect" to the lower interface. Inputs "accept" or "refuse" from the lower interface causes the SCN to output "CONrsp(+)" (i.e., positive connection response) or "CONrsp(-)" (negative connection response) to the upper interface. After the connection is established, an input "Data" from the upper interface causes the SCN to output "data" to the lower interface. To release the connection, an input "Reset" from the upper interface cause the SCN to output "abort" to the lower interface. Notice that the uppercase (or lowercase) first letter indicates that the message is related to the upper (lower) interface.

The data portion specifies other functions of the SCN by two types of variables. Variables of the first type are called the *memory variable* that will store temporary values. In the SCN, the memory variable TryCount stores the number of unsuccessful connection attempts and has values from 0 to 2, and the memory variables ReqQos and FinQos store the levels of requested and final quality of service and have values from 0 to 3. Variables of the second type are called *parameter variables* that will store the parameter values of inputs. Let X(y) denote an input X which has a parameter y, then the parameter variable denoted by X.y will store the value of parameter y. This definition can be extended to multiple parameters. In the SCN, the parameter variables CONreq.qos, accept.qos and data.qos will store the values of the parameter "qos" of inputs "CONreq(qos)", "accept(qos)" and "data(qos)" respectively. They have values from 0 to 3.

The EFSM can be represented by the Data Flow Digraph of Figure 1,

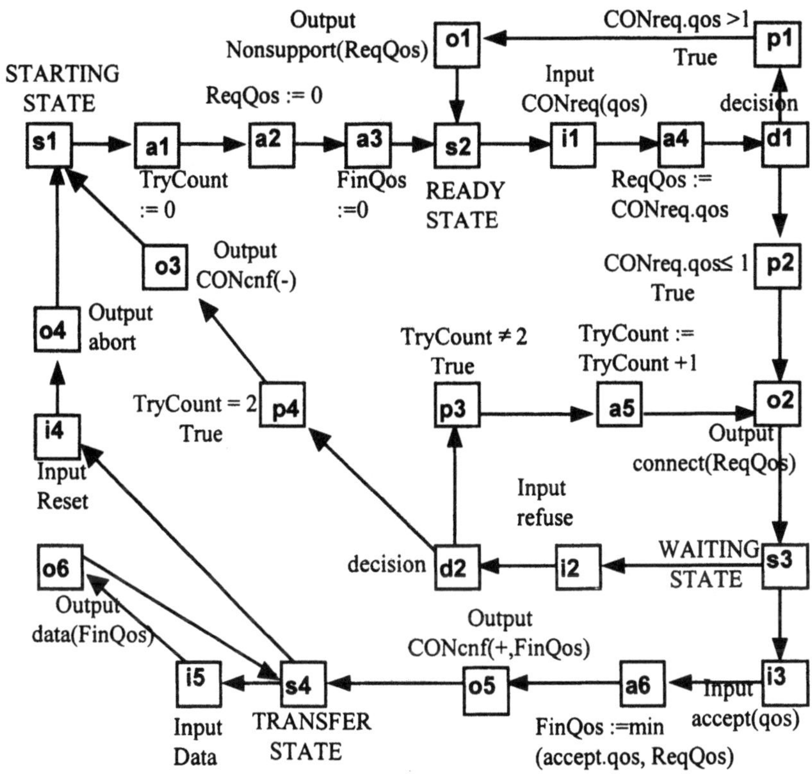

Figure 1. The Data Flow Digraph of the SCN protocol

where the states, inputs, outputs and actions are represented by the *state nodes*, *input nodes*, *output nodes and action nodes* which enclose symbols that start with "s", "i", "o" and "a" respectively. Each predicate of the EFSM is represented by one predicate-decision node (i.e., *decision node*) and two predicate-outcome nodes (i.e., *predicate nodes*) which specify the two possible outcomes[4]. The decision and predicate nodes enclose symbols that start with "d" and "p" respectively. The execution sequence of nodes is indicated by edge directions. For example, consider Figure 1. In the input node "i1", the value of variable CONreq.qos will be input according to the statement "Input CONreq(qos)." Then, in decision node "d1", the value of CONreq.qos is decided to see whether it is larger than 1 or not. Then, the EFSM may either arrive at either the predicate node p1 or predicate node p2

[4] In [10][11], a predicate is represented by a predicate node and the results of the predicates are represented by edges leaving that node.

which represents the two possible outcomes "CONreq.qos >1" and "CONreq.qos ≤1."

In Figure 1, a *path* is a sequence of nodes that are connected by edges. A *tour* is a special path which starts and ends at the same node. A path can be used to generate a test sequence by considering only the inputs and outputs in the path. For example, the path [s1, a1, a2, a3, s2, i1, a4, d1, p1, o1, s2] is used to generate the test sequence [Input CONreq(qos), Output Nonsupport(qos)] (or simply [CONreq(qos)/Nonsupport(qos)]) by considering only input "i1" and output "o1. " In reality, the parameter "qos" must be given a value which makes predicate p1 hold true. For example, [CONreq(0)/Nonsupport(0)] is not executable because predicate p1 will not hold true, but [CONreq(2)/Nonsupport(2)] is executable. An *executable path* is a path that allows parameter values to be assigned which causes each predicate in the path to be true in order to generate the *executable test sequence*. Not all paths are executable. For example, the path [s1, a1, a2, a3, s2, i1, a4, d1, p2, o2, s3, i2, d2, p4, o3, s1] is not executable. It is because that the value of variable TryCount becomes "1" in node a1. In order to pass through predicate node p4 which requires that the value of variable TryCount be equal to 2, the variable should increase its value in the intermediate nodes. However, the only node that can increment the value of TryCount is node a5, and that node is not in the path.

All-Use and IO-df-chain are two important criteria in selecting a test path from the Data Flow Digraph [11]. The first criterion claims that we should trace each variable from where it is *defined* (i.e., its value is first assigned) to where that value is *used*. We first describe where and when a variable is defined. At an input node, a variable is defined by an input statement. For example, at input node i1 of Figure 1, variable CONreq.qos is defined by the input statement "Input CONreq(qos)" which gives variable CONreq.qos a value. At an action node, a variable is defined by a computation statement which gives the variable a value. For example, at action node a1, variable TryCount is defined by the statement "TryCount :=0" where variable TryCount is assigned the value of 0. We then describe where and when a variable is used. At a predicate node, a variable is used in the predicate statement where the value of the variable will determine the result of the predicate. For example, in Figure 1, variable CONreq.qos is used at predicate node p1 because the value of CONreq.qos will determine whether the predicate "CONreq.qos >1" is true or not. At an action node, a variable is used in the computation statement where the variable value will determine the value of another variable. For example, at action node a6, variable ReqQos is used in the statement "FinQos := min(accept.qos, ReqQos)" where the value of ReqQos determines the value of variable FinQos. At an output node, a variable is used in the output statement where the value of the variable is output. For example, at node o2, variable ReqQos is used in the output statement "Output connect(ReqQos)." Table 1 lists the variables defined and used in Figure 1.

Table 1. Variables defined and used at the nodes of the Data Flow Digraph of Figure 1

Variable	Node where it is defined	Node where it is used
CONreq.qos	i1	a4 p1 p2
accept.qos	i3	a6
TryCount	a1 a5	a5 p3 p4
ReqQos	a2 a4	a6 o2 o1
FinQos	a3 a6	o5 o6

In tracing a variable X from node J (where X is defined) to node K (where X is used), the All-use criterion claims that the variable cannot not be redefined in the tracing process. Thus, to trace such an association, we must construct the *define-clear-use path*[5] which connects node J to node K in such a way that the first node is the only node of the path where X is defined. If such a define-clear-path exists, we say that variable X and nodes J and K form an association called a *define-use pair* (or *du-pair*) du(J, K, X). For example, consider Figure 1 and Table 1. Variable ReqQos and nodes a4 and o2 form a du-pair du(a4, o2, ReqQos) because it can be traced by the define-clear-use path [a4, d1, p2, o2], where variable ReqQos is defined exclusively at node a4 and used at node o2. It is possible that many define-clear-use paths can trace the same du-pair. But such a path may not exist at all so that not all associations of variables and nodes form du-pairs. A path (or test sequence) of the Data Flow Digraph is said to satisfy the All-Use criterion if all possible du-pairs can be traced.

The second criterion (i.e., the IO-df-chain criterion) claims that we should trace the data flow from an input node J (where a variable X is defined) to an output node K (where a variable Y is used), where X and Y can be either the same or different variables. Variables X and Y are input and output parameter values that are controlled and observed by the testing system respectively. Certainly, the value of variable X may not directly affect the value of variable Y, but it may affect indirectly. That is, X may affect another variable that in turn affects variable Y, and so on. For example, consider the path [i1, a4, d1, p2, o2] of Figure 1. At node i1, the EFSM receive an input "CONreq" which has a parameter "qos," the value of which is stored by the parameter variable CONreq.qos. At node a4, the value of CONreq.qos affects the value of variable ReqQos. At node o2, the value of ReqQos affects the parameter value of output "connect." As a result, we can control the parameter value of the input "CONreq" to observe

[5] the definition is extended from the define-clear path of [10][11].

the parameter value of the output "connect." Obviously, a sequence of define-clear-use paths must be connected to trace this data flow.

By definition, a sequence of define-clear-use paths $p_1, p_2, ..., p_{r-1}, p_r, ..., p_s$ (for variables $X_1, X_2, ..., X_{r-1}, X_r, ..., X_s$ respectively) which starts from an input node J and ends at an output node K form an *input-output chain* (i.e., *io-chain)* io(J, X_1, K, X_s). The chain [$p_1, p_2, .. .p_{r-1}, p_r ..., p_s$] can be used to trace the data flow from an input node J where variable X_1 is input to an output node K where the value of variable X_s is output, because the value of variable X_r is determined by the value of variable X_{r-1} through a computation statement in the node where p_{r-1} and p_r intersect, i.e., variable X_{r-1} affects X_r at the intersecting node (with the restriction that an action can has only one statement so that X_{r-1} cannot affect other variables). For example, [i1, a4] is a define-clear-use path for variable CONreq.qos which starts from an input node, and [a4, d1, p2, o2] is a define-clear-use path for variable ReqQos which ends at an output node. The two paths are connected into an io-chain io(i1, CONreq.qos; o2, ReqQos) which is [i1, a4, d1, p2, o2]. A path (or test sequence) of the Data Flow Digraph is said to satisfy the IO-df-chain criterion if all possible io-chains are traced.

3. THE METHOD FOR THE FIRST CRITERION

In Section 2, we have described the All-Use and IO-df-chain criteria that are guidelines for selecting the test paths from the Data Flow Digraph. In this section, we will propose a method which automatically generates the executable test sequence that satisfies the first criterion. Our method involves four steps.

The first step involves the conversion of the Data Flow Digraph (Figure 1)
into the Behavior Machine Digraph (Figure 2) where executable paths can be directly generated. In general, a node J (Figure 1) is converted into nodes J_1, J_2, J_3, ... (Figure 2)[6] by embedding parameter values. If all parameter values are embedded, the Behavior Machine Digraph will be very large. Hence, we only embed parameter values that enable the removal of decision nodes. For example, consider node i1 (i.e., CONreq(qos)) of Figure 1. The parameter qos has values from 0 to 3, so that we can convert node "CONreq(qos)" into four nodes "CONreq(0)", "CONreq(1)", "CONreq(2)", "CONreq(3) ." However, because either "CONreq(0)" or "CONreq(1)" can make predicate p2 true (and either "CONreq(2)" or "CONreq(3)" can make predicate p1 true), we only create node "CONreq(1)" (i.e., node $i1_1$) and node

[6] Without loss of generality, notations J_1, J_2, J_3, ... of the Behavior Machine Digraph represent the nodes converted from the node J of the Data Flow Digraph in this paper.

"CONreq(2)" (i.e., node $i1_2$). Then we remove the decision node d1 from the digraph since no decisions are required there. Because every predicate in the digraph is true, any path of the digraph can be used to generate an executable test sequence. In Figure 2, the Behavior Machine is represented by fine lines and text, but it includes some bold characters that will be used in the next step for tracing the data flow. In general, a bold character "$\mathbf{X_{def}}$" (or $\mathbf{X_{use}}$) is put on nodes J_1, J_2, J_3, .. of Figure 2 if variable X is defined (used) at node J of Figure 1. For example, a bold character $\mathbf{C_{def}}$ (an abbreviation of CONreq.qos_{def}) is put on nodes $i1_1$ and $i1_2$ of Figure 2, because variable CONreq.qos is defined at node i1 of Figure 1 (see Table 1.)

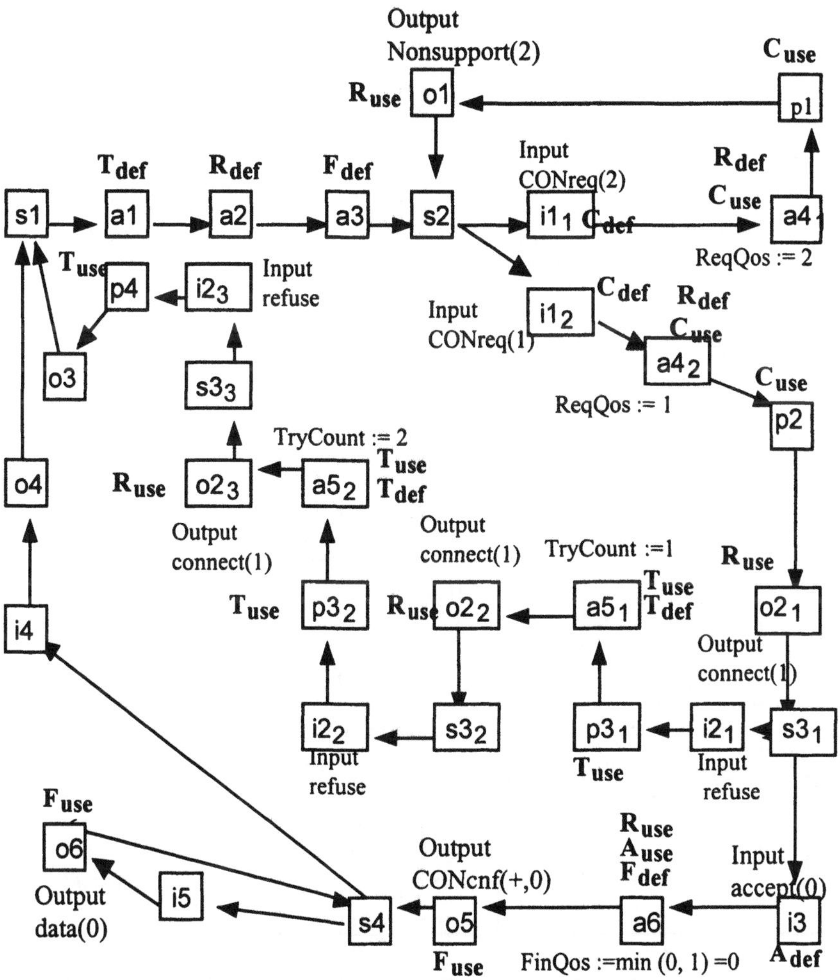

Remarks:
C: CONreq.qos **A**: accept.qos **T**:TryCount **R**: ReqQos **F**:FinQos
def (use): variable defined (used) d at that node

Figure 2. The Behavior Machine Digraph of the Data Flow Digraph of Figure 1 (only new statements other than that of Figure 1 are indicated)

The second step involves finding executable test paths of Figure 2 that can trace the du-pairs of Table 1. Consider the du-pair du(J, K, X), where variable X is defined at node J and used at node K. As described in Section 2, this du-pair is traced by a define-clear-use path of Figure 1 which starts from node J and ends at node K and does not redefine the variable in the intermediate nodes. In Figure 2, such an executable define-clear-use path corresponds to specific paths which start from nodes J_1, J_2,..., J_r ,...,J_s and end at nodes K_1,K_2,...,K_p,... ,K_q where the symbols "X_{def}" can only be seen in the starting nodes. These specific paths of Figure 2 can be constructed from shortest paths of Figure 2 where nodes which have the symbol X_{def} (except the starting nodes) and their adjacent edges are (temporarily) removed. For example, to trace the du-pair du(i1, a4, CONreq.qos) of Table 1 where "CONreq" is defined at node i1 and used at node a4, we find the specific shortest paths which starts at nodes $i1_1$ and $i1_2$ and ends at nodes $a4_1$ and $a4_2$ of Figure 2. These shortest paths are used to produce the executable define-clear-use paths [$i1_1$, $a4_1$] and [$i1_2$, $a4_2$] that can trace the du-pair du(i1, a4, CONreq.qos). The complete executable define-clear-use paths of Figure 2 for tracing the du-pairs of Table 1 are shown in Table 2.

Table 2. Executable test paths of Figure 2 for tracing the du-pairs of Table 1

Label	du-pairs of Table 1	Executable define-clear-use paths of Figure
1	du(i1,a4, CONreq.qos)	[$i1_1$, $a4_1$] or [$i1_2$, $a4_2$]
2	du(i1,p1, CONreq.qos)	[$i1_1$, $a4_1$, p1]
3	du(i1,p2, CONreq.qos)	[$i1_2$, $a4_2$, p2]
4	du(i3, a6, accept.qos)	[i3, a_6]
5	du(a1, a5, TryCount)	[a1, a2, a3, s2, $i1_2$, $a4_2$, p2,$o2_1$, $s3_1$, $i2_1$, $p3_1$, $a5_1$]
6	du(a1, p3, TryCount)	[a1, a2, a3, s2, $i1_2$, $a4_2$, p2,$o2_1$, $s3_1$,$i2_1$,$p3_1$]
7	du(a5, p3, TryCount)	[$a5_1$, $o2_2$, $s3_2$, $i2_2$, $p3_2$]
8	du(a5, p4, TryCount)	[$a5_2$, $o2_3$, , $s3_3$, $i2_3$, p4]
9	du(a4, o2, ReqQos)	[$a4_2$,p2,$o2_1$] or 2, p2, $o2_1$, $s3_1$, $i2_1$, $p3_1$, a51,$o2_2$] or [$a4_2$,p2,$o2_1$,$s3_1$,$i2_1$,$p3_1$,$a5_1$,$o2_2$,$s3_2$,$i2_2$, 2,$a5_2$,$o2_3$]
10	du(a4, a6, ReqQos)	2, p2, $o2_1$, $s3_1$, i3, a6]
11	du(a4, o1, ReqQos)	[$a4_1$, p1, o1]
12	du(a6, o5, FinQos)	[a6, o5]
13	du(a6, o6, FinQos)	[a6, o5, s4, i5, o6]

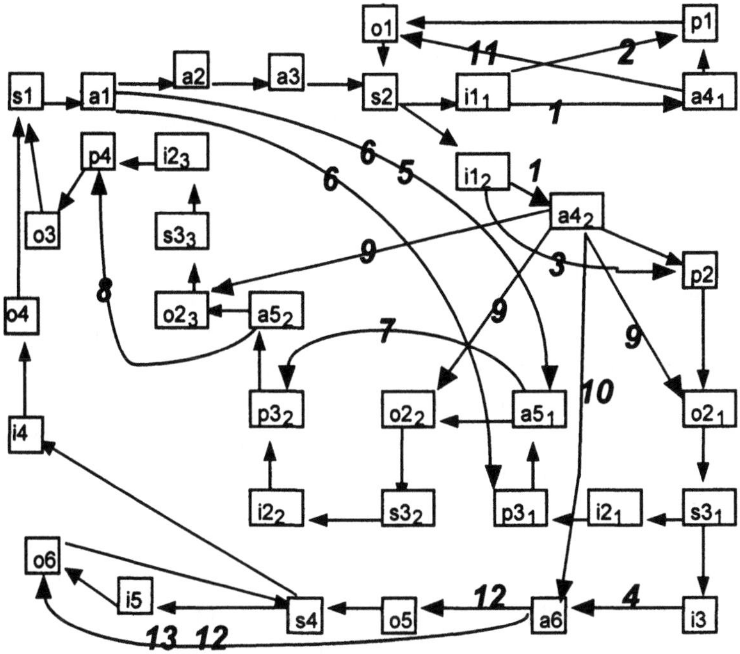

Figure 3. The SelectUse Digraph constructed from the Behavior Machine Digraph of Figure 2 by embedding the define-clear-use paths of Table 2 as bold edges

The third phase involves constructing a SelectUse Digraph of Figure 3 by embedding the executable test paths of Table 2 into the Behavior Machine Digraph of Figure 2. Generally, an executable define-clear-use path from node J to node K is embedded as a bold edge from node J to node K, and a bold label is put on the path to indicate the du-pair traced by the path. Cost is assigned to the edge according the number of inputs that it contribute to the final test sequence when that edge is traversed in the test path, because the length of a test sequence is decided by the number of inputs that it has. Thus, a fine edge is assigned a cost of 1 if it leaves an input node, and a bold edge is assigned a cost which is the number of inputs contained in the define-clear-use path represented by the bold edge. For example, consider the executable test path [a1, a2, a3, s2, $i1_2$, $a4_2$, p2,$o2_1$, $s3_1$, $i2_1$, $p3_1$, $a5_1$] of Table 2 that traces the du-pair du(a1, a5, TryCount), which has the label "5." Because the path starts at node a1 and ends at node $a5_1$, it is embedded as a bold edge from node a1 to $a5_1$. The bold edge is labeled with "5" and assigned a cost of 2. An executable define-clear-use path that can trace multiple du-pairs is embedded as an edge that has multiple labels. For example, the define-clear-use path described above can trace two du-pairs of Table 2, and we put labels "5" and "6" on it. For many bold edges that share the same label, only one edge needs to be included in the final test path because these bold edges represent the executable test paths that trace the same du-pair. As a result, the Selecting Chinese Postman Tour of the SelectUse Digraph (which is a minimum-cost tour where each type of label appears at least once) can be used to generate a minimum-length executable test sequence that traces all the du-pairs.

The fourth step involves using the Selecting Chinese Postman Algorithm proposed in [1] to find the Selecting Chinese Postman Tour of the SelectUse Digraph in order to generate the executable test sequence. The algorithm contains two phases. The first phase replicates (or deletes) each edge of the SelectUse digraph, resulting in a *minimum-cost Selecting Symmetric Augmentation* (MCSSA) which satisfies the properties that i) each node has the same number of entering and leaving edges and ii) each type of label appears in the digraph at least once. The MCSSA can be obtained by solving a system of integer programming equations formulated from the SelectUse Digraph. The integer programming problem is solved by a branch and bound algorithm which iterates to improve an initial solution [8]. When the problem size is not very large, the branch and bound algorithm can find the optimal solution. Otherwise, the branch and bound algorithm will at least obtain a significantly improved solution if we can make the algorithm run at long time. In the second phase, we check whether the MCSSA is an Euler Digraph or a collection of disjoint Euler Digraphs [7]. In the former case, an Euler Tour of the Euler Digraph is an Selecting Chinese Postman Tour of the SelectIO Digraph. In the latter case, a tour will be used to connect these disconnected Euler Digraphs into an Euler Digraph, so that the Euler Tour algorithm can be applied. For Figure 3, the Lindo package [8] has solved the integer programming equations in less than one second and results in an Euler Digraph, resulting in the first case so that that the optimal tour is found. The Selecting Chinese Postman Tour of Figure 3: [s1, a1, a2, a3, s2, $i1_1$, ***1***, $a4_1$, ***11***, o1, s2, $i1_1$, ***2***, p1, o1, s2, $i1_2$, ***1***, $a4_2$, ***10***, a6, ***(13,12)***, o6, s4, i4, o4, s1, a1, a2, a3, s2, $i1_2$, ***3***, p2, $o2_1$, $s3_1$, i3, ***4*** ,a6, ***12***, o5, s4, i4, o4, s1, a1, a2, a3, s2, $i1_2$, ***1***, $a4_2$, ***9***, $o2_2$, $s3_2$, $i2_2$, $p3_2$, $a5_2$, $o2_3$, $s3_3$, $i2_3$, p4, o3, s1, a1, ***(6, 5)***, $a5_1$, *7*, $p3_2$, $a5_2$, ***8***, p4, o3, s1, a1, a2, a3, s2, $i1_2$, ***3***, p2, o21, s31, i3, ***4***, a6, ***12***, o5, s4, i4, o4, s1] is used to generate the minimum-length executable test sequence [$i1_1$, o1, $i1_1$, o1, $i1_2$, $o2_1$, i3, o5, i5, o6, i4, o4, $i1_2$, $o2_1$, i3, o5, i4, o4, $i1_2$, $o2_1$, $i2_1$, $o2_2$, $i2_2$, $o2_3$, $i2_3$, o3, $i1_2$, $o2_1$, $i2_1$, $o2_2$, $i2_2$, $o2_3$, $i2_3$, o3, $i1_2$, $o2_1$, i3, o5, i4, o5] which traces all the du-pairs of Table 1.

4. THE METHOD FOR THE SECOND CRITERION

In Section 3, we have proposed a method to produce an executable test sequence which satisfies the first criterion by finding the Selecting Chinese

Postman Tour of the SelectUse Digraph. In this Section, we are going to extend the method to generate an executable test sequence which satisfies the second criterion, i.e., the IO-df-chain criterion which requires tracing each io-chain at least once. The extended method contains two steps.

The first step involves finding all io-chains of Figure 1. In the SelectUse Digraph of Figure 3, we consider input nodes J_1, J_2, J_3, ..., (converted from node J where variable X is defined) to output nodes K_1, K_2, K_3, ... (converted from node K where variable Y is used). As described in Section 2, an io-chain is constructed from a sequence of define-clear-use paths so as to connect an input node to an output node (but notice that a sequence of define-clear paths defined in [11] does not necessary form an input-output-chain). We call those bold edges which leave input nodes J_1, J_2, J_3, ..., that are the define-clear-use paths for variable X as the (X-J) edges, and those bold edges which enter output nodes K_1, K_2, K_3, ..., that are the define-clear-use paths for variable Y as the (Y-K) edges. A path which involves only bold edges is a bold path. In the SelectUse Digraph, the shortest bold path from these (X-J) edges to those (Y-K) edges can be used to generate the sequence of define-clear-use paths which can trace io-chain io(J, X, K, Y). Those specific shortest paths of the SelectUse Digraph can be obtained from shortest paths of the SelectUse Digraph where fine edges are (temporarily) removed. For example, the bold edge (i.e., the define-clear-use paths [$i1_1$, *1*, $a4_1$]) and another bold edge (i.e., [$a4_1$, *11*, o_1]) compose an io-chain io(i1,CONreq.qos, o1, ReqQos) = [$i1_1$, *1*, $a4_1$, $a4_1$, *11*, o_1]. The complete io-chains for Figure 1 obtained from these specific shortest paths of Figure 3 are shown in Table 3, where each io-chain is given a label.

Table 3. The io-chains for the Data Flow Digraph of Figure 1

Label	input-output-chains (Figure 1)	Composed define-clear-use paths (Figure 3)	Corresponding Executable Paths (Figure 2)
a	io(i1,CONreq.qos, o1, ReqQos)	[$i1_1$, *1*, $a4_1$, *11*, o1]	[$i1_1$, $a4_1$, p1, o1]
b	io(i1,CONreq.qos, o2, ReqQos)	[$i1_2$, *1*, $a4_2$, *9*, $o2_1$]	[$i1_2$, $a4_2$, p2, $o2_1$]
		[$i1_2$, *1*, $a4_2$, *9*, $o2_2$]	[$i1_2$, $a4_2$, p2, $o2_1$, $s3_1$, $i2_1$, $p3_1$, a51, $o2_2$]
		[$i1_2$, *1*, $a4_2$, *9*, $o2_3$]	[$i1_2$, $a4_2$, p2, $o2_1$, $s3_1$, $i2_1$, $p3_1$, $a5_1$, $o2_2$, $s3_2$, $i2_2$, $p3_2$, $a5_2$, $o2_3$]
c	io(i1,CONreq.qos, o5, FinQos)	[$i1_2$, *1*,$a4_2$, *10*,a6,*12*, 05]	[$i1_2$, $a4_2$, p2, $o2_1$, $s3_1$, i3, a6, o5]
d	io(i1,CONreq.qos, o6, FinQos)	[$i1_2$, *1*, $a4_2$, *10*, a6,*13*,06]	[$i1_2$, $a4_2$, p2, $o2_1$, $s3_1$, i3, a6, o5, s4, i5, o6]

The second step involves constructing a SelectIO Digraph of Figure 4 by embedding the executable test paths of Table 3 to the Behavior Machine Digraph of Figure 2. Generally, an executable test path listed in Table 3 which starts from an input node J and ends at output node K is embedded as a bold edge from vertex J to vertex K. The label which represents the io-chain traced by the executable path is put on the edge (see Table 3 and Figure 4.) Costs are assigned to the edges of the SelectIO Digraph similar to the process described in the third step of Section 3 for assigning the cost to the edges of the SelectUse Digraph. A Selecting Chinese Postman Tour of the SelectIO digraph can be used to generate an minimum-cost executable test sequence that checks each io-chain. The tour is obtained using the algorithm described in the four step of Section 3. The Selecting Chinese Postman Tour of Figure 4: [s1, a1, a2, a3, s2, $i1_1$, ***a***, o1, s2, $i1_2$, ***b***, $o2_1$, $s3_1$, $i2_1$, $p3_1$, $a5_1$, $o2_2$, $s3_2$, $i2_2$, $p3_2$, $a5_2$, $o2_3$, $s3_3$, $i2_3$, p4, o3, s1, a1, a2, a3, s2, $i1_2$, ***(c, d)***, o6, s4, i4, s1] is used to generate the executable test sequence [$i1_1$, o1, $i1_2$, $o2_1$, $i2_1$, $o2_2$, $i2_2$, , $o2_3$, $i2_3$, o3, $i1_2$, $o2_1$, i3, o5, i5, o6, i4, o4] which traces all the io-chains of Table 3.

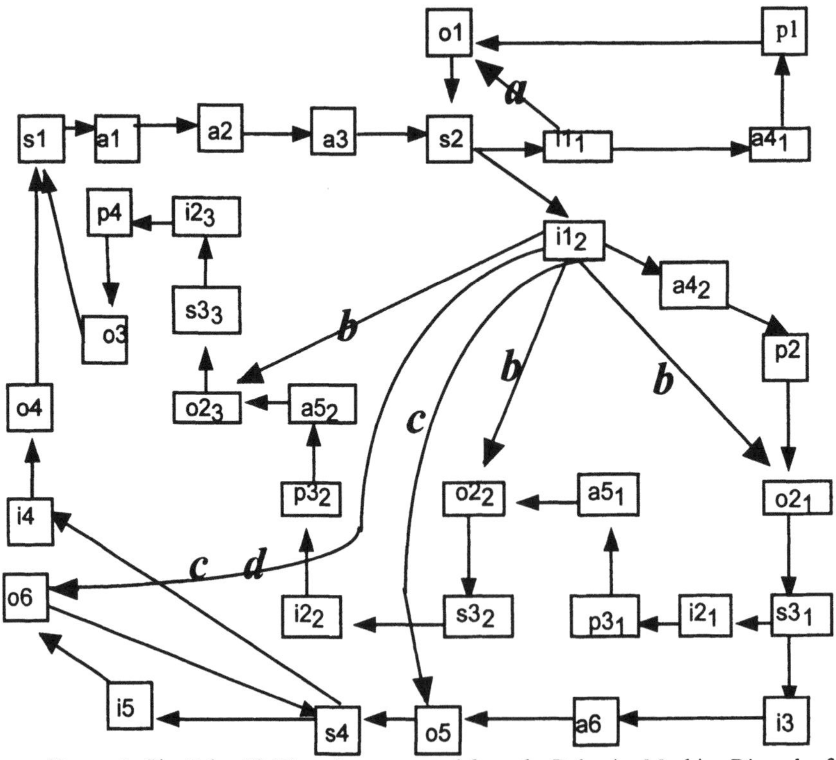

Figure 4. The SelectIO Digraph constructed from the Behavior Machine Digraph of Figure 2 by embedding the io-chains of Table 3 as bold edges

5. CONCLUSIONS

In this paper, we have proposed a method to automatically generate the executable test sequence from the protocol specification for verifying two data flow criteria, based on finding the Selecting Chinese Postman Tour of the SelectUse and SelectIO Digraphs constructed from the Data Flow Digraph of the protocol. The first phase of the Selecting Chinese Postman Algorithm involves solving a system of integer programming equations so as to find an augmentation. In our experiences of solving such equations [1][2], a linear programming version of the formulation always yield integer results. We want to check whether these equations satisfy a specific property so that the linear programming approach can be applied.

Our method of minimizing the test sequence length can be easily extended to minimizing the test sequence cost, by assigning cost to the edges of the digraph. And the method can be combined with the duplexE digraph method to generate a synchronizable and executable test sequence [3].

REFERENCES

[1] W. H. Chen, "Test sequence generation from the protocol data portion based on the Selecting Chinese Postman algorithm," *Information Processing Letters*, Vol. 65, 1998.

[2] W. H. Chen, "Executable test sequence for the protocol control and data portions," *Proc. of IEEE Int'l Conference on Communications*, New Orleans, U. S. A., 2000.

[3] W. H. Chen and H. Ural, "Synchronizable test sequence based on multiple UIO sequences," *IEEE/ACM Trans. on Networking*, Vol. 3, No. 2, 1995.

[4] K. T. Cheng and A. S. Krishnakumer, "Automatic functional test generation using the extended finite state machine model," *Proc. IEEE Design Automation Conference*, 1993.

[5] R. Dssouli, K. Saleh, E. Aboulhamid, A. En-Nouaary and C. Bourhfir, "Test development for communication protocols: towards automation," *Computer Networks*, Vol. 31, 1999.

[6] E. Kwast, "Automatic test generation for protocol data aspects, " *Proc. IFIP Int'l Symp. on Protocol Specification, Testing, and Verification*, 1992.

[7] J. A. Mchugh, *Algorithmic Graph Theory*, Prentice-Hall, Englewood Cliffs, NJ, 1990.

[8] L. Schrage, *LINDO 5.0 User's Manual*, Scientific Press., 1991.

[9] M. Schwartz, *Telecommunication Networks: Protocols, Modeling and Analysis*, Addison-Wesley Publishing Company, 1987.

[10] H. Ural, "Test sequence selection based on static data flow analysis," *Computer Communications*, Vol. 10, No. 5, 1987.

[11] H. Ural, K. Saleh and A. Williams, "Test generation based on control and data dependencies within system specification in SDL," *Computer Communications*, Vol. 23, 2000.

[12] C. J. Wang and M. T. Liu, "Generating test cases for EFSM with given fault models," *Proc. IEEE INFOCOM*, March 1993.

A METHOD TO GENERATE CONFORMANCE TEST SEQUENCES FOR FSM WITH TIMER SYSTEM CALL

Takanori Mori†, Kohei Tokuda†,
Harumasa Tada†, Masahiro Higuchi‡ and Teruo Higashino†
†*Dept. of Info. and Math. Sci., Osaka Univ., Toyonaka, Osaka 560-8531, Japan*
{t-mori, k-tokuda, tada, higashino}@ics.es.osaka-u.ac.jp
‡*School of Sci. and Eng., Kinki Univ., Higashiosaka, Osaka 577-8502, Japan*
higuchi@ee.kindai.ac.jp

Abstract In this paper, we propose a method to generate conformance test sequences for communication protocols modeled as FSM with timers. The test sequences generated by the proposed method can detect any single fault of timer commands or destination states in the transitions on protocol machines. For each single timer command fault, we give sufficient conditions that a given IUT is (or is not) equivalent to its specification. Based on these sufficient conditions, we give a method for generating test sequences. For each destination state fault, we give a test sequence generation method based on Wp-method. To show the usefulness of this method, we have developed a system for generating test sequences, and applied it to DHCP (Dynamic Host Configuration Protocol). As a result, we have generated the test sequences efficiently.

Keywords: conformance testing, test case generation, embedded system, FSM, timer

1. INTRODUCTION

Recently, complex systems consisting of two or more cooperating components are commonly used. In such complex systems, interactions between components cannot be observed and controlled from their environment, usually[1]. Conformance testing for complex systems are discussed in a lot of papers[2, 3, 4].

Communication protocols which use timer functions provided by operating systems (OS) also can be viewed as such complex systems. In this paper, we discuss conformance testing of such protocols. A system considering in this paper consists of a FSM based protocol machine and timers. A protocol machine is modeled as a deterministic finite state machine (DFSM) which uses timer functions of OS through system calls. The interactions between a protocol machine and timers cannot be observed and controlled from their environment.

In conformance testing for specific subsystems of complex systems, a given complex system is divided into two components; (i) *Spec* : the subsystem of the test target, and (ii) *Context* : the subsystems in the complex system other than *Spec*. The context is assumed to be correctly implemented. A system consisting of IUT (Implementation Under Test) and Context is called SUT (System Under Test). Such testing is called embedded testing[5] or gray box testing[2].

For embedded testing, we must consider the following features of complex systems; (i) although the IUT does not conform to its specification *Spec*, the behavior of the resulting SUT may conform to *Spec·Context*, and (ii) a single fault of IUT may cause multiple faults of SUT.

Embedded testing has been mainly investigated by assuming that each component of SUT is modeled as DFSM[2, 3, 4]. Although timers can be modeled as DFSM, the numbers of states are usually large. We will consider the above features of timers.

The remainder of this paper is organized as follows. In Section 2, we describe the formal model of FSM with timers. In Sections 3 and 4, we discuss a test sequence generating method. In Section 5, we explain an experiment of applying the proposed method to DHCP.

2. COMMUNICATION PROTOCOLS WITH TIMER SYSTEM CALL

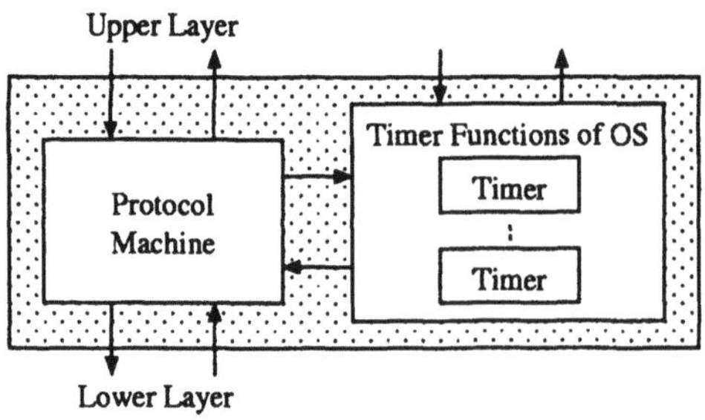

Figure 1. Communication protocols with timer system call

In this paper, we consider a class of communication protocols shown in Figure 1. In this model, a communication entity consists of a finite state protocol machine and multiple timers. The interactions between the protocol machine and timers are (i) timer commands (enabling or disabling timers) from the protocol machine to timers, and (ii) timeout notifications from timers to the protocol machine. Assume that the timer function is provided by operating systems. The timer commands are given as system calls, e.g. `set_timer` or `del_timer` in Linux. The timeout notifications are given as signals. These interactions cannot be observed and controlled from their environment.

2.1. Timer

In general, operating systems can manage multiple timers. A user process can enable or disable individual timer through system calls. Here, the timer period from enabling a timer to expiring the timer is called the *timer expiring time.* The timer expiring time is specified as a parameter of the system call. After enabling a timer, if the timer expiring time elapses without disabling or re-enabling the timer, then the timer expires and produces a timeout signal.

We assume that each timer can be identified by a timer number and the timer expiring time is fixed for individual timer. A specification of timers is defined by a vector T of timer expiring times. $v[i]$ denotes the i-th element of a vector v. $T[i]$ denotes the timer expiring time of timer i. In order to describe a state of timers, we introduce a timer value vector $\bar{\tau}$. $\tau[i]$ denotes the current value of timer i. $\tau[i]$ has an integer value $(0 \leq \tau[i] \leq T[i])$ or $\perp$. $\tau[i] = \perp$ means timer i is not active. We assume that $\perp > x$ and $\perp - x = \perp$, for all $x \in \mathcal{N}$ ($\mathcal{N}$: the set of non-negative integers). When timer i is enabled, $\tau[i]$ is set to $T[i]$. When timer i is disabled, $\tau[i]$ is set to $\perp$. The values of all $\tau[i]$ decrease one by one simultaneously every time unit. When $\tau[i]$ becomes zero, timer i produces a timeout signal and $\tau[i]$ becomes $\perp$. We define $\mathcal{I}[i]$ as the set $\{0, 1, \ldots, T[i], \perp\}$. $\mathcal{T}$ denotes the set of timer value vectors $\{\bar{\tau} \mid \tau[i] \in \mathcal{I}[i]\}$.

If there exists a timer whose value is zero, a unit time never elapses. If there exist two or more timers whose values are zero, the timers produce timeout signals in the increasing order of their timer numbers. If a timer i whose value is zero is re-enabled or disabled, $\tau[i]$ is set to $T[i]$ or $\perp$, respectively. Such a timer does not produce a timeout signal.

2.2. Protocol machine

A protocol machine is modeled as a Mealy deterministic finite state machine (DFSM) and defined as the following 6-tuple (Q, X, n, Y, H, s_0).

Q	:	a finite set of states.
X	:	a finite set of external input symbols.
n	:	the number of timers.
Y	:	a finite set of external output symbols.
H	:	a finite set of transitions $(u, v, x, y, \bar{p})$. $u, v \in Q$: source state, destination state. $x \in (X \cup \{1, \ldots, n\})$: input. $y \in Y$: external output. $\bar{p} \in \langle \mathtt{S}, \mathtt{D}, \mathtt{N} \rangle^n$: timer command vector.
s_0	:	an initial state.

A state transition is executed when a protocol machine receives an input (an external input or a timeout signal). State transitions caused by external inputs and timeout signals are called external input transitions and timeout transitions, respectively. As protocol machines are deterministic, the current state and an input can determine the destination state and outputs (an external output and a timer command vector) uniquely. We assume that state transitions are executed instantaneously. A timer command vector is a vector $\bar{p} = (p[1], p[2], \ldots, p[n])$. $p[i]$ denotes a command for timer i. The command is either S, D or N. S is a command enabling a timer. D is a command disabling a timer. N denotes null operation for a timer.

We assume that protocol machines are reduced. We also assume that each protocol machine has a reliable reset feature. If a protocol machine receives a reset input, the protocol machine is reset to its initial state and disables all timers. If a protocol machine receives an undefined input, the protocol machine is reset and produces an error output. With such error transitions, protocol machines are completely specified.

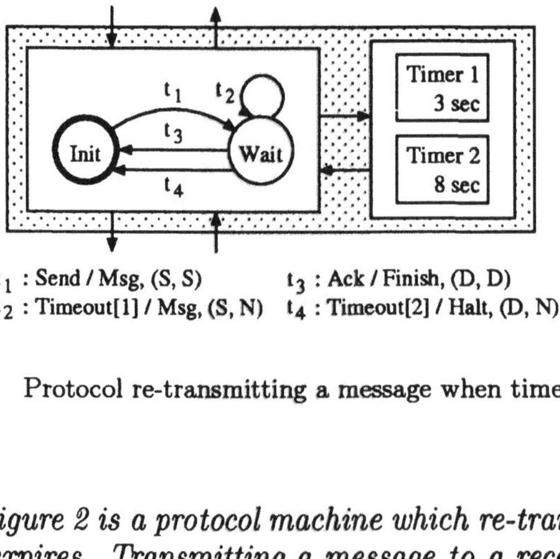

Figure 2. Protocol re-transmitting a message when timeout occurs

Example 1 *Figure 2 is a protocol machine which re-transmits messages when a timer expires. Transmitting a message to a receiver, the protocol machine enables both Timer 1 and Timer 2. The protocol machine re-transmits the message every expiration of Timer 1 until receiving an Ack message. Receiving a timeout signal of Timer 2, the protocol machine stops re-transmitting the message. In Figure 2,* Timeout[1] *and* Timeout[2] *denote timeout signals of Timer 1 and Timer 2, respectively.*

2.3. FSM with timers

For a protocol machine $M = (Q, X, n, Y, H, s_0)$ and a timer specification T, if n equals to the dimension of T, $M \cdot T$ denotes the system consisting of M and T. We use a pair $\xi = \langle s, \bar{\tau} \rangle$ of a state and a timer value vector to describe an entire state of $M \cdot T$ where $s \in Q$ and $\bar{\tau} \in T$. Such a pair is called a *composite state* of $M \cdot T$. The set of all composite states of $M \cdot T$ is denoted by Q_T. The initial composite state of $M \cdot T$ is $\xi_0 = \langle s_0, \perp^n \rangle$, where s_0 is the initial state of M, and x^n $(x \in \mathcal{N} \cup \perp)$ denotes a vector that all elements of the vector are x.

Given a composite state $\langle s, \bar{\tau} \rangle$, if there exists an integer i $(1 \leq i \leq n)$ such that $\tau[i] = 0$ and $\tau[j] > 0$ for $1 \leq j < i$, then $\langle s, \bar{\tau} \rangle$ is said to be *timer i expiring state.* $\Gamma[i]$ denotes the set of timer i expiring states. $\langle s, * \rangle$ denotes the set of composite states $\{\langle s, \bar{\tau} \rangle \mid s \in Q, \bar{\tau} \in T\}$. The timer value vector part of a composite state ξ is denoted by $\tau(\xi)$. For the set of composite states Ξ, $\tau(\Xi)$ denotes the set of timer value vectors $\{\tau(\xi) \mid \xi \in \Xi\}$.

We use H_T to denote the set of composite state transitions of $M \cdot T$. H_T is the set of composite state transitions $\eta = (\xi, \xi', x, y)$. There exist three kinds of transitions.

1. external input transitions :
   For an external input transition $(s, s', x, y, \bar{p}) \in H$, there exist composite state transitions $(\langle s, \bar{\tau} \rangle, \langle s', \bar{\tau}' \rangle, x, y)$ on $M \cdot T$ where $\tau[i] > 0 (1 \leq i \leq n)$ and

$$\tau'[i] = \begin{cases} T[i] & (p[i] = \mathtt{S}) \\ \tau[i] & (p[i] = \mathtt{N}) \\ \perp & (p[i] = \mathtt{D}) \end{cases}$$

2. timeout transitions :
   For a timeout transition $(s, s', \text{Timeout}[k], y, \bar{p}) \in H$, there exist composite state transitions $(\langle s, \bar{\tau} \rangle, \langle s', \bar{\tau}' \rangle, -, y)$ on $M \cdot T$ where $\langle s, \bar{\tau} \rangle \in \Gamma[k]$ and

$$\tau'[i] = \begin{cases} T_i & (p[i] = \mathtt{S}) \\ \tau[i] & (i \neq k \wedge p[i] = \mathtt{N}) \\ \perp & (p[i] = \mathtt{D} \vee (i = k \wedge p[i] = \mathtt{N})) \end{cases}$$

3. time elapse transitions :
   There exist composite state transitions $(\langle s, \bar{\tau} \rangle, \langle s, \bar{\tau} - 1^n \rangle, -, -)$ on $M \cdot T$ where $\tau[i] > 0 (1 \leq i \leq n)$.

H_t denotes the set of composite state transitions on $M \cdot T$ corresponding to $t \in H$. Also H_e denotes the set of the time elapse transitions.

For a transition $\eta = (\xi, \xi', x, y)$, $IO(\eta)$ denotes the input/output part of η, i.e. $IO(\eta) = (x/y)$. $\rho(\eta)$ and $\delta(\eta)$ denotes the source composite state and the destination composite state of η, respectively.

A composite state transition string is simply called a string. Given a string $r = \eta_1 \eta_2 \cdots \eta_m \in {H_T}^*$, r is said to be executable from a composite state $\xi_1 \in Q_T$, if there exist composite states $\xi_2, \ldots, \xi_{m+1} \in Q_T$ such that $\rho(\eta_k) = \xi_k$ and $\delta(\eta_k) = \xi_{k+1}$ for $k(1 \leq k \leq m)$.

Example 2 *A string "$\eta_1\eta_2\eta_3\eta_4\eta_5\eta_6$" is executable from the initial composite state of the protocol machine shown in Figure 2 where $\eta_1 \in H_{t_1}$, $\eta_2, \eta_3, \eta_4 \in H_e$, $\eta_5 \in H_{t_2}$ and $\eta_6 \in H_{t_3}$.*

We extend IO and δ to accept a string or a set of strings. For a string r, $\delta(r)$ denotes the destination composite state of the last transition of r. $TS(\xi)$ denotes the set of strings which are executable from ξ, and $TS_n(\xi)$ denotes the set of strings in $TS(\xi)$ whose length are n. $\mathcal{R}(\xi)$ denotes the set of composite states $\delta(TS(\xi))$. We may explicitly express the target machine M in our notations, like $TS_M(\xi)$ or $\mathcal{R}_M(\xi)$.

2.4. Related work

Timed automata[6, 7] are known as a model for describing and analyzing protocols which deal with time dependent behaviors. A timed automaton has several timers. The timer values increase by time elapse, and we can give timing constraints for transitions by simultaneous inequality on timer values. In state transitions, the timer value can be reset to zero. Although we do not give the proof, a timed automaton can simulate our FSM with timers. It means that the procedures for analyzing timed automata are also effective for analyzing FSM with timers.

3. TEST SEQUENCES DETECTING SINGLE FAULT

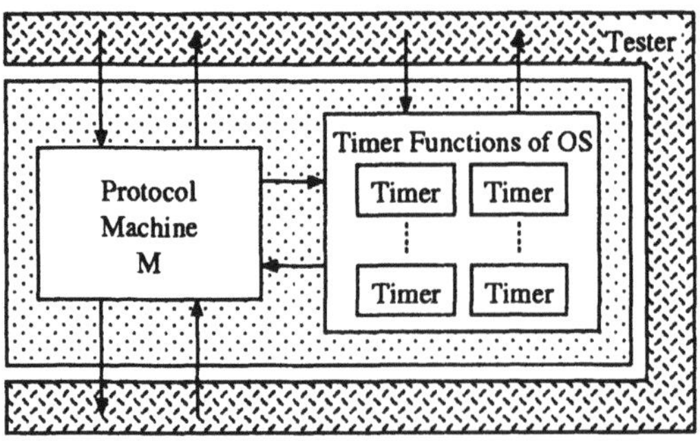

Figure 3. Test architecture

We consider a test architecture for FSM with timers (see Figure 3). In this architecture, the tester can measure time through timer system calls although the tester cannot observe and control the interactions between the protocol machine M and timers. We assume that the tester receives the timeout signals from timers after receiving all outputs from M. We use this assumption to construct test sequences in Section 3.4.

3.1. Fault model

We introduce a fault model for transitions of the protocol machines as follows.

- timer command faults :
  An element of timer command vector of a state transition differs from that of the specification.
- destination state faults :
  A destination state of a state transition differs from that of the specification.
- external output faults :
  An external output of a state transition differs from that of the specification.

In this paper, we focus to detect single faults. For a given specification $M = (Q, X, n, Y, H, s_0)$ and T, each faulty implementation can be denoted by $M[t'/t] = (Q, X, n, Y, H', s_0)$ and T such that $H' = H \setminus \{t\} \cup \{t'\}$ where t' differs from t only in one element of the timer command vector, the destination state or the external output.

3.2. IO equivalent implementations

In a complex system, although an implementation I does not conform to the specification S, a composed machine $I \cdot C$ (C is a context) may conform to $S \cdot C$ [2, 3]. It is also true in our model of FSM with timers. It means that some implementations which do not conform to the specification cannot be externally distinguished from the specification in context. We introduce the following IO equivalent relation.

Definition 1 *Given two states ξ_1 of A_T and ξ_2 of B_T, ξ_1 and ξ_2 are said to be IO equivalent, written $\xi_1 \equiv \xi_2$, if $IO(TS(\xi_1)) = IO(TS(\xi_2))$.*

Definition 2 *A_T and B_T are said to be IO equivalent, written $A_T \equiv B_T$, if the initial composite states of A_T and B_T are IO equivalent.*

Given a protocol machine M and an implementation $M[t'/t]$ which contains a single timer command fault or destination state fault, $M[t'/t] \cdot T$ may be IO equivalent to $M \cdot T$. On the other hand, if an implementation $M[t'/t]$ which contains a single external output fault, $M[t'/t] \cdot T$ is not IO equivalent to $M \cdot T$ except the case that the faulty transition is not executable, since our tester can observe external outputs.

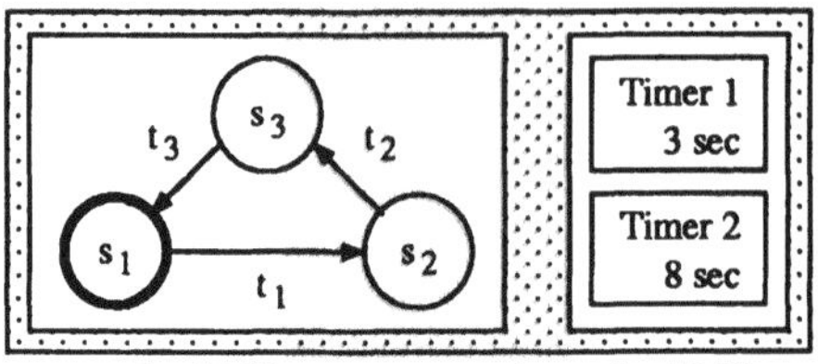

Figure 4. A faulty implementation IO equivalent to the specification

Example 3 *For the protocol machine M shown in Figure 4, let us consider a faulty implementation $M[t_1'/t_1]$ in which the transition t_1 is implemented as t_1'.*

The initial state of M is $\langle s_1, (\perp, \perp)\rangle$. In addition, after moving to s_1 by transition t_3, $(\tau_1, \tau_2) = (\perp, \perp)$. Thus, $(\tau_1, \tau_2) = (\perp, \perp)$ holds whenever M is in s_1. It also holds for $M[t_1'/t_1]$.

Let us consider strings which are executable from $\langle s_1, (\perp, \perp)\rangle$ in M and $M[t_1'/t_1]$. After executing t_1 or t_1', M is in $\langle s_2, (3, \perp)\rangle$ and $M[t_1'/t_1]$ is in $\langle s_2, (3, 8)\rangle$. For the both composite states, timer 1 will expire after three time elapse transitions. By the timer expiration, transition t_2 is executed, $M \cdot T$ is in $\langle s_3, (\perp, 8)\rangle$ and $M[t_1'/t_1] \cdot T$ is also in $\langle s_3, (\perp, 8)\rangle$. Furthermore, for the strings corresponding the above behaviors, $IO(\eta_1\eta_2\eta_3\eta_4\eta_5) = IO(\eta_1'\eta_2'\eta_3'\eta_4'\eta_5')$ where $\eta_1 \in H_{t_1}$, $\eta_1' \in H_{t_1'}$, $\eta_2, \eta_3, \eta_4, \eta_2', \eta_3', \eta_4' \in H_e$ and $\eta_5, \eta_5' \in H_{t_2}$. Thus $M \cdot T \equiv M[t_1'/t_1] \cdot T$.

3.3. Fault detecting sequence

For every implementation $M[t'/t]$, the equivalence between $M \cdot T$ and $M[t'/t] \cdot T$ can be decided by composing two finite automata which accept $IO(TS_{M \cdot T}(\xi_0))$ and $IO(TS_{M[t'/t] \cdot T}(\xi_0))$ and deciding their equivalence. We can compose the above two automata in a similar way to the composition of the region automata from timed automata[6].

Since protocol machines are assumed to be deterministic and completely specified, if $M \cdot T$ is not IO equivalent to $M[t'/t] \cdot T$, there exist at least one IO sequence *io* which is executable from the initial composite state of $M \cdot T$ and is not executable from that of $M[t'/t] \cdot T$. We can adopt such a sequence *io* as a fault detecting sequence.

3.4. Translating IO sequences to test sequences

Let *io* be an IO sequence obtained in the previous section. Since *io* is merely an IO sequence of $M \cdot T$, we have to translate *io* to a test sequence observable from our tester. Since one or more consecutive timeout transitions occur after at least one time elapse transition, a string of $M \cdot T$ should be an element of the set denoted by a regular expression $\{(E^*F)|(E^+G^+)\}^*E^*$, where E denotes the set of time elapse transitions, F and G denote the set of external input transitions and that of timeout transitions, respectively. The translation is as follows.

- IO sequences corresponding to E^*F :

$$(-/-)^n(x/y) \Longrightarrow \begin{cases} (x/y) & n=0 \\ (\mathrm{Set}(n)/-)(\mathrm{WE}/\mathrm{TO})(x/y) & n>0 \end{cases}$$

  Set(n) means that the tester sets a timer to n. WE means that the tester Waits Enough until an output comes. TO means a timeout signal corresponding to Set(n). If the tester receives a timeout signal after setting a timer to n, we can guarantee that n time units have passed.

- IO sequences corresponding to E^+G^+ :

$$(-/-)^n(-/-)(-/y_1)\cdots(-/y_m) \Longrightarrow$$
$$\begin{cases} (\mathrm{Set}(1)/-)(\mathrm{WE}/y_1)\cdots(\mathrm{WE}/y_m)(\mathrm{WE}/\mathrm{TO}) & n=0 \\ (\mathrm{Set}(n)/-)(\mathrm{WE}/\mathrm{TO}) & \\ \quad (\mathrm{Set}(1)/-)(\mathrm{WE}/y_1)\cdots(\mathrm{WE}/y_m)(\mathrm{WE}/\mathrm{TO}) & n>0 \end{cases}$$

  We treat E^+ as E^*E. From the assumption, the tester can recognize the timeout signal for Set(1) after observing all outputs $y_1 \cdots y_m$ of timeout transitions.

- IO sequences corresponding to E^* :

$$(-/-)^n \Longrightarrow (\mathrm{Set}(n)/-)(\mathrm{WE}/\mathrm{TO}) \quad n>0$$

Example 4 *For the protocol machine shown in Figure 2, suppose that the transition t_1 is implemented as t_1'.*

$$\begin{aligned} t_1 &= (\mathrm{Init}, \mathrm{Wait}, \mathrm{Send}, \mathrm{Msg},(\mathtt{S}, \mathtt{S})) \\ t_1' &= (\mathrm{Init}, \mathrm{Wait}, \mathrm{Send}, \mathrm{Msg},(\mathtt{N}, \mathtt{S})) \end{aligned}$$

The following IO sequence is a test sequence detecting the fault.

$$(\mathrm{Send}/\mathrm{Msg})(\mathrm{Set}(2)/-)(\mathrm{WE}/\mathrm{TO})(\mathrm{Set}(1)/-)(\mathrm{WE}/\mathrm{Msg})(\mathrm{WE}/\mathrm{TO})$$

The specification accepts the above IO sequence, while the faulty implementation cannot accept it.

4. EFFICIENT TEST SEQUENCE GENERATION

In general, if we enumerate all faulty implementations, check their IO equivalence and generate the corresponding test sequences, the set of the test sequences can distinguish every faulty implementation with a single fault. However, this method may be a obstacle to generating test sequences rapidly. To reduce the time necessary for test sequence generation, we try to deal with multiple faults together and/or generate test sequences by analyzing a smaller FSM than $M \cdot T$.

Since external output faults can be immediately detected by executing the faulty transitions and observing the external outputs, we will focus on detecting timer command faults and destination state faults. We begin with enumerating the reachable states of $M \cdot T$ and generate strings which lead $M \cdot T$ from the initial composite state to each reachable composite state.

4.1. Test sequences for timer command faults

Assume that the specification is given as $M = (Q, X, n, Y, H, s_0)$. An faulty implementation $M[t'/t]$ contains a single timer command fault for timer i on a transition $t = (u, v, x, y, \bar{p}) \in H$.

We will discuss sufficient conditions for $M \cdot T \equiv M[t'/t] \cdot T$ and $M \cdot T \not\equiv M[t'/t] \cdot T$ which can be checked by analyzing smaller machines. To consider such conditions, we introduce the following relation between composite states.

Definition 3 *Given two composite states ξ and ξ', we write $\xi{<}i{>}\xi'$ if ξ and ξ' differ only in the value of timer i.*

For $\xi \in Q_T$ and appropriate transitions $\eta \in H_t$ and $\eta' \in H_{t'}$ i.e. $\rho(\eta) = \rho(\eta') = \xi$, $\delta(\eta){<}i{>}\delta(\eta')$ or $\delta(\eta) = \delta(\eta')$ holds. Let us consider two composite states ξ and ξ' such that $\xi{<}i{>}\xi'$, a transition z other than the timeout transition of timer i and two transitions $\eta, \eta' \in H_z$ which are executable at ξ and ξ', respectively. If the timer command for timer i of z is either `S` or `D`, $\delta(\eta) = \delta(\eta')$ holds. On the other hand, if the timer command for timer i of z is `N`, $\delta(\eta){<}i{>}\delta(\eta')$ still holds. It means that only strings consisting of the transitions corresponding to transitions whose timer command for timer i is `N` can distinguish such ξ and ξ'.

By the above consideration, we introduce a sub-protocol machine $M_{\mathtt{N}} = (Q, X, n, Y, H_{\mathtt{N}}, s_0)$ where $H_{\mathtt{N}} = \{z = (s, s', a, b, \bar{p}) \mid z \in H \wedge p[i] = \mathtt{N} \wedge a \neq \text{Timeout}[i]\}$.

We introduce the following equivalence between strings.

Definition 4 *Assume that the sets of state transitions of M and $M[t'/t]$ are $H = \{t, t_1, \ldots, t_k\}$ and $H' = \{t', t_1, \ldots, t_k\}$, respectively. Two strings $q(= q_1 \cdots q_m)$ and $r(= r_1 \cdots r_n)$ of $(TS_{M \cdot T}(\xi_0) \cup TS_{M[t'/t] \cdot T}(\xi_0))$ are said* *to be* $\{t,t'\}$*-equivalent, written* $q \stackrel{t,t'}{=} r$*, if (1)* $m=n$*, (2)* $q_i \in (H_{t_j} \cup H'_{t_j}) \Leftrightarrow r_i \in (H_{t_j} \cup H'_{t_j})$ *for* $i(1 \le i \le m)$*,* $j(1 \le j \le k)$*, (3)* $q_i \in (H_t \cup H'_{t'}) \Leftrightarrow r_i \in (H_t \cup H'_{t'})$ *for* $i(1 \le i \le m)$*.*

We extend the relation "$\stackrel{t,t'}{=}$" to the relation between the sets of strings.

Definition 5 *Two sets of strings A, B are said to be* $\{t,t'\}$*-equivalent, written* $A \stackrel{t,t'}{=} B$*, if* $^{\forall}a \in A\ ^{\exists}b \in B\ (a \stackrel{t,t'}{=} b) \wedge {}^{\forall}b \in B\ ^{\exists}a \in A\ (b \stackrel{t,t'}{=} a)$*.*

Since composite state transitions are translated to IO sequences uniquely and composite state transitions corresponding to t or t' are translated to the same IO sequence, the following lemma holds.

Lemma 1 *For* $M \cdot T$ *and* $M[t'/t] \cdot T$*, if* $TS_{M \cdot T}(\xi_0) \stackrel{t,t'}{=} TS_{M[t'/t] \cdot T}(\xi_0)$*, then* $M \cdot T \equiv M[t'/t] \cdot T$*.*

In the following, we will consider a condition to decide $TS_{M \cdot T}(\xi_0) \stackrel{t,t'}{=} TS_{M[t'/t] \cdot T}(\xi_0)$.

For the strings whose length are one, the following lemma holds.

Lemma 2 *Given two composite states* ξ *of* $M \cdot T$ *and* ξ' *of* $M[t'/t] \cdot T$*,* $TS_{M \cdot T,1}(\xi) \stackrel{t,t'}{=} TS_{M[t'/t] \cdot T,1}(\xi')$ *if* $\xi = \xi' \vee \xi {<}i{>} \xi' \wedge (\xi, \xi' \notin \Gamma[i])$*.*

Proof Assume that $\xi {<}i{>} \xi' \wedge (\xi, \xi' \notin \Gamma[i])$. Since the executability of external input transitions does not depend on the values of timers, if an external input transition is executable at ξ, then the corresponding transition is also executable at ξ'. Next, since $\xi {<}i{>} \xi'$ and timeout transition caused by timer i is not executable at both ξ and ξ', the timeout transition which executable at ξ is $\{t,t'\}$-equivalent to that of ξ'. Both the values of timer i at ξ and ξ' are greater than zero. Hence if the time elapse transition is executable at ξ, then the transition is also executable at ξ'. The case of $\xi = \xi'$ can be shown in a similar way. □

Table 1 describes a necessary and sufficient condition for $TS_{M \cdot T}(\xi_0) \stackrel{t,t'}{=} TS_{M[t'/t] \cdot T}(\xi_0)$. $\Phi(u)$ denotes the set of timer value vectors at state u i.e. $\tau(\mathcal{R}(\xi_0) \cap \langle u, * \rangle)$. For example, the first row denotes that $M[t'/t]_{\mathbf{N}} \cdot T$ cannot reach $\Gamma[i]$ after executing a transition corresponding to t' at $\langle u, \bar{\tau} \rangle$ $(\bar{\tau} \in \Phi_{M[t'/t] \cdot T}(u) \wedge \tau[i] \neq \bot)$.

Lemma 3 *The followings hold if and only if "condition" holds for every element* $\bar{\tau}$ *in "timer value vector" of the row according to the case of* $p[i]$ *and* $p'[i]$ *in Table 1.*

1. $TS_{M \cdot T}(\xi_0) \stackrel{t,t'}{=} TS_{M[t'/t] \cdot T}(\xi_0)$.
2. *For each* $r \in TS_{M \cdot T}(\xi_0)$ *and corresponding* $r' \in TS_{M[t'/t] \cdot T}(\xi_0)$*,* $\delta(r) = \delta(r')$ *or* $\delta(r) {<}i{>} \delta(r') \wedge \delta(r), \delta(r') \notin \Gamma_i$*.*

Table 1. Necessary and sufficient condition for $TS_{M\cdot T}(\xi_0) \stackrel{t,t'}{=} TS_{M[t'/t]\cdot T}(\xi_0)$

$p[i]$	$p'[i]$	necessary and sufficient condition		
		timer value vector	condition	
S	N	$\bar{\tau} \in \Phi_{M[t'/t]\cdot T}(u)$	$\tau[i] \neq \bot$	$\mathcal{R}_{M[t'/t]_{\mathbf{N}}\cdot T}(\delta(\eta)) \cap \Gamma[i] = \emptyset$ $(\eta \in H_{t'} \wedge \rho(\eta) = \langle u, \bar{\tau} \rangle)$
			$\tau[i] = \bot$	$\mathcal{R}_{M_{\mathbf{N}}\cdot T}(\delta(\eta)) \cap \Gamma[i] = \emptyset$ $(\eta \in H_t \wedge \rho(\eta) = \langle u, \bar{\tau} \rangle)$
N	S	$\bar{\tau} \in \Phi_{M\cdot T}(u)$	$\tau[i] \neq \bot$	$\mathcal{R}_{M_{\mathbf{N}}\cdot T}(\delta(\eta)) \cap \Gamma[i] = \emptyset$ $(\eta \in H_t \wedge \rho(\eta) = \langle u, \bar{\tau} \rangle)$
			$\tau[i] = \bot$	$\mathcal{R}_{M[t'/t]_{\mathbf{N}}\cdot T}(\delta(\eta)) \cap \Gamma[i] = \emptyset$ $(\eta \in H'_{t'} \wedge \rho(\eta) = \langle u, \bar{\tau} \rangle)$
N	D	$\bar{\tau} \in \Phi_{M\cdot T}(u)$	$\tau[i] \neq \bot$	$\mathcal{R}_{M_{\mathbf{N}}\cdot T}(\delta(\eta)) \cap \Gamma[i] = \emptyset$ $(\eta \in H_t \wedge \rho(\eta) = \langle u, \bar{\tau} \rangle)$
D	N	$\bar{\tau} \in \Phi_{M\cdot T}(u)$	$\tau[i] \neq \bot$	$\mathcal{R}_{M[t'/t]_{\mathbf{N}}\cdot T}(\delta(\eta)) \cap \Gamma[i] = \emptyset$ $(\eta \in H'_{t'} \wedge \rho(\eta) = \langle u, \bar{\tau} \rangle)$
D	S	$\bar{\tau} \in \Phi_{M\cdot T}(u)$	$\mathcal{R}_{M[t'/t]_{\mathbf{N}}\cdot T}(\delta(\eta)) \cap \Gamma[i] = \emptyset$ $(\eta \in H'_{t'} \wedge \rho(\eta) = \langle u, \bar{\tau} \rangle)$	
S	D	$\bar{\tau} \in \Phi_{M\cdot T}(u)$	$\mathcal{R}_{M_{\mathbf{N}}\cdot T}(\delta(\eta)) \cap \Gamma[i] = \emptyset$ $(\eta \in H_t \wedge \rho(\eta) = \langle u, \bar{\tau} \rangle)$	

The *if part* of Lemma 3 can be shown in induction using Lemma 2. By Lemma 1 and Lemma 3, if $M\cdot T$ and $M[t'/t]\cdot T$ satisfy the conditions shown in Table 1, $M \cdot T \equiv M[t'/t] \cdot T$ holds. The conditions in Table 1 can be checked by exploring a smaller FSM. Unfortunately, we have to compose the automaton for $M[t'/t]\cdot T$ in the case of $p[i] =$S and $p'[i] =$N even with Lemma 3.

In the case that the "condition" does not hold, a string r which leads $M\cdot T$ or $M[t'/t]\cdot T$ from $\langle u, \bar{\tau} \rangle \in \mathcal{R}(\xi_0) \cap \langle u, * \rangle$ to a timer i expiring state is obtained. By adding the transfer string from the initial state to $\langle u, \bar{\tau} \rangle$ in front of r, a string r_0 can be obtained. r_0 leads $M \cdot T$ or $M[t'/t] \cdot T$ from ξ_0 to a timer i expiring state.

Next, we will consider a sufficient condition for $M \cdot T \not\equiv M[t'/t] \cdot T$.

Lemma 4 *For $M\cdot T$ and $M[t'/t]\cdot T$, if there exist two $\{t, t'\}$-equivalent executable strings r of $M \cdot T$ and r' of $M[t'/t] \cdot T$, and if two composite states $\delta(r) = \langle s, \bar{\tau} \rangle$ and $\delta(r') = \langle s, \bar{\tau}' \rangle$ satisfy either one of the following conditions, then $M \cdot T \not\equiv M[t'/t] \cdot T$ holds.*

- $\langle s, \bar{\tau} \rangle \in \Gamma[i] \wedge \langle s, \bar{\tau}' \rangle \notin \Gamma[i] \wedge \tau[j] > 0 \ (i < j \leq n)$.
- $\langle s, \bar{\tau} \rangle \notin \Gamma[i] \wedge \langle s, \bar{\tau}' \rangle \in \Gamma[i] \wedge \tau[j] > 0 \ (i < j \leq n)$.

Proof Suppose that one of the above conditions holds. If the value of timer i is zero at $\delta(r)$ or $\delta(r')$, the values of other timers are not zero. Hence, an output caused by timeout of timer i is observed. □

Based on the above lemmas, we can summarize the test sequence generating procedure as follows.

step I According to the types of faults, check the condition in Table 1. If the condition holds, we do not generate a test sequence. Otherwise, we obtain the above mentioned string r_0.

If r_0 and corresponding r_0' satisfy the condition of Lemma 4, we can adopt r_0 as a fault detecting string. Otherwise, go to step II.

step II By the way described in Section 3.3, we can obtain a fault detecting string.

4.2. Test sequences for destination state faults

Since the specification of protocol machines is assumed to be reduced, we can use existing test sequence generating methods, e.g. Wp-method [8], UIOv-method[9], to obtain strings which distinguish each state. We choose Wp-method for test sequence generation. However, if a string contains timeout transitions, the string may not be executable. Hence, we must decide whether the string is executable.

Given specification of a protocol machine $M = (Q, X, n, Y, H, s_0)$, we consider to generate a test sequence for detecting destination faults on $t = (u, v, x, y, \bar{p}) \in H$. We generate test sequences step by step. In the first step, we generate strings considering only external input transitions.

step 1 Find a state transition sequence which distinguishes v (the destination state of t) from $w_1 \in (Q \setminus \{v\})$ and contains only external input transitions in M. If such a state transition sequence r exists, we can adopt a string $Tr(u) \cdot \eta_t \cdot R$ as a fault detecting string where η_t is a composite state transition corresponding to a target transition t, $Tr(u)$ is a transfer string which leads $M \cdot T$ from the initial composite state to a composite state $\langle u, \bar{\tau} \rangle$ in which η_t is executable, and R is a string corresponding to r.

step 2 Let $w_2 \in (Q \setminus \{v\})$ be a state which cannot be distinguished from v by any string consisting of external input transitions. Find a state transition sequence $t_1 t_2 \cdots t_n \in H^*$ which distinguishes v from w_2 and decide whether there exists a string in $H_e^* H_{t_1} H_e^* H_{t_2} \cdots H_e^* H_{t_n}$ executable in $M \cdot T$. Recall that H_e is the set of time elapse transitions. If such a string exist, we can generate the fault detecting string by adding some appropriate transfer string.

step 3 If any string in $H_e^* H_{t_1} H_e^* H_{t_2} \cdots H_e^* H_{t_n}$ is not executable, there must be unexecutable timeout transitions t_k in the string. So as shown in Figure 5, if timer i is active in some composite state $\langle s_k, \bar{\tau} \rangle$ and there exists a cycle $c_1 \cdots c_m$ consisting of transitions whose commands for timer i is N, then we try to add the cycles in front of the unexecutable transition $t_k = (s_k, s_{k+1}, \mathrm{Timeout_i}, y, \bar{p})$.

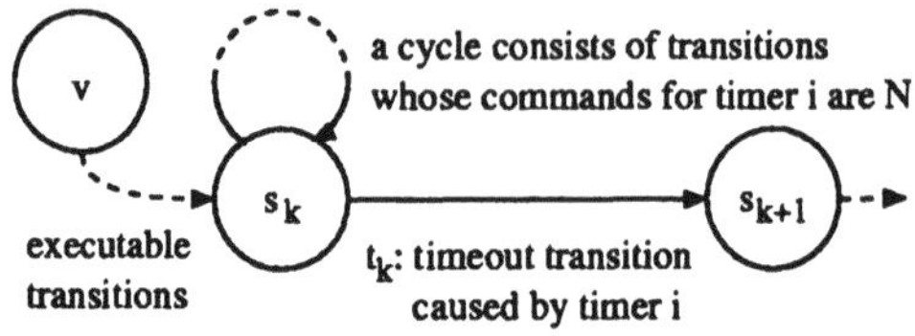

Figure 5. String extension

We also check whether there exists a string in $H_e^* H_{t_1} \cdots H_e^* H_{t_{k-1}}$ $\{H_e^* H_{c_1} \cdots H_e^* H_{c_m}\}^+ H_e^* H_{t_k} \cdots H_e^* H_{t_n}$ executable in $M \cdot T$.

step 4 For a state $w_3 \in (Q \setminus \{v\})$ at which we cannot find a string in the above steps, we can obtain a fault detecting string in the way described in Section 3.3.

In step 1, only a specification M is analyzed. In step 2 and step 3, a specification with timers $M \cdot T$ is analyzed. In step 4, the system consisting of a faulty implementation and timers $M[t'/t] \cdot T$ is analyzed.

5. AN EXPERIMENT

We developed a test sequence generating system based on the proposed method, and applied the system to DHCP (Dynamic Host Configuration Protocol)[10].

5.1. Using simultaneous inequality

In composing a finite automaton of $M \cdot T$, the state space explosion problem arises for protocols containing timers with large $T[i]$ values, e.g. two or more $T[i] = 1000$ timers.

To avoid this problem, we introduce simultaneous inequality to express the set of timer value vectors. The inequalities are classified into the following two types; (i) an inequality $\tau[i] \leq C$ or $\tau[i] \geq C$ specifies the upper limit or the lower limit C (a constant value) of timer i, (ii) an inequality $\tau[i] - \tau[j] \leq C$ specifies the upper limit C of the difference of two timer values $\tau[i]$ and $\tau[j]$. Simultaneous inequality consisting of the above types' inequalities can be solved in $O(lm)$ (l : the number of inequalities, m : the number of variables) [11]. For using such simultaneous inequality, we have made a small modification on our test sequence generating procedure.

A similar method is introduced in verification of timed automata [7].

5.2. A sample protocol

DHCP provides a framework for passing configuration information to hosts on TCP/IP networks. DHCP is built on a client-server model, where designated DHCP servers allocate network addresses and deliver configuration parameters to clients dynamically. A client requests the use of an address for some period of time (lease time). Both the client and server use timers to watch the lease time. In DHCP, the time unit is a second. The shortest lease time is an hour. A client re-transmits a message based on timeout. The first interval of re-transmitting is four seconds. The interval becomes twice every re-transmission.

If the lease time can be specified by a client, we cannot describe DHCP in our model. Hence, we limit the lease time to an hour, i.e. 3600 time units. In our model, the number of states is 11 and the number of state transitions is 74. The number of timers is nine where five timers are used to watch the interval of re-transmitting messages. The rest of the timers are used to watch the lease time.

5.3. The result

As a result, we can decide the IO equivalence between the specification and faulty implementation, and generate test sequences for non IO equivalent implementations in a reasonable processing time. The experiment is done on a PC (CPU : Pentium III 600MHz, Memory : 128MB). The computing time and memory space are about seven minutes and 2MB, respectively.

In Table 2, we show the total number of the implementations containing a single fault (test item), the number of IO equivalent implementations, and that of generated test sequences (non IO equivalent implementations). These tables also tell the numbers of faulty implementations shown to be IO equivalent or generated test sequences in every step, which correspond to each step described in Section 4.

Table 2. Result applying our method to DHCP

	timer command
test item	1332
equivalent	457
test sequence	875
step I	1332
step II	0

	destination state
test item	740
equivalent	0
test sequence	740
step 1	705
step 2, 3	35
step 4	0

For every timer command faults, the equivalence was determined and the test sequence was generated for non IO equivalent fault in step I. About 34% of faults are equivalent to $M \cdot T$. Most of these faults are the one that the timer command `N` for non-active timers at source states of target transitions are implemented as `D`.

For the destination state faults, about 95% of the test sequences are generated in step 1, and no test sequence is generated in step 4.

There is no test sequence which is generated by enumerating all reachable states. Test sequences for DHCP are generated efficiently.

6. CONCLUSIONS

In this paper, we have proposed a method to generate conformance test sequences for FSM with timers. The test sequences generated by this method can detect any single fault of timer commands or destination states. We have developed a test sequence generating system and applied it to DHCP. As a result, the test sequences was generated efficiently.

Acknowledgments

This work is partially supported by International Communications Foundation (ICF), Japan.

REFERENCES

[1] B. S. Bosik and M. U. Uyar, "Finite State Machine Based Formal Methods in Protocol Conformance Testing : From Theory to Implementation", Computer Networks and ISDN Systems, Vol.22, No.1, pp. 7–33 (1991).

[2] L. P. Lima Jr and A. R. Cavalli, "A Pragmatic Approach to Generating Test Sequences for Embedded Systems", In Proc. IFIP 10th International Workshop on Testing of Communicating Systems (IWTCS'97), pp. 288–307 (1997).

[3] A. Petrenko, N. Yevtushenko and G. v. Bochmann, "Fault Models for Testing in Context", In Proc. Joint International Conference on 9th Formal Description Techniques and 16th Protocol Specification, Testing, and Verification (FORTE/PSTV'96), pp. 163–178 (1996).

[4] A. Petrenko and N. Yevtushenko, "Fault Detection in Embedded Components", In Proc. IFIP 10th International Workshop on Testing of Communicating Systems (IWTCS'97), pp. 272–287 (1997).

[5] ISO 9646, "Information Technology, Open System Interconnection, Conformance Testing Methodology and Framework", ISO/IEC 9646 (1991).

[6] R. Alur and D. L. Dill, "A Theory of Timed Automata", Theoretical Computer Science, Vol. 126, No. 2, pp. 183–235 (1994).

[7] R. Alur, "Timed Automata", In Proc. 11th International Conference on Computer-Aided Verification (CAV'99), LNCS 1633, pp. 8–22 (1999).

[8] S. Fujiwara, G. v. Bochmann, F. Khendek, M. Amalou and A. Ghedamsi, "Test Selection Based on Finite State Models", IEEE Trans. on Soft. Eng., Vol. 17, No. 6, pp. 591–603 (1991).

[9] W. Y. L. Chan, S. T. Vuong and M. R. Ito, "An Improved Protocol Test Generation Procedure Based on UIOs", In Proc. ACM SIGCOMM'89, pp. 283–294 (1989).

[10] R. Droms, "Dynamic Host Configuration Protocol", RFC 2131, Bucknell University (1997).

[11] T. H. Cormen, C. E. Leiserson and R. L. Rivest, "Introduction to Algorithms", The MIT Press, pp. 539–543 (1990).

Part Seven

Formal Methods in Software Development II

A TOOL FOR GENERATING SPECIFICATIONS FROM A FAMILY OF FORMAL REQUIREMENTS

Jan Bredereke

Universität Bremen, FB 3 · P.O. box 330 440 · D-28334 Bremen · Germany
jan.bredereke@web.de · www.tzi.de/~brederek

Abstract. We are concerned with the maintenance of formal requirements documents in the application area of telephone switching. We propose a specification methodology that avoids some of the so-called feature interaction problems from the beginning, and that converts some more into type errors. We maintain all the variants and versions of such a system together as one family of formal specifications. For this, we define a formal feature combination mechanism. We present a tool which checks for feature interaction problems, which extracts a desired family member from the family document, and which generates documentation on the structure of the family. We also report on a quite large case study.

Keywords: tools; formal requirements; maintenance; feature interaction problems; telephone switching; Object-Z.

1 INTRODUCTION

Digital telephone switching systems already comprise hundreds of features, and the market forces the providers to incorporate an ever increasing number of new services and features into the switches. In recent years, *feature interaction (FI) problems* have become a serious obstacle to adding more features to these systems [1–3]. Adding one more feature, even when it behaves as desired for itself, may cause undesired behaviour of other features, even when these behave as desired for themselves. The number of possible feature combinations has become so large that it is not possible anymore to check all combinations manually.

One standard example of such a feature interaction occurs between the Terminating Call Screening (TCS) and the Call Forwarding (CF) feature (Fig. 1). Terminating Call Screening allows a subscriber C to specify a list of callers by whom he never wants to be called. Call Forwarding allows a subscriber B to specify another destination to which all calls to B should be forwarded. Now suppose that C has subscribed to TCS, and has put A on his screening list. Furthermore suppose that B has subscribed to CF and forwards all calls to C. Finally, A happens to attempt to call B. Will the phone of C ring? Should it ring? This depends very much on *how we define the notion of "caller"* for the two features. When CF consists of two calls "glued" together, then B is the caller for C, the screening condition is not satisfied, and the phone will ring. When we take into account that C probably does not want to be disturbed by A, the phone should not ring. We will come back to this example later.

The point we want to make at the above example is that feature interactions already arise in the requirements documents. When these documents are a *complete* description of the behaviours (and of other interesting properties), then even *all* feature interaction

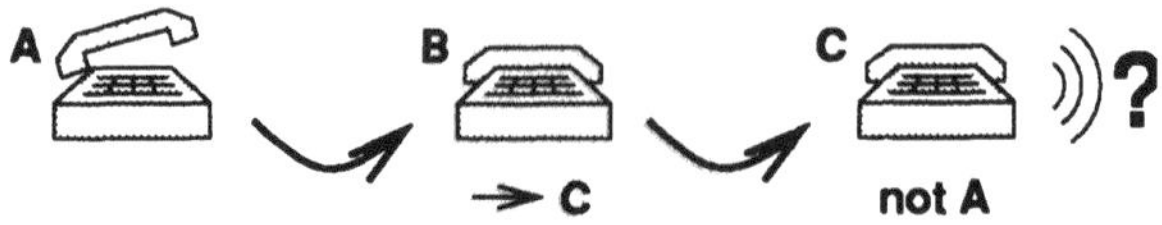

Fig. 1. Feature interaction between Terminating Call Screening and Call Forwarding.

problems are at least inherently present in the requirements documents. Therefore, they should be tackled already there.

The notion of complete requirements leads us to formal, i.e., precise, requirements. Considerable work about formal requirements specifications exists. But an aspect of them that deserves more attention is the maintenance of formal requirements documents. How can we handle many changes to them over time, and how can we handle a large number of variants for these documents when different customers of a successful product have different needs?

In a recent paper[4], we introduced the idea of *families of formal requirements*. Our approach proposes to keep all requirements variants together, and to encapsulate variant choices. It enables information hiding for requirements documents, similar to how it is known for programs. Furthermore, it encourages to plan ahead for future members of the family. Another core idea is to let the original specifier document explicitly as much of the information as possible that another specifier of an incremental feature will need. The specifier of an incremental feature can use this explicit information either for avoiding feature interaction problems, or for at least detecting feature interaction problems, such that they can be resolved.

A family of requirements needs a means for incremental specification. One standard means is refinement. In refinement approaches, a base system is specified first, which is then extended step by step by specifying further parts. The formalisms are constructed such that the interesting properties of previous steps are preserved by subsequent steps. For example, some approaches allow to add further possible behaviour to a behaviour description, while all existing behaviour remains possible. Other approaches allow to impose incremental, explicit constraints on an (implicit) behaviour description, which preserves all previous constraints due to the nature of the logical "and". While these formal mechanisms are useful for many applications, they are not yet sufficient in our application area of telephone switching. In our experience with telephone switching requirements, most *new features change the behaviour of the base system or of other features*. "Change" here means that something really becomes different than it was before. For example, the Call Forwarding feature stops to connect to the dialled number and thus restricts the base system behaviour, and it starts connecting to the forwarded-to number, extending the base system behaviour. A telephony feature typically is *non-monotonous*.

Various feature operators have been proposed to allow non-monotonous changes. They allow to express the desired changes. If their definition and use reflect the current practice of modelling telephone switching features, feature interaction problems occur in such a formal setting in the same way as in an informal setting. A lot of research work has been done on a formal analysis of the changes [1–3, 5]. Our experience is that this indeed works to some extent, but that the exponential growth of complexity not only beats the human specifiers, but also beats the automated tools, when it comes to real-sized systems with hundreds of features. In particular, feature operators allowing arbitrary syntactic modifications of individual source lines lead to unmanageable feature interaction problems, since the complete system must be analyzed for the consequences.

We therefore attempt to reduce the complexity by using modularization in the information hiding sense for the requirements documents. We support it by a feature construct which allows for non-monotonous changes but which does not allow for arbitrary changes. In our experience, planning ahead for a careful modularization and for future potential modifications reduces the number and complexity of the dependences between features.

In Section 2, we briefly introduce to the formal specification language we use, and we present a suitable, new feature construct for it. Section 3 describes our tool support for managing features. Section 4 demonstrates the application of the approach and the tool in a case study from telephone switching, and Section 5 discusses how we avoided or detected feature interaction problems. Section 6 compares our approach to related work, and Section 7 summarizes this paper.

2 A SPECIFICATION METHODOLOGY SUPPORTING FAMILIES OF FORMAL REQUIREMENTS

In this section, we give a very brief overview of the formal specification language we use, concentrating on the means to structure a specification document. Another prerequisite is a suitable specification style, which we describe. We then add in a first step the concept of families to this language, and in a second step the concept of non-monotonous, but well-structured, changes.

2.1 Base Language CSP-OZ

We have chosen the formal specification language CSP-OZ [6, 7] as a base for adding our feature constructs. This language is suitable for telephone switching, since it supports communication events explicitly, and it is is suitable for specifying requirements, since it allows to specify the communication events that are observable externally, without prescribing any internal structure. CSP-OZ is a combination of the process algebra CSP (Communicating Sequential Processes) [8] and of Object-Z [9]. CSP-OZ allows to specify communication and control aspects in the CSP part, and data aspects separately in the Object-Z part. Such a separation has proven useful in the communications area. The Object-Z data part is in turn based on the language Z [10, 11], which is a formalization of set theory. Object-Z provides the notions of abstract data types and state invariants which are not present in CSP. Numerous algebraic laws have been derived for the refinement of behaviour in CSP and for the refinement of data in Z, providing a sound theoretical base to the inheritance operators we use.

The inheritance operator of CSP-OZ, in its basic form, allows to specify refinement. Optionally it can be supplemented by a renaming operator. If the latter is used, it effectively allows to change the behaviour of an inherited class completely. CSP-OZ inheritance thus offers either pure refinement or powerful, but uncontrolled, changes.

The Z language was defined for many years by Spivey's reference manual [11]. Currently, the International Standardization Organization (ISO) finalizes a standard for Z [10]. Besides providing a more detailed, well-structured definition of the syntax and semantics of Z, the standard also provides a few extensions to Spivey's Z. One of these extensions concerns the structuring of a specification document, and it supports our work considerably. We therefore already base our work on the forthcoming Z standard. Fortunately, the definition of CSP-OZ is also already based on the new Z standard.

A Z (and thus also a CSP-OZ) specification consists of paragraphs of formal text. A paragraph can be, for example, one type definition or one schema definition. The new standard allows to put a *specification* and *sections* on top of paragraphs. A specification consists of sections, which in turn consist of paragraphs. The meaning of a Z specification is the set of named theories established by its sections. Each section adds one named theory to the set established by its predecessors. A named theory associates a section name with a set of models.

As a consequence, each section has a self-contained formal meaning. Any initial sequence of sections can be taken as the requirements for a variant of the specified system. Since each further section can only add more constraints on the system (and new declarations), this allows for a constraint-oriented, incremental style of specification.

But there is no means for non-monotonous extensions. Z sections have a *parents* construct such that a section can be an extension of specifically named sections. The only way to get a modified version of a system is to copy, rename, and edit the changed sections, and also to copy and rename all sections that have these sections as a parent. If a section needs to be modified that is at the leaves of the hierarchy of parent sections, this can amount to a large part of the document being duplicated.

2.2 Specification Style

For requirements, we prefer a constraint-oriented specification style. Adding one (small) constraint after the other helps us to concentrate on one aspect of the system at the time, without accidentally being overly restrictive on other aspects.

When we perform non-monotonous extensions, it is important that the changes happen in a controlled way. Therefore, we choose that a feature may modify another feature at the level of Z sections. Leaving out an entire feature and introducing a new feature with a slight modification is too coarse-grained. If the feature is complex, this results in a considerable duplication of code. Allowing a feature to modify the specification text of another feature in an arbitrary way, e.g., by changing an integer value or by substituting an arithmetic operator, is too fine-grained. It is hard to understand the consequences of such changes, in particular if many of them are applied at the same time. This easily leads to feature interaction problems.

We distinguish between the *essential behaviour* and the *changeable behaviour* of a feature. Some parts of a feature are essential for its nature. Other parts are provided only in order to make the requirements specification complete. For example, some of the behaviour restrictions might only be made to make the behaviour predictable for the user. If the changeable behaviour is modified, this can never be an undesired feature interaction. When a specifier needs to modify a feature that was written some time ago by another person, he will find it difficult to find out which part of the behaviour is essential. Therefore, we require the original specifier of a feature to document which behaviour is essential and which behaviour is changeable, by using an appropriate language construct.

Furthermore, we distinguish between the requirements of the system and the requirements of its environment. This is important for any software development contract. It is the developer's duty to implement the former, and it is the customer's duty to ensure the latter, in order to make the system actually work. Formally, we collect both parts into one CSP-OZ class each, using inheritance. We then compose them by parallel composition into a description of the aspects of the world which are relevant to the contract.

2.3 Extension of CSP-OZ: Families with Monotonous Increments Only

In order to support families, we change the top-level structure of the language. We remove specifications from the syntax, and we add *families*, *feature chapters*, and *family member chapters* instead.

A family consists of feature chapters and of family member chapters. Feature chapters are (named) chapters of (named) sections. Family member chapters are named chapters that contain nothing but a features list. Informally, each family member chapter defines one specification in plain CSP-OZ. We can extract one family member from a family document by collecting all its features, and by concatenating all the sections of these features, resulting in a specification in plain CSP-OZ.

A detailed definition of the syntax and semantics of the language extension can be found in the manual [12].

The inheritance operator of standard CSP-OZ can be supplemented by a renaming operator, effectively allowing to change the behaviour of an inherited class completely. We therefore ban the use of the renaming option in our extension of CSP-OZ.

Semantics. The set *Features* contains all feature chapters' names. The set *Sections* contains all mappings from feature chapter names to sets of section names: $Sections == Features \twoheadrightarrow\!\!\!\!\!\!+\; \mathbb{P}\, Name$. The meaning of a feature chapter can be determined by applying the function *Sections* to the feature chapter's name. The meaning of a family is a mapping from family member names to meanings of plain CSP-OZ specifications.

$$[\![\cdot]\!]^{\mathcal{F}} : Family \nrightarrow (Name \twoheadrightarrow\!\!\!\!\!\!+\; \mathbb{P}\, Theory)$$

That is, each (pure) family member chapter is essentially a complete CSP-OZ specification, except that all of its actual sections have been moved into feature chapters, which can be shared with other family members.

Shorthand Notations. Optionally, a family member chapter may contain a feature chapter body. This is transformed syntactically into a separate feature chapter and a "pure" family member chapter. Such a feature chapter body can be convenient to specify the composition of all items from the different features in a suitable way.

CSP-OZ classes are larger units than mere (Z language) paragraphs. Therefore we found it sometimes convenient to use classes for structuring instead of sections. This is true in particular when each section contains only one class. Therefore, we allow a class in the places where a section can occur, too. When determining the semantics, we transform such a class into a section heading followed by this class, and followed by whatever other formal paragraphs may follow. The name of this implicit section is the name of the class. (Please note that family members, features, and sections each have separate name spaces.) The parents sections names of the section are the names of the classes inherited by this class.

2.4 Extension of CSP-OZ: Families with Non-Monotonous Increments

Sections are our unit of increment for monotonous increments, and they are also our unit of modification for non-monotonic changes. We distinguish between the essential and the changeable behaviour of a feature, and we specify this information explicitly by using different kinds of section. The type rules on the new constructs defined below will allow us to exploit the information for formal checks.

But one more kind of section is not sufficient. Both the essential and the changeable behaviour constraints need to be composed. The composition operators need to be grouped into sections too, and these operators specify neither essential nor changeable behaviour. They require a third kind of section.

We therefore add two more kinds of section. We put the essential properties of a feature into the normal kind of section. The two new kinds of section have the same syntax as the normal kind, except for the new keywords default_section and collect_section. *Default sections* are different insofar that they can be referenced by a *remove clause*, and they serve to express changeable properties. A remove clause is a another new construct that may appear in feature chapters. When the family member comprises a feature that contains a remove clause, the named default sections are disregarded when determining the set of sections. A *collect section* is special in that it adapts its list of parent sections automatically to losses of default sections due to remove clauses. A collect section should be used only to collect and combine properties defined in other sections.

We define six *type constraints* on these different kinds of sections which ensure their intended use, and which flag errors that may constitute feature interaction problems. A basic rule is that only changeable properties can be removed: 1) A remove clause may name default sections only. 2) The parent of a normal, i.e., essential, section must not be a default section, except if it is from a different feature or family member chapter.

The above exception is necessary to allow one feature to use a specific, non-essential version of another feature as a base. Note: a type checker may issue a warning if there is more than one remove clause for one default section in the features of a family member.

By further rules, parents clauses and remove clauses are restricted to sensible section names with respect to any specific family member: 3) For each family member m, we construct the set $\mathit{SectionNames}(m)$ of names of all its feature's normal sections, default sections, and collect sections. For each family member m, all remove clauses of its features must name sections from the set $\mathit{SectionNames}(m)$ only. 4) We also construct a reduced set $\mathit{RSectionNames}(m)$ from $\mathit{SectionNames}(m)$ by removing those default sections that are named in a remove clause of this family member. For any normal or default section in $\mathit{RSectionNames}(m)$, its set of parents sections must be a subset of $\mathit{RSectionNames}(m)$.

Finally, the varying set of defined items in collect sections cannot be referenced in property-defining sections: 5) Neither normal nor default sections may have a collect section as a parent. 6) The parents lists of collect sections must contain sections only that are either from the same feature or that are collect sections.

We also introduce a special kind of inheritance operator, called "default_properties", that is ignored in case the inherited class has been removed. We extend the syntax accordingly, and we add two more type constraints for the new inheritance operator: 7) The default properties operator may appear in a collect section only. 8) The class name in the operator must be a valid class name in the entire family document; but for any specific family member, the default properties operators in its sections may reference class names from outside of these sections.

A detailed definition of the syntax and semantics of this part of the language extension can be found, again, in the manual [12].

Semantics. The meaning of a remove clause is the set of its names of default sections. The function *Remove* maps feature chapter names to the sets of names of default sections which are named in a remove clause in this feature chapter: $Remove : Features \twoheadrightarrow\!\!\!\!\!\!\shortmid \; \mathbb{P}\, Name$.

The features list of a family member chapter determines the features that are used to construct its CSP-OZ specification. The features list is a subset of the set *Features*. Its meaning is the union of all the sets of section names of the features listed, minus the set of default sections that are removed by the features listed.

$$\begin{aligned}[\![fe_1 \ldots fe_n]\!]^{\mathcal{FL}} = {} & (Sections(fe_1) \cup \ldots \cup Sections(fe_n)) \setminus \\ & (Remove(fe_1) \cup \ldots \cup Remove(fe_n))\end{aligned}$$

The sections of a family member chapter's specification are all the sections $[\![fe_1 \ldots fe_n]\!]^{\mathcal{FL}}$. The only modification is that we remove all section names in ($Remove\ fe_1 \cup \ldots \cup Remove\ fe_n$) from the sets of parents sections. The meaning of the specification is then constructed in the usual way from these sections.

Shorthand Notations. When classes are used for structuring instead of sections, a class is transformed into a collect section instead of a normal section exactly if it contains a "default_properties" operator. A class is transformed into a default section exactly if there exists a "default_properties" operator somewhere in the same feature that references its name.

3 THE TOOL

Our approach is supported by a tool, called *genFamMem 2.0*, which

- extracts specifications in plain CSP-OZ from a family document,
- detects feature interactions by
    - performing the additional type checks for families like the constraints on normal, default, and collect sections, on "remove", and on the "default_properties"/ "inherit" operators,

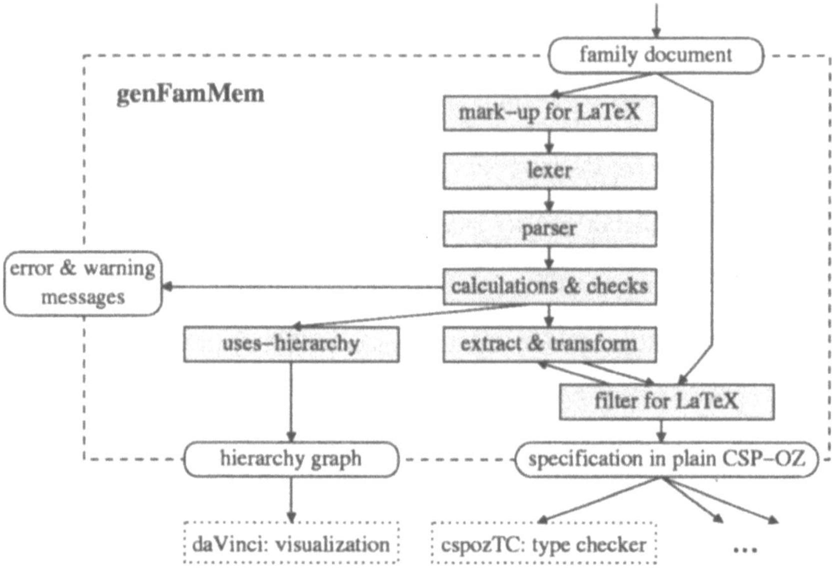

Fig. 2. Structure of the genFamMem tool.

  - issuing some heuristic warnings, for example when two different remove clauses remove the same class, and
- helps avoiding feature interactions by generating documentation on the structure of the family.

The tool contains a lexer/parser for our extension of CSP-OZ which is written using the lex/yacc parser generator [13], and C code. The tool comprises about 8500 lines of commented source code. Figure 2 shows its module structure. The tool has a modular structure that reflects the well-structured definition of the Z semantics in the forthcoming standard. The standard defines the syntax and the semantics in nine phases, the first ones being mark-up, lexing, and parsing, and the last one being the semantic relation. The separate mark-up phase solves the problem that Z provides many more special symbols than normal keyboards can provide, by encoding them as sequences of characters. The standard defines two mark-ups, LaTeX markup and email mark-up.

The mark-up module takes a family document for a first pass and translates LaTeX control sequences into CSP-OZ characters like "$\mathbb{F}$", "$\nrightarrow$" etc. It could be replaced by mark-up modules for other mark-ups. The next steps are the lexical and the syntactic analysis. The semantic calculation module takes the information from the syntactic analysis, and the name of a family member, and it calculates which sections should be included into the generated specification. It also performs the type checks described above, and flags any type errors in the family document structure, indicating feature interaction problems. When no type errors occurred, the tool extracts the actual specification document in a second pass, while adjusting the parents clauses, turning the default properties clauses into appropriately reduced inherit clauses, and generating the

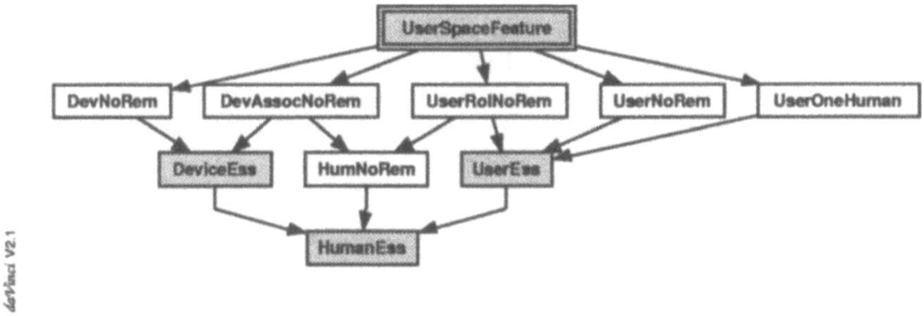

Fig. 3. The uses-hierarchy of the sections of feature UserSpace from the case study [16, 17].

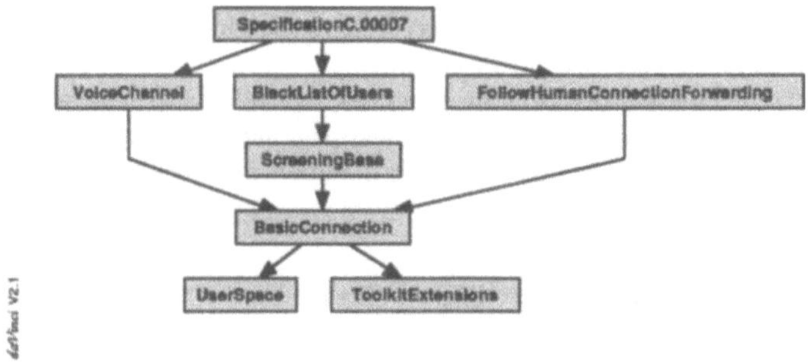

Fig. 4. The uses-hierarchy of family member SpecificationC from the case study, abstracted to features.

final family member collect section (discussed below in Section 4.4). After that, we type-check the resulting requirements document in plain CSP-OZ using von Garrel's CSP-OZ type checker cspozTC [14].

Besides generating and checking family members, the tool genFamMem also has another mode of operation, in which it generates documentation on the structure of the family. It extracts the parents relations of all sections and the inheritance relations of all classes, it extracts the association of sections to features, and then it constructs and outputs a *uses-hierarchy graph*. The uses-hierarchy graph describes which section needs which sections as prerequisites. This graph then is visualized graphically using the tool daVinci [15]. We have the choice to see the uses-hierarchy of all sections of the family (which is large!), of one family member only, or of one feature only. Furthermore, we have the choice to abstract sections to features, such that we see the uses-hierarchy of all features of the family, or of all features of one family member.

For example, Figure 3 presents the uses-hierarchy graph of the sections of feature UserSpace from our case study [16, 17]. Plain grey boxes denote essential sections, white boxes denote changeable sections, and double border boxes denote collect sections. Since the uses-relation is transitive, redundant arrows are not shown, such as the one from DevAssocNoRem to HumanEss. Figure 4 shows the uses-hierarchy of family member SpecificationC from our case study, abstracted to features.

Our extensions can be defined not only for CSP-OZ, but also for plain Z and for CSP_Z. The latter is a dialect of CSP that uses Z for data definitions, but does not offer object-based constructs like classes and inheritance. Our tool already supports Z and CSP_Z, through switches which optionally disable the more advanced language constructs. We have not conducted experiments with families of Z or CSP_Z specifications yet, but plan to do so.

The tool is available without charge through its Web home page [18].

4 CASE STUDY

We specified the requirements for a telephone switching system in a case study [16, 17], using our extension of CSP-OZ. The case study currently comprises about 40 pages of commented formal specification, with about 50 sections in nine features, including the base system. We demonstrate the new structuring means using excerpts from two simple features and from the base switching system.

The family of requirements from the case study can be used to generate requirements specifications with different combinations of features. Such a requirements specification can then be the starting point for a software development process delivering an implementation for this family member.

4.1 The Base Telephone Switching System

The base concepts of telephony are at the bottom of the class inheritance hierarchy. The class HumanEss introduces the set of humans, modelled as the variable humans. Classes are written in CSP-OZ as a box, with the class name in the heading. We write the state space of the class in a contained, unnamed box. The (essential) class UserEss introduces the set of user roles users of humans and also the association of humans to user roles userRoles:

section HumanEss [Humans]

parents

...

HumanEss

...

$humans : \mathbb{F}\, humanIds$

section UserEss [User Roles]

parents HumanEss

...

UserEss

inherit HumanEss

...

$users : \mathbb{F}\, userIds$

$userRoles : humanIds \rightarrow \mathbb{F}\, userIds$

The class DeviceEss introduces the set of terminal devices devices and also the association of humans to devices possessedDevices. The section and class DeviceEss is very similar to UserEss, we do not reproduce it here.

In order to get a plain old telephone system, we specify next that the elements of these sets are all in a one-to-one relation, and that these sets are static. For example, the association between humans and devices should never change:

`default_section` **DevAssocNoRem** **[Telephone Device Associations neither Added nor Removed]**

```
DevAssocNoRem
inherit HumanEss, DeviceEss, ...

  ∀ h : humans •
      possessedDevices(h) = initialPossessedDevices(h)
```

The specification describes furthermore how a connection is set up and torn down. Due to space limitations, we leave out this rather large description [16, 17] here.

4.2 A Black List Feature

A black list feature protects its subscriber from calls of users that she has put on a black list. We specify the black list feature in two steps. First, we specify a feature ScreeningBase that only serves as a "hook" which allows to block connections according to some criteria. The variable connectionScreen is defined in section ConScreenEss, it describes the pairs of users which shall never be connected. We specify this screening as a kind of missing resource for a connection. A default section ConNoScreen of this feature then states that the screening condition is never satisfied.

In a second step, we specify one specific kind of black list feature BlackListOfUsers, including means to add users to the list and to delete them again. Due to the size of this definition [16, 17], we omit it here. But we present how we couple the feature blackListUsr to the general screening above:

`feature` **BlackListOfUsers**

...

`remove` ConNoScreen

`default_section` **BlackListUsrOnly** **[User Black List is the only Screening Feature]**

```
BlackListUsrOnly
inherit ..., ConScreenEss

  ∀ u, v : users •
      ((connectionScreen(u, v) = connectionScreenBlock)
      ⇔ ((u, v) ∈ blackListUsr))
```

`collect_section` **BlackListOfUsersFeature** [**Collection of all the Requirements**]
parents ..., BlackListUsrOnly, ...

BlackListOfUsersFeatureSys
```
inherit ...
default_properties ..., BlackListUsrOnly, ...
```

BlackListOfUsersFeatureEnv
```
...
```

4.3 A Call Forwarding Feature

The feature FollowHumanConnectionForwarding allows a human to move to another telephone device and register himself there, in order to have all of his calls forwarded there. We specify this feature mainly by lifting the restriction of a static association between humans and devices introduced above.

`feature` **FollowHumanConnectionForwarding**

...
`remove` DevAssocNoRem

In order to facilitate the implementation of these requirements, the feature also demands that a registration may be performed only when no connection to the human exists. Again, we omit the details [16, 17] due to space limitations.

4.4 Composition of a Complete System

The above two features, and some more, can be combined into the specification of the requirements of a complete system:

`familymember` **SpecificationC**

`features` ..., BlackListOfUsers, FollowHumanConnectionForwarding

The collect sections of the listed features are composed in a single collect section of the family member. The structure of this collect section always follows a fixed pattern: one class inherits all requirements on the system, a second class inherits all requirements on the environment, and in a third, final class the first two classes are composed by a parallel composition operator:

$$main = \text{SpecificationCEnv} \parallel \text{SpecificationCSys}$$

Since this final collect section has such a fixed structure, it can be omitted, the semantics definition of the language then derives it from the features list. The entire family member chapter then consists of the features list only. Accordingly, our genFamMem tool fills in this collect section when generating a family member in plain CSP-OZ.

Similarly, we do not need to specify the parents sections of a section under certain circumstances, they are then derived from the inheritance operators. We omit the formal details [12] of these shorthand notations here.

Performing the Generation. We use the tool genFamMem to extract the above family member from the family document. The command

```
genFamMem SpecificationC case-study.tex > SpecificationC.tex
```

extracts SpecificationC and performs the type checks, and the command

```
genFamMem --uhier-familymember SpecificationC --only-features \
          case-study.tex > SpecificationC.daVinci
```

extracts the uses-hierarchy graph shown a few pages above in Figure 4.

5 AVOIDING AND AUTOMATICALLY DETECTING FEATURE INTERACTIONS

The goal of our approach is to either 1) avoid feature interaction problems in requirements documents from the beginning, or 2) to help detect them, such that they can be resolved, by

- allowing to maintain all variants of a requirements document together, by
- using a specification style that encapsulates variant choices, and by
- allowing/urging the original specifier to document as much of the information as possible that another specifier of an incremental feature will need.
  The information provided formally by a specification written in CSP-OZ with our extension comprises
  - the *dependences* among the features, through the hierarchy of parents sections sets and of class inheritance, and
  - what is *the core of a feature*, by the distinction between default sections, which are changeable, and normal sections, which are essential.

  This information can help the maintenance specifier to avoid problems, and it can be used for automated type checks, as performed by our generator tool. The detected feature interactions then can be resolved by the specifier.

Therefore, thanks to the additional explicit information in the document, our approach converts some of the feature interaction problems into *type errors*.

5.1 Example of Feature Interaction Avoidance

In our case study, we explicitly introduce the notions of telephone devices, humans, and user roles, even though we then state a static association between these for the Plain Old Telephone Service. Nevertheless in the definition of the screening feature presented above, we consequently must specify that user roles are screened, not devices. Furthermore, we encapsulated this variant choice by specifying separately a screening base feature and a user screening feature. We also specified another variant, a device screening feature not presented in this paper, which also builds on the screening base feature.

Similarly, our case study comprises two kinds of call forwarding features. One registers a human with another device, it is presented above. The other feature transfers a user role to another human. An example use case would be a help desk employee who transfers his help desk user role to a colleague while having a coffee break. His boss can still reach him in his role as an employee, even though he is not disturbed by customers.

We can now come back to the standard example of a feature interaction between Terminating Call Screening (TCS) and Call Forwarding (CF) from the introduction. Due to our separation of notions, for each combination of forwarding/screening features it is clear whether the phone should ring. A subscriber of BlackListOfUsers will be protected against a connection from a screened user from all devices, at the device where she is registered currently. A subscriber of BlackListOfDevices will not expect to be protected from humans that are able to use different devices. We just don't have the ambiguous notion of "caller" anymore.

5.2 Example of Feature Interaction Detection and Resolution

Our case study does not have any feature interaction problems between Terminating Call Screening and Call Forwarding, and therefore we cannot demonstrate any feature interaction detection using this standard example. Nevertheless, our type rules defined above point us to some remaining problems.

For example, both screening features "remove" the same changeable section ConNoScreen of the screening base feature (compare Section 4.2, first paragraph). Each feature then states in one new, essential section when no connection must be possible. The type rules flag this double removal as a warning. A manual analysis shows immediately that the two replacing, simple sections are not contradictory. But a further look reveals that both features also have a changeable section each that states that its respective feature is the only screening feature, such that otherwise no screening happens. This of course is a contradiction in the general case.

After we have detected a feature interaction problem, we need to resolve it. As in most cases, the harder part was to detect the problem before a customer notices it, and the relatively easy part is to find a specific remedy. We specify this remedy as just another feature which we select whenever both of the above features are included. It contains a section that states that screening happens exactly when either of the two screening features demands it. Furthermore, it removes both conflicting sections. We are explicitly allowed to do this since both sections are marked as changeable.

5.3 Conventional Further Consistency Checks

After we have generated one specific requirements document, all the conventional methods and tools can be applied which check the consistency of formal requirements. We used the CSP-OZ type checker cspozTC [14] to find conventional type errors ("bugs"), but they are of no interest here.

6 RELATED WORK

Related work investigates "feature" constructs for different specification and programming languages, in particular in the context of the recently terminated FIREworks project [19]. The aim of this project is to define a feature construct for several languages, and then to evaluate it in case studies. The approach bases on the superimposition idea of Katz [20], where we specify a base system first, and where we then specify textual increments by a suitable operator. For example Plath and Ryan [21] do this, they define a feature construct for CSP. They use a state-oriented specification style, and their feature operator allows to add or replace transitions at states. They don't define type rules for their operator, instead they detect feature interactions by model-checking the combined systems. One problem of the superimposition idea is that, even though the textual increments have a defined interface, the base system does not. Therefore, a superimposition can invalidate arbitrary assumptions about the base system and thus cause feature interaction problems. We try to reduce these problems by encapsulating one self-contained property each in one section, and by either including or excluding it entirely.

Also related is the foreground/background approach of Hall [22]. He separates the description of a feature into its foreground behaviour, which is "essential" behaviour in our terms, and its background behaviour, which is "changeable" by another feature's foreground behaviour without causing a warning by his analysis tools. His goal is to perform feature interaction detection with *less* results; i.e., his tools do not report "spurious" interactions. His approach does not provide multiple layers of features, as ours does, even though he mentions a generalization to an "n-ground" approach. We use a hierarchical, constraint-oriented structuring of features, and we are in particular concerned about avoiding feature interactions. Hall attempts to cope with current telephony system architectures, while we investigate a more suitable architecture.

7 SUMMARY

We are concerned with the maintenance of formal requirements documents. In the application area of telephone switching, so-called feature interaction problems have become a serious obstacle to evolve these systems according to market needs.

We propose a specification methodology that either 1) avoids feature interaction problems in requirements documents from the beginning, or 2) helps to detect them, such that they can be resolved. The original specifier of a feature must document additional information on what is the core of the feature, and what are the dependences on other features. Keeping all the variants and versions of a system as a single family enables us to exploit this information.

For specifying families, we present in this paper (and in full detail in the manual [12]) the syntax and semantics of a formal feature combination mechanism. It is an extension to the specification language CSP-OZ. In our formalism, some feature interactions appear as type errors.

In this paper, we also present a tool which checks for feature interaction problems, which extracts and transforms a desired family member from the family document, and which generates documentation on the structure of the family. The tool is available freely.

We finally report on a quite large case study. We show how a standard example of feature interactions does not occur due to the modular structure we use, and we show how our type checker detects a further feature interaction, which we then resolve.

Several issues are worth further research. More experience with our methodology is important; we plan to extend our case study further. Our approach should be applicable with other formalisms than CSP-OZ as well, as long as the base formalism used allows for a constraint-oriented specification style and provides a means for incremental refinement. Our tool even already supports the languages CSP_Z and Z. Finally, the relationship between families of requirements and families of programs needs to be investigated, with respect to a reuse-oriented implementation process.

REFERENCES

1. L. G. BOUMA, NANCY GRIFFETH, AND KRISTOFER KIMBLER. Special issue: Feature interactions in telecommunications systems. *Comp. Networks* **32**(4) (April 2000).
2. MUFFY CALDER AND EVAN MAGILL, editors. "Feature Interactions in Telecommunications and Software Systems VI". IOS Press, Amsterdam (May 2000).
3. KRISTOFER KIMBLER AND L. G. BOUMA, editors. "Feature Interactions in Telecommunications and Software Systems V". IOS Press, Amsterdam (September 1998).
4. JAN BREDEREKE. Families of formal requirements in telephone switching. In Calder and Magill [2], pages 257–273.
5. JAN BREDEREKE. "Communication Systems Design With Estelle – On Style, Efficiency, and Analysis". PhD thesis, University of Kaiserslautern, Shaker Verlag, Aachen, Germany (August 1997).
6. CLEMENS FISCHER. Combination and implementation of processes and data: from CSP-OZ to Java. PhD thesis, report of the Comp. Sce. dept. 2/2000, University of Oldenburg, Oldenburg, Germany (April 2000).
7. CLEMENS FISCHER. CSP-OZ: a combination of Object-Z and CSP. In HOWARD BOWMAN AND JOHN DERRICK, editors, "Formal Methods for Open Object-Based Distributed Systems (FMOODS'97)", volume 2, pages 423–438. Chapman & Hall (July 1997).
8. A. W. ROSCOE. "The Theory and Practice of Concurrency". Prentice-Hall (1997).
9. GRAEME SMITH. "The Object-Z Specification Language". Kluwer Academic Publishers (2000).
10. ISO Panel JTC1/SC22/WG19, Final Committee Draft, CD 13568.2. "Z Notation" (August 1999).
11. JOHN MICHAEL SPIVEY. "The Z notation: a reference manual". Series in Computer Science. Prentice-Hall, New York, 2nd edition (1995).
12. JAN BREDEREKE. "genFamMem 2.0 Manual – a Specification Generator and Type Checker for Families of Formal Requirements". University of Bremen (October 2000). URL http://www.tzi.de/~brederek/genFamMem/.
13. LEVINE, MASON, AND BROWN. "Lex & Yacc". O'Reilly & Associates Inc (1992).
14. JENS VON GARREL. Parsing, Typechecking und Transformation von CSP-OZ nach Jass. Masters thesis, University of Oldenburg, Germany, Dept. of Comp. Sci. (July 1999).
15. MATTIAS WERNER. "daVinci V2.1.x Online Documentation". University of Bremen (June 1998). URL: http://www.tzi.de/~davinci/doc_V2.1/.
16. JAN BREDEREKE. Modular, changeable requirements for telephone switching in CSP-OZ – revision 2.0. Tech. Rep. IBS-00-1, University of Oldenburg, Oldenburg, Germany (2001). *To appear.*
17. JAN BREDEREKE. Modular, changeable requirements for telephone switching in CSP-OZ. Tech. Rep. IBS-99-1, University of Oldenburg, Oldenburg, Germany (October 1999).
18. JAN BREDEREKE. "genFamMem 2.0 Home Page". University of Bremen (2000). URL http://www.tzi.de/~brederek/genFamMem/.
19. STEPHEN GILMORE AND MARK RYAN, editors. "Proc. of Workshop on Language Constructs for Describing Features", Glasgow, Scotland (15–16 May 2000). ESPRIT Working Group 23531 – Feature Integration in Requirements Engineering.
20. SHMUEL KATZ. A superimposition control construct for distributed systems. *ACM Trans. Prog. Lang. Syst.* **15**(2), 337–356 (April 1993).
21. MALTE PLATH AND MARK RYAN. Defining features for CSP: Reflections on the feature interaction contest. In Gilmore and Ryan [19], pages 55–69.
22. ROBERT J. HALL. Feature combination and interaction detection via foreground/background models. *Comp. Networks* **32**(4), 449–469 (April 2000).

PATTERNS AND RULES FOR BEHAVIOURAL SUBTYPING

Heike Wehrheim
Universitat Oldenburg
Fachbereich Informatik
Postfach 2503, D–26111 Oldenburg, Germany
wehrheim@informatik.uni-oldenburg.de

Abstract Subtyping relations for object-oriented formalisms describe relationships between super- and subclasses which satisfy the substitutability requirement imposed on types and their subtypes. *Behavioural* subtyping is concerned with subtypes for *active* classes with an explicit dynamic behaviour, specifiable for instance by object-oriented formal methods combining state-based with behavioural formalisms.

In this paper we develop syntactic *patterns* and semantic *rules* for the state-based part of a subclass which guarantee that the subclass is a behavioural subtype of its superclass. This allows to check for subtypes *without* computing the behavioural semantics of the class at all. Our results are similar to the ones linking data refinement in state-based methods with failure-divergence refinement in CSP. In contrast to classical data refinement, subtyping has to cope with additional new operations in the subclass.

Keywords: Object-orientation, refinement, behavioural subtyping, Z, CSP.

1. INTRODUCTION

Inheritance is one of the key features for the success of object-oriented analysis and design methods as well as programming languages. Inheritance is employed to re-use already existing class descriptions and extend them with additional attributes and methods, thus creating a new subclass from an existing superclass. Subtyping for object-oriented formalisms [1, 6, 12] studies relationships between super- and subclasses with the aim of clarifying when a subclass can be said to be *type substitutable* for its superclass. Type substitutability means that "a subtype instance can be used wherever a supertype instance is expected" [17]. Subtyping is also studied in the context of *active* classes with a behavioural semantics (as for instance obtained when using LOTOS or CSP as a specification language). Active classes have an explicit specification of the dynamic behaviour which describes synchronisation constraints of the class or certain protocols which have to be obeyed. In this paper, we use the term behavioural subtyping to refer to subtyping for this type of classes. A number of proposals for *behavioural subtyping* relations have been made (among others [16, 2, 4, 13]; see conclusion for a discussion of related work), with for instance Petri nets or labelled transition systems as semantic basis. The starting point of this paper are three subtyping relations from [9] with proven substitutability properties, two of them also adequate for shared mutable objects. The formal basis for these two relations are transition systems and their (CSP) failures and divergences.

Here we will be concerned with *checking* subtype relationships among classes specified in object-oriented formal methods combining state-based with behavioural aspects. The formal method which we have in mind (and which we use throughout our examples) is CSP-OZ [7], a combination of CSP [10] and Object-Z [15], but the theory we develop will be independent of this particular choice. Two aspects of CSP-OZ are however of importance: first, the fact that the state-based part of CSP-OZ has a *behavioural semantics* (and thus we have to use *behavioural* subtyping relations), and second, that the semantics of the combination is obtained by separately computing the semantics of the Object-Z and CSP part, and afterwards joining them by CSP parallel composition. This second point allows us to look at state-based and behaviour part in isolation when checking for subtypes.

A straightforward approach to check subtyping relationships among classes is the following: given two class descriptions, compute their behavioural semantics and compare the classes according to some chosen relation. However, since the semantics of CSP-OZ (and most other such types of integrated formal methods) has to be constructed by hand, this approach is in general impracticable. Instead we are interested in developing simple, at the best *syntactic, patterns*, which allow to avoid the explicit check and instead guarantee subtypes *by construction*. Such patterns or rules have to give conditions on attributes and operations in subclasses, which, when followed, immediately lead to subtypes.In this paper we will solely be interested in subtyping patterns for the state-based part (which, due to the compositional semantics, can without change be combined with patterns for the behaviour part). Our investigations are similar to the work about the relationship between data refinement rules in state-based formalisms and failure-divergence refinement [11, 18]. There, it has been studied whether data refinement induces failure-divergence refinement when the state-based formalism is given a behavioural semantics. The basic approach taken here is the same: we assume to have a behavioural semantics for a state-based formalism, have a set of conditions on two class specifications and show that the conditions induce a behavioural subtyping relationship between the classes (soundness of rules).

We illustrate our approach by a small example of a buffer which is extended by means of inheritance in three different ways yielding three different kinds of subtypes.

2. CLASSES

We start with the specification of the buffer. We use CSP-OZ [7] syntax in the specifications. The following class specifications will *not* contain any CSP behaviour descriptions since we are only interested in patterns for the state-based part. Thus the classes almost look like Object-Z classes. The specified buffer can hold elements (in a finite bag) of a basic type *Element* up to some arbitrary number *n*. Elements may be put into the buffer and retrieved by means of the methods *put* and *get*, and an attribute *alive* is used to indicate whether some specific instance of class *Buffer* is alive or has already been deleted.

$[Element]$

$Buffer(n : \mathbb{N})$

method $put : [\, el? : Element \,]$
method $get : [\, el! : Element \,]$

State:
$contents : \mathrm{fbag}\ Element$
$alive : \mathbb{B}$

$\#contents \leq n$

Init:
$contents = [\![\,]\!]$
$alive = \mathrm{true}$

enable_*put*:
$el? : Element$

$\#contents < n \land alive$

effect_*put*:
$\Delta(contents)$
$el? : Element$

$contents' = contents \uplus [\![el?]\!]$

enable_*get*:

$\#contents \geq 1 \land alive$

effect_*get*:
$\Delta(contents)$
$el! : Element$

$el! \sqsubseteq contents$
$contents' = contents \cup [\![el!]\!]$

The class is parameterised in the length of the buffer. The first two lines define the *interface* of the class: it has methods *put* and *get*. Next the attributes and their initial values are defined. The application of methods *put* and *get* is described by enable and effect predicates, the enabling predicate specifying a *guard* for the application of the method, the effect predicate the possible results of applying the method.

In general we represent a class C by a *state schema*, an *initialisation schema* and a number of *operation schemas*: $C = (State, Init, (\texttt{enable\_}Op_j)_{j\in J}, (\texttt{effect\_}Op_j)_{j\in J})$. The components have the following meaning: *State* describes the state space of the class. It is determined by giving a set of *attributes* *Attr* and their types together with an *invariant* *Inv* further restricting the allowed values. *Init* describes the set of initial states determined by a predicate over *Attr*. $(Op_j)_{j\in J}$ gives a family of methods or operations on the state space. Every operation Op_j has a number of input variables $(in_{j,i}?)_{1\leq i\leq k_j}$ of type $T_{in_{j,i}}$ and a number of output variables $(out_{j,i}!)_{1\leq i\leq l_j}$ of type $T_{out_{j,i}}$. Two schemas are given for every operation (an omitted schema in the CSP-OZ specification corresponds to *true*):

1 an enabling schema $\texttt{enable\_}op_j$ whose predicate ranges over the current state *Attr* and input variables $in_{j,i}?$. It acts as a *guard* to the operation execution, and

2 an effect schema $\texttt{effect\_}op_j$ with predicate ranging over current state *Attr*, next state *Attr'*, input variables $in_{j,i}?$ and output variables $out_{j,i}!$ which describes the postcondition of the operation (where *Attr'* refers to the new state of attributes after execution of the operation).

We use the following abbreviation for input and output variables:

$$\begin{aligned} In_j &\mathrel{\widehat=} [in_{j,1}? : T_{in_{j,1}}; \ \ldots ; \ in_{j,k_j}? : T_{in_{j,k_j}}] \\ Out_j &\mathrel{\widehat=} [out_{j,1}! : T_{out_{j,1}}; \ \ldots ; \ out_{j,l_j}! : T_{out_{j,l_j}}] \end{aligned}$$

Additionally we define the precondition of an operation Op_j to be $\texttt{pre\_}Op_j := \exists State', Out_j : \texttt{effect\_}Op_j$.

Next we provide the class description with a behavioural semantics in terms of labelled transition systems. The labels are taken from an alphabet Σ which contains all events $op.i_1.\ldots.i_k.o_1.\ldots.o_l$ where op is the name of a method and the rest are possible values of input and output parameters. For an event $ev = op.i_1.\ldots.i_k.o_1.\ldots.o_l$ we write $ch(ev) = op$ to refer to the method or operation name of an event. Σ_τ stands for $\Sigma \cup \{\tau\}$. A *labelled transition system* (LTS) is a tuple $T = (Q, \rightarrow, Q^{in})$ with Q a set of states, $\rightarrow$ a transition relation and Q^{in} the set of initial states. A state q is said to be *stable* if it has no outgoing τ transitions, q diverges ($q\uparrow$) if an infinite number of τ steps is possible from q, and $next(q)$ is the set of all events which are executable from q.

The basic idea of the semantics is to view a class as a *process* which is continuously executing enabled operations and thus changes its state. Initially it is in one of the states of *Init*. The symbol $\perp$ is used to indicate one particular *divergent* state: The class diverges if an enabled operation is applied outside its precondition. In the following we take a schema to stand for a set of *bindings* and for a binding $q = \langle\!| a_1 \Rrightarrow v_1, \ldots, a_n \Rrightarrow v_n |\!\rangle$ and a schema S we write $S(q)$

if q is a valid binding of schema S. The operator $\cup$ is used to join bindings: $\langle\!| a_1 \Rrightarrow v_1, \ldots, a_n \Rrightarrow v_n |\!\rangle \cup \langle\!| b_1 \Rrightarrow w_1, \ldots, b_m \Rrightarrow w_m |\!\rangle = \langle\!| a_1 \Rrightarrow v_1, \ldots, a_n \Rrightarrow v_n, b_1 \Rrightarrow w_1, \ldots, b_m \Rrightarrow w_m |\!\rangle$ assuming disjointness of the sets $\{a_1, \ldots, a_n\}$ and $\{b_1, \ldots, b_m\}$. For a state q, the notation $q\prime$ is used to describe the same binding but with components decorated with $'$, the notation $q \restriction_{State}$ is a binding which only gives values to variables in q which are declared in *State*.

Definition 2.1
The operational semantics of a class C *is defined as* $[\![C]\!] = (Q_C, \rightarrow_C, Q_C^{in})$, *where*

- $Q_C = State \cup \{\bot\}$, $Q_C^{in} = Init$ *and*
- $\rightarrow_C$ *is defined as* $\bot \xrightarrow{\tau}_C \bot$ *and*
    - $q_1 \xrightarrow{ev}_C q_2$ *if* $ev = op_j.i_1.\ldots.i_{k_j}.out_1.\ldots.out_{l_j}$ *and* $\mathtt{enable\_}op_j(q_1 \cup \langle\!| in_n \Rrightarrow i_n, 1 \le n \le k_j |\!\rangle)$ *and* $\mathtt{effect\_}op_j(q_1 \cup q_2\prime \cup \langle\!| in_n \Rrightarrow i_n, out_m \Rrightarrow o_m, 1 \le n \le k_j, 1 \le m \le l_j |\!\rangle)$,
    - $q \xrightarrow{ev}_C \bot$ *if* $ev = op_j.i_1.\ldots.i_{k_j}.out_1.\ldots.out_{l_j}$ *and* $\mathtt{enable\_}op_j(q_1 \cup \langle\!| in_n \Rrightarrow i_n, 1 \le n \le k_j |\!\rangle)$ *but not* $\mathtt{pre\_}op_j(q_1 \cup \langle\!| in_n \Rrightarrow i_n, 1 \le n \le k_j |\!\rangle)$.

Note that we use state $q_2\prime$ here since the effect predicate uses primed variables to refer to the state after execution.

The dynamic behaviour of class *Buffer*(2) with an instantiation of type *Element* with the set $\{1, 2\}$ is depicted by the transition system in Figure 1.

Next we present three extensions of the basic class *Buffer* by inheritance: the first adds a possibility of explicitly deleting the buffer, the second adds a new method checking for emptiness and the third a method which allows to insert two elements at once.

```
FinBuffer(n : ℕ)
  inherit Buffer(n)
  method del

  enable_del
  ─────────────
  #contents = 0 ∧ alive

  effect_del
  Δ(alive)
  ─────────────
  alive = false
```

```
EmptyCheckBuffer(n : ℕ)
  inherit Buffer(n)
```

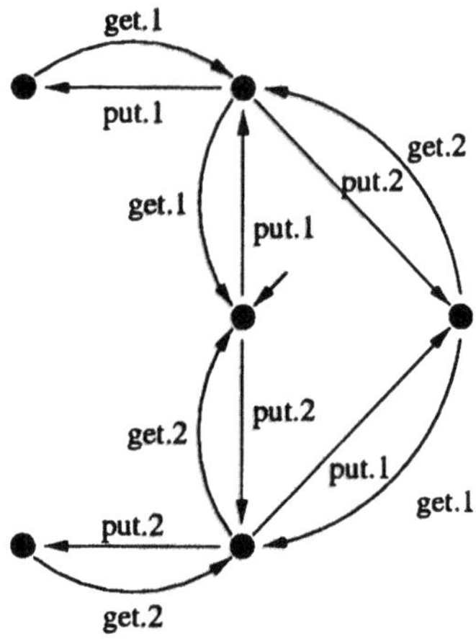

Figure 1. Dynamic behaviour of superclass *Buffer*(2)

$\texttt{method}\ empty : [b! : \mathbb{B}]$

$\texttt{enable\_}empty$

$alive$

$\texttt{effect\_}empty$

$\Delta()$

$b! : \mathbb{B}$

$b! = (contents = \varnothing)$

$2Buffer(n : \mathbb{N})$

$\texttt{inherit}\ Buffer(n)$

$\texttt{method}\ put2 : [el1? : Element;\ el2? : Element]$

$\texttt{enable\_}put2$

$el1? : Element;\ el2? : Element$

$alive \wedge \#contents < n - 1$

$\texttt{effect\_}put2$

$\Delta(contents)$

$el1? : Element;\ el2? : Element$

$contents' = contents \uplus \{el1, el2\}$

Conceptually inheritance is defined as conjunction on schemas: given an abstract class $A = (AState, AInit, (\texttt{enable\_}AOp_j)_{j \in J}, (\texttt{effect\_}AOp_j)_{j \in J})$ and specialised class $C = (CState, CInit, (\texttt{enable\_}COp_i)_{i \in I}, (\texttt{effect\_}COp_j)_{j \in J})$ assuming C is constructed from A by means of inheritance, then $J \subseteq I$ and

$$\begin{aligned} CState &= AState \wedge ExtState \\ CInit &= AInit \wedge ExtInit, \end{aligned}$$

$$\begin{array}{llll}
\texttt{enable_}COp_j & = & \texttt{enable_}AOp_j \wedge \texttt{enable_}ExtOp_j & \forall j \in J, \\
\texttt{effect_}COp_j & = & \texttt{effect_}AOp_j \wedge \texttt{effect_}ExtOp_j & \forall j \in J, \\
\texttt{enable_}COp_i & = & \texttt{enable_}ExtOp_i \quad \forall i \in I \setminus J, & \\
\texttt{effect_}COp_i & = & \texttt{effect_}ExtOp_i \quad \forall i \in I \setminus J. &
\end{array}$$

ExtState (*ExtInit* etc.) is the schema which is explicitly written in the subclass and *AState* (etc.) is inherited.

3. SUBTYPING

Subtyping on ordinary types is usually demanded to fulfill the requirement of *type substitutability*: subtypes should be usable wherever supertypes are expected without any noticeable differences [17]. When lifting this concept to classes, the first idea is to use the inheritance hierarchy for defining subtyping relations. This however does not fit the requirement of substitutivity anymore: inheritance is first of all concerned with *re-use* of class descriptions and allows fundamental changes to the behaviour of superclasses in a subclass. The requirement of substitutivity suggests that instead some sort of refinement should be used as a subtyping relation.

Since we are working in the context of CSP, our notion of refinement is failure-divergence refinement. In the process algebra CSP process refinement is used to compare an implementation against a specification with respect to their failures and divergences: the implementation has to be more deterministic than the specification, i.e. has less failures and divergences. Failures are sets of pairs $(\sigma, X) \in \Sigma^* \times \mathbf{2}^\Sigma$ with the meaning that a process has such a pair in its failure set if it can execute the trace σ and may thereby reach a stable state (no τ-transitions possible) in which none of the events from X are possible. Divergences are sets of traces σ such that a process may diverge after the trace, i.e. execute an infinite number of invisible actions. Failures and divergences can for instance be computed from the transition system of a process, but due to lack of space we refrain from giving a formal definition here. Buffer(2) for instance has failures (with maximal refusals) $(\varepsilon, \{get.1, get.2\}), (put.1, \{get.2\}), (put.1 ^\frown put.2, \{put.1, put.2\}), \ldots$ and an empty set of divergences.

Refinement then simply is inclusion of failure and divergence sets: A labelled transition system I is a *failure-divergence refinement* of an lts S ($S \sqsubseteq_{\mathcal{FD}} I$) iff $failures(I) \subseteq failures(S)$ and $divergences(I) \subseteq divergences(S)$.

According to this definition none of the subclasses of *Buffer* are refinements (e.g. *FinBuffer* has a failure (del, Σ) which is not in *Buffer*'s failure set). Nevertheless refinement will be the basis for our three behavioural subtyping relations. The relations additionally have to cope with the fact that the subclass may introduce new methods. The comparison of sub- with superclass has to "ignore" these methods or at least relate new methods to old ones. For this, we use three different operators: restriction, concealment (next) and substitution (below), and define them by their effect on failures and divergences.

Definition 3.1 *Let $T = (Q, \rightarrow, Q^{in})$ be a labelled transition system and $N \subseteq \Sigma$.*

- *Restriction: $failures(T \setminus_r N) = \{(\sigma, X) \mid \exists Y \bullet (\sigma, Y) \in failures(T), \sigma \in (\Sigma \setminus N)^*, X \subseteq Y \cup N\}$,*
- *Concealment: $failures(T \setminus_c N) = \{(\sigma, X) \mid \exists (\sigma', Y) \in failures(T), \sigma = \sigma' \upharpoonright (\Sigma \setminus N) \wedge X \subseteq Y \cup N\}$.*

Divergences are unchanged by restriction and concealment.

Restriction forbids the execution of events in N, concealment makes N-events invisible. Note that concealment differs from classical hiding in process algebras. Hiding may usually introduce new non-stable states, whereas concealment does not.

The basic idea of substitution is to *map* new methods onto sequences of old methods: the state changes achieved by new method should be the same as that achieved by a consecutive execution of the old methods. This is the "dynamic behaviour" counterpart of Liskov and Wing's extension maps [12].

Definition 3.2 *Let $(\sigma, X) \subseteq \Sigma \times 2^{\Sigma}$ be a pair of trace and refusal, and $N \subseteq \Sigma$ a set of events (the new events to be mapped onto old ones). Furthermore let $f_N : N \rightarrow \Sigma^*$ be a function, the* substitution *function. We define $f_N(\sigma, X)$ (by overloading the function f_N) to be $(f_N(\sigma), X \cup N)$ where $f_N(\sigma)$ is inductively defined by $f_N(\varepsilon) = \varepsilon$ and*

$$f_N(a \frown \sigma) = \begin{cases} a \frown f_N(\sigma) & \text{if } a \notin N \\ f_N(a) \frown f_N(\sigma) & \text{else} \end{cases}$$

Superclass and subclass are then compared by applying concealment, restriction or substitution to the new methods before checking for failure and divergence inclusion (see [9]).

Definition 3.3 *Let A, C be labelled transition systems, A standing for the super- and C for the subclass, and let N be the set of events over new methods.*

- *C is a* weak subtype *of A (denoted $A \sqsubseteq_{wst}^{N} C$) iff $A \sqsubseteq_{\mathcal{FD}} C \setminus_r N$.*
- *C is a* safe subtype *of A according to a substitution function f_N (denoted $A \sqsubseteq_{sst}^{f_N} C$) iff $\forall (\sigma, X) \in failures(C) : f_N(\sigma, X) \in failures(A)$ and $\forall \sigma \in divergences(C) : f_N(\sigma) \in divergences(A)$.*
- *C is an* optimal subtype *of A (denoted $A \sqsubseteq_{ost}^{N} C$) iff $A \sqsubseteq_{\mathcal{FD}} C \setminus_c N$.*

Coming back to our four buffers: *FinBuffer* is a weak subtype of *Buffer* and *EmptyCheckBuffer* is a weak as well as an optimal subtype. *FinBuffer* is not an optimal subtype of *Buffer* since $[\![FinBuffer]\!] \setminus_c \{del\}$ has a failure (ε, Σ) which *Buffer* does not have. Finally *2Buffer* is a safe subtype (mapping *put*2 to $put ^\frown put$) as well as a weak subtype of *Buffer* but no optimal subtype.

In order to check for subtypes we first had to construct the transition systems, compute failures/divergences for the classes and afterwards apply the subtyping definition. Next we look at *patterns* which give us the possibility of obtaining a subtype *by construction* and *rules* which allow to *locally* check for subtypes without computing the semantics of classes at all.

4. PATTERNS AND RULES

We refer to *patterns* as conditions on the specification of a subclass which can be checked (almost) completely syntactically, without looking at the semantics at all. *Rules* will be conditions on classes which can be checked by looking at the semantics of methods in isolation, i.e. without computing the transition system semantics of a class but by comparing the meaning of methods of sub- and superclass.

Besides the patterns and rules discussed here, it is always possible to employ *data refinement rules* (specific to CSP-OZ [8]): any subclass which is a refinement of its superclass is a subtype in the sense of all three relations. All of the following patterns and rules can thus be combined with a valid data refinement.

Patterns. We present two patterns here, one for weak and one for optimal subtyping. The patterns are conditions on the changes and extensions allowed in the subclass. We again assume $A = (AState, AInit, (AOp_j)_{j \in J})$ to be the super- and, letting $J \subseteq I$, $C = (CState, CInit, (COp_i)_{i \in I})$ to be the subclass.

P1 $\forall j \in J$: $\texttt{enable\_}COp_j = \texttt{enable\_}AOp_j$ and $\texttt{effect\_}COp_j = \texttt{effect\_}AOp_j$, i.e. the specialised subclass does not change the old methods at all. There are no restrictions on new methods or attributes.

P2 **P2.A** $\forall j \in J : \texttt{enable\_}COp_j = \texttt{enable\_}AOp_j$ and $\texttt{effect\_}COp_j = \texttt{effect\_}AOp_j$,

P2.B $\forall n \in I \setminus J$: the predicate of schema $\texttt{effect\_}COp_n$ ranges solely over *Attr*, *ExtAttr* and *ExtAttr'*, i.e. no changes to old attributes are allowed,

P2.C $\forall n \in I \setminus J : \texttt{enable\_}COp_n \Rightarrow \texttt{pre\_}COp_n$.

Here = stands for syntactical equality. Pattern P1 is a pattern for a weak subtype, e.g. *FinBuffer* may be constructed from *Buffer* using pattern P1, it leaves all old methods unchanged. Pattern P2 is valid for *EmptyCheckBuffer*: the new method *empty* does not change old attributes and its precondition holds whenever the method is enabled. We however cannot construct *FinBuffer* from *Buffer* using pattern P2 since *FinBuffer*'s effect predicate for operation *del* changes a value of an old attribute (*alive*). Condition P2.C is essentially used to avoid new divergences in the subclass. It is the only part of the patterns which is not purely syntactic, but this is necessary since we want to take divergence into account.

The following main theorem now states the soundness of the patterns for the subtyping relations.

Theorem 1
If C is a subclass of A obeying pattern P1 then $A \sqsubseteq_{wst} C$.
If C is a subclass of A obeying pattern P2 then $A \sqsubseteq_{ost} C$.

Note that such a relationship cannot be guaranteed by inheritance alone. Inheritance allows all kinds of changes in the subclass. Due to lack of space we only show the proof for the second part of the theorem. For the proof we show that, when constructed by a pattern, subclass and superclass are even more close to each other than defined by subtyping: the superclass *simulates* the subclass in a certain sense. Hence in the following we define a simulation relation inducing optimal subtyping. The relation is in spirit a forward simulation. We do not consider backward simulations here since we are not interested in completeness. In fact, it is immediately clear that the patterns are not complete: there are as well valid subtypes which cannot be constructed according to the patterns.

For the simulation definition we assume that the semantics of Definition 2.1 is used for classes, with the consequence that there are no τ-transitions except from $\perp$ to $\perp$.

Definition 4.1 *Class A* simulates *C* concealing *N*, $A \preceq_{(N)} C$, *if there is a relation* $\rho \subseteq Q_A \times Q_C$ *such that*

1 $\forall q_C \in Q_C^{in}\ \exists q_A \in Q_A^{in} : (q_A, q_C) \in \rho$,
2 $\forall (q_A, q_C) \in \rho$ *the following holds:*

- $stable(q_C) \Rightarrow stable(q_A)$, $next(q_C) \setminus N = next(q_A)$, $q_C\uparrow \Rightarrow q_A\uparrow$,
- $\forall a \notin N$:
  $q_C \xrightarrow{a}_C q'_C \Rightarrow \exists q'_A : q_A \xrightarrow{a}_A q'_A \wedge (q'_A, q'_C) \in \rho$,
  $\forall a \in N$:
  $q_C \xrightarrow{a}_C q'_C \Rightarrow (q_A, q'_C) \in \rho$.

The idea behind simulation for optimal subtypes is that for the N-steps of the subclass we do not require a matching step of the superclass, instead states q'

reached after executing new methods from some state q must be simulated by the matching partner of q.

Theorem 2 $A \preceq_{(N)} C \Rightarrow A \sqsubseteq_{ost}^{N} C$.

The remaining part for proving (the second part of) Theorem 1 is to show that subclasses constructed by patterns can be simulated by the superclasses. We again use N to refer to the set of all events over new methods.

Theorem 3 *If C is a subclass of A obeying pattern P2 then* $A \preceq_{(N)} C$.

Rules. We proceed with giving rules for the last subtyping relation, safe subtyping. Here, we actually have to compare the semantics of new methods with that of (sequences of) old methods. This requires a little more technical overhead. We explain this by means of *2Buffer*. *2Buffer* is a safe subtype of *Buffer* since the new method *put*2 can be explained in terms of the old method *put*: the effect of *put*2 is the same as that of two *put*'s. However, the substitution function f_N that we need does not map a new method/operation to a sequence of operations but *events* to *sequences of events*. Besides transformation of operations we also need some means for transforming inputs and outputs of a new operation onto the inputs and outputs of the old operations. For this, we use (an adaption of) the technique of input and output transformers as for instance advocated in [5]. An input or output transformer simply is a schema that relates inputs (or outputs) of operations. Looking at *2Buffer* the input transformer should relate *el*1? and *el*2? with the two inputs *el*? of two *put*'s, the output transformer is empty. In order to be able to distinguish the inputs of the two *put*'s we define another decoration on schemas, that decorate input and output variables with numbers.

For an operation schema $S = [a_1 : T_1; \ \ldots; \ a_n : T_n, in? : T_{in}, out! : T_{out} \mid p]$ we define S^k, $k \in \mathbb{N}$, to be $[a_1 : T_1; \ \ldots; \ a_n : T_n, in^k? : T_{in}, out^k! : T_{out} \mid p[in?/in^k?, out!/out^k!]]$ (and similar for more than one input or output variable). The decoration is used to distinguish variables with the same name coming from operations at different positions in the substituting sequence. The input transformer from *put*2 to $put^1 \mathbin{;} put^2$ can thus be formulated as:

```
IT
el1? : Element; el2? : Element
el¹! : Element; el²! : Element
-----
el1? = el¹! ∧ el2? = el²!
```

$$el1? : Element;\ el2? : Element$$
$$el^1! : Element;\ el^2! : Element$$
$$el1? = el^1! \land el2? = el^2!$$

It transforms inputs of *put*2 into outputs which can than be fed into $put^1 \mathbin{;} put^2$ (using the Z pipe operator $\gg$). Conversely, an output transformer from a sequence of old operations $AOp_1^1 \mathbin{;} AOp_2^2$ to a new operation *COp* would take the outputs of $AOp_1^1 \fatsemi AOp_2^2$ as inputs and transform them into outputs of *COp*. The output transformer from $put^1 \fatsemi put^2$ to *put* is an empty schema: no declarations and the predicate is true.

To simplify matters, we only consider substitutions which map one operation to a sequence of two operations in the sequel. This can however easily be generalised to an arbitrary number of operations.

Definition 4.2 *An* input transformer *from COp to* $AOp_1^1 \fatsemi AOp_2^2$ *is a schema IT s.t.*

- *all input variables from COp are declared in IT,*
- *for every input variable* $x^i?$ *of type T from operation* AOp_i^i, $i = 1, 2$, *there is one output variable* $x^i! : T$ *in IT,*
- *no other variables are declared in IT and*
- *the predicate of IT is a conjunction of equalities of the form* $x? = y!$ *such that every output variable occurs exactly once.*

Conversely, an output transformer *from* $AOp_1^1 \fatsemi AOp_2^2$ *to COp is a schema OT in which all output variables from COp are declared, for every output variable* $x^i! : T$ *from* AOp_i^i, $i = 1, 2$, *there is one input variable* $x^i? : T$, *no other variables are declared in OT, and the predicate of IT is a conjunction of equalities of the form* $x? = y!$ *such that every output variable occurs exactly once.*

Figure 2 illustrates the use of input and output transformers: execution of *COp* should be the same as that of $AOp_1 \fatsemi AOp_2$, and the input and output transformers define the valid relationships between the input and outputs.
We now again assume *A* to be the super- and *C* to be the subclass. The following rule formalises Liskov and Wings's extension maps [12] in our setting.

Extension Rule: The extension rule *R* holds between *A* and *C* if all old methods remain unchanged in the subclass (or are changed according to data refinement rules) and there is an *extension map* $E : N_{op} \to O_{op}^*$ such that the following holds: for every $COp \in N_{op}$, either *COp* does not change attributes at all and $E(COp) = \varepsilon$, or else *COp* does change attribute values, $E(COp) = AOp_1 \frown AOp_2$ and there is an input transformer *IT* from *COp* to $AOp_1 \fatsemi AOp_2$, an output transformer *OT* from $AOp_1 \fatsemi AOp_2$ to *COp* and the following conditions are furthermore met.

R.A $\forall CState, \forall In \bullet \mathtt{enable\_}COp \Rightarrow IT \gg (\mathtt{enable\_}AOp_1{}^1 \land \mathtt{pre\_}AOp_1{}^1)$,

R.B $\forall CState, \forall In \bullet \mathtt{enable\_}COp \Rightarrow \mathtt{pre\_}COp$,

R.C $\forall CState, CState', \forall In \bullet \mathtt{enable\_}COp \land (IT \gg \mathtt{effect\_}AOp_1{}^1)$
$\Rightarrow IT \gg (\mathtt{enable\_}AOp_2{}^{2\prime} \land \mathtt{pre\_}AOp_2{}^{2\prime})$,

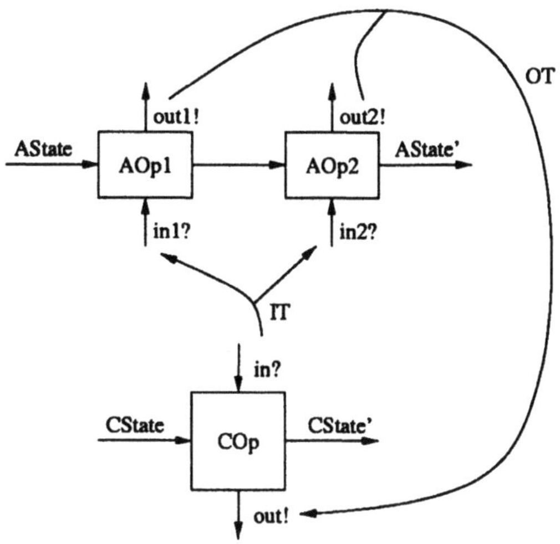

Figure 2. Input and output transformer

R.D $\forall \mathit{CState}, \mathit{CState}', \forall \mathit{In}, \mathit{Out} \bullet \mathtt{enable\_}\mathit{COp} \wedge \mathtt{effect\_}\mathit{COp}$
$\Rightarrow \mathit{IT} \gg (\mathtt{effect\_}\mathit{AOp}_1{}^1 \mathbin{;} \mathtt{effect\_}\mathit{AOp}_2{}^2) \gg \mathit{OT}.$

The conditions have the following meaning: R.A states that whenever *COp* is enabled so is AOp_1 and its precondition is also fulfilled; R.B guarantees that *COp* does not introduce new divergences; R.C states that after execution of AOp_1 the second operation of the sequence AOp_2 is enabled and defined. Finally R.D guarantees that the effect of executing *COp* is the same as that of $AOp_1 \mathbin{;} AOp_2$.

An extension map and the corresponding input and output transformers *IT*, *OT* induce a substitution function $f^E_{IT,OT} : N \rightarrow \Sigma^*$. This function has to transform events into sequences of events; the extension map tells us which operation to map onto which sequence of operations and the input and output transformers tell us the values of parameters. Additionally we need the *ordering* among inputs and outputs as defined in the interface declaration of the class. Let *COp* and AOp_1, AOp_2 be defined by

$$\begin{aligned}
&\mathtt{method}\ COp : [cin_1? : T_1, \ldots, cin_k : T_k,\\
&\qquad cout_1 : T_{k+1}, \ldots, cout_l : T_{k+l}]\\
&\mathtt{method}\ AOp_1 : [ain_{1,1}? : T_{1,1}, \ldots, ain_{1,k_1} : T_{1,k_1},\\
&\qquad aout_{1,1} : T_{1,k_1+1}, \ldots, aout_{1,l_1} : T_{1,k_1+l_1}]\\
&\mathtt{method}\ AOp_2 : [ain_{2,1}? : T_{2,1}, \ldots, ain_{2,k_2} : T_{2,k_2},\\
&\qquad aout_{2,2} : T_{2,k_2+1}, \ldots, aout_{2,l_2} : T_{2,k_2+l_2}]
\end{aligned}$$

Given this declaration we associate with an input transformer *IT* two functions $\mathcal{IT}^i, i = 1, 2$, mapping input values of *COp* onto input values of AOp_1, AOp_2, respectively.

$$\mathcal{IT}^i(i_1, \ldots i_k) = (w_1, \ldots, w_{k_i})$$

with $w_j = i_m$, $1 \leq j \leq k_i$, $1 \leq m \leq k$, if $cin_m? = ain^i_j!$ is an equation in *IT*. For output transformers *OT* we have a single function that maps ordered outputs of $AOp_1 \fatsemi AOp_2$ to outputs of *COp*.

$$\mathcal{OT}(v_{1,1}, \ldots v_{1,l_1}, v_{2,1}, \ldots, v_{2,l_2}) = (o_1, \ldots, o_l)$$

with $o_j = v_{i,m}$ if $cout_j! = aout^i_m?$ is an equation in *OT*. Then $f^E_{IT,OT}$ is defined to be
$f^E_{IT,OT}(op.i_1. \ldots .i_k.o_1. \ldots .o_l) =$
$E^1(op).\mathcal{IT}^1(i_1, \ldots, i_k).v_1. \ldots .v_{l_1} \frown E^2(op).\mathcal{IT}^2(i_1, \ldots, i_k).v_{l_1+1}. \ldots .v_{l_1+l_2}$
such that $\mathcal{OT}(v_1, \ldots, v_{l_1}, v_{l_1+1}, \ldots, v_{l_1+l_2}) = (o_1, \ldots, o_l)$.

Looking at *2Buffer* we find that E defined by $E(put2) = put \frown put$ is an extension map; for instance $\#contents < n - 1$ implies $\#contents < n$.

The following Lemma is the key to correctness of the extension rule:

Lemma 4.1
Let $A = (AState, AInit, (\texttt{enable\_}AOp_j)_{j \in J}, (\texttt{effect\_}AOp_j)_{j \in J})$ *be the super- and* $C = (CState, CInit, (\texttt{enable\_}COp_i)_{i \in I}, (\texttt{effect\_}COp_i)_{i \in I})$ *the subclass. Let* $E : N_{op} \rightarrow O^*_{op}$ *be an extension map with input and output transformers IT, OT for every new operation. Let* $T_C = [\![C]\!]$ *and* $T_A = [\![A]\!]$ *be the transition systems. Then the following holds:*

$$\begin{aligned} &\forall q_1 \in Q_C : q_1 \xrightarrow{ev} q_2 \text{ with } ch(ev) \notin N_{op} \text{ a new operation} \\ &\quad \Rightarrow \exists q_3 \in Q_A : q_1 \mid_{AState} \xrightarrow{ev_1} q_3 \xrightarrow{ev_2} q_2 \mid_{AState} \\ &\qquad \text{such that } f^E_{IT,OT}(ev) = ev_1 \frown ev_2 \,. \end{aligned}$$

The proof is again omitted. Analogous to the other subtyping relations, we have a simulation relation which induces safe subtyping.

Definition 4.3 *Let* $f_N : N \rightarrow \Sigma^*$ *be a substitution function. Class A* simulates *C* changing *N* by f_N, $A \preceq_{f_N} C$, *if there is a relation* $\rho \subseteq Q_A \times Q_C$ *such that*

1. $\forall q_C \in Q^{in}_C \; \exists q_A \in Q^{in}_A : (q_A, q_C) \in \rho$,
2. $\forall (q_A, q_C) \in \rho$ *the following holds:*
   - $stable(q_C) \Rightarrow stable(q_A)$, $next(q_C) \setminus N = next(q_A)$, $q_C\uparrow \Rightarrow q_A\uparrow$,
   - $\forall a \notin N$:
     $q_C \xrightarrow{a}_C q'_C \Rightarrow \exists q'_A : q_A \xrightarrow{a}_A q'_A \wedge (q'_A, q'_C) \in \rho$.
     $\forall a \in N$:
     $q_C \xrightarrow{a}_C q'_C \Rightarrow \exists q'_A : q_A \xrightarrow{f(a)} q'_A \wedge (q'_A, q'_C) \in \rho$.

As before we get two results relating simulation with subtyping and patterns with simulation.

Theorem 4 $A \preceq_{f_N} C \Rightarrow A \sqsubseteq^{f_N}_{sst} C$.
If C is a subclass of A obeying the extension rule R then $A \preceq_{f_N} C$.

5. DISCUSSION

In this paper we presented subtyping patterns and rules for active classes with a behavioural semantics. Patterns are simple, mostly syntactic conditions which, when applied during construction of subclasses via inheritance, guarantee correct subtypes. Rules are conditions which can be locally checked on the semantics of operations.

Related work. First of all there is a huge amount of work on behavioural subtyping relations for object-oriented formalisms. They differ in the semantic models used as a basis and in the defined relations. A rough classification could divide them into bisimulation and refinement-based approaches. Bisimulation based approaches easily achieve some form of substitutivity since they use an *equivalence* relation for subtyping. We consider this to be too strong since subtyping in its basic nature is not symmetric. Refinement-based proposals on the other hand in most cases only discuss substitutability but do not give a formal proof. The relation most often used in this context is *extension*, developed for the process algebra LOTOS. Extension allows extension of functionality, but not only on new methods but also on old methods.

The discussion of related work concerning the specific topic of this paper can be split in two parts. The first one concerns the large amount of work on simulation relations for refinement, especially those showing the tight connections between data refinement in state-based and process refinement in behaviour-based systems [11, 18]. The ideas of these papers are clearly the basis for our simulation relations which adapt this basic concept to the case of behavioural subtyping relations.The second area are patterns for subtyping relations. The work of van der Alst and Basten [16] presents transformation rules on Petri nets which construct subtypes, where the subtyping relations are based on branching bisimilarity. Rudkin [14] studies subtyping patterns for LOTOS, i.e. specific LOTOS operators that induce subtypes. The relation chosen there is again extension.

Patterns which are similar to ours are also employed in UNITY's superposition operator [3]. This is not an object-oriented setting, nevertheless the goal is similar: extend a program in a conservative way, not disturbing any of the properties holding beforehand. In fact, the idea of conservative extension appears in many theories and behavioural subtyping is just a form of conservative extension in object-orientation.

REFERENCES

[1] P. America. Designing an object-oriented programming language with behavioural subtyping. In J.W. de Bakker, W.P. de Roever, and G. Rozenberg, editors, *REX Workshop: Foundations of Object-Oriented Languages*, number 489 in LNCS. Springer, 1991.

[2] H. Bowman, C. Briscoe-Smith, J. Derrick, and B. Strulo. On behavioural subtyping in LOTOS. In H. Bowman and J. Derrick, editors, *FMOODS '97*, pages 335 – 351. Chapman & Hall, 1997.

[3] K.M. Chandy and J. Misra. *Parallel program design: a foundation*. Addison Wesley, 1988.

[4] E. Cusack. Refinement, conformance and inheritance. *Formal Aspects of Computing*, 3:129 – 141, 1991.

[5] J. Derrick and E. Boiten. Non-atomic refinement in Z. In J. Woodcock and J. Wing, editors, *FM'99*, number 1709 in LNCS, pages 1477–1496. Springer, 1999.

[6] K. K. Dhara and G. T. Leavens. Forcing behavioral subtyping through specification inheritance. In *ICSE '96*, pages 258–267. IEEE Computer Society Press, March 1996.

[7] C. Fischer. CSP-OZ: A combination of Object-Z and CSP. In H. Bowman and J. Derrick, editors, *FMOODS '97*, volume 2, pages 423–438. Chapman & Hall, 1997.

[8] C. Fischer and S. Hallerstede. Data-Refinement in CSP-OZ. Technical Report TRCF-97-3, University of Oldenburg, June 1997.

[9] C. Fischer and H. Wehrheim. Behavioural subtyping relations for object-oriented formalisms. In T. Rus, editor, *AMAST 2000*, number 1816 in LNCS, pages 469–483. Springer, 2000.

[10] C. A. R. Hoare. *Communicating Sequential Processes*. Prentice-Hall, 1985.

[11] M.B. Josephs. A state-based approach to communicating processes. *Distributed Computing*, 3:9–18, 1988.

[12] B. Liskov and J. Wing. A behavioural notion of subtyping. *ACM Transactions on Programming Languages and Systems*, 16(6):1811 – 1841, 1994.

[13] O. Nierstrasz. Regular types for active objects. In O. Nierstrasz and D. Tsichritzis, editors, *Object-oriented software composition*, pages 99 – 121. Prentice Hall, 1995.

[14] S. Rudkin. Inheritance in LOTOS. In K. Parker and G. Rose, editors, *FORTE '92*, pages 409–424. Elsevier, 1992.

[15] G. Smith. *The Object-Z Specification Language*. Kluwer Academic Publisher, 2000.

[16] W.M.P. van der Aalst and T. Basten. Life-cycle inheritance - a Petri-net-based approach. In P. Azema and G. Balbo, editors, *ATPN '97*, number 1248 in LNCS, pages 62–81. Springer, 1997.

[17] P. Wegner and B. Zdonik. Inheritance as an incremental modification mechanism or What like is and isn't like. In *ECOOP'88*, volume 322 of *LNCS*. Springer, 1988.

[18] J. C. P. Woodcock and C. C. Morgan. Refinement of state-based concurrent systems. In *VDM Symposium 1990*, volume 428 of *LNCS*, pages 340–351. Springer-Verlag, 1990.

Part Eight

Theories of Verification

VERIFICATION OF DENSE TIME PROPERTIES USING THEORIES OF UNTIMED PROCESS ALGEBRA

Matti Luukkainen

University of Helsinki, Department of Computer Science

PO Box 26, FIN-00014 University of Helsinki, FINLAND

mluukkai@cs.helsinki.fi

Abstract The article shows how untimed process algebraic methods can be used in verifying a category of dense time properties from a restricted class of timed automata. The work is based on the known results that if we restrict ourselves to a certain class of timed automata, then most of the interesting dense time properties are verifiable using discrete time methods. Those results are refined here in order to fit the process algebraic context, and furthermore, reduction strategies based on behavior equivalences, compositionality and abstraction are indicated. The usability of the method is evaluated with two case studies.

Keywords: Formal verification, Timed automata, Process algebra, Compositionality

1. INTRODUCTION

Because of the state explosion problem, a useful principle in algorithmic verification is that *the information unnecessary for a verification task should be abstracted away.* Especially, we would like to see in the context of systems specified with timed automata when it is sufficient to use discrete time algorithms but still say something about the system behavior in dense time.

In Henzinger et al., 1992 it was shown that for systems specified with timed transition systems, a large class of dense time properties can be verified by using discrete time algorithms. These results were generalized to timed automata in Bošnački, 1999. In this article we show how the methods used in standard untimed process algebra, and the related tools, can be used in verifying this category of dense time properties from the timed automata descriptions. We have refined the previously known results in order to fit the process algebraic context, and furthermore, some efficient reduction strategies based on behavior equivalences, abstraction and compositionality are advocated.

2. TIMED AUTOMATA

Informally, a timed automaton (Alur and Dill, 1994) is just a transition system extended with a set of a special type of variables, namely *clocks* which are used to control the executability and urgency of transitions with respect to real time.

Let $\mathcal{X}$ be a finite set of non-negative valued variables, namely clocks. Let Φ be a set of predicates on clocks, defined as the smallest set satisfying: $\phi ::= true \mid x \sim c \mid \phi \wedge \phi$, where $x \in \mathcal{X}$, $c \in \mathbb{N}$ and $\sim \in \{<, \leq, =, \geq, >\}$.

A timed automaton is a 6-tuple (S, A, X, R, I, s_0), where

- S is a finite set of control states,
- A is a finite set of transition labels or actions,
- $X \subseteq \mathcal{X}$ is a finite set of real-valued clocks,
- $R \subseteq S \times A \times \Phi \times 2^{\mathcal{X}} \times S$ is the transition relation,
- $I : S \rightarrow \Phi$ is a function that labels control states with time progress conditions,
- $s_0 \in S$ is the initial control state.

In a transition $(s, a, \phi, Y, s') \in R$, s is the source and s' the destination control state, $\phi \in \Phi$ is the guard of the transition. and $Y \subseteq X$ denotes the set of clocks that are reset to zero during the transition.

A valuation $v \in \mathcal{V}$ is a function that assigns a non-negative real value to each of the clocks, i.e. for every $x \in X : v(x) \in \mathbb{R}^{\geq 0}$. If X denotes a set of clocks, valuation $v[X := 0]$ is defined as follows: if $x \in X$ then $v[X := 0](x) = 0$, else $v[X := 0](x) = v(x)$. If $t \in \mathbb{R}^{\geq 0}$, and v is a valuation, then $v + t$ is a valuation such that $(v + t)(x) = v(x) + t$ for all the clocks x.

A *state* of a timed automaton is its control state and a valuation of the clocks. Let $Q \in S \times \mathcal{V}$ be the set of states of a timed automaton. A timed automaton can evolve in two ways, either executing a discrete transition or letting time pass without changing the control state. Possible transitions are formulated by the two rules below.

- Let (s, a, ϕ, Y, s') be a transition. The state (s, v) has a *discrete transition* labeled with a to (s',v'), if $\phi(v)$ evaluates to true and $v' = v[Y := 0]$.
- Let $t \in \mathbb{R}^{\geq 0}$. The state (s, v) has a *time transition* labeled with t to $(s, v + t)$ if for all $t' \leq t$ the time progress condition $I(s)(v + t')$ is true.

As can be noticed, the *state space* of a timed automaton can be defined with a transition system. Its alphabet consists of the transition labels of the timed automaton (the discrete transitions) and positive real values (the time transitions). Formally, given a timed automaton $TA = (S, A, X, R, I, s_0)$, its state space is defined by a transition system $S_{TA} = (S \times \mathcal{V}, A \cup \mathbb{R}^{\geq 0}, R^t, (s_0, v_0))$, where R^t is the smallest set containing all discrete and timed transitions of the timed automaton (which are given by the two rules above), and v_0 is a valuation that assigns the value zero to all the clocks.

The most natural way of modeling a system is to describe each of the components with a separate timed automaton. The behavior of the entire system is then defined by the timed automaton resulting from the parallel composition (denoted with $||$) of the component automata.

Now we will give an automata theoretic semantics of timed automata. A *timed word* is a pair $(\overline{\sigma}, \overline{t})$, where $\overline{\sigma} = \sigma_1, \sigma_2 \ldots$ is an infinite sequence of transition labels $\sigma_i \in A$, and $\overline{t} = t_1, t_2 \ldots$ is an infinite, monotone and progressive sequence of time values. The intuition behind the timed word is that a transition labeled with σ_i has occurred at time t_i from the beginning of the observation. All the possible timed words with time domain $\mathbb{T}$ are denoted $TW_{\mathbb{T}}$.

If we add to the definition of timed automaton *a set of accepting states* $F \subseteq S$, then it can be thought of as a *timed Büchi-automaton* which accepts a set of timed words.

An *execution* of timed automaton TA which *corresponds* to the timed word $(\overline{\sigma}, \overline{t})$ is an infinite sequence

$\overline{q} = q_1 \xrightarrow{t_1} q_1' \xrightarrow{\sigma_1} q_2 \xrightarrow{t_2-t_1} q_2' \xrightarrow{\sigma_2} q_3 \xrightarrow{t_3-t_2} q_3' \xrightarrow{\sigma_3} q_4 \ldots,$

in the S_{TA}, so each q_i, q_i' are of the form (s, v). The intuition behind the execution is that at the ith step, the automaton first lets time progress for $t_i - t_{i-1}$ units and then executes a transition labeled with σ_i.

A timed automaton TA *accepts* the timed word $(\overline{\sigma}, \overline{t})$ if it has an execution $\overline{q}$ corresponding to that word, starting from the initial state (s_0, v_0) and which contains infinite occurrences of states (s_i, v_i) such that $s_i \in F$. Language accepted by a timed automaton TA is defined as usual:

$[\![TA]\!] = \{(\overline{\sigma}, \overline{t}) \mid TA$ has an accepting computation corresponding to the word $(\overline{\sigma}, \overline{t})\ \}$.

In the automata theoretic setting, the question whether a system S satisfies a requirement φ (denoted $S \models \varphi$) can be formulated as follows. The system and the property are defined as two timed Büchi-automata TA_S and TA_φ. The automata define two sets of timed words $[\![TA_S]\!]$ and $[\![TA_\varphi]\!]$. Now the question whether $S \models \varphi$ is equivalent to the inclusion $[\![TA_S]\!] \subseteq [\![TA_\varphi]\!]$. The inclusion can be replaced by the emptiness check of $[\![TA_S]\!] \cap [\![TA_{\overline{\varphi}}]\!]$, where $TA_{\overline{\varphi}}$ is a timed Büchi-automata that defines the negation of the property φ.

Intersection $[TA_1] \cap [TA_2]$ of the languages accepted by two timed Büchi-automata equals the language accepted by the product automaton $[TA_1 \times TA_2]$ (Alur and Dill, 1994). In the special case where all the states of the other automata, e.g. TA_1 are accepting, the product automaton $TA_1 \times TA_2$ corresponds to the parallel composition $TA_1 \parallel TA_2$, where the accepting states are those in which the component corresponding to TA_2 is in accepting state. Since the automaton TA_S that describes the system usually has all the states as accepting ones, the verification can be done by checking whether $[TA_S \parallel TA_{\overline{\varphi}}] = \emptyset$.

3. VERIFICATION

It is known that several interesting verification problems are decidable for timed automata. Essentially the method of showing the decidability, and also the way of conducting the verification is the construction of a *region graph*, which is a finite abstraction of the state space of a timed automaton.

The dense time verification methods based on finite abstractions are computationally quite hard. Furthermore, the decidability of the dense time methods does not allow the clock values to be compared against arbitrary real values, but instead just integers. A question that now arises is, whether we could use discrete time verification algorithms in deducing dense time properties.

In Henzinger et al., 1992 it was shown that for systems specified with timed transition systems, a large class of dense time properties can be verified by using discrete time algorithms. These results were generalized to timed automata in Bošnački, 1999.

We have taken these known results as our starting point in developing our verification methodology. Since the theory in Henzinger et al., 1992 and Bošnački, 1999 is quite involved, we just rephrase here the most interesting consequences from our point of view. See the full version of this paper (M. Luukkainen, 2001) for full treatment and motivation for the result.

Let Π be a set of timed words. Let us denote with $\mathbb{Z}(\Pi)$ a subset of timed words of Π, which consists only of those words where the observation times t_i are natural numbers, formally $\mathbb{Z}(\Pi) = \Pi \cap TW_{\mathbb{N}}$. In other words, if $[TA]$ characterizes the behavior of an automaton TA within dense time semantics, $\mathbb{Z}(TA)$ would be its *interpretation in the discrete time model.* Now the question that we are interested in is, what can be concluded about the dense time behavior of an automaton TA by verifying $\mathbb{Z}(TA) \subseteq \mathbb{Z}(TA_{\varphi})$, or $\mathbb{Z}(TA) \cap \mathbb{Z}(TA_{\overline{\varphi}}) = \emptyset$, for a property φ. Thus, we want to know what dense time properties of systems can be deduced by using discrete time methods.

Firstly, let us call a timed automaton where relations only of the form $\leq, =, \geq$ are used as a *weakly constrained* timed automaton. Analogously let us call a timed automaton where only $<, >$ or the trivial constraint *true* are used as a *strongly constrained* timed automaton.

Now the following theorem (Henzinger et al., 1992; Bošnački, 1999) describes the relation between discrete and dense time verification.

Theorem 1 *Let TA_1 and TA_2 be two weakly constrained and TA_3 a strongly constrained timed Büchi automata.*

1 $[TA_1] \subseteq [TA_3]$ *if and only if* $\mathbf{Z}(TA_1) \subseteq \mathbf{Z}(TA_3)$.

2 $[TA_1] \cap [TA_2] = \emptyset$ *if and only if* $\mathbf{Z}(TA_1) \cap \mathbf{Z}(TA_2) = \emptyset$.

3 $\mathbf{Z}(TA_1) \cap \mathbf{Z}(TA_2) = \mathbf{Z}(TA_1 \parallel TA_2)$.

The above theorem guarantees the following: if a system is described with weakly constrained timed automaton TA_S and the negation of the correctness requirements is given as another weakly constrained automaton, say $\overline{\varphi}$, the verification task whether $S \models \varphi$ can now be reduced to checking whether the language accepted by the parallel composition, of TA_S and $TA_{\overline{\varphi}}$ is empty in the discrete time domain. Since the result of the parallel composition itself is just one timed automaton (with a set of accepting states), we can now concentrate on checking the language emptiness of a single timed automaton in the discrete time interpretation.

As stated in the previous section, a timed automaton TA accepts a timed word $(\overline{\sigma}, \overline{t})$ if and only if the corresponding state space S_{TA} has a path which corresponds to the timed word, and the path visits an accepting state infinitely often. So, in order to check the emptiness of TA one could observe whether the condition holds in the transition system S_{TA}. But since S_{TA} is inherently an infinite structure, even in the discrete time interpretation that we now only consider, it is not practical for verification purposes.

In the discrete time domain, the reason for infiniteness is the fact that clock values can potentially increase without bounds. In every timed automaton there exists the greatest value x_c against which a clock $x \in X$ is compared. When the value of x is greater than x_c, the further increase is no more meaningful to the behavior of the system. So, keeping this in mind, we can modify the rules for transitions of timed automata in the following way (the first rule is unaltered):

- Let (s, a, ϕ, Y, s') be a transition. The state (s, v) has a *discrete transition* labeled with a to (s',v'), if $\phi(v)$ evaluates to true and $v' = v[Y := 0]$.
- Let $t \in \mathbb{R}^{\geq 0}$. The state (s, v) has a *time transition* labeled with t to (s, v'),

where $v'(x) = v(x) + t$ if $x \leq x_c$, else $v'(x) = x_c$,
if for all $t' \leq t$ the time progress condition $I(s)(v + t')$ is true.

Now we can define the *reduced state space* RS_{TA}, a finite structure that can be used to derive the information about the executions of the timed automaton TA in the discrete time interpretation.

Definition 2 *Given a timed automaton* $TA = (S, A, X, R, I, s_0)$, *its reduced state space is defined by a transition system* $RS_{TA} = (S \times \mathcal{V}, A \cup \mathbb{R}^{\geq 0}, R^t, (s_0, v_0))$, *where* R^t *is the smallest set containing all discrete and timed transitions (given by the two rules above) of the timed automaton in the discrete time domain.*

The next theorem summarizes two important properties of transition systems corresponding to reduced state spaces. The proofs can be found in the full version of the paper (M. Luukkainen, 2001).

Theorem 3 *Let TA and TA′ be two timed automata.*

1. *There is a path* $\overline{q}_S = q_1 \xrightarrow{t_1} q_1' \xrightarrow{\sigma_1} q_2 \xrightarrow{t_2} q_2' \xrightarrow{\sigma_2} \ldots$ *in* S_{TA} *if and only if there is a corresponding path* $\overline{q}_{RS} = q_1 \xrightarrow{1} q_{1_1} \xrightarrow{1} \ldots q_{1_{t_1-1}} \xrightarrow{1} q_1' \xrightarrow{\sigma_1} q_2 \xrightarrow{1} q_{2_1} \xrightarrow{1} \ldots q_{2_{t_2-1}} \xrightarrow{1} q_2' \xrightarrow{\sigma_2} \ldots$ *in* RS_{TA}.
2. $RS_{TA} \mid\mid RS_{TA'}$ *is strongly bisimilar to* $RS_{TA \mid\mid TA'}$.

The first part gives the needed evidence that the use of RS_{TA} is actually enough to determine whether the language accepted by TA is empty. The second part indicates that the transitions corresponding to time progress behave just like ordinary synchronizations between the automaton, so from the semantical point of view those do not differ from ordinary transitions. The theorem also allows building the reduced state space of a system $TA_1 \mid\mid \ldots \mid\mid TA_n$ in a compositional manner by first building the reduced state spaces of the individual components TA_i and afterwards combining those with parallel operator.

With the above results, we can now conclude the following:

Corrollary 4 *Let* TA_1, TA_2 *and TA be weakly constrained timed automata.*

1. *The language accepted by TA is nonempty if and only if* RS_{TA} *contains a cycle which is reachable from the initial state, and contains at least one* 1*-transition, and a control state component* $s \in F$.
2. *If TA is built from several parallel components, like* TA_1 *and* TA_2, *then the emptiness of TA can be checked by observing the existence of such a cycle from* $RS_{TA_1} \mid\mid RS_{TA_2}$.

The existence of a 1-transition in the cycle is required in order to ensure that the corresponding timed word is progressive.

These results show how the verification of dense time properties of a weakly constrained timed automaton can be reduced to cycle detection from a finite discrete structure which is suitable for algorithmic verification. However, some further optimizations can be achieved by noticing that all the state information except the fact whether a state is accepting or not, is actually superfluous for the verification task. So, the structure that is needed greatly resembles the labeled transition systems that are used as semantical models for process algebra descriptions. Actually, our structure, the reduced state space, can be seen as a labeled transition system where the states corresponding to acceptance states are augmented with atomic propositions f.

Next we will rephrase a result concerning the truth of Linear temporal logics preserving reductions for labeled transition systems where states can be augmented with a set of atomic propositions (Kaivola, 1996). Let us call this structure *Augmented labeled transition system.*

In the theorem, the hide-operator well known in process algebra is also used. The semantics of hide is that it simply replaced all the listed transition labels with τ-labels, denoting an internal action. The theorem also refers to the CFFD (Chaos Free Failures Divergences) semantical model (Valmari and Tienari, 1995), which is a semantics inducing an equivalence relation and the related pre-order between labeled transition systems. It is a variant of the Failures Divergences model well known from the CSP community.

Theorem 5 *Let $ALTS_i = (S_i, A_i, R_i, s_{0i}, L_i)$ be alts's for $i \in \{1, 2\}$, where L_i is a mapping from S_i to the set of atomic propositions true in that state.*

1. *If $ALTS_1$ and $ALTS_2$ are CFFD-equivalent, then for every formula φ of Linear Temporal Logic without 'nexttime'-operator (LTL' in short): $ALTS_1 \models \varphi$ if and only if $ALTS_2 \models \varphi$.*
2. *If φ is a formula of LTL' that does not refer to actions in the set A', then: $ALTS_1 \models \varphi$ if and only if* (**hide** A' **in** $ALTS_1$) $\models \varphi$.
3. *Let A'_1 and A'_2 be such that $A'_1 \cap A_2 = \emptyset$ and $A'_2 \cap A_1 = \emptyset$, and furthermore let φ be such a LTL' formula that does not refer to actions in $A'_1 \cup A'_2$. Now* (**hide** $A'_1 \cup A'_2$ **in** $(ALTS_1 \parallel ALTS_2)$) $\models \varphi$ *if and only if*
   (**hide** A'_1 **in** $ALTS_1$) $\parallel$ (**hide** A'_2 **in** $ALTS_2$) $\models \varphi$.
4. *If $ALTS_1 \sqsubseteq_{\mathrm{CFFD}} ALTS_2$ then for a LTL' formula φ if $ALTS_2 \models \varphi$ then also $ALTS_1 \models \varphi$.*

This result thus covers the general case with a finite set of state propositions, but we need only one of them, f, which is true for the acceptance states only.

Since the nonexistence of an acceptance cycle can be expressed in LTL' with the negation of the formula $\Box\Diamond f \land \Box\Diamond executed(1)$, we can use the above result in order to reduce the size of the structure before the cycle detection. Since the property we are verifying refers only to atomic proposition f and to the transition label 1, it is possible to hide all the other labels. Theorem 3 allows the compositional construction of reduced state space, and because of part 3 of Theorem 5, hiding can be 'pushed' inside the components for the transitions labels that are internal to one component. Then before making the composition, thanks to part 1 of Theorem 5, these components may be minimized with respect to an equivalence relation that is stronger or equal to the CFFD, for example one suitable is the version of weak bisimulation equivalence that takes into account the divergence information. The next corollary summarizes this, as well as the fact that a CFFD-abstraction of a reduced state space can be used to guarantee the nonexistence of an acceptance cycle.

Corrollary 6 *Let TA and $TA_{\overline{\varphi}}$ be weakly constrained timed automata, where the latter corresponds to negation of the property φ, and let RA_{TA} and $RA_{TA_{\overline{\varphi}}}$ be the reduced state spaces of the above. Furthermore let RS_{min} be an augmented labeled transition system which is CFFD-equivalent with* **hide** $A_{TA} \cup A_{TA_{\overline{\varphi}}}$ **in** $RS_{TA} \parallel RS_{TA_{\overline{\varphi}}}$. *Now the following are equivalent:*

1 TA satisfies the property φ,

2 $RS_{TA} \parallel RS_{TA_{\overline{\varphi}}}$ does not have any accepting cycles containing 1 *transitions,*

3 RS_{min} does not have any accepting cycles containing 1 *transitions.*

Furthermore, let RS' be an augmented labeled transition system, for which the relation $RS_{min} \sqsubseteq_{\mathrm{CFFD}} RS'$ holds. If RS' does not have any accepting cycles containing 1 *transitions then TA satisfies the property φ.*

The last part of the corollary is an interesting consequence of part 4 of Theorem 5. It allows the possibility of replacing some components in the reduced state space with ones that are 'bigger' with respect to CFFD pre-order, and if the reduced state space with this 'bigger' version (often called *abstraction*) does not contain cycles, it is guaranteed that the concrete system does not contain a cycle either.

As the following theorem shows, the nonexistence of an acceptance cycle can be verified when the time progress transitions are hidden.

Theorem 7 *Let TA and $TA_{\overline{\varphi}}$ be weakly constrained timed automata, and $RS' =$* **hide** $A_{TA} \cup A_{TA_{\overline{\varphi}}} \cup \{1\}$ **in** $RS_{TA} \parallel RS_{TA_{\overline{\varphi}}}$. *If RS' does not contain any accepting cycles then TA satisfies the property φ.*

Thus, the cycle detection might at first be done for a reduced state space that has no visible transitions at all, and only if a cycle is found, then the time progress transition 1 is unhidden and the existence of the cycle is re-checked.

If only some of the system components actually refer to the progress of time, some further optimization can be achieved. This is formalized with the following theorem.

Theorem 8 *Let TA_1 be an arbitrary timed automaton, TA_2 a timed automaton that allows time to progress in every state, and RS_{TA_1}, RS_{TA_2} the corresponding reduced state spaces. Let us denote with RS'_{TA_2} the alts which is obtained from RS_{TA_2} by removing all the* 1 *transitions. Now $RS_{TA_1} \parallel RS_{TA_2} =_{\text{CFFD}} RS_{TA_1} \parallel RS'_{TA_2}$.*

4. FISCHER'S PROTOCOL

Fischer's mutual exclusion protocol uses only one shared variable in guaranteeing the mutual exclusion, but it requires correct timing from the processes trying to enter their critical sections.

Every client of the protocol is given a unique identifier *id* which is a natural number greater than zero. When a client wants to enter the critical section, it first waits until the value of the shared variable is zero which indicates that nobody is currently in the critical section. When detecting that, the client writes its own *id* to the shared variable. This operation has to be done within T time units from the moment when the value of the shared variable was noted as zero. After that, the client waits for T' time units, where $T' > T$. If the shared variable still contains the *id* of the client, it is allowed to enter the critical section. In other cases, the mutex-protocol is started from the beginning.

Every process as well as the memory of the system are modeled as timed automata. The automata P_i and Mem, modeling process i and the shared memory are defined in Figure 1. As usual, we assume that all the states in the automata describing the system are accepting.

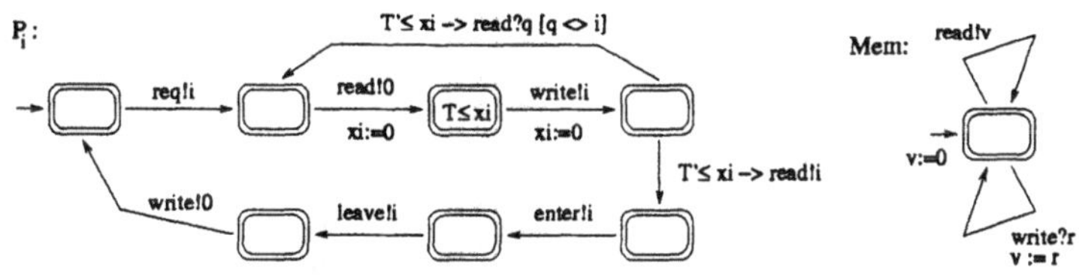

Figure 1. Fischer's protocol

A timed automaton describing the behavior of the system is obtained with the parallel composition ($P_1 \mid\mid\mid \cdots \mid\mid\mid P_n$) $\parallel$ Mem, where $\mid\mid\mid$ denotes the full interleaving operator.

The negation of the mutual exclusion property, which states that there is maximally one process at a time in the critical section, is captured by the automaton $\overline{Mutex}$ which is defined in Figure 2.

The protocol satisfies the mutual exclusion property if we can show that the intersection of the languages defined by ($P_1 \mid\mid\mid \cdots \mid\mid\mid P_n$) $\parallel$ Mem and $\overline{Mutex}$ is empty.

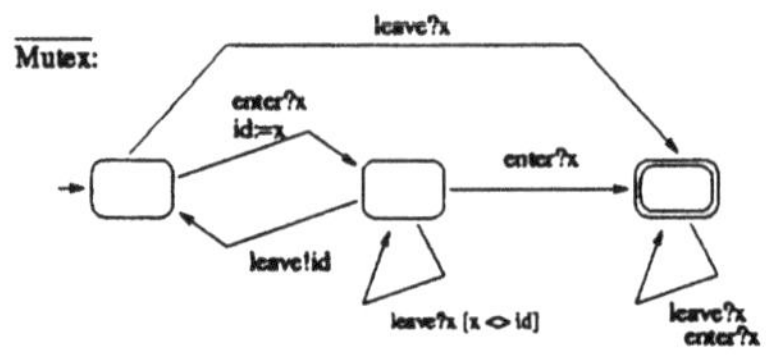

Figure 2. Automaton defining the negation of the mutual exclusion property

In order to do the verification we produced the transition systems $RS_{\overline{Mutex}}$, RS_{Mem} and RS_{P_i} for every $i \in \{1, \ldots, n\}$. The results with parameter values $T = 1, T' = 2$ and $n = 2$ are shown in Figure 3.

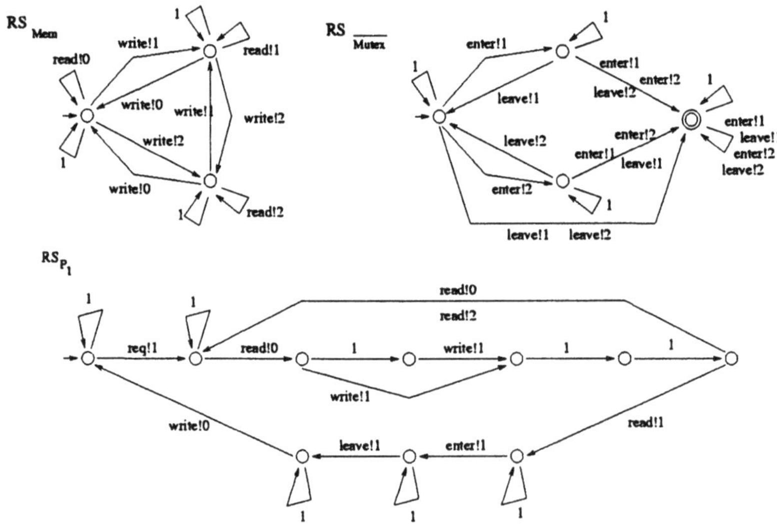

Figure 3. Reduced state spaces for Fischer's protocol

Then the next step was building the transition systems

$System := red(red(\textbf{hide } read, write, req, 1 \textbf{ in } red(RS_{P_1}) \mid\mid\mid \ldots \mid\mid\mid red(RS_{P_n}))$
$\qquad \mid\mid red(\textbf{hide } 1 \textbf{ in } RS_{Mem}))$

$Test := red(\textbf{hide } enter, leave \textbf{ in } System \mid\mid red(\textbf{hide } 1 \textbf{ in } RS_{\overline{Mutex}}))$,

where *red* denotes a reduction procedure that preserves CFFD-equivalence. Transition systems *System* and *Test* with the example parameters are shown in Figure 4.

Now because of Corollary 6 and Theorem 8 it follows that if *Test* does not have any accepting cycles then the intersection of the languages defined by $(P_1 \mid\mid\mid \ldots \mid\mid\mid P_n) \mid\mid Mem$ and $\overline{Mutex}$ is empty, and thus the protocol satisfies the mutual exclusion property. Since all the automata

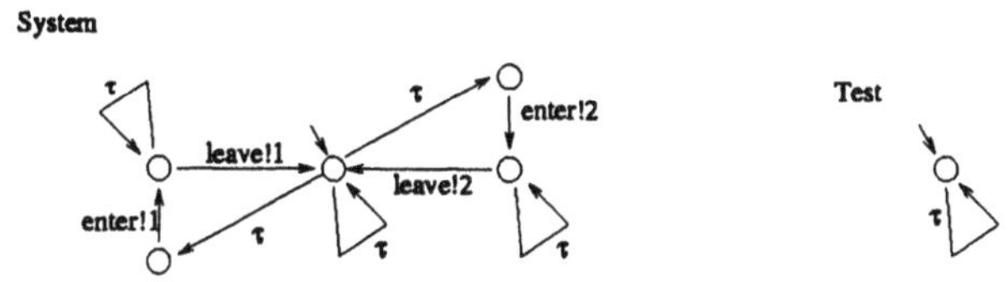

Figure 4. Transition systems *System* and *Test* in the example case

are weakly constrained, the verification results hold for the dense time domain.

We verified the system with several parameter value combinations. The ARA (Advanced Reachability Analysis) toolset (Valmari et al., 1993) was used for the verification. The sizes of the largest transition systems that were encountered during the verification are shown in Table 1. For the compositional style of building the state space it is often the case that not the final result, but some intermediate step is the one that produces the largest transition system, thus the one that restricts the size of the systems that the tool can handle. So, for the compositional method, the largest intermediate result is shown in the table. Besides the number of client processes, the time parameter values have a great impaction the size of the verification model. This fact is also reflected in Table 1.

n	T	T'	compositional	noncompositional
3	0	1	294	415
3	1	2	496	777
3	4	5	1882	3351
4	0	1	1417	1899
4	1	2	3013	4559
4	4	5	19340	33399
5	0	1	6901	7205
5	1	2	18838	27769
5	4	5	200011	-
6	0	1	34124	43854
6	1	2	120254	-
7	0	1	172115	-
8	0	1	156564	-

Table 1. Sizes of the reduced state spaces with the various parameter combinations. Note that the 8 client case was done with an optimized version of the protocol clients, where the *enter*, *req* and *leave* transitions were not used, but instead just the information whether the client is in its critical section or not.

Let us now demonstrate what happens if the property to be verified does not hold. Consider the protocol in case of two clients and parameter values $T = 1$, and $T' = 2$ for P_1 but $T = T' = 1$ for P_2. The transition systems *System* and *Test* are shown in Figure 5.

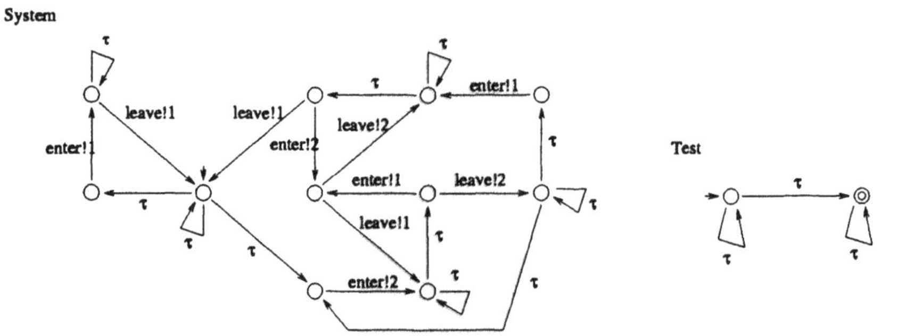

Figure 5. *System* and *Test* in the case of an erroneous parameter value in the second entity

As is seen, the *Test* now contains an accepting cycle, so the property might not hold. The next step is to built *System'* and *Test'* which are just as the above two, except the transition label 1 is not hidden. The result *Test'*, which is shown in Figure 6, indeed contains an accepting cycle, with a 1-transition, so Corollary 6 now says that the this version of the protocol does not guarantee the mutual exclusion.

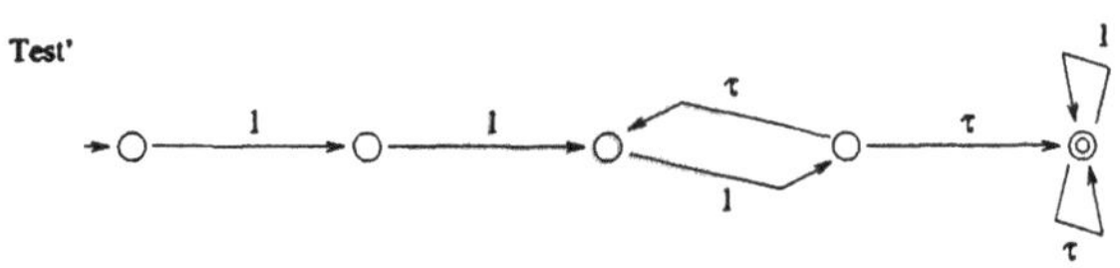

Figure 6. *Test'* showing the existence of an accepting cycle with progress of time

It is impossible to see the reason for the error from *Test'*, and the only way to produce a useful counter-example is trying to preserve more transition information in the transition system from which the cycle is to be detected.

In comparison with other reported studies on Fischer's protocol, our compositional method performs quite well. The Discrete Time SPIN (Bošnački and Damms, 1998) was able to do the verification up to 6 client processes, and the Real Time Spin (Tripakis and Courcoubetis, 1996) with 4 clients and UPPAAL with 8 clients. The tool Kronos has been reported to manage with 5 clients (Daws et al., 1996), and UPPAAL with 8 (Larsen et al., 1995). With the convex-hull over approximation technique (Daws and Tripakis, 1998) both UPPAAL and Kronos can handle the protocol with 9 clients. In the above case studies, the timing parameters T and T' used were not reported, but since those methods, except Discrete Time SPIN, rely on region or zone techniques, it is presumable that the results are not as sensitive to the values of time parameters as ours.

5. BOUNDED RETRANSMISSION PROTOCOL

The bounded retransmission protocol, or BRP in short (D'Argenio et al., 1997), is used to implement a reliable file transfer service using an unreliable point-to-point data link, such as a telephone line. The underlying data link offers unreliable delivery of small, fixed size data *packets*. The size of the packets is far smaller than a typical file which has to be transferred. Thus, the file has to be fragmented into *frames*, which are smaller than a data packet of the link.

The formal specification of the protocol consists of four timed automata: *Sender*, *Receiver*, Ch_{SR}, Ch_{RS} where the latter two describe the behavior of the underlying channel. The description of the entire system is parallel composition of the automata.

Due to the restricted length of the paper, only the verification result is reported here, and we suggest that reader see the full version of the paper (M. Luukkainen, 2001) to get a more detailed view of the protocol and how the verification was done.

The parameter describing the length of the file in the *Sender* automaton turned out to be problematic from the verification point of view. In fact, there has to be an upper limit in the length of the transmitted file in order to make *Sender* finite stated. And even if a limit is set, it has a great impact on the size of the generated transition systems. Therefore, we used an abstraction *Sender'* of the sender automaton. In *Sender'* the actual length of the transferred file is not remembered, but instead it is 'guessed' while sending the file.

Let us denote the sender with k as the maximum length of files as $Sender_k$. Since for all values of k the reduced state space RS_{Sender_k} of the actual sender automaton is CFFD refinement of the transition system $RS_{Sender'}$ of the abstraction, i.e. $RS_{Sender_k} \sqsubseteq_{\text{CFFD}} RS_{Sender'}$, the Corollary 6 guarantees, that if there are no acceptance cycles in the system with the abstract sender, the property holds for the actual version of the system. So, with the use of an abstraction of the sender, we now gain two things. Firstly there is a substantial saving in the size of the generated transition systems, since already the size of $Sender_3$ is greater than that of the abstraction. Even if the effect of increasing k appears to be linear, see the Table 2, it is advantageous to eliminate this source of state explosion. The second, even more impressive advantage is that the verification generalizes to all the possible file lengths.

	2	4	6	8	10	Abstraction
Sender	51	107	163	219	275	65
Model	1156	2393	3628	4864	6100	1433

Table 2. Effect of the maximum file length to the size of sender and to the verification model in case the channel delay is 1 and number of retransmissions is 1

The correctness requirements of the protocol are twofold. The automaton defining *the correspondence of indications given to clients of the protocol* was quite straightforward to construct. The verification itself was then done in the same way as in the case of Fischer's protocol.

From the verification point of view, the other requirement *the frames are received in correct order, and no frames are dropped if an error is not indicated* was more problematic. To verify this property, the theory of *data independence* (Wolper, 1986) was exploited. The data independence guarantees that, if a transfer protocol does not alter or use in its logic the user data, it is enough to consider just a small number of different types of user data frames, even if the domain of the frames is extremely large. So, with the help of this we formulated the corresponding property automaton and carried out the verification in the same way as was reported for Fischer's protocol.

The verification was done with several choices of values for the parameters D_{min}, D_{max} and MAX where the first two define the transmission delay of a frame, and the last describes the maximum number of retransmission trials for a frame. Table 3 summarizes the sizes of the largest reduced states spaces during the verification.

D_{min}	D_{max}	MAX	verifying indications	verifying data transfer
0	1	1	1433	1993
0	1	3	5415	7411
1	2	1	2108	4499
1	2	3	8915	11693
2	3	1	2958	3748
2	3	3	12967	16378

Table 3. Sizes of the largest encountered reduced state spaces with the various parameter combinations.

In D'Argenio et al., 1997 verification of the protocol with UPPAAL was reported. The main difference compared to our results is that in the study of D'Argenio et al., 1997 the verification was done only for files of length 1, 2 and 3 frames, and the correctness of data transfer was not verified. In the UPPAAL study the channel delay was fixed to one, and as in our case, the maximum number of retransmissions between 1 and 3 were considered.

In order to compare the efficiency of our method versus UPPAAL, we ran our verification with a Sender process where the maximum length of transmitted files was fixed to 3. With the number of retransmissions 1, 2 and 3 the running times with our method were 2, 4 and 13 seconds and with UPPAAL the corresponding measures were 61, 156 and 340 seconds. Our method also outperforms the Discrete Time SPIN verification reported in Bošnački and Damms, 1998.

6. CONCLUSIONS

In the article we showed how the untimed process algebra theory and tools can be used in verifying a category of dense time properties from a restricted class of timed automata. The work was based on the known result that by restricting ourselves to a class of timed automata where only the operators $\leq, =$, and $\geq$ are used in the clock comparisons, most of the interesting dense time properties are verifiable using discrete time methods. Those results were refined in order to fit the process algebraic context, and furthermore, some reduction strategies based on behavior equivalences, abstraction and compositionality were indicated. It is in particular the possibility for the compositional generation of state spaces that makes the process algebraic method efficient for the timed systems, as well.

The experimental results obtained from the verification of Fischer's mutual exclusion protocol and Bounded retransmission protocol indicated that the efficiency of our method is of the same magnitude as for the other available timed methods. Especially in the BRP case study our method clearly worked faster than UPPAAL in the study reported in D'Argenio et al., 1997. An interesting point in our BRP study was the possibility of using an abstract version of the sender process that allowed us to conduct the verification for arbitrary long files. Thus, with our method, we were able to obtain a more complete verification result than the UPPAAL, since there the maximum length of files had a fixed upper bound of just 3 frames.

REFERENCES

Alur, R. and Dill, D. (1994). A Theory of Timed Automata. *Theoretical Computer Science*, 126:183–235.

Bošnački, D. (1999). Digitization of Timed Automata. In *Fourth International Workshop on Formal Methods for Industrial Critical Systems*.

Bošnački, D. and Damms, D. (1998). Integrating Real-Time in Spin: a Prototype Implementation. In *Proceedings of IFIP Joint Conference on Formal Description Techniques and Protocol Specification, Testing and Verification*, pages 423–439. Kluwer Academic Publishers.

D'Argenio, P., Katoen, J.-P., Ruys, T., and Tretmans, J. (1997). The Bounded Retransmission Protocol Must Be in Time. In *Third International Workshop of Tools and Algorithms for the Construction and Analysis of Systems*, number 1217 in Lecture Notes in Computer Science, pages 416–431. Springer-Verlag.

Daws, C., Olivero, A., Tripakis, S., and Yovine, S. (1996). The Tool KRONOS. In *Hybrid Systems III*, number 1066 in Lecture Notes in Computer Science, pages 208–219. Springer-Verlag.

Daws, C. and Tripakis, S. (1998). Model Checking of Timed Reachability Properties Using Abstraction. In *Fourth International Workshop of Tools and Algorithms for the Construction and Analysis of Systems*, Lecture Notes in Computer Science. Springer-Verlag.

Henzinger, T., Manna, Z., and Pnueli, A. (1992). What Good are Digital Clocks? In *Proceedings of the 20th International Conference on Automata, Languages and Programming*, number 623 in Lecture Notes in Computer Science, pages 545–558. Springer-Verlag.

Kaivola, R. (1996). *Equivalences, Preorders and Compositional Verification for Linear Time Temporal Logic and Concurrent Systems*. Report A-1996-1. PhD Thesis, University of Helsinki, Department of Computer Science.

Larsen, K. G., Pettersson, P., and Yi, W. (1995). Model-Checking for Real-Time Systems. In *Proc. of Fundamentals of Computation Theory*, number 965 in Lecture Notes in Computer Science, pages 62–88.

M. Luukkainen (2001). Verification of dense time properties using theories of untimed process algebra. www.cs.helsinki.fi/u/mluukkai/full.ps.

Tripakis, S. and Courcoubetis, C. (1996). Extending Promela and Spin with Real Time. In *Second International Workshop of Tools and Algorithms for the Construction and Analysis of Systems*, Lecture Notes in Computer Science. Springer-Verlag.

Valmari, A., Kemppainen, J., Clegg, M., and Levanto, M. (1993). Putting Advanced Reachability Analysis Techniques Together: the "ARA" Tool. In *Proceedings of Formal Methods Europe*, number 670 in Lecture Notes in Computer Science, pages 597–616. Springer-Verlag.

Valmari, A. and Tienari, M. (1995). Compositional Failure-Based Semantic Models for Basic LOTOS. *Formal Aspects of Computing*, 7:440–468.

Wolper, P. (1986). Expressing Interesting Properties of Programs in Propositional Temporal Logic. In *Proceedings of the 13th ACM Symposium on Principles of Programming Languages*, pages 184–193.

TESTING LIVENESS PROPERTIES

Approximating Liveness Properties by Safety Properties

Ulrich Ultes-Nitsche and Simon St James
University of Southampton
Department of Electronics and Computer Science
Southampton, SO17 1BJ, United Kingdom
{uun,spstj99r}@ecs.soton.ac.uk

Abstract Satisfaction within fairness establishes an abstract notion of truth under fairness in linear-time verification. It can be argued that the idea of an implementation meeting the intersection of the system behaviour and the requirement specification forms a desirable basis for testing systems. Unfortunately this intersection of the behaviours of the system and the requirements specification is not accessible directly and cannot be computed practically. It is, however, possible to derive some information of this intersection from a high-level (abstract) specifation of the system. We will present in this paper an abstraction-based approach that will enable us to derive some test cases, even for liveness properties.

Keywords: Test case generation, satisfaction within fairness, liveness properties, behaviour abstraction.

1. INTRODUCTION

In [12], the concept of satisfaction within fairness is presented and adopted to testing systems. The basic idea is to test a system not against requirement specifications (properties) on their own but against the intersection of the requirments specification with the behaviour of the system. A similar observation was made by Petrenko [10] at the same conference (the observation was made in his presentation). Such an approach is more suited towards verification than towards testing. The reason being the accessability of the above mentioned intersection. We simply have no direct access to an exhaustive system behaviour description of the implemented system. However, we normally will have a high-level system specification on which the implementation is based upon.

We will give in this paper a definition of a safety property that approximates a liveness property. Using this safety approximation will enable us to test for violation of liveness properties in principle. The established definition will be similar to the notion of *satisfaction within fairness* in verification [6; 7; 12]. In order to overcome practical limitation of the introduced concepts, we will adopt an abstraction result for satisfaction within fairness that will make the results of this paper more applicable.

2. LINEAR-TIME VERIFICATION

We review linear satisfaction of properties very briefly in this section. Let $\mathcal{A}$ be the finite set of actions that a system can perform. By $\mathcal{A}^\omega$ we denote the set of all infinitely long sequences of actions in $\mathcal{A}$. A computation c of the system is an element of $\mathcal{A}^\omega$, i.e. $c = a_1a_2a_3\ldots$ such that all a_i are in $\mathcal{A}$. The behaviour B of the system is then a subset of $\mathcal{A}^\omega$.[1] A property P, from an abstract point of view, is itself a subset of $\mathcal{A}^\omega$, representing all correct behaviours with respect to the intuitive meaning of the property [1]. Hence the system, represented by its behaviour B, satisfies property P (written: $B \models P$) if and only if all computations in B are also in P. This *linear-time* satisfaction of properties is defined by

$$B \models P \iff B \subseteq P.$$

3. EXHAUSTIVE TESTING

In contrast to the satisfaction relation for linear-time temporal verification, when testing a system with respect to a property P (P, in this context, can be called a requirement specification), we consider finite behaviours: We can only observe a system for a finite amount of time, during which we only see a prefix of its (potentially) infinite behaviour. Let us define formally the relation that describes testing a system exhaustively:

Let $\mathcal{A}^*$ denote the set of all finitely long sequences of actions in $\mathcal{A}$. For a subset S of $\mathcal{A}^\omega$ let $pre(S)$ denote the set $\{w \in \mathcal{A}^* \mid \exists x \in \mathcal{A}^\omega : wx \in S\}$ of finite prefixes of behaviours in S. For an element s of $\mathcal{A}^\omega$, let $pre(s) = pre(\{s\})$. B satisfies P with respect to exhaustive testing, written $B \vdash P$ in this paper, if and only if

$$pre(B) \subseteq pre(P).$$

Typically, properties are classified into two groups, which reflect either absence of errors or satisfactory progress. These two property classes are called *safety* and *liveness* properties respectively [1]:

P is a liveness property if and only if

$$\forall w \in \mathcal{A}^* : \exists x \in \mathcal{A}^\omega : wx \models P.$$

[1] In terms of formal language theory, c is an ω-word and B is an ω-language on the alphabet $\mathcal{A}$.

P is a safety property if and only if

$$\forall x \in \mathcal{A}^\omega : (x \not\models P \Rightarrow \exists w \in pre(x) : \forall z \in \mathcal{A}^\omega : wz \not\models P).$$

Note that any property is the intersection of a safety and a liveness property [1]. Also note that P is a liveness property if and only if $pre(P) = \mathcal{A}^*$ (recall that P is a subset of $\mathcal{A}^\omega$) [2]. Therefore $pre(B)$ is always a subset of $pre(P)$ for a liveness property P. In other words: *We cannot test liveness properties,* because they are always trivially satisfied (wrt. exhaustive testing).

In the case of safety properties, however, exhaustive testing is as precise as linear-time verification:

Lemma 1 *P is a safety property if and only if $B \vdash P \iff B \models P$, for all B.*

Proof *"if:" We assume $B \vdash P \Rightarrow B \models P$ and take the counterpositive: $B \not\models P \Rightarrow B \not\vdash P$, for all B. Let $x \in \mathcal{A}^\omega$ such that $x \not\models P$. Then, by the above assumption, $x \not\vdash P$. Let $w \in pre(x)$ such that $w \notin pre(P)$. Then for all $z \in \mathcal{A}^\omega$, $wz \not\models P$. As x was chosen arbitrarily, P is a safety property.*

"only if:" Let P be a safety property. Since $B \models P$ always implies $B \vdash P$, we only have to show $B \vdash P \Rightarrow B \models P$. We prove the counterpositive version: $B \not\models P \Rightarrow B \not\vdash P$. Assume $B \not\models P$. Then there must be an $x \in B$ such that $x \not\models P$. Because P is a safety property, there must be a prefix w of x such that for all $z \in \mathcal{A}^\omega$, $wz \not\models P$. Therefore $w \notin pre(P)$. Because $w \in pre(B)$, we have $B \not\vdash P$, completing the proof. □

4. APPROXIMATING LIVENESS PROPERTIES BY SAFETY PROPERTIES

We are going to define to a behaviour B and a property P a safety property S_{BP}. Testing S_{BP} exhaustively will establish an approximation to the linear-time satisfaction of P by B. Note that P can be any property, e.g. a liveness property.

First, we need a dual operation to constructing prefix sets: the Eilenberg-limit. Let L be a subset of $\mathcal{A}^*$. The Eilenberg-limit $lim(L)$ [3; 11] is defined by

$$lim(L) = \{x \in \mathcal{A}^\omega \mid |pre(x) \cap L| = \infty\}.$$

We define the safety property to B and P by

$$S_{BP} = lim(pre(B \cap P)).$$

First we prove that S_{BP} is indeed a safety property:

Lemma 2 *S_{BP} is a safety property.*

Proof *Let $x \in \mathcal{A}^\omega$. If $x \not\models S_{BP}$, then there must be a prefix $w \in pre(x)$ that is not in $pre(B \cap P)$ (otherwise $x \models S_{BP}$). Then, for all $z \in \mathcal{A}^\omega$,*

$wz \notin \lim(pre(B \cap P))$ and thus $wz \not\models S_{BP}$. Hence S_{BP} is a safety property.
□

Note that $pre(lim(pre(B \cap P))) = pre(B \cap P)$. Hence

$$B \vdash S_{BP} \iff pre(B) \subseteq pre(B \cap P).$$

This relation also occurs in the context of linear-time properties with an abstraction notion of fairness [7; 12]. We use the following result to interpret the definition of S_{BP} with respect to exhaustive testing:

Lemma 3 *Let $B \subseteq \mathcal{A}^\omega$ and $P \subseteq \mathcal{A}^\omega$ be the behaviour of a system and the property it should meet respectively. Then $B \vdash S_{BP}$ if and only if $B \cap P$ is a dense set in B [7; 12].*

We use in this lemma the following definition of a metric space (known as the Cantor topology) (for topological notions see [4]):

Let $common(a, b)$ designate the longest common prefix of two computations a and b in $\mathcal{A}^\omega$. We define the metric $d(a, b)$ by

$$\forall a, b \in \mathcal{A}^\omega, a \neq b : d(a, b) = \frac{1}{|common(a, b)| + 1}$$

$$\forall b \in \mathcal{A}^\omega : d(b, b) = 0.$$

How can we interpret the obtained results? Lemma 3 tells us that testing safety property S_{BP} exhaustively is equivalent to checking whether $B \cap P$ is a dense set in B. However, in topological terms, $B \cap P$ being a dense set in B is an approximation of $B \subseteq B \cap P$ and therefore an approximation of linear time satisfaction $B \models P$: A system whose behaviour satisfies that $B \cap P$ is a dense set in B, but not that $B \models P$, must be observed for an infinite amount of time in order to detect $B \not\models P$ [12]. This distinction is therefore practically irrelevant, since systems cannot be observed for an infinite amount of time. Because S_{BP} is a safety property, we can indeed test it. As P can even be a liveness property, we have found a way to check the (approximate) satisfaction of a (liveness) property by testing a related safety property exhaustively.

Lemma 3, establishing a very interesting theoretical result about the testability of liveness properties, is unfortunately not too practical without additional effort: As we need access to $B \cap P$ in the definition of S_{BP}, we must be able to construct the intersection of the implementation of a system (B) and the property (P). But the implementation will in practice not be accessible in such a way that this intersection could be constructed.

Having mentioned above that the relation $pre(B) \subseteq pre(B \cap P)$ is a satisfaction relation in verification under fairness constraint (called *satisfaction within fairness*), we can use an abstraction result for this relation that will allow us to use the results of this section at least partly (in general we will lose some information, but not all).

5. SATISFACTION WITHIN FAIRNESS AND BEHAVIOUR ABSTRACTION

In this section we use *behaviour abstraction* [6; 7]. Before we define it this, we need a few preliminary results and definitions. First of all we will write $B \Vdash P$ for saying that B satisfies P within fairness:

$$B \Vdash P \iff pre(B) \subseteq pre(B \cap P).$$

Let $\mathcal{A}^\infty = \mathcal{A}^* \cup \mathcal{A}^\omega$. The leftquotient [3] of $M \subseteq \mathcal{A}^\infty$ by $w \in \mathcal{A}^*$ is then the set

$$cont(w, M) = \{v \in \mathcal{A}^\infty \mid wv \in M\}.$$

Lemma 4 *$lim(L) \Vdash P$ if and only if*

$$\forall w \in pre(lim(L)) : \exists x \in cont(w, lim(L)) : wx \in P \ [7].$$

We can now give a definition of behaviour abstraction and explore its relation to satisfaction within fairness. It is defined by alphabetic language homomorphisms extended to ω-languages [5]:

Definition 5 *$h : \mathcal{A}^\infty \to \mathcal{A}'^\infty$ is an abstraction homomorphism if and only if the following conditions hold:*

- $h(\mathcal{A}) \subseteq \mathcal{A}' \cup \{\varepsilon\}$
  (action renaming and hiding)
- $\forall v, w \in \mathcal{A}^*, x \in \mathcal{A}^\omega : h(vw) = h(v)h(w)$ *and* $h(vx) = h(v)h(x)$
  (compatibility with concatenation)
- $h(\mathcal{A}^\omega) \subseteq \mathcal{A}'^\omega$
  (undefined for ω-words that would be taken to finite words).

Note that abstraction homomorphisms are partial mappings since they are not defined on ω-words that would be taken to finitely long words. Also, note that they do not increase the length of words; that is, we always have that $|h(w)| \leq |w|$. For an abstraction homomorphism h, we call $lim(h(L))$ the abstraction of concrete behaviour $lim(L)$. For prefix-closed regular L, $lim(h(L)) = h(lim(L))$ [7]. We will consider subsequently only behaviours that are the Eilenberg-limit of a prefix-closed regular language L (i.e. $B = lim(L)$, $L = pre(L)$, and L is regular). That means we are considering behaviours that can be represented by finite-state labelled transition systems.

In general, a property satisfied within fairness on the abstraction is not necessarily satisfied within fairness on the concrete behaviour in a corresponding way. We need an additional requirement to be satisfied by the abstraction homomorphism [8; 9]:

Essential information about properties that are satisfied within fairness by $lim(L)$ is contained in the sets $cont(w, L)$, for $w \in L$. At the abstract level, we obviously have access to $cont(h(w), h(L))$, but we really need $h(cont(w, L))$ in order to ensure that properties satisfied within fairness by the abstraction will also be satisfied within fairness by the concrete system in a corresponding way. Thus, we need to find conditions under which $cont(h(w), h(L))$ can be used instead of $h(cont(w, L))$.

In general, $h(cont(w, L))$ is a proper subset of the set $cont(h(w), h(L))$. In order to obtain sufficient information about $h(cont(w, L))$ from considering $cont(h(w), h(L))$, one would be tempted to require equality of the two sets. Those homomorphisms are *continuation closed*, since computation of the continuation (leftquotient) and the abstraction commute. However, this is stronger than needed. Indeed, since we are dealing with satisfaction within fairness, it is sufficient that the computations in $cont(h(w), h(L))$ "eventually" become computations in $h(cont(w, L))$:

Definition 6 *h is weakly continuation-closed on L if and only if, for all* $w \in \mathcal{A}^*$, *there exists* $v \in cont(h(w), h(L))$ *such that* $cont(v, cont(h(w), h(L))) = cont(v, h(cont(w, L)))$.

One should not worry about this rather technical definition. Its essence of this definition revealed by the following lemma:

Lemma 7 *Let L be regular. If* $h(L)$ *does not contain maximal words,*[2] *then the condition*

$$lim(h(L)) \Vdash P \quad \textit{if and only if} \quad lim(L) \Vdash h^{-1}(P)$$

holds if and only if h is weakly continuation-closed on L [6].

Weakly continuation-closed homomorphisms are therefore precisely the class of homomorphisms that preserve satisfaction within fairness.

What is the essence of this result? To test whether the concrete behaviour $lim(L)$ (i.e. the implementation's behaviour) satisfies property $h^{-1}(P)$[3] with the help of a safety approximation $S_{lim(L)h^{-1}(P)}$ — a test that is practically impossible directly — we can test instead whether the abstract behaviour $lim(h(L))$ satisfies abstract property P with the help of $S_{lim(h(L))P}$ which can be constructed effectively. Since we can turn an observed (finite) behaviour of the system under test on-the-fly into its corresponding abstract behaviour by application of abstraction homomorphism h, we can then immediately test whether

[2] $w \in L$ is a maximal word in L if and only if $cont(w, L) = \{\varepsilon\}$. This restriction can be removed easily by introducing additional "dummy" actions [6]. However, we avoid introducing additional notation here.
[3] Note that $h^{-1}(P)$ can even be a liveness property.

the obtained trace is in $pre(S_{lim(h(L))P})$. If we tested the on-the-fly abstraction of the system traces exhaustively against $S_{lim(h(L))P}$ would immediately establish $lim(L) \Vdash h^{-1}(P)$. We therefore have established an indirect way, with the help of abstraction, for testing even liveness properties of a system.

6. CONCLUSIONS

We have presented how liveness properties can be approximated by safety properties. The prefixes of these safety properties form then test cases that allow to check whether a system nearly meets some lieveness properties. "Nearly" means in this context that if the liveness property were not satisfied (wrt. linear-time verification), then it would only be detectable by observing the system for an infinite amount of time and would therefore be practically irrelevant.

Since constructing the safety property mentioned above requires construction of the intersection of the system implementation and the original property, the described approach is not yet feasible. However an abstraction result as presented can be used that allows to gain at least some benefit from the defined safety approximation. The abstraction result in the previous section requires the concrete behaviour (i.e. the implementation) to be representable by a finite-state labelled transition system and the abstraction to be weakly continuation-closed. These are two conditions that will not necessarily be satisfied in practice, or we may simply not know whether they are satisfied when testing the system. However, if the system is an implementation derived from some high-level specification where the high-level specification has a finite-state labelled transition system semantics, we can simply assume that the high-level specification is a weakly-continuation closed abstraction of the implementation and that the implementation itself also has a finite-state labelled transition system semantics. If the assumption is false, we will have lost information and our testing will inaccurate, even though it may seem to be exhaustive. We will still be able to spot some problems with respect to meeting progress properties that will not be testable without the presented approach.

The claim made is oviously quite vague, even though it is underpinned by interesting theoretical results, and needs solid experimental justification. Experiments using the concepts introduced in this paper will be part of our future work. We will therefore aspire co-operation with researchers who work on the practical application of system testing. Any comments and suggestions are welcome.

REFERENCES

[1] B. Alpern and F. B. Schneider. Defining liveness. *Information Processing Letters*, 21(4):181–185, October 1985.

[2] B. Alpern and F. B. Schneider. Recognizing safety and liveness. Technical Report TR 86-727, Department of Computer Science, Cornell University, January 1986.

[3] S. Eilenberg. *Automata, Languages and Machines*, volume A. Academic Press, New York, 1974.

[4] J. L. Kelley. *General Topology*. Van Nostrand, Princeton, 1955.

[5] U. Nitsche. Propositional linear temporal logic and language homomorphisms. In A. Nerode and Y. V. Matiyasevich, editors, *Proceedings of the 3rd International Symposium on Logical Foundations of Computer Science (LFCS'94)*, volume 813 of *Lecture Notes in Computer Science*, pages 265–277, Saint Petersburg, Russia, 1994. Springer Verlag.

[6] U. Nitsche and P. Ochsenschläger. Approximately satisfied properties of systems and simple language homomorphisms. *Information Processing Letters*, 60:201–206, 1996.

[7] U. Nitsche and P. Wolper. Relative liveness and behavior abstraction (extended abstract). In *Proceedings of the 16th ACM Symposium on Principles of Distributed Computing (PODC'97)*, pages 45–52, Santa Barbara, CA, 1997.

[8] P. Ochsenschläger. Verifikation kooperierender Systeme mittels schlichter Homomorphismen. Arbeitspapiere der GMD 688, Gesellschaft für Mathematik und Datenverarbeitung (GMD), Darmstadt, Oktober 1992.

[9] P. Ochsenschläger. Verification of cooperating systems by simple homomorphisms using the product net machine. In J. Desel, A. Oberweis, and W. Reisig, editors, *Workshop: Algorithmen und Werkzeuge für Petrinetze*, pages 48–53. Humboldt Universität Berlin, 1994.

[10] A. Petrenko and N. Yevtushenko. On test derivation from partial specifications. In *Proceedings of the IFIP TC6/WG6.1 International Conference on Formal Description Techniques for Distributed Systems and Communication Protocols and Protocol Specification, Testing and Verification (FORTE/PSTV'2000)*, pages 85–102, Pisa, Italy, October 2000. Springer Verlag.

[11] W. Thomas. Automata on infinite objects. In J. van Leeuwen, editor, *Formal Models and Semantics*, volume B of *Handbook of Theoretical Computer Science*, pages 133–191. Elsevier, 1990.

[12] U. Ultes-Nitsche. Satisfaction up to liveness — automatic verification based on exhaustive testing. In *Proceedings of the IFIP TC6/WG6.1 International Conference on Formal Description Techniques for Distributed Systems and Communication Protocols and Protocol Specification, Testing and Verification (FORTE/PSTV'2000)*, pages 237–248, Pisa, Italy, October 2000. Springer Verlag.

SVL: A SCRIPTING LANGUAGE FOR COMPOSITIONAL VERIFICATION

Hubert Garavel, Frédéric Lang*
INRIA Rhône-Alpes, 655, avenue de l'Europe, F-38330 Montbonnot St. Martin, France
Hubert.Garavel@inria.fr, Frederic.Lang@inria.fr

Abstract Compositional verification is a way to avoid state explosion for the enumerative verification of complex concurrent systems. Process algebras such as LOTOS are suitable for compositional verification, because of their appropriate parallel composition operators and concurrency semantics. Extending prior work by Krimm and Mounier, this article presents the SVL language, which allows compositional verification of LOTOS descriptions to be performed simply and efficiently. A compiler for SVL has been implemented using an original compiler-generation technique based on the Enhanced LOTOS language. This compiler supports several formats and tools for handling Labeled Transition Systems. It is available as a component of the CADP toolbox and has been applied on various case-studies profitably.

Keywords: Abstraction, Bisimulation, Compositional Verification, Concurrency, Coordination Language, E-LOTOS, Enumerative Verification, Labeled Transition System, LOTOS, Model-Checking, Process Algebra, Reachability Analysis, Specification, Validation.

1. INTRODUCTION

Enumerative verification (also called *reachability analysis* or *model-checking*) is a popular technique for verifying concurrent systems. Roughly speaking, it is a "brute force" technique, which consists in exploring and checking all states and transitions reachable by a concurrent system. This technique is confronted to the *state explosion* problem, which occurs when the number of states grows exponentially as the number of concurrent processes in the system increases. To avoid or reduce state explosion, various approaches have been proposed, among which:

*This work has been partially supported by DYADE, the BULL/INRIA joint venture for advanced reasearch symbolic verification, on-the-fly verification, partial orders, symmetries, data flow analysis, and compositional verification.

This article deals with the latter approach (also known as *compositional reachability analysis* or *compositional minimization*). This approach assumes that the concurrent system under study can be expressed as a collection of communicating sequential processes, the behaviors of which are modeled as finite state machines or labeled transition systems (LTSs, for short). The sequential processes are composed in parallel, either in a flat or hierarchical manner.

In its simplest form [9, 23, 27, 32, 28, 29, 30], compositional verification consists in replacing each sequential process by an *abstraction*, simpler than the original process but still preserving the properties to be verified on the whole system. Quite often, abstracting a process is done by minimizing its corresponding LTS modulo an appropriate equivalence or preorder relation (e.g., a bisimulation relation, such as strong, branching or observational equivalence). If the system has a hierarchical structure, minimization can also be applied at every intermediate level in the hierarchy. Clearly, this approach is only possible if the parallel composition is "compatible" with LTS minimization: in particular, this is the case with the parallel composition operators of most process algebras, for which bisimulation is a congruence (see [32] for a discussion on this issue).

Although this simple form of compositional verification has been applied successfully to some complex systems (e.g., [11, 4] in the case of the LOTOS language [20]), it may be counter-productive in some other cases: generating the LTS of each process separately may lead to state explosion, whereas the generation of the whole system of concurrent processes might succeed if processes constrain each other when composed in parallel.

This issue has been addressed by refined compositional verification approaches [15, 6, 31, 7, 8, 18, 22, 5, 14], which allow to generate the LTS of each separate process by taking into account *interface constraints* (also known as *environment constraints* or *context constraints*). These constraints express the behavioral restrictions imposed on each process by synchronization with its neighbor processes. Taking into account the environment of each process allows to eliminate states and transitions that are not reachable in the LTS of the whole system. Depending on the approach, interface constraints can be either written by the user or generated automatically.

The refined approach to compositional verification has been implemented in two tools, namely the TRACTA tool [14] and the PROJECTOR/DES2AUT tools [22]. The latter tools are part of the CADP protocol engineering toolbox [10] and have been applied to industrial case-studies [22, 24, 19]. Although positive, these experiments revealed various issues and shortcomings that prevented compositional verification to be used on a larger scale, especially in industrial projects. To solve these problems, we designed a scripting language named SVL, which can be seen as a process algebra extended with operations on LTSs, e.g., minimization (also called reduction), abstraction, comparison, deadlock/livelock detection, etc. We implemented a compiler for this language, with the goal of making compositional verification easier for non-experts.

This article is organized as follows. Section 2 gives a few preliminary definitions. Section 3 briefly presents the principles and limitations of the DES2AUT tool of [22]. Section 4 defines the syntax and semantics of the SVL language. Section 5 introduces high-level features of SVL, which allow sophisticated strategies for compositional verification. Section 6 describes the implementation of the SVL 2.0 compiler. Finally, Section 7 gives concluding remarks and lists directions for future work.

2. DEFINITIONS

Labelled Transition Systems are the natural model for action-based specification languages, especially process algebras such as LOTOS [20]. Formally, an LTS is a tuple $M = (S, A, T, s_0)$, where S is the set of *states*, A the set of *actions* (or *labels*), $T \subseteq S \times A \times S$ the *transition relation*, and $s_0 \in S$ the *initial state*. A transition $(s, a, s') \in T$ indicates that the system can evolve from state s to state s' by performing action a. In enumerative verification, there are essentially two ways to represent an LTS:

- An *explicit* LTS is defined in extension, by enumerating all its states and transitions. Practically, there exist several formats to store explicit LTSs in computer files. The CADP verification tool set uses three such formats: AUT, a simple textual format, BCG (*Binary Coded Graphs*), a compact binary format based upon dedicated compression algorithms, and SEQ, a human-readable format for displaying sequences of transitions produced by verification tools to explain why a given property is not verified. There exist other LTS formats, for instance the FC2 format used by the FC2TOOLS [3], with which the CADP tools are interfaced.

- An *implicit* LTS is defined in comprehension by giving its initial state s_0 and its successor function $succ : S \rightarrow 2^T$ defined by $succ(s) = \{(s, a, s') \mid (s, a, s') \in T\}$. A generic representation of implicit LTSs is provided by the language-independent environment OPEN/CÆSAR [13] embedded in CADP. OPEN/CÆSAR offers primitives for accessing the initial state of an LTS and for enumerating the successors of a given state, as well as various data structures (state tables, stacks, etc.), allowing on-the-fly verification algorithms to be implemented concisely. A number of input languages are connected to OPEN/CÆSAR, including LOTOS and the EXP formalism of CADP, which describes implicit LTSs as sets of explicit LTSs combined together using LOTOS parallel composition and hiding operators. Similarly, the FC2 format allows to describe networks of communicating automata, which can be processed by the FC2TOOLS.

The tools available in CADP and FC2TOOLS allow to perform the usual operations on (explicit or implicit) LTSs in several complementary ways, as summarized in the following table:

ALDÉBARAN	Reduction, comparison, deadlock/livelock detection
BCG_IO	Conversion from one explicit LTS format to another
BCG_LABELS	Hiding and renaming of labels
BCG_MIN	Reduction
BCG_OPEN	Implicit LTS view of an explicit LTS
CÆSAR.ADT, CÆSAR	LTS generation from a LOTOS description
CÆSAR.OPEN	Implicit LTS view of a LOTOS description
EVALUATOR	Deadlock/livelock detection
EXHIBITOR	Deadlock detection
EXP2FC2	Conversion from EXP format to FC2
EXP.OPEN	Implicit LTS view of an EXP description
FC2EXPLICIT, FC2IMPLICIT	Reduction, comparison, deadlock/livelock detection
GENERATOR	Explicit LTS generation from an implicit LTS
PROJECTOR	Abstraction (see Section 3 below)

3. THE DES2AUT TOOL

DES2AUT is a tool performing compositional generation of an LTS from a composition expression (called *behavior*) written in the DES language [22]. Given a behavior, the DES2AUT tool generates and minimizes its LTS compositionally modulo a relation R (e.g., strong or branching bisimulation) specified on the command-line. The DES language is defined by the following grammar, where B, B_0, B_1, B_2 are non-terminal symbols denoting behaviors and F, P, $G_1, \ldots, G_n$ are terminal symbols denoting respectively file prefixes, LOTOS process identifiers, and LOTOS gate identifiers:

$$\begin{array}{rcll} B & ::= & F\texttt{.aut} \mid F\texttt{.exp} \mid F\texttt{.lotos} \mid \textbf{Proc}(F\texttt{.lotos}, P\texttt{[}G_1,\ldots,G_n\texttt{]}) & (1) \\ & \mid & \textbf{hide } G_1,\ldots,G_n \textbf{ in } B_0 & (2) \\ & \mid & B_1 \texttt{ |[}G_1,\ldots,G_n\texttt{]| } B_2 & (3) \\ & \mid & B_1 \texttt{ -|[}G_1,\ldots,G_n\texttt{]| } B_2 \mid B_1 \texttt{ -|[}G_1,\ldots,G_n\texttt{]|? } B_2 & (4) \end{array}$$

The syntactic constructs above denote:

(1) an explicit LTS, either contained in a file in AUT format, or generated from an implicit LTS given as an EXP description, an entire

LOTOS description, or a particular process P defined in a LOTOS description (this process being instantiated with actual gate parameters $G_1, \ldots, G_n$);

(2) the behavior B_0 in which gates $G_1, \ldots, G_n$ are hidden using the LOTOS hiding operator;

(3) the behaviors B_1 and B_2 executing in parallel with synchronization and communication on gates $G_1, \ldots, G_n$ according to the LOTOS parallel composition semantics;

(4) the behavior B_1 restricted by synchronizing it on the set of gates $G_1, \ldots, G_n$ with B_2 (B_2 is considered as an automaton, the regular language of which expresses interface constraints); the resulting behavior is called the *abstraction* of B_1 w.r.t. B_2. The "?" symbol, if present, indicates that the user is unsure that the interface constraints expressed in B_2 are sound w.r.t. the real environment of B_1: as a consequence, validity checks are performed when the resulting LTS is composed in parallel with other LTSs.

Although the usefulness of the DES2AUT tool has been established on significant case-studies [22, 24, 19], its applicability is limited by several practical shortcomings:

- It only supports a single verification strategy: for each parallel composition operator, the operands are minimized first, then combined in parallel. Therefore, operands can only be combined two by two according to the evaluation order defined by the algebraic parallel composition operators; it is not possible to combine more than two components simultaneously. In some cases, this leads to state explosions that could be avoided otherwise. In the following example,

  `(User1.aut ||| User2.aut) |[`*G*`]| Medium.aut`

  generating the interleaved combination of "`User1.aut`" and "`User2.aut`" first — independently from "`Medium.aut`" — is likely to produce a larger state space than if "`Medium.aut`" was taken into account.

  So far, this problem has been circumvented by using other CADP tools (namely ALDÉBARAN and EXP.OPEN), which allow several LTSs to be combined together using LOTOS parallel composition operators. Practically, this is tedious because the DES and EXP languages accepted by DES2AUT and ALDÉBARAN–EXP.OPEN, respectively, are incompatible in syntax and semantics, and because many auxiliary DES and EXP files are needed to describe a compositional verification scenario this way. The problem is even worse as there are often many possible scenarios corresponding to various decompositions of the system into sub-systems, which requires a trial-and-error approach to find the most effective decomposition.

- The DES2AUT tool relies on ALDÉBARAN to perform LTS minimization and does not support other bisimulation tools such as BCG_MIN and FC2TOOLS. Moreover, it only accepts the AUT format, which is much less compact than the BCG format, so that compositional verification often aborts due to a lack of disk space.
- When a compositional verification scenario fails (e.g., if LTS generation, minimization, or parallel composition aborts at some point, because of a lack of memory or disk space, for instance), localizing and understanding the reason of the problem is difficult, as DES2AUT does not provide sufficient debugging information.

4. THE SVL LANGUAGE

To address these problems, we propose a scripting language for compositional verification. This language, named SVL, contains all the features of the DES language, as well as new ones that provide increased flexibility in the verification task.

To define the abstract syntax of SVL, we first introduce several terminal symbols: F denotes a file prefix, $G, G_1, \ldots, G_n$ denote LOTOS gate identifiers, $L, L_1, \ldots, L_n$ denote (UNIX-like) regular expressions on labels considered as character strings, and P denotes a LOTOS process identifier. We also introduce the following non-terminal symbols:

```
 op  ::=  <= | == | >=
 par ::=  |[G1,...,Gn]| | ||| | ||
 E   ::=  aut | bcg | exp | fc2 | lotos | seq
 R   ::=  branching | strong | observational | safety | tau*.a | ...
 M   ::=  std | fly | bdd
 T   ::=  aldebaran | bcg_min | evaluator | exhibitor | fc2tools | ...
 U   ::=  user | <empty>
```

where *op* denotes an equivalence or preorder relation, *par* a LOTOS parallel operator, E a file extension, R a bisimulation relation, M an algorithmic method to compute bisimulations ("**std**" meaning a standard partition refinement algorithm such as Paige-Tarjan or Kanellakis-Smolka, "**fly**" meaning the Fernandez-Mounier on-the-fly algorithm, and "**bdd**" meaning an algorithm based on Binary Decision Diagrams), and T a tool.

The two main non-terminal symbols are S (*statements*), and B (*behaviors*). They are defined by the following grammar, where '[' and ']' delimit optional clauses in the grammar, and where '`[`' and '`]`' denote the terminal bracket symbols. An SVL program is a list of statements "$S_1; \dots ; S_n$".

$$
\begin{array}{rcll}
S & ::= & \texttt{"}F.E\texttt{"} = B_0 & (S1) \\
 & | & \texttt{"}F.E\texttt{"} = R \textbf{ comparison } [\textbf{using } M] \; [\textbf{with } T] \; B_1 \; op \; B_2 & (S2) \\
 & | & \texttt{"}F.E\texttt{"} = \textbf{deadlock } [\textbf{with } T] \textbf{ of } B_0 & (S3) \\
 & | & \texttt{"}F.E\texttt{"} = \textbf{livelock } [\textbf{with } T] \textbf{ of } B_0 & (S4) \\
B & ::= & \texttt{"}F.E\texttt{"} \mid \texttt{"}F.\texttt{lotos"} : P\texttt{[}G_1, \dots, G_n\texttt{]} & (B1) \\
 & | & \textbf{hide } L_1, \dots, L_n \textbf{ in } B_0 & (B2) \\
 & | & B_1 \; par \; B_2 & (B3) \\
 & | & \textbf{rename } L_1 \to L'_1, \dots, L_n \to L'_n \textbf{ in } B_0 & (B4) \\
 & | & \textbf{generation of } B_0 & (B5) \\
 & | & R \textbf{ reduction } [\textbf{using } M] \; [\textbf{with } T] \textbf{ of } B_0 & (B6) \\
 & | & U \textbf{ abstraction } B_1 \textbf{ sync } G_1, \dots, G_n \textbf{ of } B_2 & (B7)
\end{array}
$$

An SVL behavior denotes either an explicit or an implicit LTS, contrary to the DES language, in which every implicit LTS is immediately converted to an explicit one. Formally, the semantics of a behavior B is given by a denotation function $[\![B]\!]_\sigma$, the result of which is either an (explicit or implicit) LTS file, a LOTOS file, a LOTOS process instantiation, or an EXP composition expression (i.e., a set of explicit LTSs combined together with hiding and parallel composition operators). The subscript σ denotes a set of file extensions, which denote all acceptable formats in which the result of $[\![B]\!]_\sigma$ should be produced. The value of σ is determined by the context in which B will be used, and namely by the tools that will be applied to B, given that certain tools require certain formats for their inputs and outputs (for instance, the FC2TOOLS only handle LTSs in the FC2 format). Hence, format conversions may be required at some places but, for efficiency reasons, should be avoided as much as possible. σ always contains at least an explicit LTS format; if it contains more than one, a preferred LTS format noted $pref(\sigma)$ is selected in the following preference order: `bcg`, then `aut`, then `fc2`, then `seq`. A dedicated type-checking is done on SVL programs, mainly to distinguish between explicit and implicit LTSs: certain constraints on σ (listed below) must be satisfied, otherwise type-checking errors are reported. The semantics of behaviors is the following:

($B1$) denotes a behavior contained in a file (or a LOTOS process instantiation). If it is an implicit LTS file (resp., a LOTOS file), then `exp` or `fc2` (resp., `lotos`) must belong to σ. If $E \in \sigma$ then $[\![B]\!]_\sigma$ returns B else it converts the LTS contained in B to $pref(\sigma)$ using BCG_IO or EXP2FC2. LOTOS process instantiations are handled similarly as LOTOS files.

($B2$) denotes the label hiding of an implicit or explicit LTS using the ALDÉBARAN, BCG_LABELS, and EXP.OPEN tools. This generalizes the LOTOS hiding operator by allowing regular expressions on labels. Contrary to the DES language, in which all implicit LTSs are converted into explicit ones before hiding, SVL semantics preserves implicit LTSs as long as possible (according to the definition of EXP composition expressions). If σ contains **exp** or **fc2** and all labels $L_1, \ldots, L_n$ are LOTOS gates and B_0 is a parallel composition of behaviors, then $[\![B]\!]_\sigma$ returns the EXP composition expression **hide** $L_1, \ldots, L_n$ **in** $[\![B_0]\!]_{\{\mathtt{exp,aut,bcg,fc2,seq}\}}$, else $[\![B]\!]_\sigma$ returns the conversion to $pref(\sigma)$ of $[\![B_0]\!]_{\{\mathtt{bcg}\}}$ in which labels matching one of $L_1, \ldots, L_n$ are hidden.

($B3$) denotes the parallel composition of B_1 and B_2. σ must contain **exp** or **fc2**. $[\![B]\!]_\sigma$ returns the EXP composition expression $[\![B_1]\!]_{\sigma'}$ *par* $[\![B_2]\!]_{\sigma'}$ where $\sigma' = \{\mathtt{exp}, \mathtt{aut}, \mathtt{bcg}, \mathtt{fc2}, \mathtt{seq}\}$.

($B4$) denotes the label renaming of an explicit LTS using the BCG_LABELS tool, which supports UNIX-like regular expression matching and substring replacement. $[\![B]\!]_\sigma$ returns the conversion to $pref(\sigma)$ of $[\![B_0]\!]_{\{\mathtt{bcg}\}}$ in which labels are renamed as specified by the rules $L_1 \to L'_1, \ldots, L_n \to L'_n$.

($B5$) denotes the conversion from an implicit LTS to an explicit one, which is computed using OPEN/CÆSAR compilers and GENERATOR. In SVL, such conversions must be requested explicitly using the "**generation**" operator, unlike the DES language in which such conversions are automatic and cannot be avoided. Let α be $[\![B_0]\!]_{\sigma \cup \{\mathtt{exp,fc2,lotos}\}}$. If α is already an explicit LTS, then $[\![B]\!]_\sigma$ returns α else it returns the conversion of α to an explicit LTS of format $pref(\sigma)$.

($B6$) denotes the reduction of an LTS modulo an equivalence relation R, using an algorithmic method M and a tool T (ALDÉBARAN, BCG_MIN, or FC2TOOLS). For short, we abbreviate this operator to *RMT-reduction*. $[\![B]\!]_\sigma$ returns the conversion to format $pref(\sigma)$ of the RMT-reduction of $[\![B_0]\!]_{\sigma^T}$, where σ^T is the set of the accepted input formats for reduction with tool T.

($B7$) denotes the abstraction w.r.t. interface constraints, which is computed using OPEN/CÆSAR compilers and PROJECTOR. $[\![B]\!]_\sigma$ returns the conversion to format $pref(\sigma)$ of the abstraction of $[\![B_2]\!]_{\{\mathtt{bcg,exp,fc2,lotos}\}}$ w.r.t. the interface $[\![B_1]\!]_{\{\mathtt{aut}\}}$. The "**user**" keyword has the same meaning as the "?" symbol of the DES language.

The statements have the following effects:

($S1$) stores in file F.E (where $E \neq \mathtt{lotos}$) either $[\![B_0]\!]_{\{E\}}$ if E denotes an explicit LTS format or $[\![B_0]\!]_{\{\mathtt{aut,bcg,exp,fc2,seq}\}}$ if E denotes an implicit LTS format.

($S2$) compares two LTSs modulo an equivalence or preorder relation R, using an algorithmic method M and a tool T (ALDÉBARAN or FC2TOOLS). Formally, it compares $[\![B_1]\!]_{\sigma^T}$ and $[\![B_2]\!]_{\sigma^T}$, where σ^T denotes the set of possible input formats accepted by T. The result stored in file $F.E$ (where $E \notin \{\texttt{exp}, \texttt{lotos}\}$) is a (set of) distinguishing path(s) if the comparison returns false or an empty path otherwise.

($S3$) detects deadlocks using a tool T (ALDÉBARAN, EVALUATOR, EXHIBITOR, or FC2TOOLS) in $[\![B_0]\!]_{\sigma^T}$, where σ^T is the set of possible input formats accepted by T. The result stored in file $F.E$ (where $E \notin \{\texttt{exp}, \texttt{lotos}\}$) is a (set of) path(s) leading to (one or all) deadlock state(s), if any, or an empty path otherwise.

($S4$) checks for livelocks in a way similar to statement ($S3$).

5. META-OPERATIONS

SVL has so-called *meta-operations*, which allow various compositional reduction strategies to be written concisely; this has proven to be useful in many case-studies. Meta-operations extend the syntax of SVL behaviors as follows:

$$B \;::=\; \ldots$$
$$\mid \; A \; R \; \textbf{reduction using} \; M \; \textbf{with} \; T \; \textbf{of} \; B_0$$

where the attribute A is defined by:

$$A \;::=\; \textbf{leaf} \mid \textbf{root leaf} \mid \textbf{node}$$

Informally, meta-operations have the effect of propagating RMT-reductions automatically at various places in the algebraic term B_0. Depending on the value of A, they have different semantics:

- "**leaf**" means that an RMT-reduction must be applied to all leaves "$F.E$" but also to all subterms of B_0 that generate an explicit LTS (i.e., to all abstraction, renaming, hiding, generation, and reduction operators). Leaf reduction is not propagated through generation and reduction operators. Moreover, to maximize the use of the EXP format of CADP and to avoid converting EXP descriptions into explicit LTSs (unless requested by the user), RMT-reduction is not applied to hiding operators whose operands are parallel compositions of LTSs.
- "**root leaf**" is similar to "**leaf**" except that B_0 itself is also RMT-reduced at the top-level.
- "**node**" is similar to "**leaf**" except that RMT-reductions are also applied to all parallel composition operators in the abstract tree of B_0. This emulates the reduction strategy of the DES2AUT tool and is mainly implemented for backward compatibility purpose.

Formally, "*A R* **reduction using** *M* **with** *T* **of** *B*" is expanded into $\mathcal{E}_{RMT}(B, A)$ where $\mathcal{E}_{RMT}$ is defined as follows and where the shorthand notation $\mathcal{R}_{RMT}(B)$ stands for an RMT-reduction of B:

$\mathcal{E}_{RMT}(B, A) =$ **if** $A =$ "**root leaf**" **then** $\mathcal{R}_{RMT}(\mathcal{E}_{RMT}(B, \textbf{leaf}))$
  else case B **in**
    rename ... **in** $B_0 \rightarrow \mathcal{R}_{RMT}($**rename**... **in** $\mathcal{E}_{RMT}(B_0, A))$
    | U **abstraction** B_1 ... **of** $B_2 \rightarrow$
      $\mathcal{R}_{RMT}(U$ **abstraction** B_1 ... **of** $\mathcal{E}_{RMT}(B_2, A))$
    | B_1 *par* $B_2 \rightarrow$
      if $A =$ "**node**" **then** $\mathcal{R}_{RMT}(\mathcal{E}_{RMT}(B_1, A)$ *par* $\mathcal{E}_{RMT}(B_2, A))$
      else $\mathcal{E}_{RMT}(B_1, A)$ *par* $\mathcal{E}_{RMT}(B_2, A)$
    | **hide** $L_1, \ldots, L_n$ **in** $B_0 \rightarrow$ **if** $\mathcal{E}_{RMT}(B_0, A)$ *has the form* "B_1 *par* B_2" *and*
      all L_i *are* LOTOS *gates* **then hide** $L_1, \ldots, L_n$ **in** $\mathcal{E}_{RMT}(B_0, A)$
      else $\mathcal{R}_{RMT}($**hide** $L_1, \ldots, L_n$ **in** $\mathcal{E}_{RMT}(B_0, A))$
    | **otherwise** $\rightarrow \mathcal{R}_{RMT}(B)$

After meta-operation expansion, all behaviors containing meta-operations are replaced by simple behaviors (as defined in Section 4). Since the application of $\mathcal{E}_{RMT}$ may create superfluous operations, an optimization function called $\mathcal{O}$ is applied to the resulting behaviors in order to increase efficiency. For instance, $\mathcal{O}$ merges nested hiding operators into a single hiding operator. $\mathcal{O}$ also replaces $\mathcal{R}_{RMT}(\mathcal{R}_{R'M'T'}(B))$ by $\mathcal{R}_{R'M'T'}(B)$ if the relation R' is weaker or equal to the relation R (noted $R' \sqsubseteq R$), since an RMT-reduction is useless after an $R'M'T'$-reduction. At last, $\mathcal{O}$ suppresses an RMT-reduction applied to the operand of a hiding (resp. abstraction) operator if the result of this hiding (resp. abstraction) itself is to be reduced modulo the same relation R.

$\mathcal{O}(B) =$ **case** B **in**
    "$F.E$"[:P[...]] $\rightarrow B$
    | B_1 *par* $B_2 \rightarrow \mathcal{O}(B_1)$ *par* $\mathcal{O}(B_2)$
    | **generation of** $B_0 \rightarrow$ **generation of** $\mathcal{O}(B_0)$
    | **rename** ... **in** $B_0 \rightarrow$ **rename** ... **in** $\mathcal{O}(B_0)$
    | U **abstraction** B_1 ... **of** $B_2 \rightarrow U$ **abstraction** $\mathcal{O}(B_1)$ ... **of** $\mathcal{O}(B_2)$
    | **hide** $L_1, \ldots, L_n$ **in** $B_0 \rightarrow$
        if B_0 *has the form* "**hide** $L_{n+1}, \ldots, L_{n+m}$ **in** B_0'" **then**
          $\mathcal{O}($**hide** $L_1, \ldots, L_{n+m}$ **in** $B_0')$
        else hide $L_1, \ldots, L_n$ **in** $\mathcal{O}(B_0)$
    | $\mathcal{R}_{RMT}(B_0) \rightarrow$ **case** B_0 **in**
        $\mathcal{R}_{R'M'T'}(B_0')$ *where* $R' \sqsubseteq R \rightarrow \mathcal{O}(\mathcal{R}_{R'M'T'}(B_0'))$
        | **hide** $L_1, \ldots, L_n$ **in** $\mathcal{R}_{R'M'T'}(B_0')$ *where* $R = R' \rightarrow$
          $\mathcal{O}(\mathcal{R}_{RMT}($**hide** $L_1, \ldots, L_n$ **in** $B_0'))$
        | U **abstraction** B_1 ... **of** $\mathcal{R}_{R'M'T'}(B_2)$ *where* $R = R' \rightarrow$
          $\mathcal{O}(\mathcal{R}_{RMT}(U$ **abstraction** B_1 ... **of** $B_2))$
        | **otherwise** $\rightarrow \mathcal{R}_{RMT}(\mathcal{O}(B_0))$

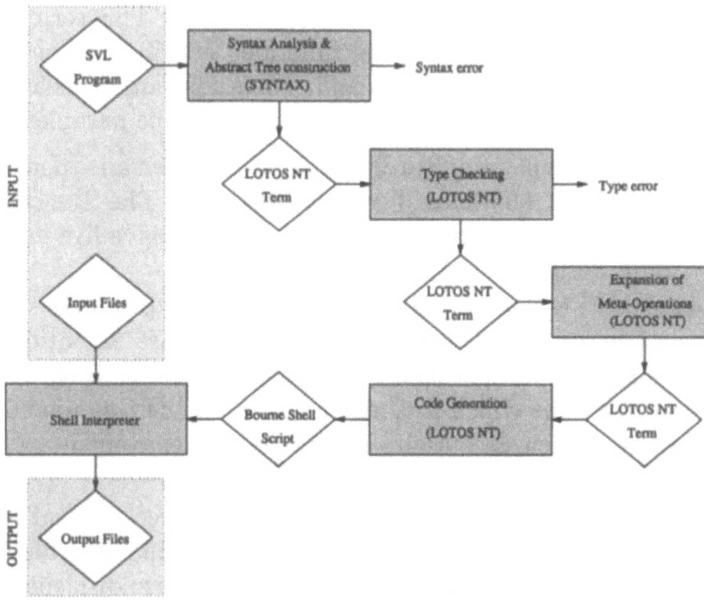

Figure 1. Architecture of SVL 2.0

6. THE SVL 2.0 COMPILER

A compiler for the SVL language has been implemented within the CADP toolbox. This compiler, named SVL 2.0, includes five phases, as shown in Figure 1. It was developed using an original compiler construction technology also used for two other compilers developed by the VASY team of INRIA:

- The lexical analysis, syntax analysis, and abstract tree construction phases (1,100 lines of code) are implemented using the SYNTAX compiler generator [2]. SYNTAX has similar functionalities as LEX and YACC, enhanced with a well-designed automatic error recovery mechanism.
- Type checking, expansion of meta-operations, and code generation (2,900 lines of code) are implemented as a set of data types and functions written in the LOTOS NT language [16, 26]. Inspired by the standardization work on Enhanced LOTOS [21], LOTOS NT combines the theoretical foundations of process algebras with features borrowed from both functional and imperative languages (such as abstract data types, patterns, assignments, loops, ...) suitable for a wider industrial acceptance of formal methods. The TRAIAN compiler [25] translates the LOTOS NT code into a set of C types and functions. The generated C code is augmented with about 200 lines of hand-written C code.

- SVL generates a script written in Bourne shell. This script starts by including a set of predefined functions (1,750 lines of Bourne shell), then the SVL statements and expressions are translated into sequences of calls to these functions with suitable parameters.

In total, the development of SVL 2.0 took about 5 person · month, and totalizes more than 11,000 lines of generated C code. The SVL compiler is designed carefully to ease the verification task. We give five examples of such a careful design:

- When executing a verification scenario, SVL 2.0 produces simultaneously two kinds of output: a high-level, concise execution trace expressed at the abstraction level of the SVL source program, and a low-level log file which lists all the executed shell commands and the output of their execution. This file is very convenient to locate and understand run-time errors.
- During its execution, the generated script produces several temporary files (explicit LTSs, EXP files, hiding and renaming files, etc.) needed by the various tools of CADP. To minimize disk space consumption, SVL removes temporary files as soon as possible and uses as much as possible the BCG format for representing explicit LTSs, because the BCG format is more compact than other LTS formats.
- Several convenient compiler options are implemented, e.g., "`-debug`", which prevents intermediate files from being erased, "`-expand`", which produces the SVL program obtained by expanding all meta-operations, and "`-script`", which generates the Bourne shell script without executing it.
- As much as possible, the generated script tries to recover from errors automatically by using "expert" knowledge about the verification process. For instance:
  - When a reduction fails because the combination of relation, tool, and method specified in the source SVL program is not available, SVL 2.0 attempts to change some of the parameters (tool and/or method). For instance, it can replace "**std**" with "**bdd**" when using ALDÉBARAN to perform strong reduction of an LTS in the EXP format since standard reduction is not available in this case.
  - If a weak reduction fails, e.g., because of memory exhaustion, SVL 2.0 tries to perform first stronger reduction(s) before trying again the weak reduction. For instance, observational reduction may be preceded by branching reduction, and branching reduction may be preceded by strong reduction. If the reduction still fails, SVL 2.0 will leave the LTS un-reduced and continue execution.

- The SVL semantics states that a "**generation**" operator is mandatory whenever an implicit LTS should be converted to an explicit LTS because such conversion is costly and can lead to state explosion. Practically, this constraint is slightly relaxed: if a "**generation**" operator is omitted, it will be inserted automatically and a warning message will be emitted.

- SVL 2.0 permits to invoke shell commands from the SVL description. This can be used for calling tools with features not implemented in SVL, for using the shell control structures to perform conditionals and loops, and for modifying the values of environment variables specifying the default tools, methods, and relations as well as options to be passed to the tools.

7. CONCLUSION

Although compositional verification has a strong potential in attacking the state explosion problem, only a few implementations have been carried out. In this respect, the compositional verification tools of CADP are of particular interest because they are widely distributed and have been applied to several industrial case-studies [4, 22, 24, 19]. However, these tools require a good level of expertise to be effective. To address this problem, we designed the scripting language SVL, which is well-adapted to compositional verification:

- SVL combines process algebra operators (parallel composition, label hiding) with operations on LTSs (e.g., minimization, abstraction, comparison, livelock/deadlock detection, label hiding and label renaming using regular expressions, etc.). It also provides high-level meta-operators, which allow sophisticated compositional verification strategies to be expressed concisely. Practically, SVL is expressive enough to supersede the two formats EXP and DES previously used in the CADP toolbox and to suppress the need for hand-written MAKEFILEs.

- SVL behaves as a tool-independent coordination language (in the same way as EUCALYPTUS [12] is a tool-independent graphical user interface). Due to its concise syntax and well-chosen default options, SVL relieves the user from the burden of launching verification tools manually: it invokes the relevant tools with appropriate options and allocates/deletes temporary files automatically. SVL supports several verification tools (ALDÉBARAN, BCG_MIN, FC2TOOLS) both for *explicit* (enumerative) and *implicit* (on-the-fly) verification. It supports several LTS formats (AUT, BCG, FC2, SEQ) though using as much as possible the BCG format, which allows significant savings in both disk space and access time. Switching from one tool or one format to another can be done simply by changing a few words in the SVL description.

- SVL is extensible in two ways. Its modular design will allow new tools and new formats to be integrated easily. Moreover, as Bourne shell commands can be directly invoked from an SVL description, the user can easily describe specific processings and benefit from the high-level constructs of the Bourne shell (**if** and **case** conditions, **while** and **for** loops, etc.).

For this language, we have fully implemented the SVL 2.0 compiler which is developed using an original approach to compiler construction, combining the SYNTAX compiler generator and LOTOS NT, a variant of the forthcoming E-LOTOS standard.

SVL 2.0 has reached a good level of stability and maturity. It is distributed as a component of CADP and available on four different UNIX and WINDOWS platforms. It has been used in ten case-studies profitably, in particular for the compositional verification of a dynamic reconfiguration protocol for agent-based applications [1]. As regards future work, three directions can be identified:

- SVL could be enhanced with common sub-expressions detection. At present, the user can always avoid redundant computations by storing their results in intermediate, named files, which can be reused later. For instance, the statement:

  "a.bcg" = **leaf strong reduction of** $((B_1 \; |[G]| \; B_0) \; || \; (B_2 \; |[G]| \; B_0))$

  can be evaluated more efficiently as:

  "b.bcg" = **strong reduction of** B_0
  "a.bcg" = ((**strong reduction of** B_1) |[G]| "b.bcg") ||
  ((**strong reduction of** B_2) |[G]| "b.bcg")

  Ideally, this optimization could also be performed automatically.

- The SVL language could be enriched with additional operators, e.g., to model-check temporal logic formulas, and additional meta-operators, such as recursive propagation of hiding (so as to hide labels as soon as possible) or recursive abstractions. More applications are needed to determine which extensions are practically useful.

- The SVL language and related tools should be extended to support the new parallel composition operators [17] introduced in E-LOTOS and LOTOS NT; these operators are more expressive and user-friendly than LOTOS ones and would thus contribute to make compositional verification easier.

Acknowledgements

We would like to thank the former and current members of the INRIA/VASY team involved in the SVL project, namely: Mark Jorgensen and Christophe Discours, who implemented the early versions (1.0–1.6) of the SVL compiler, Ghassan Chehaibar and Charles Pecheur, who used and gave valuable feedback about the DES2AUT and SVL 1.* tools, Marc Herbert and Stéphane Martin, who tested the improved CADP tools on several architectures carefully, and Radu Mateescu, who experimented the SVL 2.0 compiler on several case-studies.

We are also grateful to Ji He (University of Stirling) for her reports about the DES2AUT tool, to Laurent Mounier (Univ. Joseph Fourier, Grenoble) for his pertinent explanations regarding DES2AUT in particular and compositional verification in general, and to Solofo Ramangalahy, and to the anonymous referees for their useful comments about this paper.

REFERENCES

[1] M. Aguilar Cornejo, H. Garavel, R. Mateescu, N. de Palma. Specification and Verification of a Dynamic Reconfiguration Protocol for Agent-Based Applications. In *Proc. of DAIS'01*. Kluwer Academic Publishers.

[2] P. Boullier, P. Deschamp. Le système SYNTAX : Manuel d'utilisation et de mise en œuvre sous Unix. `http://www-rocq.inria.fr/oscar/www/syntax`, 1997.

[3] A. Bouali, A. Ressouche, V. Roy, and R. de Simone. The Fc2Tools set: a Toolset for the Verification of Concurrent Systems. In *Proc. of CAV'96*, LNCS vol. 1102.

[4] G. Chehaibar, H. Garavel, L. Mounier, N. Tawbi, and F. Zulian. Specification and Verification of the PowerScale Bus Arbitration Protocol: An Industrial Experiment with LOTOS. In *Proc. of FORTE/PSTV'96*. IFIP.

[5] K. H. Cheung. *Compositional Analysis of Complex Distributed Systems.* Ph.D. thesis, Hong Kong Univ. of Science and Technology, 1998.

[6] S. C. Cheung and J. Kramer. Enhancing Compositional Reachability Analysis with Context Constraints. In *Proc. of FSE'93*. ACM.

[7] S. C. Cheung and J. Kramer. Compositional Reachability Analysis of Finite-State Distributed Systems with User-Specified Constraints. In *Proc. of FSE'95*. ACM.

[8] S. C. Cheung and J.Kramer. Context Constraints for Compositional Reachability. *ACM Transactions on Software Engineering Methodology*, 1996.

[9] J.-C. Fernandez. *ALDEBARAN : un système de vérification par réduction de processus communicants.* Thèse de Doctorat, Univ. Joseph Fourier (Grenoble), 1988.

[10] J.-C. Fernandez, H. Garavel, A. Kerbrat, R. Mateescu, L. Mounier, and M. Sighireanu. CADP (CÆSAR/ALDEBARAN Development Package): A Protocol Validation and Verification Toolbox. In *Proc. of CAV'96*, LNCS vol. 1102.

[11] J.-C. Fernandez, H. Garavel, L. Mounier, A. Rasse, C. Rodríguez, and J. Sifakis. A Toolbox for the Verification of LOTOS Programs. In *Proc. of ICSE'92*. ACM.

[12] H. Garavel. An Overview of the Eucalyptus Toolbox. In *Proc. of the Int. Workshop on Applied Formal Methods in System Design*. Univ. of Maribor, Slovenia, 1996.

[13] H. Garavel. OPEN/CÆSAR: An Open Software Architecture for Verification, Simulation, and Testing. In *Proc. of TACAS'98*. LNCS vol. 1384.

[14] D. Giannakopoulou. *Model Checking for Concurrent Software Architectures.* Ph.D. thesis, Imp. College of Science, Technology and Medicine — Univ. of

London, 1999.

[15] S. Graf and B. Steffen. Compositional Minimization of Finite State Systems. In *Proc. of CAV'90*, LNCS vol. 531.

[16] H. Garavel and M. Sighireanu. Towards a Second Generation of Formal Description Techniques – Rationale for the Design of E-LOTOS. In *Proc. of FMICS'98*. CWI.

[17] H. Garavel and M. Sighireanu. A Graphical Parallel Composition Operator for Process Algebras. In *Proc. of FORTE/PSTV'99*. IFIP.

[18] S. Graf, B. Steffen, and G. Lüttgen. Compositional Minimisation of Finite State Systems using Interface Specifications. *Formal Aspects of Computation*, Sept. 1996.

[19] H. Garavel, C. Viho, and M. Zendri. System Design of a CC-NUMA Multiprocessor Architecture using Formal Specification, Model-Checking, Co-Simulation, and Test Generation. *STTT*. Springer-Verlag, July 2001.

[20] ISO/IEC. LOTOS — A Formal Description Technique Based on the Temporal Ordering of Observational Behaviour. International Standard 8807, ISO — Information Processing Systems — OSI, September 1988.

[21] ISO/IEC. Information Technology - Enhancements to LOTOS (E-LOTOS). Juan Quemada, editor, FDIS 15437 ballot, May 2001.

[22] J.-P. Krimm and L. Mounier. Compositional State Space Generation from LOTOS Programs. In *Proc. of TACAS'97*. LNCS vol. 1217.

[23] J. Malhotra, S. A. Smolka, A. Giacalone, and R. Shapiro. A Tool for Hierarchical Design and Simulation of Concurrent Systems. In *Proc. of the Workshop on Specification and Verification of Concurrent Systems*. British Computer Society, 1988.

[24] C. Pecheur. Advanced Modelling and Verification Techniques Applied to a Cluster File System. In *Proc. of ASE'99*. IEEE Computer Society.

[25] M. Sighireanu, X. Bouchoux, C. Chaudet, H. Garavel, M. Herbert, F. Lang, and B. Vivien. TRAIAN: A Compiler for E-LOTOS/LOTOS NT Specifications. `http://www.inrialpes.fr/vasy/traian`, 2000.

[26] M. Sighireanu. LOTOS NT User's Manual (Version 2.1). INRIA projet VASY. `ftp://ftp.inrialpes.fr/pub/vasy/traian/manual.ps.Z`, 2000.

[27] K. K. Sabnani, A. M. Lapone, and M. U. Uyar. An Algorithmic Procedure for Checking Safety Properties of Protocols. *IEEE Transactions on Communications*, Sept. 1989.

[28] K. C. Tai and V. Koppol. Hierarchy-Based Incremental Reachability Analysis of Communication Protocols. In *Proc. of the Int. Conf. on Network Protocols*. IEEE, 1993.

[29] K. C. Tai and V. Koppol. An Incremental Approach to Reachability Analysis of Distributed Programs. In *Proc. of the Int. Workshop on Software Specification and Design*. IEEE, 1993.

[30] A. Valmari. Compositional State Space Generation. In *Proc. of Advances in Petri Nets*. LNCS vol. 674, 1993.

[31] W. J. Yeh. *Controlling State Explosion in Reachability Analysis*. Ph.D. thesis, Software Engineering Research Center (SERC) Laboratory, Purdue Univ., Dec. 1993.

[32] W. J. Yeh and M. Young. Compositional Reachability Analysis Using Process Algebra. In *Proc. of SIGSOFT'91*. ACM.

Part Nine

Invited Papers

ON FORMAL TECHNIQUES IN PROTOCOL ENGINEERING — EXAMPLE CHALLENGES

Dines Bjørner
Informatics and Mathematical Modelling, Technical University of Denmark
Building 322, Richard Petersens Plads, DK–2800 Lyngby, Denmark
URI: www.imm.dtu.dk/~db, E–Mail: db@imm.dtu.dk

Abstract In this invited paper — for the 2001 *FORTE* Conference on *FOR*mal Description *TE*chniques for Protocol Specification, Testing and Verification — we intend to illustrate a number of methodological issues that the practising software engineer might wish to consider when developing software for interacting, distributed software, more precisely, for such software which significantly contains interaction protocols. We "attach" our "guided tour" of software engineering using formal techniques, and following a *triptych* of *domain engineering*, *requirements engineering*, and *software design*, to two example "challenges". The particular examples are the *domain* and *requirements* for an "inverse" smart card application, that is, of a *domain* where ordinary citizens gather, refer to and update ("change") accumulated information and money resources and of *requirements* to a "really" smart card system which is the primary carrier of (ie. repository for) such information, and the (electronic) monies (themselves). The other example is that of "market" transactions; that is, a *domain* of interactions between government ($\mathcal{G}$) institutions, businesses ($\mathcal{B}$), and citizens ($\mathcal{C}$). That is with $\mathcal{G} \leftrightarrow \mathcal{B} \leftrightarrow \mathcal{C}$ sequences of interactions — ie. "protocols", as well as some thoughts on *requirements* (an "extended E–Business system").

We outline the problems while adhering to our approaches to both *informal* and *formal software engineering*, to the development *triptych*, and indicate some of the very many *principles* and *techniques* inherent in any good *method*. By showing two different examples we illustrate the "impossibility" of one simpe, straightforward *method*.

Thus the paper presents a capsule view of some of our view of the Theory & Practice of Software Engineering.[1]

[1]This is also the title of: [1].

1. INTRODUCTION

Protocol engineering, a sub-discipline of software engineering, as automotive engineering is of mechanical engineering, is a complex professional field. Protocol designers understood — earlier than software engineers in general — the importance of formal specification and verification.

Perhaps protocol engineers focused more on abstractly specifying intricate protocols and verifying some of their properties, than on developing, in general, provably correct software that satisfied the protocol specifications ? This paper is about the latter.

In this presentation we wish to alert the field of protocol engineering to a number of more general software engineering development principles and techniques whose use in other fields of software engineering has been of some use.

Our two main points are those of the software engineer vigilantly following and adhering to a long check list of *method principles* and *techniques*, and that protocol development step back to include careful considerations of the actual world, ie. the *domain*, in which protocols serve, to better understand as many of the issues that ultimately lie behind a great number of protocol properties.

For software engineering to eventually label itself a professional engineering activity it must, it seems, rely, as do other engineering professions, on high specialisation — hence our insistence on referring to *protocol engineering*. Specialisation is characterised in terms of varieties of principles, techniques and tools whose conditional application follows careful analyses. Our field is slowly accumulating and enunciating these principles, techniques and tools. It may take a whole generation more to get them into general use.

1.1. "Formal" Methods

1.1.1 A Caveat. The label *' "formal" method'* has been affixed to such *techniques* as use *"formal" specification* and *"formal" verification*. Here we first double quote the term *'method'*, secondly we "reduce" the scope of *'formal'* from *'method'*, to *'specification &[2] verification'*. Double–quoting expresses that we do not quite mean what we write: *Methods* cannot be "formal", but some of the *techniques* "they" imply may.

[2]The '&' connective is here used in its conventional meaning: Its two operands with '&' form one concept, not two !

1.1.2 ***"What is a Method ?".*** Here the double quoted *slanted* display line title text designates a rhetorical question. Our answer is: A *method* is a set of analysis and synthesis *principles* for selecting and applying a number of *techniques* and *tools* in order efficiently to construct an efficient artifact. Here those artifacts are software.

Principles are to be adhered to, ie. followed, by humans. Hence they cannot be formalised, ie. are not " formal". Some *techniques* and tools (addressing such techniques) can be said to be formal. Specification languages are *tools,* and certainly can be formal: Have formal semantics, based on formal syntax, but never a formal pragmatics: Pragmatics is what leads outside formality: Has to do with the socio–economic context in which any utterance is made. Hence *methods* cannot be "formal", but crucial aspects of *techniques* and hence their associated *tools* can.

1.2. A Triptych of Software Engineering

Before *software* can be *designed* we must understand its *requirements,* hence they must (first ?) be engineered. Before *requirements* can be developed we must understand the (application) *domain,* hence it must (first ?) be engineered. Our approach is to both *informally* and *formally describe* all three: *Domains, requirements* and *software* — perhaps — in many `stages` and `steps` of "refinement". And to relate these `phases`, `stages` and `steps`.

1.3. Problem Synopses

Engineering is *"walking the bridge between science and technology": "Constructing artifacts (here computing systems, notably software) based on scientific insight,"* and *"investigating technology in order to ascertain its possible scientific value."*

In this paper, as in so much else, we, engineers, are "driven" (forward, or is it sometimes "backwards") by the promises of the hard technologies.

In the case of this paper it is the "technologies" of "smart-cards" and "Internet", and their "derivatives" that first "drive' us.

1.3.1 **The Citizen Smart Card:** *The Technology Incentive.* Imagine, a few years from now, that low power, mass storage, fast computing & optical communication is being facilitated by credit card sized (incl. thin) cards — which we shall henceforth refer to as The Citizen Smart Cards.

The Citizen Smart Card is capable, we conjecture, of holding, typically all of a citizens accumulated health–care information, and more, from a full lifetime: Information about all the health states, temperatures, blood pressure, blood sugar, cholesterol content, etc., all the original ECGs ever measured, all the original X–Rays ever taken, special photos taken before and after cosmetic (plastic) surgery, all the patient medical records ever recorded during hospital and clinic visits, your family doctors' notes about your health. All of this time stamped, etc.; all of a citizens accumulated financial information, again for any full lifetime, as well as all financial instruments currently holdable by any such citizen — ie. the monies themselves suitable transport timetables as well as commuter (daily, weekly, monthly or seasonal) tickets.

Add to the above that the first two kinds of information held on The Citizen Smart Card, in principle, is held nowhere else: No copies ![3] The citizen is in sole and full possession of all *own* information !

The *development brief* now being posed, as our first example, is: Now what does the above imply in a world where we forget this card, where we loose it, where card terminals break down, etc.? Which citizen/card/machine interaction protocols does this call for ?

We will try "edge" towards an answer by examining and modelling, also formally, the actual world of these three information source in the actual *domain*, and then suggest properties that a protocol and its supporting machinery *required* to satisfy.

1.3.2 E–Business: *The Technology Incentive.* The Internet is here. Here and there, *contractual transactions* take place over the net. *Merchandise* (ie. manifest, tangible goods, instantly tradeable, transferable) and (intangible, time–interval providable) *services* are offered, are sold, are sought, are bought, etc.

So far two categories of *traders: Buyers* and *sellers* have been identified: *Businesses* ($\mathcal{B}$) and *citizens* (or consumers, $\mathcal{C}$). But in general there are three, the above and *government institutions* ($\mathcal{G}$).

The question is now: How to generalise the notions of $\mathcal{B}2\mathcal{C}$ and $\mathcal{B}2\mathcal{B}$ transactions to arbitrary *pairings* of traders (buyers and sellers) from the three categories; and how to extend these transactions to such which involve "middlemen": *Agents* and *brokers ?*

The *development brief* now being posed, as our second example, is — in the *domain* — to investigate the full triangle of pairwise interactions; identify "primitive" such transactions, such which enter into most transaction sequences; explore possible behaviours; and — as *requirements* — to identify suitable protocols.

[3] That is: No copies, in a logical sense. Physically, instead of The Citizen Smart Card storing bulks of information it may instead refer to more stationary repositories where (sole) originals are kept. Still, the idea is that the citizen is in full control.

1.4. Structure of Paper

In Section 2 we informally and formally describe the two domains of the citizen information "repository", and "the market". Based on that, we can, in Section 3, outline The Citizen Smart Card and E–Business. requirements — emphasising their protocol aspects. In those sections, and in Section 4, we review a number of method principles and techniques. Space limitations do not permit a thorough treatment — so we shall refer, instead, to published papers and a forthcoming text book on methodical software engineering based on the use of formal techniques.

2. DOMAIN ENGINEERING

Domain engineering is not the same as knowledge engineering. The latter could be said to be "a subset of" the former. In knowledge engineering, briefly, computing scientists and knowledge engineers try model the knowledge of stake–holders of the domain in such a way that the knowledge models are "directly" computable. In domain engineering we do not consider computable aspects only. We wish to model as much of the domain that can be described. We refer to [2] for details.

2.1. The Citizen Smart Card: A Domain

2.1.1 Three Sub–Domains: Informal Sketches. We treat three diverse information and resource "gathering" and "dispensing" domains: citizen health information, financial information and securities, and transport services and tickets.

Patient Medical Records (PMRs) and Journals (PMJs) By a Patient Medical Record, PMR, we mean a collection of documents pertaining to one particular health case of one particular citizen — whether that citizen sought help from medical professionals or not. At any one time several PMRs may be in effect: The citizen may be analysed or undergoing treatments for more than one disease or similar. By a Patient Medical Journal, PMJ, we mean the collection of all PMRs, over the entire life–time of that patient, "up till now !".

We refer to PMRs, not necessarily as manifest, tangible paper, instrument output strips or X–ray files, documents, but, in addition to such, also as what the citizens (the patient, relatives, working place colleagues), in the past remembered, or now remembers. We certainly do not have electronic PMRs in mind but do not a–priori rule them out either.

In other words: A PMJ represents all that is knowable, by whatever means, about a citizen's health.

A PMJ may be subject to a number of operations: (1) creation, ie. opening of new PMRs, and (2) closing, ie. "filing away" of existing PMRs. Individual PMRs may be subject to a number of operations: (3) entering and update of administrative information: Patient name, dates, addresses, etc.; (4) entering and update, incl. "deletion" of "raw", undigested (ie., further unanalysed) medical information obtained from the patient, "previous" PMRs, relatives, etc., or obtained as analysis results from medical professionals (or other), or in the form of analysis originals, or from, analytical (medico-technical) instruments (ECGs, blood analysis, X—Rays, etc.); (5) entering and update, incl. "deletion" of diagnostics; (6) entering and update of treatment plans; (7) entering and update of treatment actions; (8) entering and update of how the patient responds to treatment actions. Whereby we are back to item (4).

In addition we can inquire about PMR/PMJ information: search, correlate, and compute over such information — as do medical professionals: When analysing, when diagnosing, etc.

That's basically "all !"

We refer to two kinds of operations and of two kinds of queries: those which update PMJs with no input provided (O), those which update PMJs with input provided (IO), those which query PMJs with no input provided (Q), and those which update PMJs with input provided (IQ).

Financial Services We treat only banking and securities trading.

• **Banking** Today you would be expected to have a number of bank accounts (Acc): `Savings & Loan` (S&L), `Demand/Deposit`, `Cheque`, etc. Each of these show a balance, either positive or negative. Each such balance represents `credit` or `debit` in the bank. One may perform the following operations related to accounts: `open`, `close` (new, resp. old account), `save`, `borrow`, `repay` (in preparation for loan, establish loan, repay loan — in installments), `withdraw`, `deposit`, `change credit` (latter stands for extending or lowering credit limit), `statement` (list of transactions "since last"), etc.

The *development brief* calls for an analysis of the bank/customer interaction interface today — with a view towards keeping the banks, but maintaining all information and all balances (whether negative or positive) — that is: "the real monies" — on The Citizen Smart Card in future.

• **Securities Trading** Securities trading is an important financial service: Enterprises and governments offer securities (stocks, respectively bonds, Sec), for sale. Enterprises and people buy and sell stocks and bonds. Stocks and bonds are traded on the "floor" of stock exchanges. Brokers and traders act as middlemen between buyers and sellers. The following are typical securities trading operations: initial stock offering; placing and withdrawing orders to buy, respectively sell; effect, as it were "simultaneously" a batch of one or more buy/sell orders; transfer of securities instruments: The stock, bonds and payments. *&c.*

The *development brief* calls for an analysis of the client/broker interaction interface today — with a view towards keeping the broker and securities exchanges, but maintaining all information and all securities instruments (stocks, bonds) on The Citizen Smart Card in future.

Public Transport We shall look at Public Transport only from the very narrow perspective of potential passengers (incl. commuters). Commuters possess knowledge of time and fare tables. Commuters enter, pay for (cash or by showing seasonal tickets), and leave transport vehicles (taxis, busses, street cars, metro trains, canal boats, etc.).

The *development brief* calls for an analysis of the commuter/transport system interaction interface today — with a view towards keeping the transport, but maintaining all information (time and fare tables), and all monies and tickets (InfoTick) on The Citizen Smart Card in future.

2.1.2 Analysis and Formal Model. We decompose our analysis and models into data (ie. type) and function (incl. process) models.

Data Analysis The unifying quantity, whether PMR, PMJ, Sec, or InfoTick, will be referred to as R. Operations on (ie. functions that change) PMR, PMJ, Sec, or InfoTick, will be referred to a O. Queries on PMR, PMJ, Sec, or InfoTick, will be referred to a Q. Extensions of these are operations or queries which update with special information or query wrt. such information I.

Formal Domain Data Model

type
$\quad \mathrm{R}, \mathrm{I}$
$\quad \mathrm{O} = \mathrm{R} \to \mathrm{R}$
$\quad \mathrm{Q} = \mathrm{R} \to \mathrm{I}$
$\quad \mathrm{IO} = \mathrm{I} \times \mathrm{R} \to \mathrm{R}$
$\quad \mathrm{IQ} = \mathrm{I} \times \mathrm{R} \to \mathrm{I}$

Process Analysis Whether a PMR or a PMJ: moving around the health-care system, being tested, taking medicine, having a fever, etc.; or a bank account or a securities instrument (Acc, Sec): depositing and withdrawing cash into and from an account (etc.), buying or selling stocks and bonds (etc.), etc.; or a transport repository (InfoTick): obtaining new time table and fare information, boarding and leaving a bus (train, etc.); all amounts to updating or querying the repository.

Formal Process Model Each repository is like a process. Each such process is, non–deterministically, subject to any of a great variety of operations or queries.

```
type
   Idx = {| 1..n |}
   Θ = Idx →m R
value
   θ:Θ
   system(n) ≡ || { citizen(θ(i)) | i:Idx }
   citizen(r) ≡
      let oq = io ⊓ o ⊓ iq ⊓ q in
      cases oq of
         io → let io:IO,i:I in citizen(io(i,r)) end
         o → let o:O in citizen(o(r)) end
         iq → let iq:IQ,i:I in iq(i,r) ; citizen(r) end
         q → let q:Q in q(r) ; citizen(r) end
      end end
```

The system of all (n) citizens proceed, in the domain, as a set of non–interfering parallel processes. Each citizen non–deterministically (internal choice, ⊓) perform an information–based update (io), a "standard" (a simple) update (o), an information–based query (iq), or a (simple) query (q).

To paraphrase: That's all !

Discussion We remind the reader that the narration and formalisation above is of a domain, not of requirements to software. And we remind the reader that we are focusing on the repository of information and certain (here the financial) resources.

In this example the formal domain model is very trivial. It expresses only the very basics. In the domain all is possible: Humans err, are not consistent, etc. One reason for seeking computing and communication support is to help avoid mistakes and to secure consistency.

The protocol issue was highlighted: We have modelled the non–deterministic handling, ie. arbitrary interleaving of the operations and queries of this information and resource repository. We have abstracted which these operations and queries were in specific.

The next example will be in contrast to the present.

2.2. E–Business: A Domain

2.2.1 Government, Business and Citizen Transactions. In the case of the government, business and citizen infrastructure interactions we can postulate the following *domain*, ie. not necessarily computer & electronic communication supported, transactions:

$\mathcal{G}2\mathcal{G}$: Government institutions, $\mathcal{G}$, buy services from other $\mathcal{G}$, and $\mathcal{G}$ sell services to other $\mathcal{G}$. The $\mathcal{G}$ pay with monies (obtained through taxes, etc.), respectively offer free services, in return. $\mathcal{G}2\mathcal{B}$: $\mathcal{G}$ buy services, or request taxes, from businesses ($\mathcal{B}$), and pay, respectively offer free services, in return. $\mathcal{G}2\mathcal{C}$: $\mathcal{G}$ buy services (hire), or request taxes, from citizens ($\mathcal{C}$), and pay, respectively offer free services, in return. $\mathcal{B}2\mathcal{G}$: Businesses ($\mathcal{B}$) buy services from $\mathcal{G}$, and pay $\mathcal{B}$ for these either through having already paid taxes or by paying special fees. $\mathcal{B}2\mathcal{B}$: $\mathcal{B}$ buy merchandise or services from other $\mathcal{B}$, and $\mathcal{B}$ offer merchandise or services to other $\mathcal{B}$. $\mathcal{B}$ usually pay for these outright. $\mathcal{B}2\mathcal{C}$: $\mathcal{B}$ buy services from citizens: ie. hire temporary or permanent staff (employment), and $\mathcal{B}$ pay for these through salaries. $\mathcal{C}2\mathcal{G}$: Citizens ($\mathcal{C}$) obtain services from $\mathcal{G}$ (passport, drivers licence, etc., health–care, education, safety and security, etc.) and $\mathcal{C}$ pay for these either by paying special fees or through having already paid taxes. $\mathcal{C}2\mathcal{B}$: $\mathcal{C}$ buy merchandise from $\mathcal{B}$, and $\mathcal{C}$ pay for this. $\mathcal{C}2\mathcal{C}$: Two or more $\mathcal{C}$ together enter into political "grass–root" organisations, or leisure–time hobby club activities, or just plainly arrange meetings (incl. BBQ parties); and the two or more $\mathcal{C}$ "pay" for this by being "together".

2.2.2 Traders: Buyers and Sellers. Above we have stressed that also government (institutions) are part of the more general concept of `E-Business`, some aspects of contractual obligations, and a seeming "symmetry" between partners to any such contract (ie. buy, sell, etc.). As such we have stressed that "The Market" consists of buyers and sellers, whom we, as one, refer to as traders.

2.2.3 Traders: Agents and Brokers. An *agent*, to us, while we are still only discussing the *domain*, is a trader that acts (in a biased manner) on behalf of usually one other trader (either a buyer, or a seller), vis—a–vis a number of other traders (sellers, respectively buyers), in order to secure a "best deal". A *broker*, to us, while we are still only discussing the *domain*, is a trader that acts (in a neutral manner) on behalf one or more buyers and one or more sellers in order to help them negotiate a "deal."

2.2.4 Schematic Transactions. Sequences of contractual transactions can be understood in terms of "primitive" transactions:

A buyer `inquires` as to some merchandise or service. A seller may respond with a `quote`. A buyer may `order` some merchandise or service. A seller may `confirm` an order. A seller may `deliver` an order. A buyer may `accept` a delivery. A seller may send an `invoice`. A buyer may `pay` according to the invoice. A buyer may `return`, within warranty period, a delivery. And a seller may `refund` such a return.

We have, deliberately, used the "hedge" 'may':

A trader may choose an action of `no response`, or a trader may inform that a transaction was `misdirected`, or a trader may `decline` to `quote`, `order`, `confirm`, `deliver`, `accept`, `pay` or `refund` !

2.2.5 Formalisation of Syntax.

type

Trans == Inq | Ord | Acc | Pay | Rej
Qou | Con | Del | Acc | Inv | Ref
NoR | Dec | Mis

The first two lines list the 'buyer', respectively the 'seller' initiated transaction types. The third line lists common transaction types.

U below stand for unique identifications, including time stamps (T), Sui for surrogate information, and MQP alludes to merchandise identification, quantity, price.

U, T, Su1, Su2, MQP
Inq = MQP × U
Qou = (Inq|Su1) × Inf × U
Ord = Qou|Su2 × U
Con = Ord × U
Del = Ord × U
Acc = Del × U
Inv = Del × U
Pay = Inv × U
Rej = Del × U
Ref = Pay × U
NoR = Trans × U
Dec = Trans × U

Mis = Trans × U
value
obs_T: U → T

In general we model, in the *domain,* a "subsequent" transaction by referring to a complete trace. of unique, time stamped transactions. Thus, in general, a transaction "embodies" the transaction it is a manifest response to, and time of response.

Figure 1 attempts to illustrate possible transaction transitions between buyers and sellers.

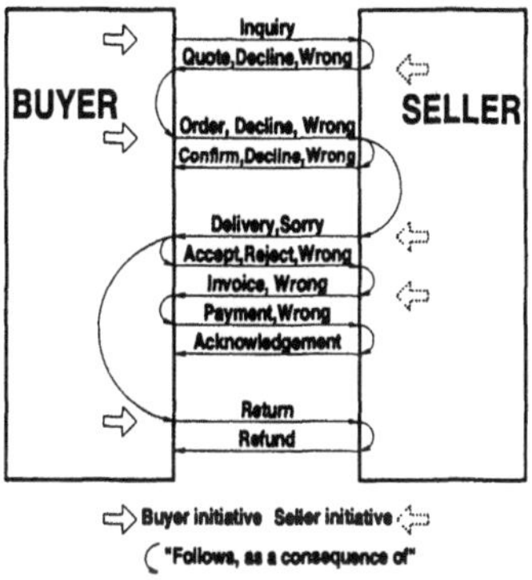

Figure 1. Buyer / Seller Protocol

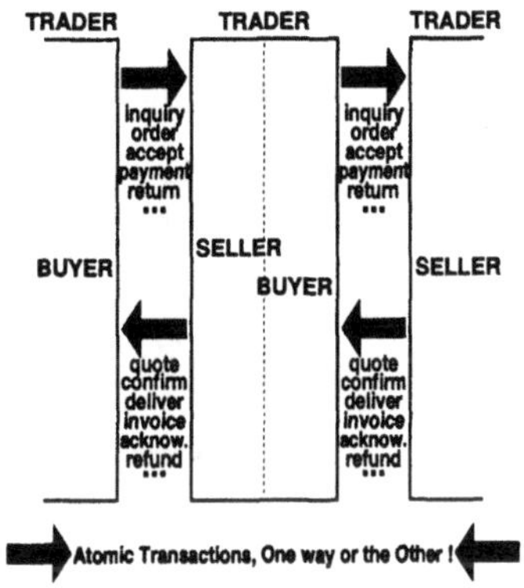

Figure 2. Trader=Buyer+Seller

2.2.6 Transaction Sequences. Figure 3 attempts to show the possible sequences of transactions as related to one "trade": From inquiry through refunding, that is: For one pair of buyer/seller.

2.2.7 "The Market". Figure 2 attempts to show that a trader can be both a buyer and a seller. Thus traders "alternate" between buying and selling, that is: Between performing 'buy' and performing 'sell' transactions.

Figure 4 attempts to show "an arbitrary" constellation of buyer and seller traders. It highlights three supply chains. Each chain, in this example, consists, in this example, of a "consumer", a retailer, a wholesaler, and a producer.

2.2.8 Formalisation of Process Protocols. "The Market" consist of n traders, whether buyers, or sellers, or both; whether additionally agents or brokers. Each trader τ_i is able, potentially to com-

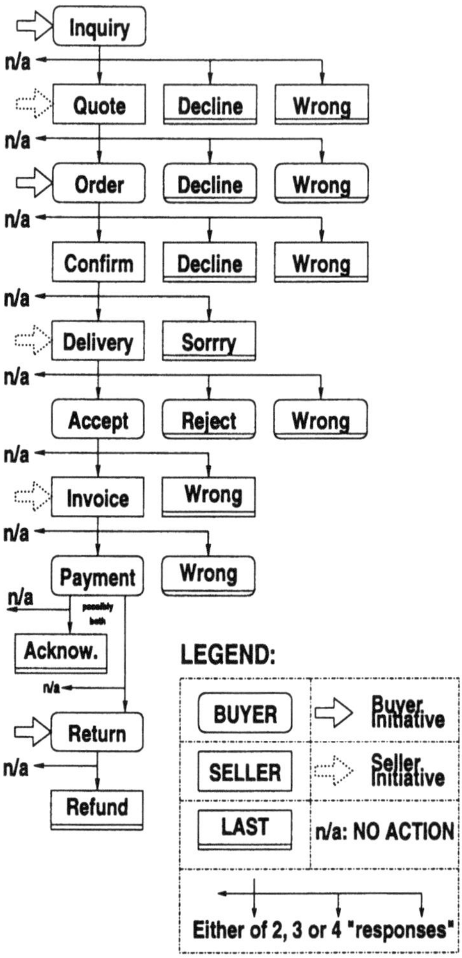

Figure 3. Transaction Sequence "Chart"

municate with any other trader $\{\tau_1, \ldots, \tau_{i-1}, \tau_{i+1}, \ldots, \tau_n\}$. We omit treatment of how traders come to know of one another. And focus only on the internal and external non–determinism which is always there, in the *domain*, when transactions are selected, sent and received.

Our model is in a variant of CSP, but expressed "within" RSL [3].

type
  Θ
  Idx = {| 1..n |}
value

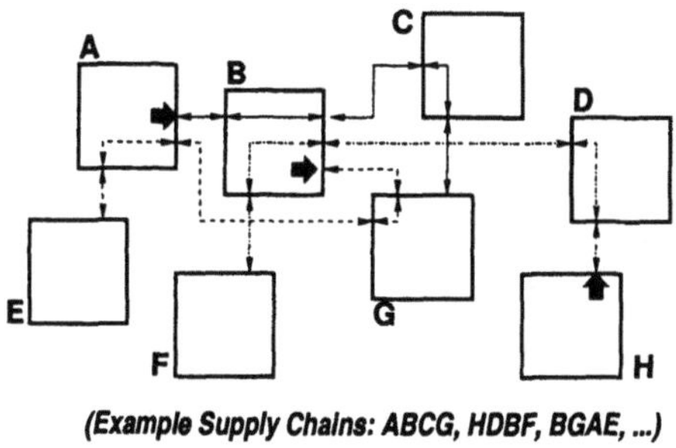

Figure 4. A Network of Traders and Supply Chains

```
sys: (Idx m→ Θ) × n:Nat → Unit
sys(mθ,n) ≡ || { tra(i)(mθ(i)) | i:Idx }

tra: i:Idx → Θ →
        in {tc[j,i]|j:Idx•i≠j} out {tc[i,j]|j:Idx•i≠j} Unit
tra(i)(θ) ≡ tra(i)(nxt(i)(θ))

nxt: i:Idx → Θ →
        in {tc[j,i]|j:Idx•i≠j} out {tc[i,j]|j:Idx•i≠j} Θ
nxt(i)(θ) ≡
   let choice = rcv ⊓ snd in
   cases choice of
      rcv → receive(i)(θ), snd → send(i)(θ) end end
```

The *system* is the parallel combination of n *traders*. *Traders* communicate over channels: *tc[i,j]* — from *trader i* to *trader j*. Each *trader* is modelled as a process which "goes on forever", but in steps of *next* state transitions. The *next* state transition non—deterministically (internal choice, ⊓) "alternates" between expressing willingness to *receive*, respectively desire to *send*. In "real life", ie. in the *domain*, the choice as to which transactions are taken is *non–deterministic*. And it is an *internal choice*. That is: The *choice* is not influenced by the *environment*.

```
receive: i:Idx → Θ → in {tc[j,i]|j:Idx•i≠j} Θ
receive(i)(θ) ≡
   ⊓ {let msg=tc[j,i]? in
         update_rcv_state(msg,j)(θ) end|j:Idx}

update_rcv_state: i:Idx × Θ → Θ
```

update_rcv_state is not a protocol function. *update_rcv_state* describes the deposit of *msg* in a repository of received messages. If *msg* is a response to an earlier sent transaction, *msg_o*, then *update_rcv_state* describes the removal of *msg_o* from a repository of sent messages. *remove_set_msg* models the situation where no response (*nor*) is (ever) made to an earlier sent message. Once the internal non-deterministic choice ($\sqcap$) has been made: Whether to *receive* or *send*, the *choice* as to whom to 'receive from' is also *non-deterministic*, but now *external* ($\square$). That is: *receive* expresses willingness to receive from any other *trader*. But just one. As long as no other *trader j* does not send anything to *trader i* that *trader i* just "sits" there, "waiting" — potentially forever. This is indeed a model of the real world, the *domain*. A subsequent *requirement* may therefore, naturally, be to provide some form of *time out*. A re–specification of *receive* with time out is a *correct implementation* of the above.

send: i:Idx → Θ → **in** {tc[i,j]|j:Idx•i≠j} Θ
send(i)(θ) ≡
  let choice = ini ⊓ res ⊓ nor **in**
  cases choice **of**
    ini → send_initial(i)(θ),
    res → send_response(i)(θ),
    nor → remove_received_msg(θ) **end end**

Either a *trader*, when communicating a transaction chooses an *initial (ini)* one, or chooses one which is in *response (res)* to a message received earlier, or chooses to *not respond (nor)* to such an earlier message The choice is again non–deterministic internal. In the last case the state is updated by *non–deterministically internal choice* (not shown) *removing* the, or an earlier *received message.*

Note that the above functions describe the internal as well as the external non–determinism of protocols. We omit the detailed description of those functions which can be claimed to not be proper protocol description functions — but are functions which describe more the particular *domain* at hand: Here "The Market".

send_initial: i:Idx → Θ → **out** {tc[i,j]|j:Idx•i≠j} Θ
send_initial(i)(θ) ≡
  let choice = buy ⊓ sell **in**
  cases choice **of**
    buy → send_init_buy(i)(θ),
    sell → send_init_sell(i)(θ) **end end**

```
send_response: i:Idx → Θ → out {tc[i,j]|j:Idx•i≠j} Θ
send_response(i)(θ) ≡
    let choice = buy ⊓ sell in
    cases choice of
        buy → send_res_buy(i)(θ),
        sell → send_res_sell(i)(θ) end end
```

In the above functions we have, perhaps arbitrarily chosen, to distinguish between *buy* and *sell* transactions. Both *send_initial* and *send_response* functions — as well as the four auxiliary functions they invoke — describe aspects of the protocol.

```
send_init_buy: i:Idx → Θ → out {tc[i,j]|j:Idx•i≠j} Θ
send_init_buy(i)(θ) ≡
    let choice = inq ⊓ ord ⊓ pay ⊓ ret ⊓ ... in
    let (j,msg,θ') = prepare_init_buy(choice)(i)(θ) in
    tc[i,j]!msg ; θ' end end

send_init_sell: i:Idx → Θ → out {tc[i,j]|j:Idx•i≠j} Θ
send_init_sell(i)(θ) ≡
    let choice =  quo ⊓ con ⊓ del ⊓ inv ⊓ ... in
    let (j,msg,θ') = prepare_init_sell(choice)(i)(θ) in
    tc[i,j]!msg ; θ' end end
```

prepare_init_buy is not a protocol function, nor is *prepare_init_sell.* They both assemble an initial buy, respectively sell message, *msg,* a target trader, *j,* and update a send repository state component.

```
send_res_buy: i:Idx → Θ → out {tc[i,j]|j:Idx•i≠j} Θ
send_res_buy(i)(θ) ≡
    let (θ',msg)=sel_update_buy_state(θ),
          j=obs_trader(msg) in
    let (θ'',msg') = response_buy_msg(msg)(θ') in
    tc[i,j]!msg'; θ'' end end

send_res_sell: i:Idx → Θ → out {tc[i,j]|j:Idx•i≠j} Θ
send_res_sell(i)(θ) ≡
    let (θ',msg)=sel_update_sell_state(θ),
          j=obs_trader(msg) in
    let (θ'',msg') = response_sell_msg(msg)(θ') in
    tc[i,j]!msg'; θ'' end end
```

sel_update_buy_state is not a protocol function, neither is *sel_update_-sell_state.* They both describe the selection of a previously deposited, buy, respectively a sell message, *msg*, (from it) the index, *j*, of the trader originating that message, and describes the update of a received messages repository state component. *response_sell_msg* and *response_buy_msg* both effect the assembly, from *msg*, of suitable response messages, *msg'*. As such they are partly protocol functions. Thus, if *msg* was an *inquiry* then *msg'* may be either a *quote*, *decline*, or a *misdirected* transaction message. Etcetera.

2.2.9 Discussion. In this example, in contrast to the E–Business example, the protocol aspect was quite "pronounced". Again we remind the reader that we have, so far only described facets of the domain. Next we shall deal with requirements.

2.3. Discussion: Domain Engineering

The two example cases differ widely: The Citizen Smart Card "resulted" in a "small" domain model. The E–Business model, in contrast, was quite substantive. The latter included quite an elaborate, yet, in essence "free" protocol: Any sequence of transactions will occur in the domain, and there were many variants. The former seems, in the domain, to have an utterly uninteresting "protocol" — if such a term could indeed be used !

3. REQUIREMENTS ENGINEERING

We illustrate a novel technique of "deriving" significant features of requirements from the domain description. The idea is that the domain descriptions (and ours were necessarily rather cursory) "point" to the necessity of certain domain requirements. Interface and machine requirements will only be cursorily covered. We refer to [4, 5, 6] and the forthcoming [1] for details.

3.1. Domain Requirements

Domain requirements "construction" proceeds, it is suggested, by deploying the following techniques: *Projection* — where only those parts of the domain model which is of interest are "carried" forward to requirements. *Instantiation* — where abstract types and functions defined only through their signatures are made more specific. *Extension* — where new domain possibilities, afforded through computing and communication, but hitherto not feasible without, are "added" to the requirements (although "really" a domain issue). *Initialisation* — where the whole issue of how to initialise the context and state configuration components is dealt with. Ie., the input, vetting and update of configuration data. We shall mostly be concerned with *projection* and *instantiation.*

3.1.1 The Citizen Smart Card.

Projection Synopsis We *project* information and securities (incl. money) resources onto The Citizen Smart Card as mentioned earlier. Figure 5 attempts to show one possible rendition of The Citizen Smart Card. The Citizen Smart Card is to store all information and be an electronic purse.

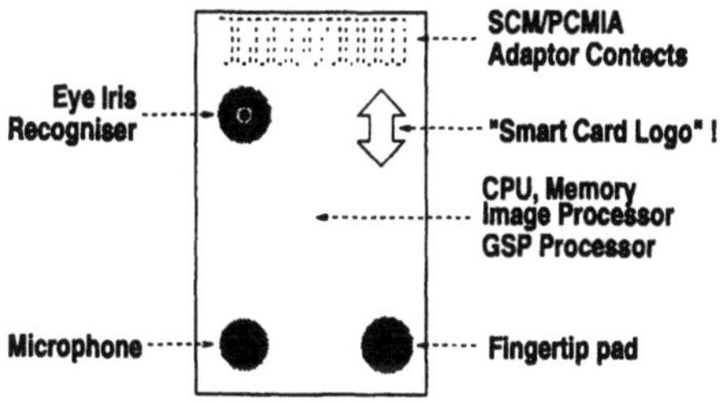

Figure 5. A Smart Card

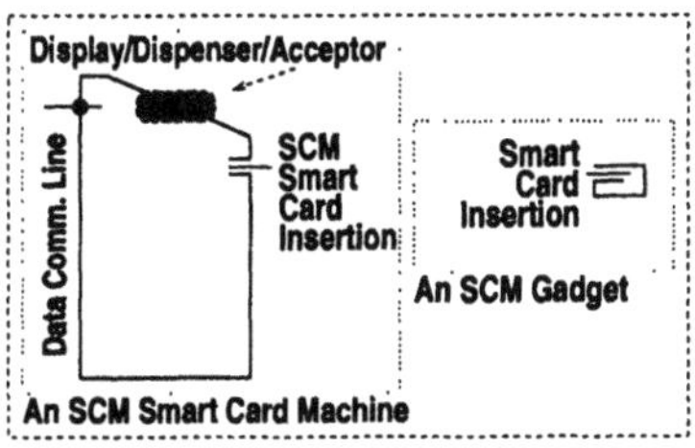

Figure 6. Smart Card Terminals

Instantiation Synopsis We omit consideration of how the repository information and electronic (purse) securities, ie. R, are represented.

Extension Synopsis Here is where some considerations of machine requirements (notably dependability issues) are to be made first. We refer forward to the machine requirements for the The Citizen Smart Card.

We *extend* as now outlined: To complement The Citizen Smart Card there must be means of "feeding" (health–care, financial and transport services) information onto the card and "moving" electronic money to and from the card. Figure 6 attempts to show two possible such means: An ETM/ATM–like machine, as shown to the left, and a "gadget" as indicated to the right. The ETM/ATM–like machines are well–known. The "gadgets" provide the means of writing and reading information to screens and from for example medico–electronics instruments, and bus and train ticketing gadgets.

Over time the citizen applies the card to a variety of such machines and gadgets (ie. terminals, SCM). Figure 7 attempts to show, as temporally determined, such a sequence of terminals: The Citizen Smart Card was first issued by SCMa, and last used in connection with SCMn.

In-between The Citizen Smart Card was "applied" to SCMb, ..., SCMi, SCMj, SCMk, ..., and SCMm.

Ignore, for the moment, the arrows between SCMs.

But observe that the trace of SCM usage portrays a virtual, a logic usage: Two or more SCMx and SCMy (where x and y in a sense stands for temporal order), may be the same physical terminal. Thus there may be loops. In fact: They all may be the same terminal. The card holder is "provincial", never gets around !

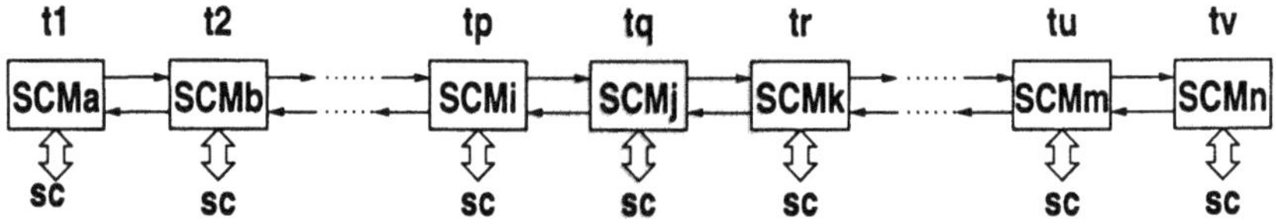

Figure 7. Trace of Terminal Use

Failures of Operation We can now discuss some possibilities: (i) The Citizen Smart Card may be lost. (ii) The Citizen Smart Card may be stolen. (iii) The smart card terminals may break down. *&c.*

We discuss possible protocols to be observed for each of these three situations.

(i) The Citizen Smart Card may be lost. We assume each card to have a unique identifier. We assume each SCM to remember all the uniquely time–stamped and card identified transactions it has been engaged in for all applied cards. By, for example, letting The Citizen Smart Card remember the identity of the SCM it was last applied to, a link can be made from the SCM it is currently applied to to that "previous" SCM, which, in turn, can then provide a link to the "current" SCM. If the holder of the The Citizen Smart Card can remember just one of the "past" SCMs, then a new card can be issued with all the accumulated and processed information and securities instruments as per the day of loosing the card. (ii) The Citizen Smart Card may be stolen. By providing, as indicated by Figure 5, a triple identification means: Eye iris "stamp", finger print, and voice print, any high degree of privacy can be obtained. A stolen card otherwise leads to the same re–establishment protocol as a lost card. (iii) The smart card terminals may break down. Figure 8 attempts to show a failing terminal SCMj. By further providing "leap frog" links, forward and backward, skipping one, as shown, two and/or more SCMs, together with copying also "neighbouring" transactions, a suitable redundancies can, again, help ensure recovery of data. And so forth.

Now recall that the trace of terminals used by a card may use the same SCM repeatedly, or always ! This could call, in general, for a set of SCM "buddies": SCMs that are physically different from the ("master") SCM of which they are "buddies". Now, instead of inserting links and copying transaction data to "master" SCMs — when the same as "current" SCM — they are copied, for example, to a rotating selection of "buddies". And so forth !

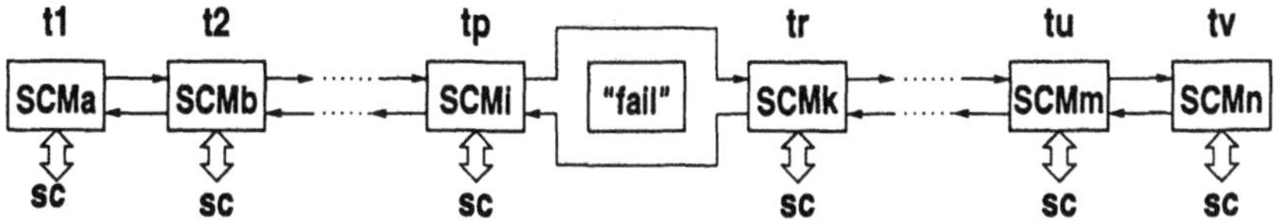

Figure 8. "Broken" Trace of Terminal use

Formalisation of Card/Terminal Interaction We leave out a formal model of the interlocking process of a card, mutually exclusively, interacting with a terminal.

3.1.2 E–Business.

Projection Synopsis We omit consideration of all operations and queries, ie. of any specific o, io, q and iq in respectively O, IO, Q and IQ. and focus only on the communication between traders. We basically ignore the "content" of any transaction, and shall instead focus on automating certain sequences of transactions.

Instantiation Synopsis Whereas the domain model of traders was a model, essentially, intrinsically, of human operations, we now try to automate as much as possible the response to received transactions. Thus, as an example: (1) If a consumer order can be delivered by the retailer, without human (retailer staff) intervention, it will be done so. (2) If a consumer order cannot be delivered by the retailer, but that retailer can re-order from a wholesaler, who can deliver — both atomic transactions without human (retailer and wholesaler staff) intervention, it will be done so. (3) And if a consumer order cannot be delivered by the retailer, but that retailer can re-order from a wholesaler, who then cannot deliver, but must re–order from producer, who can deliver — all atomic transactions without human (retailer, wholesaler and producer staff) intervention, it will be done so.

Figure 9 attempts to show the three cases listed above. There might be delays, waiting times, between order receipt and delivery and/or re–ordering, across the supply–chain.

Extension Synopsis We introduce electronic traders and brokers. They permit arbitrarily wide inquiries: Potentially to all providers (retailers, wholesalers, or producers) of specified merchandise (or services), offers ("confirmations") of merchandise (or services) to all "takers" (consumers, retailers, or wholesalers), first–come–first serve ("auction"-like) orders, *&c.* These roughly sketched domain requirements are considered extensions as they might not be humanly feasible in actual domains.

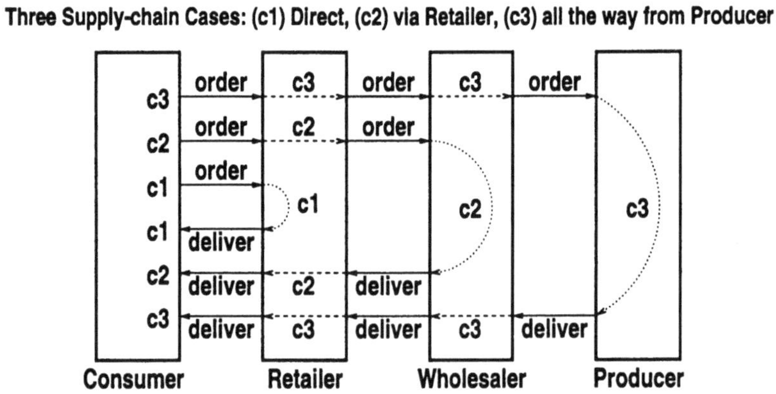

Figure 9. E–Market Supply Chain

Initialisation Synopsis Due to our projection we need only consider how traders, agents and brokers initially, and in an ongoing way, come to know of one another. We omit details — *"left as an exercise"*.

3.1.3 Discussion: Domain Requirements. We see a big difference in the two example cases: The The Citizen Smart Card case requires consideration of — as we shall later see — fairly, technologically detailed machine requirements, and in reality, as we shall see when all three kinds of requirements have been summarised, require quite an elaborate set of protocols. We are "close" to the hard technologies, and the protocol very much depends on hard technology dependability issues. The E–Business case is — most likely — to be implemented "on top" of Internet and the web; hence can rely on its underlying protocols; but requires, as we shall see, a mostly domain requirements–determined protocol.

3.2. Interface Requirements

Interface requirements deal with *"phenomena shared between users and the machine"* (software + hardware). Interface requirements thus include consideration of such issues as graphical user interfaces (GUIs), input/output in general, and user/machine dialogue sequences.

3.2.1 The Citizen Smart Card. The interface requirements have, in this case, to do with the interface between the card and its user with respect to for example: (i) *Authentication:* For example sequence or simultaneity of pressing card with designated finger for its "print" recognition, uttering a certain phonetic sequence for "voice print" recognition, while aligning, say a left eye with a minute whole through the card, to see light trough it, for an "iris print". (ii) *Transaction Processing:* For example which card and terminal "buttons/icons" etc. to press/click in order to initiate, conduct and terminate update and query actions. *&c.* Since they only have marginal relevance to protocol engineering we omit further consideration.

3.2.2 E–Business. As for The Citizen Smart Card we omit detailed consideration of trader interfaces to the electronic market support. This is "classic" web–design !

3.2.3 Discussion: Interface Requirements. The The Citizen Smart Card case raises the most non–standard questions when it comes to interface requirements — it is so hard technology, so "gadget" biased. The E–Business case, in contrast, "falls" within the, by now, "classical" field of web and Internet design.

3.3. Machine Requirements

Machine requirements has to do with determining and describing such issues as (i) performance (temporal and spatial [response times, storage and other tangible resource consumption]), (ii) dependability (availability [up time], accessability [fairness], security [authentication, etc.], fault tolerance, etc.), (iii) maintainability (perfective, adaptive and corrective), (iv) portability (hardware and software platforms: development, execution and maintenance), (v) documentation, *&c.*

3.3.1 The Citizen Smart Card. This case, of the two being treated, raises the most issues concerning machine, and in particular *dependability requirements.* They include: *Availability:* The up–time of the system of all smart card terminals. This issue has little to do with the concerns of protocol engineering — so we skip it. *Accessibility:* Provision — in the context of multiple users — of fair access to central resources. Since the entire idea of The Citizen Smart Card is "inverted" storage and computing the issue is almost mute. Anyway this issue also has little to do with the concerns of protocol engineering — so we skip it. *Security:* Ensuring protection of privacy of information and private ownership of securities instruments — in this case — leads to for example the triplet schema of eye iris, voice and finger print. There are some, however scant protocol issues related to this, but we skip them ! *Fault Tolerance:* Provision of "soft" failures: That is, piecemeal, initially negligible degradation of service in case of lost or stolen cards and in case of terminal breakdowns. We have already, in the domain requirements section, had to presuppose a possibly elaborate scheme for protection against failures of these three kinds. They are the ones that makes, in our view, for an interesting set of protocols, three of which have been mentioned: The protocol(s) of "standard" (failure–free) operation, the protocol(s) for restoring lost or stolen cards, and the protocol(s) for routing around erroneous terminals. *&c.*

3.3.2 E–Business. The *availability*, *accessibility*, and *security* requirements are assumed "taken care" of by an "underlying" Internet system. We focus just on the *fault tolerance* issue: If a trader "goes out of business", or electronically "breaks down", while many transactions, from and to many other traders are pending, then what ? Here, for example, the possible simplicity of a supply chain protocol, as indicated earlier, is at stake. A proper protocol for handling this requires back–ups, duplication, "proxies" and the like. In other words: Protocol engineering "takes over" !

3.3.3 Discussion: Machine Requirements. We have seen that machine requirements could not be kept strictly separate from domain and interface requirements; and we have seen the diversity and differing intensity with which various machine requirements issues "flavours" protocol considerations. Finally we have seen that some protocol aspects, as dictated for example by machine requirements, can be taken care of by assumed implementation platforms, while others lead to additional protocols operating, as it were, "on top of platform protocols".

3.4. Challenges

There are two challenges. They can only be formulated now, after all aspects of the requirements have been listed: Domain, interface and machine requirements.

3.4.1 Challenge I: The Citizen Smart Card Protocols. We challenge the reader to augment the The Citizen Smart Card synopses, and to informally and formally describe the protocols for handling (i) normal card/machine operation (insertion of forward and backward links and the copying of transaction data); (ii) re—establishment of a "new" card, when one is lost or stolen; and (iii) detection of and "rerouting" around failing machines.

3.4.2 Challenge II: E–Business Protocols. As for Challenge I we challenge the reader to informally and formally complete the E–Business synopses, and in particular to informally and formally describe the implied E–Business protocols.

3.5. Discussion: Requirements

The various discussion sections, in connection with domain requirements, interface requirements, and machine requirements, seems to have exhausted the issue. Let us, however, just remind the reader that one example's "thin" domain description "explodes" into a hefty protocol requirements description, while another example's "substantial" domain description "fizzles" into a slightly augmented protocol requirements description, We refer to (one) The Citizen Smart Card versus (another) E–Business.

4. FURTHER METHODOLOGY ISSUES

4.1. Abstraction & Modelling Principles and Techniques

In general [1] outlines abstraction & modelling principles and techniques applicable in any phase (domain, requirements and software design) of computing systems development: (a) property vs. model–oriented descriptions, (b) representation and operation abstraction, (c) denotations vs. computations (operational semantics), (d) hierarchies and compositions, (e) contexts, states and configurations, (f) applicative (functional), imperative, logic and parallel specification "programming", and many other specification principles and techniques. In this short paper we, unfortunately, must omit relating each and every one of these principles and techniques to protocol engineering as for example practised in this paper.

4.2. Domain Modelling Principles and Techniques

[1, 2] outline a number of domain modelling principles and techniques: Modelling (i) static and dynamic, (ii) temporal and spatial, (iii) discrete, continuous and chaotic, and (iv) tangible and intangible attributes; modelling (v–vi) stake–holders and their perspectives; and modelling (vii) the intrinsics, (viii) the support technologies, (ix) the management & organisation, (x) the rules & regulations, and (xi) the human (correct, sloppy and criminal) behaviour facets.

Some relations were: (i) The The Citizen Smart Card PMR, PMJ, and Info information repositories were basically *Static*, whereas the Acc, Sec and Tick securities instrument (ie. money) repositories were *Dynamic*. (ii) There were *Temporal* relations between E–Business and The Citizen Smart Card transactions that determined crucial properties of protocols, and there we *Spatial* properties of the domain: Citizens gathering PMJ information "from far afield", and traders trading across geographical distance — all necessitating distribution of information and processing. (iii) The examples mostly exemplified *Discrete* systems. The fact that the stock market may behave *Chaotically*, or that electro cardiograms portray *Continuous* phenomena is of no interest for the systems such as we choose to see them. (iv) Most PMR information is *Tangible*: "Feelings" of being in bad health may be indicated by informal utterings in a PMR, but are *Intangible*. Whether to order or not to order a product, and in case of ordering, then from whom, may be *Intangible* attributes, but, as for the others, are of little concern to our protocol designs.

(v) We took a rather limited view of stake–holders, focusing only on patients, bank account holders, securities buyers and sellers, and commuters, respectively only on traders (consumers, retailers, wholesaler and producers). We did not consider health–care professionals, staff of financial institutions, transport company staff (drivers, etc.), nor the specific clerks, etc. of traders. (vi) And we took in account only a rather limited perspective. Enough to focus on the essentials for this article: The protocol development.

(vii) The citizen behaviour of Section 2.1.2 as well as the trader behaviour of Section 2.2.8 can be said to reflect the very basics, the *Intrinsics* of citizens and traders vis–a–vis the issues dealt with. (viii) We did not illustrate domain *Support Technologies* — prior to the advent of The Citizen Smart Card and E–Business. Once new requirements are expressed wrt. existing The Citizen Smart Card or E–Business system, then their card and terminal, respectively their assumed Internet technologies become support technologies. (ix) Past *Management & Organisation,* past "bureaucracies" tended to prevent such "free" mechanisms as The Citizen Smart Card and E–Business. (x) The *Rules & Regulations* of government supervised health-care, financial institutions and transport was not considered at all. Some such rules may affect protocols, however. (ix) We did consider *Human Behaviour* in both case examples — as they seem to influence, very strongly several aspects of protocol design: Loosing or stealing cards, issuing trading transactions "out of order" or forgetting to respond to such transactions. We refer to [1, 2].

4.3. Requirements Modelling Principles and Techniques

We have mentioned some requirements construction principles and techniques: Projection, instantiation, extension, and initialisation. They were illustrated in Section 2. We refer, in general to [1, 4].

4.4. Software Design Principles and Techniques

Domain and some, often all, interface requirements lead to software architecture design: A *software architecture,* to us, reflects exactly the issues with which the users are immediately and tangibly concerned. And these issues have to do with the domain, and with most of the interface. Machine requirements, and remaining interface requirements lead to *program organisation* design: They are intangible as far as the end–user is concerned, and may differ between computing and communication platforms. We refer to [1, 5].

5. CONCLUSION

This paper has — obviously — been written by a person who is not a researcher nor a practitioner of *protocol engineering.* Instead the author is a researcher and practitioner of the more "general" field of *software engineering* — if one can claim that such a broad field can be justified, let alone exists !

5.1. What Have We Achieved ?

What we have tried to do is to cast new light on protocol engineering. We claim, and we have shown, we believe, that it is useful — to the extent of being necessary — to first study and document the domain from which applications (also of protocols) emerge. We have indicated how protocols emerge from requirements, and how requirements emerge from domains.

5.2. What Remains to be Done ?

The two examples were formulated as *challenges*. We have, ourselves, developed sketches and draft formalisations of some protocols, but would like to see others tackle the problem cases raised here !

Acknowledgements

The author gratefully acknowledges the kind invitation by Prof. Myungchul Kim (Inform. & Comm. Univ., Korea), Dr. Byungmool Chin (Electronics Telecomm. Res. Inst., Korea), and Dr. Sungown Kang (Korea Telecom, Korea). Thanks to them, and Prof. Sam Chanson (HKU-S&T), I was challenged into formulating my thoughts on protocols and protocol engineering as emanating from domain and requirements engineering.

References

[1] Dines Bjørner. *Software Engineering: Theory & Practice.* (Publisher is being contacted), 2002. This book represents the author's *Chef d'Œvres* — the summary of more than 25 years of research, development and teaching.

[2] Annabelle McIver and Carrol Morgan, editors. *Programming Methodology: Recent Work by Members of IFIP Working Group 2.3*, chapter Dines Bjørner: "What is a Method ?" — A Study of Some Aspects of Software Engineering. IFIP WG2.3. MacMillan, Oxford, UK, 2001. To be published.

[3] The RAISE Language Group. *The RAISE Specification Language.* The BCS Practitioner Series. Prentice-Hall, Hemel Hampstead, England, 1992.

[4] Dines Bjørner. Domains as Prerequisites for Requirements and Software *&c.* In M. Broy and B. Rumpe, editors, *RTSE'97: Requirements Targeted Software and Systems Engineering*, volume 1526 of *Lecture Notes in Computer Science*, pages 1–41. Springer-Verlag, Berlin Heidelberg, 1998.

[5] Dines Bjørner. Where do Software Architectures come from ? Systematic Development from Domains and Requirements. A Re–assessment of Software Engneering ? *South African Journal of Computer Science*, 1999. Editor: Chris Brink.

[6] Dines Bjørner. A Triptych Software Development Paradigm: Domain, Requirements and Software. Towards a Model Development of A Decision Support System for Sustainable Development. In ErnstRüdiger Olderog, editor, *Festschrift to Hans Langmaack.* University of Kiel, Germany, October 1999.

A PKI-BASED END-TO-END SECURE INFRASTRUCTURE FOR MOBILE E-COMMERCE*

Tin-Wo Cheung and Samuel T. Chanson
Department of Computer Science
Hong Kong University of Science & Technology
Clear Water Bay, Kowloon, Hong Kong
{cswo, chanson}@cs.ust.hk

Abstract The popularity of handheld mobile devices, the move towards electronic commerce, and deployment of the public key infrastructure in many parts of the world have led to the development of electronic commerce on mobile devices. The main challenge is the limited computing capacity on these devices for PKI-based end-to-end secure transactions. This paper presents a new architecture and protocol for authentication and key exchange as well as the supporting infrastructure that is suitable for the mobile environment. The system requirements and our solutions in addressing these requirements in the restrictive environment are discussed. An evaluation of the system performance is also included. The system has been implemented and is supporting some real-life applications.

Keywords: wireless security, key exchange protocols, mobile e-commerce, PKI end-to-end security, smart card, GSM Short Message Service

* This work was supported by an ISF grant from the Hong Kong Innovation and Technology Commission (AF/153/99).

1. INTRODUCTION

1.1 Background

The rapid advances in wireless mobile communication technology and the growth of electronic commerce have naturally led to the development of electronic commercial services on the wireless medium through mobile phones. In Hong Kong as well as many parts of the world, the population of mobile phone users has more than doubled the number of Internet users. The popularity and high penetration rate of the mobile phone can be attributed to its convenience, ease of use, and low cost of ownership and operation. For business transactions conducted on electronic means, security is a major concern. Both the Internet and the wireless network are public networks and considered to be insecure, where messages can be eavesdropped, captured, modified and inserted by intruders. Intruders may also impersonate as legitimate parties for personal gain. Therefore, some mechanism is needed to guarantee the confidentiality, authenticity and integrity of the transmitted messages. Internationally, the Public Key Infrastructure (PKI) is accepted as an effective means to tackle the above security problem.

In January 2000, Hongkong Post [4] started its digital certification service and became the first recognized Certification Authority (CA) in Hong Kong. The Electronic Transactions Ordinance enacted in March the same year gives electronic signatures the same legal stature as handwritten signatures. Therefore, as in a number of other countries, a sound and legal public key infrastructure has been established in Hong Kong that is ready for commercial activities on the electronic medium.

1.2 Objective of Work

Our objective is to develop a PKI-based open infrastructure that supports end-to-end secure electronic transactions through mobile phones. We are not just looking for a theoretical solution; rather the solution must be practical, efficient, and implementable. Besides the security concerns, efficiency and availability of supporting hardware products are also important. The main challenge in this project is the limited capacity of the mobile device, especially the mobile phone [10, 12, 13, 14, 15, 17]. About the only computing facilities available on the mobile phone are those offered by the SIM card which has a very slow speed processor and little memory (of the order of 16 KB). The resources on current SIM cards are not sufficient to perform general PKI-based authentication. Moreover, the wireless network is error-prone and slow compared to wired networks. We have designed a set of key exchange and authentication protocols that can run on a thin client model.

The infrastructure consists of a number of servers offering supporting services for on-line purchase and payment in addition to authentication. Security is end-to-end rather than link-by-link.

The rest of the paper is organized as follows. Section 2 describes related works by others. The requirements of our system are given in Section 3 while Section 4 provides an overview of the system architecture. Section 5 presents our key exchange and authentication protocols, and Section 6 discusses some system implementation issues. System performance is presented in Section 7, and finally Section 8 concludes the paper.

2. RELATED WORKS

For secure digital communication, the communicating parties need to be sure of the identity of one another and the contents of their messages must not be tampered with in any way even if the communication network is insecure. An authentication mechanism is needed to satisfy the first requirement and a secret key exchange mechanism is needed to achieve the second goal. This is known as the *authenticated key exchange (AK)* problem, and a variation is called the *authenticated key exchange with key confirmation (AKC)* problem in [9]. Both problems are concerned with key agreement among the communicating parties. The *AK* protocol makes sure the key can be successfully exchanged only if the identities of the communication parties are genuine. However, it does not check if the exchanged key really works. The *AKC* protocol further adds messages to verify the key being established is correct. In this paper, we shall use the term *AKC* protocol to refer to the authentication and key exchange protocol in general.

2.1 Categories of AKC protocols

There is a vast amount of literature on AKC protocols. In a survey article [20], Clark et al. categorize AKC protocols according to criteria such as the cryptographic approach used, whether a trusted third party is present, and the number of protocol messages exchanged. In this section, we propose an alternative approach and categorize the published AKC protocols based on the methodologies used in authentication and key exchange. Most existing work deals with two communicating parties only.

2.1.1 Key exchange with Implicit Authentication

This class of protocols mainly apply the Diffie-Hellman [11] key exchange algorithm for exchanging keys between the communication parties. In this case, when two parties need to establish a session key, they exchange their public keys. An identical data can be generated by each party by combining its own private key with the counterpart's public key (see Formula (A)). Since only these two parties are able to generate the secret data, proof of the ability to generate the data implicitly authenticates the participants.

The Beller-Chang-Yacobi [12] protocol is designed for the mobile environment, in which the client is a low-power portable mobile device. Let PK_j and SK_j be the public key and the corresponding private key of the base station respectively.

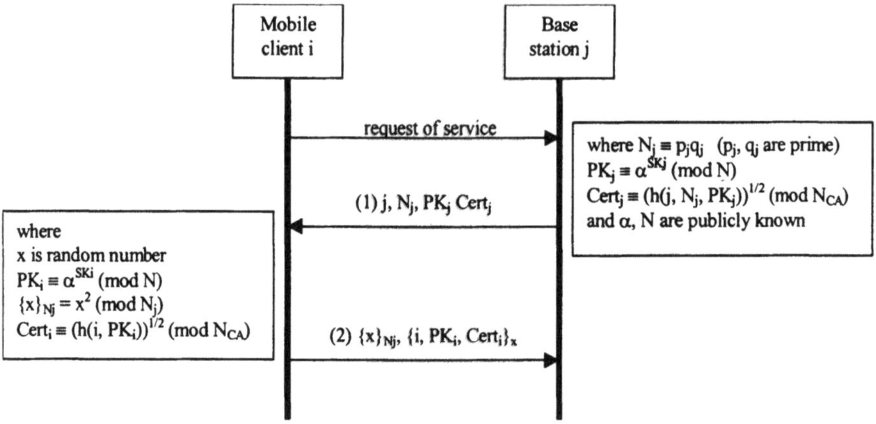

Figure 1. Beller-Chang-Yacobi Protocol

When both parties have successfully verified the certificates of their counterparts, they can calculate the value η using the Diffie-Hellman technique:

$$\eta \equiv (PK_j)^{SKi} \equiv (PK_i)^{SKj} \pmod{N} \quad \text{---- (A)}$$

A session key *sk* is calculated by symmetrically encrypting a random number x with key η, i.e., $sk = f(\eta, x)$.

The idea is similar to Diffie-Hellman in that a common session key can only be calculated by those with the corresponding private keys. The certificates ($Cert_i$ and $Cert_j$) in the protocol are for the purpose of ensuring the public keys used are valid.

Carlsen [13] and Zheng [14] have pointed out the weakness of this protocol due to the weak authenticity mechanism. The identities of both parties have not been authenticated even after message 2 has been received. This means either one can be an impersonator at this point without being detected.

2.1.2 Explicit Authentication by Challenge-Response with key exchange

The simplest way to authenticate a party is the "Challenge-Response" method. The responding party holds a secret, such as a secret symmetric key or a private key, that enables him to answer the challenge correctly. When some random data are embedded in the challenge and/or response messages, these data can be used to generate the session key after successful authentication.

The Aziz-Diffie protocol [15] is another certificate based AKC protocol. The difference between Beller-Chang-Yacobi and Aziz-Diffie is that the Aziz-Diffie protocol is based on the challenge-response mechanism. Therefore, we regard it as an authentication-oriented protocol with key-exchange functionality, while the Beller-Chang-Yacobi protocol is a key-exchange oriented protocol with implicit authentication functionality. The main design goal of Aziz-Diffie is to achieve both privacy of wireless data communication as well as mutual authentication between the communicating parties. The protocol can be summarized in the following diagram:

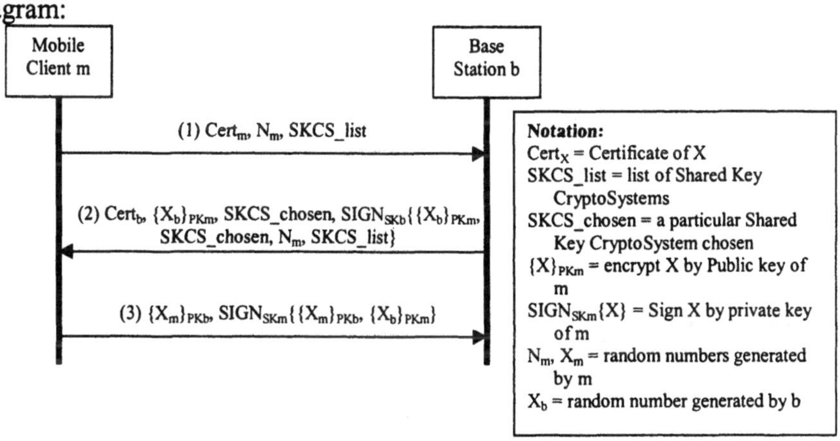

Figure 2. Aziz-Diffie Protocol

The mobile client initiates the protocol by sending N_m in message 1 as the first challenge to the base station. The signature in message 2 contains the element N_m which proves ownership of the private key of the base station b whose certificate $Cert_b$ is also passed to Client m in message 2. At the same time, the random number X_b in message 2 is encrypted by the public key of m as a challenge to the mobile client. In message 3, the mobile client responds to this challenge by decrypting it with its own private key and then encrypts it with the public key of the base station b. Since only the holder of the private key m can decrypt X_b correctly, the base station can authenticate the mobile user after receiving message 3. As a result, mutual authentication is achieved. The session key is calculated by X_b XOR X_m.

The Aziz-Diffie protocol is a typical challenge-response AKC protocol. However, many other proposals take a hybrid approach. In [10, 16], both Diffie-Hellman factors and challenge-response pairs are involved in the protocols messages. Therefore, mutual authentication as well as key establishment are achieved simultaneously.

2.1.3 Explicit Authentication by synchronised data with key exchange

Another approach to explicitly authenticate a party is by using some kind of synchronized data. The principle is basically the same as that of the challenge-response approach. The difference is instead of random numbers, synchronized data such as timestamps from synchronized clocks or counters are used. It is important to ensure the freshness of the synchronized data to avoid replay attacks. In [14], timestamps are used. However, in reality, it is difficult to guarantee the clocks are synchronized.

The Lin-Harn [17] protocol employs a chained hash technique for generating the synchronized data. Before a session key can be established, a registration process is required where the mobile client generates a group of hash chains, signs the "tail" of each chain and then sends to the base station for registration. The client can prove its identity by revealing a previous image of the registered hash chain. Due to the non-reversible property of the hash function, the ability to present a previous image in the chain implies ownership (generator) of the chain. The verifier can verify a received image by hashing it a number of times and compare the result to the 'tail' image.

In the Lin-Harn protocol, the synchronized data as well as the challenge is the last revealed image in the hash chain. The verifier accepts an image only if it is a previous image of the last revealed one. The ability to provide a valid response to this challenge implies the ability to reverse the hashing operation to recover the previous images. Since only the generator of the hash chain knows the "head" image of the chain, the authenticity requirement can be satisfied. Note that in this protocol a single message is enough for authentication.

2.2 Standardized AKC protocols

In order to make use of the AKC protocols in real life, standardization of some generally accepted protocols is essential. Normally, the standardized protocols are independent of the cryptographic algorithm. Negotiation of parameters is generally required at the beginning of the protocol.

2.2.1 Secure Socket Layer (SSL) [1]/Transport Layer Security (TLS) [7]

SSL/TLS is a well-known standard that is generally used for secure Internet web browsing. Its handshake protocol is designed to address the AKC problem between the SSL server and SSL client. The authentication mechanism is based on the challenge-response approach discussed above.

There are two important concepts of SSL: *SSL session* and *SSL connection*. A *SSL session* associates a client and a server. It is created through a handshake protocol involving negotiation of version and algorithms to be used in the session. After the handshake protocol, both the client and the server will be able to calculate a secret data for the *SSL session* called the *master secret*. Every *SSL connection* is a peer-to-peer logical connection built on top of a *SSL session* for data exchange in the application layer. In a *SSL connection*, the two peers share a set of secrets derived from the *master secret* of the *SSL session*. This set of secrets is used for encrypting the exchanged data and creating the message authentication codes (MACs) used to ensure data integrity.

The main purpose of the *SSL connection* is to enhance the efficiency of the protocol by reusing the *master secret* of a *SSL session* in different *SSL connections* between the same client and server. The handshake protocol for *SSL connection* establishment contains fewer messages and involves no public key cryptographic operations which are CPU intensive.

SSL is suitable for wired Internet environment. However, for restrictive devices such as the mobile phone or smart card, the protocol is too heavy for implementation. Moreover, a *SSL session* can only be reused for connections between the same pair of client-server. If a client tends to communicate with different servers, a new *SSL session* must be establishment for each server.

2.2.2 Wireless Transport Layer Security (WTLS) [8]

WTLS is the security layer protocol for the Wireless Application Protocol (WAP) architecture to provide confidentiality, data integrity and authenticity of the communicating applications. The design of WTLS bears a close resemblance to the design of TLS 1.0. Because of the low bandwidth and unreliability of the wireless network, WTLS is concerned with datagram support, optimized handshake and a key refreshment scheme.

One limitation of WTLS is it only provides point-to-point security in most cases. Figure 3 shows a general configuration of a WAP network. WTLS provides a secure channel between the WAP user and the WAP gateway. But it does not protect the communication all the way to the Wireless Markup Language (WML) server. Typically, a *SSL session* is established between the WAP gateway and the WML server if data security is a concern. In reality, WAP gateways are maintained by mobile operators and WML servers are owned by service providers. Therefore, WTLS normally only provides a point-to-point secure solution to the service provider (WML sever) and the end user. If end-to-end security is required, either the WAP gateway has to be maintained by a trusted party or the service provider must maintain the WAP gateway herself.

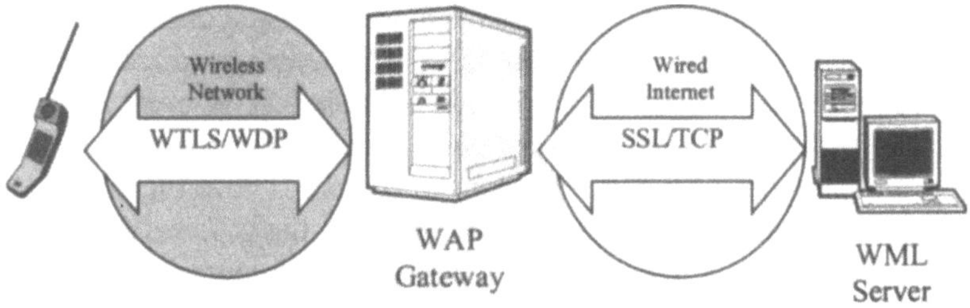

Figure 3. WAP Network configuration

2.3 Discussion

Different AKC protocols have been designed to tackle different issues, such as security of the protocol, restricted computational power of the client or privacy of the customers. However, few of the proposed solutions have considered utilizing the existing mobile system for implementing security adds-on, and as far as we know only the WTLS protocol has been implemented and deployed outside of the laboratory. The practical constraints and problems, therefore, have rarely been addressed in the literature. WTLS, however, does not provide end-to-end security between the user and the service provider.

In this work, the constraints of implementing an AKC protocol on the existing GSM mobile network using existing hardware products are discussed. A trusted third party is introduced in our AKC protocol to share the workload of certificate verification with the mobile client. Our design considerations are on security, restricted processing resources of the client, number of messages required and the size of each message packet.

3. SYSTEM REQUIREMENTS

In our system, users conduct mobile electronic transactions from a GSM dual-slot cellular phone. The phone has an additional smart card slot besides the GSM SIM card. The second smart card is chosen to be a Java Card [3] with 16K EEPROM and a cryptographic co-processor which can perform RSA cryptographic functions, triple-DES cipher, SHA-1 hashing and random number generation. All AKC protocol processing as well as transaction operations are performed on the processor chip on-board the smart card. The GSM SIM card is only responsible for providing the wireless communication channel and invoking the applications on the second card. Message packets between the phone and the base station are transmitted through SMS (short message service) of GSM. A SMS packet has a size limitation of 140 bytes.

4. OVERVIEW OF SYSTEM ARCHITECTURE

Due to the scarce resource for both memory space and computational power, the mobile equipment (ME) is incapable of verifying a X.509 digital certificate to authenticate a service provider. We have developed a server called the User Authentication Server (UAS) to act as a trusted third party to assist the mobile client to authenticate and exchange keys with the service provider, which is named the PKI End-to-end Secure Module (PESM). The following diagram shows the system architecture. Each component will now be described below.

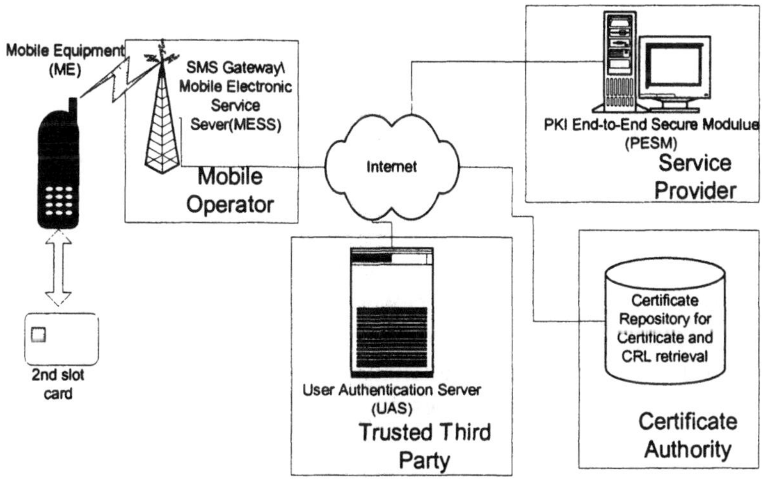

Figure 4. Overview of the system architecture

4.1 SMS Gateway and Mobile Electronic Service Server (MESS)

The SMS Gateway and MESS together act as an interface between the wireless and wired networks. MESS interprets the header of message packets and routes the packets to the proper MEs and servers. It is unable to read the message contents since they are encrypted at source.

4.2 Mobile client

The mobile client is a portable device which in our case is a dual slot GSM phone and a smart card with cryptographic functionality. Each user is required to have his own digital certificate issued by a valid Certification Authority (CA). The corresponding private key is stored in the user's second slot smart card. Moreover, we require the UAS's public key be pre-loaded on the card as well. In subsequent sections, the mobile client is abbreviated as ME (Mobile Equipment) for simplicity.

4.3 User Authentication Server (UAS)

The UAS is a centralized server that should be operated by a trusted third party. Its role is to help the ME to authenticate the party it is communicating with. First, mutual authentication is performed between the ME and UAS. Then, the UAS authenticates the PESM on behalf of the ME. Following that, a PESM session key is exchanged between the ME and PESM to establish an end-to-end secure communication channel.

4.4 PKI End-to-end Secure Module (PESM)

PESM is a server operated by the service provider. It is responsible for ensuring security at the application level, includes authentication, confidentiality, integrity and non-repudiation. For authentication, it performs the handshake protocol to authenticate the UAS or optionally authenticate the mobile client and establishes a session key. For confidentiality, it encrypts and decrypts messages sent and received from the mobile client using the established session key. Furthermore, it verifies the Message Authentication Code (MAC) of each message to guarantee integrity. For non-repudiation, it verifies the digital signature of a message if it is present.

4.5 Certificate repository

The certificate repository is a service provided by the CA which allows the public to access the issued digital certificates. It is usually implemented by a LDAP server on which object records can be searched by subjects. The UAS and PESMs will access this server from time to time to retrieve digital certificates for verification purposes.

5. PROTOCOLS

5.1 Design Criteria

Our AKC protocol is based on a 3-tiered model involving the ME, UAS and PESM. With the assistance of the UAS, the ME and PESM establish a session key between them for encryption and MAC calculation. Prior to key establishment, authentication is required between the UAS and ME, and then the UAS and PESM. For certain applications, the PESM may want to authenticate the client (ME) in order to determine the privileges allowed. If this is not necessary, the PESM normally does not need to authenticate the client. The ME and UAS, however, must authenticate the PESM to ensure they are talking to the right service provider.

The communication channel between the ME and MESS is furnished by the GSM SMS (Short Message Service) service. SMS is a datagram based packet switching channel in which message packets can be corrupted, delayed or lost. The maximum size of each message packet is 140 bytes and there is a significant delay associated with each SMS packet transmission. Furthermore, the slow processing speed of the ME and the limited memory space for program code and data on the ME's second slot smart (of the order of 16K Bytes) should also been taken into account in the design of the protocol. Therefore, the protocol should satisfy the following requirements:

- The size of each message packet on the wireless network must be less than 140 bytes
- The number of messages to be transmitted on the wireless network should be minimized
- The process on the client side must be simplified in order to minimize the code size and processing time on the second slot smart card
- The protocol must be able to tolerate the errors on the wireless transmission channel

5.2 Description of Protocols

Our protocol is divided into 2 phrases, namely: *UAS Session Establishment* and *PESM Session Establishment*. This is illustrated in the following diagram (Figure 5).

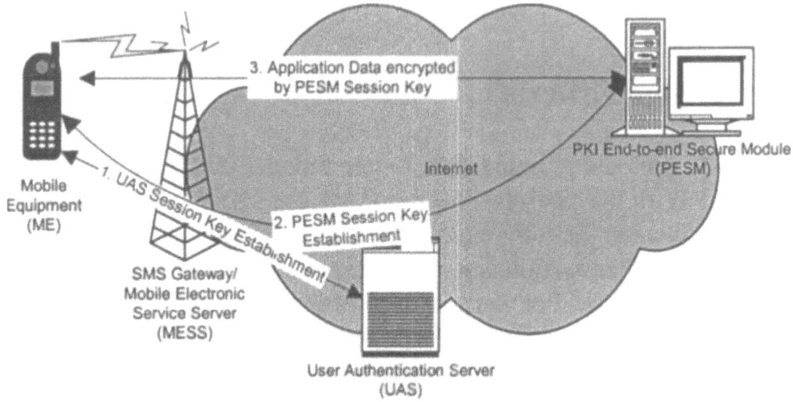

Figure 5. Steps of Key establishment

We use the following notations in describing our protocol:

Table 1. Symbols used in protocol description

Symbol	Description
ID_P	A unique identifier of entity *P*
$E_P\{x\}$	Encrypt *x* by *P*'s public key
$Cert_P$	Digital certificate of *P*
$ES_{KEY}\{x\}$	Encrypt by symmetric key block cipher in CBC mode (3DES) with the key "*KEY*"
$hash\{x\}$	Hash the value *x*
$f(x)$	Some kind of one-way function for session key diversification
$N \in_R \{0, 1\}^k$	Randomly generate *k* bits of binary data *N*
$A\|\|B$ or A,B	*A* concatenates with *B*

5.2.1 UAS Session establishment

The session is established between the ME and UAS using a two-pass AKC protocol based on a general challenge-response authentication mechanism (see Figure 6).

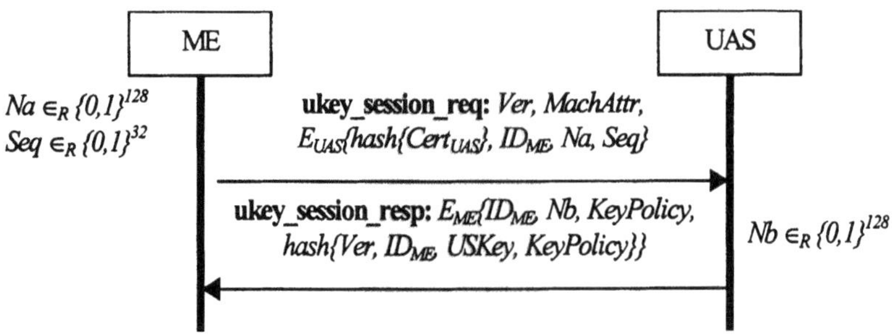

Figure 6 UAS Session Establishment

Table 2. Definition of terms used in UAS Session Establishment protocol

Items	**Description**
Ver	Version of the protocol
MachAttr	Configuration attribute of the ME (eg. Language)
Na	A random number generated by ME
Seq	A random number generated by ME as the starting sequence number of this session
Nb	A random number generated by UAS
USKey	UAS Session Key calculated from *f(Na||Nb)*
KeyPolicy	A value defining the lifetime of *USKey*

The ME initiates the establishment of a secure session with the UAS by performing the following operations:

1. Randomly generates *Na* and *Seq*.
2. Encrypts (*hash{Cert$_{UAS}$}, ID_{ME}, Na, Seq*) using UAS's public key.
3. Composes and sends *ukey_session_req* to UAS.

When UAS receives the *ukey_session_req* message, it should

1. Decrypt the message using its own private key.
2. Check if *hash{Cert$_{UAS}$}* is the fingerprint of its current certificate. If the check fails, the protocol cannot be continued since the ME does not have the correct public key of the UAS.
3. Randomly generate *Nb* and calculate *USKey* (UAS session key) from *f(Na||Nb)*.
4. Determine the lifetime of the session key and specify it in the value of *KeyPolicy*.
5. Compose and send *ukey_session_resp* to ME.

On receiving the *ukey_session_resp* message the ME verifies the validity of the message by generating its own value of *Hash{Ver, ID_{ME}, USKey, KeyPolicy}* and comparing it with the one in the message

Since only the valid UAS can decrypt *ukey_session_req* to get the value of *Na*, ME can authenticate UAS by checking the correctness of the *ukey_session_req* message. If the message is correct, ME accepts *USKey* and *KeyPolicy*.

In the above protocol, only one-way authentication of UAS is achieved. Adversaries can impersonate the ME by creating its own *ukey_session_req* message. Therefore, the UAS does not accept this newly established session yet. Instead, it stores the state parameters (i.e., *USKey, KeyPolicy* and *Seq*) of the session as a pending state and switch to the current state only after the ME has further authenticated itself in the *PESM Session Establishment* protocol that follows.

5.2.2 PESM Session Establishment

After the UAS session has been established (either in pending state or current state), the ME may start the *PESM Session Establishment* protocol in order to establish a secure communication session with the service provider. A request is sent by the ME to the UAS specifying which PESM it would like to talk with. The UAS then communicates with the target PESM on behalf of the ME.

Before the UAS can start the key exchange protocol with a PESM, it may have to interact with the PESM to find out the key exchange mode required and exchange the related certificates. If the information is already known then this step can be skipped.

The PESM can choose from two authentication modes of session key establishment when it receives an enquiry: Server Authentication and Client Authentication. Server Authentication means the PESM does not need to authenticate the ME. Otherwise, Client Authentication mode is used.

Server Authentication

If Server Authentication mode is selected, the protocol runs as follows:

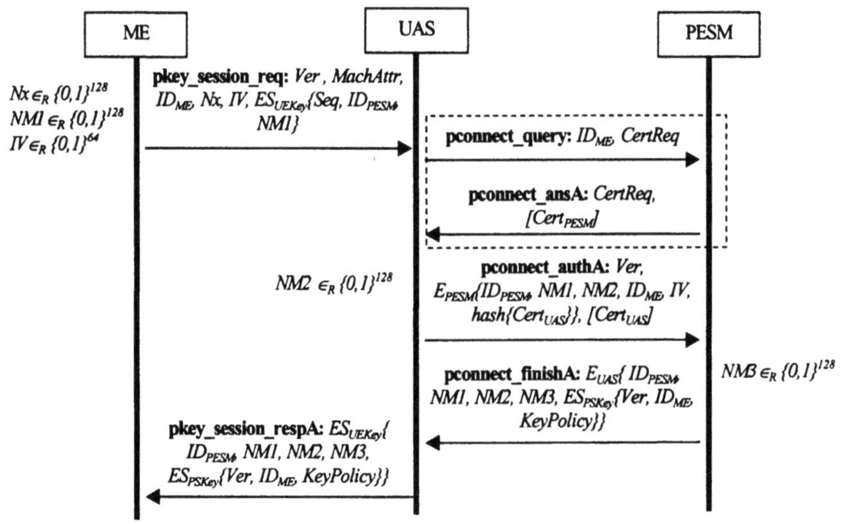

Figure 7. Protocol for PESM Session establishment in Server authentication mode

Table 3. Definition of terms used in PESM Session Establishment protocol

Items	Description
Nx	A random number generated by ME
Seq	Sequence number of this session
NM1	A random number generated by ME

NM2	A random number generated by UAS
NM3	A random number generated by PESM
IV	Initialization Vector for CBC mode encryption
UEKey	UAS Encryption Key which is calculated from *f(USKey‖Nx)*
PSKey	PESM Session Key which is calculated from *f(NM1‖NM2‖NM3)*
KeyPolicy	A value define the lifetime of the *PSKey*
CertReq	A list of CAs which is recognized by the sender. If this list is empty, it means certificate is not requested.

The ME initiates the protocol with the following actions:

1. Randomly generates *NM1* and *Nx* and calculates *UEKey* from *f(USKey‖Nx)*.
2. Increments *Seq*.
3. Encrypts (*Seq*, ID_{PESM}, *NM1*) using *UEKey*.
4. Composes and sends *pkey_session_req to* UAS.

When UAS receives the *pkey_session_req* message, it

1. Computes *UEKey* using the received *Nx* and its own *USKey*.
2. Decrypts the message using *UEKey*.
3. Checks if the value of *Seq* is valid. UAS will only accept *Seq* if it is larger than the last accepted *Seq* but falls within a certain predefined range. This mechanism is to avoid intruder's attack by replaying the *pkey_session_req* message.
4. If this message is valid, UAS switches session state from 'pending' to 'current'.
5. If UAS has no information about the mode of authentication or the certificate of PESM, a *pconnect_query* message is sent to the PESM. After receiving the *pconnect_ansA* response, UAS checks if the PESM's certificate was issued by one of the CAs listed in the non-empty *CertReq*. If it does not check out, the session cannot be established.
6. Randomly generates *NM2*.
7. Encrypts the elements in the *pconnect_authA* message using PESM's public key and sends the message to PESM.

On receiving the *pconnect_authA* message, the PESM

1. Decrypts the message using its private key.
2. Checks if the UAS certificate fingerprint in the message matches that of the certificate.
3. Randomly generates *NM3* and computes the *PSKey* by *f(NM1‖NM2‖NM3)*.
4. Determines the lifetime of the *PSKey* and assigns the value of *KeyPolicy*.
5. Encrypts (*Ver, SRN, KeyPolicy)* using *PSKey*.
6. Composes the *pconnect_finishA* message, encrypts it using the public key of UAS and sends to UAS.

On receiving the message *pconnect_FinishA*, the UAS can authenticate the PESM by checking if the values of *NM1* and *NM2* are the same as what were sent in the *pconnect_authA* message. This is, again, a simple challenge-response mechanism since the values of *NM1* and *NM2* can only be obtained by the holder of PESM's private key. After authenticating the identity of PSEM, the UAS forwards the neededl data to ME needed to calculate the PESM session key.

Note that the message type of the response by PESM indicates the mode of authentication. A *pconnect_ansA* message means server authentication is required. This step may be skipped if UAS already has this information.

Client Authentication

If the PESM requires client authentication, a *pconnect_ansB* message is sent instead of *pconnect_ansA*. The only difference in message packets contents between these two modes is that the public key of ME is used to encrypt the message sent from PESM. As a consequence, only the ME is able to calculate *PSKey*. The protocol runs as follows:

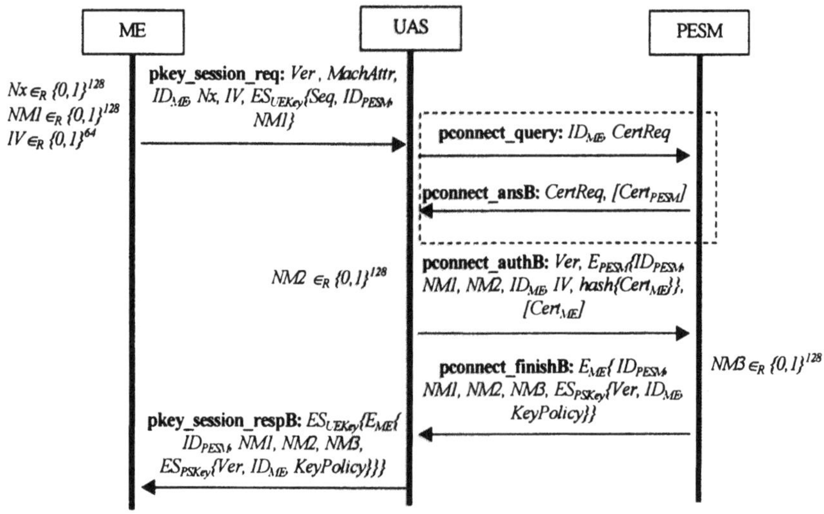

Figure 8. Protocol for PESM Session Establishment in Server authentication mode

5.3 Discussion

All authentication mechanisms in the protocols are achieved by the "challenge-response" approach. In both the *UAS and PESM Session* protocols, two messages are exchanged on the wireless network, which is the minimal number for challenge-response. However, this can achieve one-way authentication only. For mutual authentication, at least 3 messages are needed. Our approach of achieving mutual authentication with the minimum number of messages is to combine the last response with the message in the next protocol. In other words, completion of the authentication procedure is postponed until the next protocol. For example, in the *UAS Session Establishment* protocol, authentication of ME is not achieved until the start of the following *PESM Session Establishment* protocol. The same is done in the *PESM Session Establishment* protocol. Because of this, PESM and UAS should put the unauthenticated session in the pending state until confirmation can be made.

Another interesting property of this protocol is that since every protocol is a 2-pass protocol, the servers (i.e., UAS in the *UAS Session Establishment* protocol and PESM in the *PESM Session Establishment* protocol) are stateless. This is particularly important from the efficiency point of view since the servers do not need a waiting state.

6. IMPLEMENTATION ISSUES

This AKC protocol is designed to shift as much computational load as possible from the mobile device (ME) to the servers without compromising security. The infrastructure consists of a number of components as described in Section 4. In this section, some practical issues on the implementation of these components are discussed. There are 4 major entities involved in a transaction: mobile client, mobile operator, service provider and trusted third party.

As mentioned earlier, the mobile client is composed of three components: a dual slot GSM mobile phone, a GSM Subscriber Identity Module (SIM) which is a smart card with SIM Application Toolkit [21] enabled, and a second slot smart card. We have defined a set of protocols between the SIM and second slot cards so that they can communication with each other. The workflow is controlled by the second slot card. The SIM and mobile phone together provide a user interface and communication channel for the applications on the second slot card. In our implementation, the second slot smart card is a Java Card [3] which is compliant with the Java Card 2.1 standards. The card has a crypto co-processor on-board that supports public key cryptographic functions in hardware. The Java card was chosen for its flexibility and ease of programming; and the crypto engine is needed for speed in executing the RSA operations. We have tried three different brands of this type of card to test the generality of the infrastructure. The code size of the protocol on the card is about 6~7 Kbytes.

For the mobile operator, a prototype of the MESS (see Figure 4) is implemented. This is a server that interfaces the wireless network with the wired network. It is physically connected to a SMS gateway and also the

Internet. A set of message headers is defined for the MESS to route the associated messages from the SMS gateway to the appropriate destinations.

Because of the 140-byte limitation of the SMS message packet, all messages in our protocol are designed to satisfy this constraint. However, for the few cases when an application data message exceeds this limit, it will be broken down into two or more SMS packets. MESS is responsible to combine the fragmented SMS packets back to a single message according to the information in the headers.

7. PERFORMANCE

We have measured the various segments of time needed to execute the protocols on the mobile device in order to analyze the performance of the protocols and determine the system's bottleneck.

First we examine the time required for transmitting a SMS message (see Table 4). This time mainly depends on the performance of the mobile handset and the GSM SIM card. Different products can exhibit very different performance. The purpose of the study is not to compare the service provided by different mobile operators or to compare the performance of different mobile phone and SIM card products. We are only interested in gaining some ideas on the amount of time needed by each process involved in the transmission of a short message.

Table 4. Timing for SMS transmission

Operation	Time (in sec)
Fixed overhead for sending each SMS packet which is independent of packet size (depends only on the handset)	4.43 ~ 6.39
Transmission delay of 140 bytes of data	1.75
Processing delay of 140 of bytes data (handset dependent)	0.22 ~ 0.28
Total amount of time for a handset to send a 140-byte SMS packet (1.75+0.22+4.43) ~ (1.75+0.28+6.39)	6.4 ~ 8.42

As can be seen, the bulk of the time is in transmitting the SMS packet onto the wireless network from the handset no matter how small the size of the packet. Therefore, it is obvious the priority is to minimize the number of SMS messages used by the protocols rather than minimizing the packet size.

Another factor affecting the execution time of the protocol is the processing delay of the second slot smart card. Since the processing speed of the processor chip on the smart card chip is very slow compared with that of a PC, it may induce a significant delay on protocol execution. Again the processing time is highly dependent on the smart card product. We have compared three brands of Java Card with cryptographic function (see Table 5).

Table 5. Timing for smart card operations

Operation	Time (in sec)		
Invocation time of the second slot card by the handset	1.3		
Brand of Java Card	A	B	C
RSA Private Key Decryption (Key length = 1024 bits)	0.58	0.43	0.14
RSA Public Key Encryption (Key length = 1024 bits)	0.19	0.10	0.02
SHA-1 hash on 64 bytes of data	0.16	0.29	0.08
Triple-DES encryption in CBC mode on 64 bytes of data	0.28	0.04	0.008
Triple-DES encryption in ECB mode on 64 bytes of data	0.21	0.04	0.008
Random number generation (64 bytes)	0.43	0.007	0.008

For the *UAS Session Establishment* protocol, the overall processing time spent by the mobile client is about 18~22s (two SMS messages and four second slot card invocations with the associated processing time on the smart card). On the other hand, the processing time incurred by the UAS, which is implemented on a Pentium III 550MHz PC, is only about 0.1s and is negligible compared to that of the mobile client. The total time to establish a UAS session is the sum of the processing times and the queuing delay at the SMS gateway which varies from time to time depending on the traffic intensity. Since the major part of the processing time is spent on SMS transmission and invocation of the second slot card, the delay time for the *PESM Session Establishment* protocol is roughly the same as that for the *UAS Session Establishment protocol.*

We can see that performing an AKC protocol on the GSM mobile wireless is unacceptable for time critical applications. If the mobile phone starts with the *UAS Session Establishment* protocol and then the *PESM Session Establishment* protocol and lastly the application data, it would likely take about 1 minute to finish the process. A way to reduce this time is to reuse the session keys which have been previously established. In our protocol, both the UAS session key and PESM session key are kept for a period of time determined by the *KeyPolicy* value assigned by the UAS and PESM. Normally, a session key can be used for hundreds of times before refreshment since the time needed to break a 128-bit session key using the traditional method is of the order of tens of years. In this case, the AKC protocols will not need to be executed most of the time before a transaction. The application data is directly encrypted by the established session key stored in memory on the second slot smart card and transmitted. The time needed for the transaction is about 9 seconds if no digital signature is required or 18 seconds otherwise.

8. CONCLUSIONS

In this paper AKC protocols are categorized according to their authentication and key exchange approaches. We then presented a new set of AKC protocols that are practical and optimized for the GSM wireless network with SMS as the carrier. The protocols involve three parties - the mobile client, the service provider, and a trusted third party- and provide PKI-based end-to-end security between the client and the service provider. It is one of the few systems that has been implemented and an analysis of its performance has been given.

To evaluate the applicability of the protocols to electronic commerce activities, we have implemented a payment application on the infrastructure. The payment system makes use of the secure features of the protocols to protect the credit card number transmitted from the second slot smart card on the mobile client to the payment server of the merchant's acquirer bank. The application is being used in real-life for purchase and payment. The customers have found the system convenient to use and practical. The performance is acceptable and so far no security problem has been reported.

REFERENCES

[1] Secure Socket Layer (SSL) Version 3.0, http://home.netscape.com/eng/ssl3

[2] "Wireless Application Protocol Architecture Specification Version 30-Apr-1998", Wireless Application Protocol Forum, Ltd., http://www.wapforum.org

[3] Java Card API 2.1, Sun Microsystems Inc, http://java.sun.com/products/javacard

[4] Hongkong Post e-Cert, http://www.hongkongpost.gov.hk

[5] "Wireless Transaction Protocol Specification Version 05-Nov-1999", Wireless Application Protocol Forum,Ltd., http://www.wapforum.org

[6] "Specification of the SIM Application Toolkit for the Subscriber Identity Module – Mobile Equipment (SIM-ME) Interface (GSM 11.14 ver 7.1.0 Release 1998)", European Telecommunications Standards Institute

[7] T. Dierks and C. Allen, "The TLS Protocols Version 1.0", RFC 2246, 1999, ftp://ftp.isi.edu/in-notes/rfc2246.txt

[8] "Wireless Application Protocol Wireless Transport Layer Security Specification Version 18-Feb-2000" Wireless Application Protocol Forum, Ltd., http://www.wapforum.org

[9] S. Blake-Wilson, D. Johnson and A. Menezes, "Key Agreement Protocols and their Security Analysis", Sixth IMA International Conference on Cryptography and Coding, December 1997.

[10] K. H. Lee and S. J. Moon, "AKA Protocols for Mobile Communications", Proceedings of the 5th Australasian Conference on Information Security and Privacy, ACISP 2000, pp. 400-411, 2000

[11] W. Diffie and M. Hellman, "New directions in cryptography", IEEE Transactions on Information Theory, IT-22(6): pp. 644-654, November 1976

[12] M. J. Beller, L. -F. Chang and Y. Yacobi, "Privacy and Authentication on a Portable Communication System", IEEE Journal on Selected Areas in Communications, vol. 11, pp.821-829, August 1993.

[13] U. Carlsen, "Optimal Privacy and Authentication on a Portable Communications System", ACM Operating Systems Review, 28(3), 1994, pp. 16-23
[14] Y. Zheng, "An Authentication and Security Protocol for Mobile Computing", Proceedings of IFIP, pp. 249-257, Sep. 1996.
[15] Aziz and W. Diffie, "Privacy and Authentication for Wireless Local Area Networks", IEEE Personal Communications, vol. 1, pp. 25-31, 1994
[16] H. Lim, P. J. Lee, "Several practical protocols for authentication and key exchange", Information Processing Letters, vol 53, pp. 91-96, 1995
[17] H.-Y. Lin and L. Harn, "Authentication Protocols with Nonrepudiation Services in Personal Communication Systems", IEEE Communications Letters, vol. 3, no. 8, pp. 236-238, Aug. 1999
[18] "GSM System Security Study", RACAL Research Ltd, http://jya.com/gsm061088.htm, 1998
[19] R. Needham and M. Schroeder, "Using Encryption for Authentication in Large Networks of Computers", Communications of the ACM, 21(12), December 1978.
[20] J. Clark and J. Jacob, "A Survey of Authentication Protocol Literature: Version 1.0", http://www.cs.york.ac.uk/jac/papers/drareview.ps.gz, 17 November, 1997
[21] "Specification of the SIM Application Toolkit for the Subscriber Identity Module – Mobile Equipment (SIM-ME) interface (3GPP TS 11.14 version 8.5.0)", ETSI, 1999

A FAMILY OF RESOURCE-BOUND REAL-TIME PROCESS ALGEBRAS

Insup Lee
University of Pennsylvania, Philadelphia, PA (lee@cis.upenn.edu)

Jin-Young Choi
Korea University, Seoul, Korea (choi@formal.korea.ac.kr)

Hee Hwan Kwak
Synopsys Inc., Beaverton, OR (hkwak@synopsys.com)

Anna Philippou
University of Cyprus, Cyprus (annap@cs.ucy.ac.cy)

Oleg Sokolsky
University of Pennsylvania, Philadelphia, PA (sokolsky@saul.cis.upenn.edu)

Abstract *This paper describes three real-time process algebras, ACSR, PACSR and ACSR-VP. ACSR is a resource-bound real-time process that supports synchronous timed actions and asynchronous instantaneous events as well as the notions of resource, priority, exception, and interrupt. PACSR is a probabilistic extension of ACSR with resources that can fail and associated failure probabilities. ACSR-VP extends ACSR with value passing between processes and parameterized process definitions. This paper also provides three simple real-time system examples to illustrate the expressive power and analysis technique of each process algebra.*

Keywords: Real-time process algebra, probablistic process algebra, value-passing process algebra, schedulability analysis, real-time systems, resource-bound process algebra.

1. INTRODUCTION

Reliability in real-time systems can be improved through the use of formal methods for the specification and analysis of timed behaviors. Recently, there has been a spate of progress in the development of real-time formal methods. Much of this work falls into the traditional categories of untimed systems such as temporal logics, assertional methods, net-based models, automata theory and process algebras. In this paper, we provide an overview of the family of resource-bound real-time process algebras that we have developed.

Process algebras, such as CCS [12], CSP [7], Acceptance Trees [5] and ACP [2], have been developed to describe and analyze communicating, concurrently executing systems. They are based on the premises that the two most essential notions in understanding complex dynamic systems are concurrency and communication [12]. The most salient aspect of process algebras is that they support the *modular* specification and verification of a system. This is due to the algebraic laws that form a compositional proof system which enable the verification of a whole system by reasoning about its parts. Process algebras are being used widely in specifying and verifying concurrent systems.

Algebra of Communicating Shared Resource (ACSR) introduced by Lee *et. al.* [10], is a timed process algebra which can be regarded as an extension of CCS. The timing behavior of a real-time system depends not only on delays due to process synchronization, but also on the availability of shared resources. Most current real-time process algebras adequately capture delays due to process synchronization; however, they abstract out resource-specific details by assuming idealistic operating environments. On the other hand, scheduling and resource allocation algorithms used for real-time systems ignore the effect of process synchronization except for simple precedence relations between processes. ACSR algebra provides a formal framework that combines the areas of process algebra and real-time scheduling, and thus, can help us to reason about systems that are sensitive to deadlines, process interaction and resource availability.

ACSR supports the notions of resources, priorities, interrupt, timeout, and process structure. The notion of real-time in ACSR is quantitative and discrete, and is accommodated using the concept of timed actions. Executing a timed action requires access to a set of resources and takes one unit of time. Resources are serially reusable, and access to them is governed by priorities. Similar to CCS, the execution of an event is instantaneous and never consumes any resource. The notion of communication is modeled using events through the execution of complementary events, which are then converted into an internal event. As with timed actions, priorities are also used to arbitrate the choice of several events that are possible at the same time. Although the concurrency model of CCS-like process algebras is based on interleaving semantics, ACSR includes interleaving semantics for events as well as lock-step parallelism for timed actions.

The computation model of ACSR is based on the view that a real-time system consists of a set of communicating processes that use shared resources for execution and synchronize with one another. The use of shared resources is represented by timed actions and synchronization is supported by instantaneous events. The execution of a timed action is assumed to take one time unit and to consume a set of resources during the same time unit. Idling of a process is treated as a special timed action that consumes no resources. The execution of a timed action is subject to availability of the resources used in the timed action. The contention for resources is arbitrated according to the priorities of competing actions. To ensure the uniform progression of time, processes execute timed actions synchronously. Unlike a timed action, the execution of an event is instantaneous and never consumes any resource. Processes execute events asynchronously except when two processes synchronize through matching events. Priorities are used to direct the choice when several events are possible at the same time.

We have extended ACSR into a family of process algebras, GCSR [1], Dense-time ACSR [4], ACSR-VP [9], and PACSR [15]. GCSR is a graphical version of ACSR which allows the visual representation of ACSR processes. Dense-time ACSR is an extension of ACSR with dense time. ACSR-VP extends ACSR with value-passing capability so that arbitrary scheduling problems can be specified and analyzed. Probabilistic ACSR allows the modeling of resource failure with probabilities.

The rest of the paper is organized as follows. Section 2 describes the basic computation model of ACSR and explains its notions of events and timed actions. Section 3 overviews the syntax and semantics of ACSR and describes a simple scheduling example. Section 4 explains PACSR and extends the same scheduling example with probabilistic resource failure. Section 5 describes ACSR-VP and shows how parametric scheduling analysis can be done using basically the same schedule example.

2. THE COMPUTATION MODEL

In our algebra there are two types of actions: those which consume time, and those which are instantaneous. The time-consuming actions represent one "tick" of a global clock. These actions may also represent the consumption of resources, e.g., CPUs, devices, memory, batteries in the system configuration. In contrast, the instantaneous actions provide a synchronization mechanism between a set of concurrent processes.

Timed Actions. We consider a system to be composed of a finite set of serially reusable resources, denoted by $\mathcal{R}$. An action that consumes one "tick" of time is drawn from the domain $\mathbf{P}(\mathcal{R} \times \mathbf{N})$, with the restriction that each resource be represented at most once. As an example, the singleton action, $\{(r,p)\}$, denotes the use of some resource $r \in \mathcal{R}$ running at the priority level p. The action $\emptyset$ represents idling for one time unit, since no resuable resource is consumed.

We use $\mathcal{D}_R$ to denote the domain of timed actions, and we let A, B, C range over $\mathcal{D}_R$. We define $\rho(A)$ to be the set of resources used by the action A; e.g., $\rho(\{(r_1,p_1),(r_2,p_2)\}) = \{r_1,r_2\}$. We also use $\pi_r(A)$ to denote the priority level of the use of resource r in action A; e.g., $\pi_{r_1}(\{(r_1,p_1),(r_2,p_2)\}) = p_1$. By convention, if r is not in $\rho(A)$, then $\pi_r(A) = 0$.

Instantaneous Events. We call instantaneous actions *events*, which provide the basic synchronization in our process algebra. We assume a set of channels L. An event is denoted by a pair (a,p), where a is the *label* of the event, and p is its *priority*. Labels are drawn from the set $\mathcal{L} \cup \bar{\mathcal{L}} \cup \{\tau\}$, where for all $a \in L$ $a? \in \mathcal{L}$ and $a! \in \bar{\mathcal{L}}$. We say that $a?$ and $a!$ are *inverse* labels. As in CCS, the special identity label, τ, arises when two events with inverse labels are executed in parallel.

We use $\mathcal{D}_E$ to denote the domain of events, and let e, f and g range over $\mathcal{D}_E$. We use $l(e)$ and $\pi(e)$ to represent the label and priority, respectively, of the event e. The entire domain of actions is $\mathcal{D} = \mathcal{D}_R \cup \mathcal{D}_E$, and we let α and β range over $\mathcal{D}$.

The executions of a process are defined by a timed labelled transition system (timed LTS). A timed LTS M is defined as $\langle \mathcal{P}, \mathcal{D}, \rightarrow \rangle$, where (1) $\mathcal{P}$ is a set of ACSR processes, ranged over by P, Q, (2) $\mathcal{D}$ is a set of actions, and (3) $\rightarrow$ is a labeled transition relation such that $P \xrightarrow{\alpha} Q$ if the process P may perform an instantaneous event or timed action α and then behave as Q.

For example, a process P_1 may have the following behavior: $P_1 \xrightarrow{\alpha_1} P_2 \xrightarrow{\alpha_2} P_3 \xrightarrow{\alpha_3} \ldots$ That is, P_1 first executes α_1 and evolves into P_2, which executes α_2, etc. It takes no time to execute an instantaneous event. A timed action however is executed for exactly one unit of time.

3. ACSR

The following grammar describes the syntax of ACSR processes.

$$P \ ::= \ \text{NIL} \mid (a,n).P \mid A{:}P \mid P+P \mid P\|P \mid$$
$$P \,\triangle_t^a\, (P,P,P) \mid P\backslash F \mid [P]_I \mid P\backslash\!\backslash I \mid b \rightarrow P \mid C.$$

The process NIL represents the inactive process. There are two prefix operators, corresponding to the two types of actions. The process $(a,n).P$ executes the instantaneous event (a,n) and proceeds to P. The process $A{:}P$ executes a resource-consuming action during the first time unit and proceeds to P. The process $P + Q$ represents a nondeterministic choice between the two summands. The process $P \| Q$ describes the concurrent composition of P and Q: the component processes may proceed independently or interact with one another while executing events, and they synchronize on timed actions.

The scope construct, $P \triangle_t^a(Q, R, S)$, binds the process P by a temporal scope and incorporates the notions of timeout and interrupts. We call t the *time bound*, where $t \in \boldsymbol{N} \cup \{\infty\}$ and require that P may execute for a maximum of t time units. The scope may be exited in one of three ways: First, if P terminates successfully within t time-units by executing an event labeled $a!$ where $a \in L$, then control is delegated to Q, the success-handler. Else, if P fails to terminate within time t then control proceeds to R. Finally, throughout execution of this process construct, P may be interrupted by process S.

In $P \backslash F$, where $F \subseteq L$, the scope of channels in F is restricted to process P, and thus, components of P may use these labels to interact with one another but not with P's environment. The construct $[P]_I$, $I \subseteq \mathcal{R}$, produces a process that reserves the use of resources in I for itself, extending every action A in P with resources in $I - \rho(A)$ at priority 0. $P \backslash\backslash I$ hides the identity of resources in I so that they are not visible on the interface with the environment. That is, the operator $P \backslash\backslash I$ binds all free occurrences of the resources of I in P. This binder gives rise to the sets of *free* and *bound resources* of a process P. Process $b \rightarrow P$ represents the conditional process: it performs as P if boolean expression b evaluates to *true* and as NIL otherwise. Process constant C with process definition $C \stackrel{\text{def}}{=} P$ allows standard recursion.

The Structured Transition System. The informal account of behavior just given is made precise via a family of rules that define the labeled transition relations on processes. The semantics is defined in two steps. First, we develop the *unconstrained* transition system, where a transition is denoted as $P \xrightarrow{\alpha} P'$. Within "$\rightarrow$" no priority arbitration is made between actions; rather, we subsequently refine "$\rightarrow$" to define our prioritized transition system, "$\rightarrow_\pi$." The precise semantics rules are omitted but can be found in [3].

The prioritized transition system is based on *preemption*, which incorporates our treatment of synchronization, resource-sharing, and priority. The definition of preemption is straightforward. Let "$\prec$", called the *preemption relation*, be a transitive, irreflexive, binary relation on actions. Then for two actions α and β, if $\alpha \prec \beta$, we can say that "α is preempted by β." This means that in any real-time system, if there is a choice between executing either α or β, β will always be executed.

There are three cases to consider [3]: The first case is for the two timed actions, α and β, that compete for common resources. Here, the preempted action α may use a superset of β's resources. However, β uses all the resources at least at the same priority level as α. Thus, for any resource r in $\rho(\alpha) - \rho(\beta)$, the priority of r in α must be zero in order that β may preempt α since $\pi_r(B)$ is, by convention, 0 when r is not in B. Also, β uses at least one resource at a higher level. For instance, $\{(r_1,2),(r_2,0)\} \prec \{(r_1,7)\}$ but $\{(r_1,2),(r_2,1)\} \not\prec \{(r_1,7)\}$.

The second case is for the two events with the same label. Here, an event may be preempted by another event sharing the same label, but with a higher priority. For example, $(\tau,1) \prec (\tau,2)$, $(a,2) \prec (a,5)$, and $(a,1) \not\prec (b,2)$ if $a \neq b$.

The third case is when an event and a timed action are comparable under "$\prec$." Here, if $n > 0$ in an event (τ, n), we let the event preempt any timed action. For instance, $\{(r_1,2),(r_2,5)\} \prec (\tau,2)$, but $\{(r_1,2),(r_2,5)\} \not\prec (\tau,0)$.
We define the prioritized transition system "$\rightarrow_\pi$," which simply refines "$\rightarrow$" to account for preemption.

Definition 1 *The labeled transition system "$\rightarrow_\pi$" is defined as follows: $P \xrightarrow{\alpha}_\pi P'$ if and only if (1) $P \xrightarrow{\alpha} P'$ is an unprioritized transition, and (2) There is no unprioritized transition $P \xrightarrow{\beta} P''$ such that $\alpha \prec \beta$.* □

Analysis of Real-Time Systems in ACSR. Within the ACSR formalism we can conduct two types of analysis for real-time scheduling: validation and schedulability analysis. Validation shows that a given specification correctly models the required real-time scheduling discipline, such as Rate Monotonic and Earliest-Deadline-First. Schedulability analysis determines whether or not a real-time system with a particular scheduling discipline misses any of its deadlines. The validation and schedulability analysis of a real-time system can be carried out using the *equivalence* of ACSR processes.

Equivalence between ACSR processes is based on the concept of *bisimulation* [14] which compares the computation trees of two processes. Using the theory found in [12], it is straightforward to show that there exists a largest such bisimulation over "$\rightarrow_\pi$," which we denote as "$\sim_\pi$." This relation is an equivalence relation, and is a congruence with respect to ACSR's operators [3].

When comparing processes, we often find that because different objectives were pursued in formulating the two process expressions (perhaps simplicity of expressions for one, and efficiency for the other), the internal synchronization actions of the two processes are not identical. Consequently, even though the two processes may display identical "external" behavior (i.e., non-τ event labels and timed action steps), there may be τ actions in one process that do not correspond directly with τ actions in the other. (Recall that synchronization replaces the complementary event labels with a single τ event.) For those situations where matching of external behaviors is sufficient a weaker form of equivalence, *weak bisimulation* [12], is used. It is straightforward to prove the existence of a largest weak bisimulation $\approx_\pi$ over "$\rightarrow_\pi$" in a manner analogous to the case for bisimulation. Weak bisimulation, $\approx_\pi$, is an equivalence relation (though not a congruence) for ACSR that compares observable behaviors of processes.

Example. Throughout the paper, we will use a simple example from the area of schedulability analysis. The example describes a set of periodic tasks scheduled according to the Rate Monotonic (RM) scheduling policy. This policy assigns static priorities to the tasks in the inverse proportion to their periods. As a syntactic convenience, we allow ACSR processes to be parameterized by a set of index variables. Each index variable is given a fixed range of values. This restricted notion of parameterization allows us to represent collections of similar processes concisely. For example, the parameterized process

$$P_t = t < 2 \rightarrow (a_t, t).P_{t+1}, t \in \{0..2\}$$

is equivalent to the following three processes:

$$P_0 = (a_0, 0).P_1, \quad P_1 = (a_1, 1).P_2, \quad P_2 = \text{NIL}.$$

As shown in Section 5, the addition of parameterized process definition in ACSR-VP allows us to get rid of this kind of parameterization.

The example is constructed as follows. We have two tasks, $Task_1$ and $Task_2$. $Task_i$ has period p_i and execution time e_i. The deadline for each task is equal to its period. Both tasks share the same processor, modeled by the resource *cpu*. No other tasks use the processor.

Each $Task_i$ idles until it is awakened by the operating system by the $start_i$ event and starts competing for the processor. At each time unit, the task may either get access to the processor or, if it is preempted by a higher-priority task, it idles until the next time unit. Once the necessary amount (i.e., e_i) of execution time is accumulated, the task returns to the initial state and waits for the next period.

In order to detect missed deadlines, we also model the task dispatcher, which initiates the tasks according to their periods. For each $Task_i$, there is a process $Dispatch_i$, which sends the $start_i$ event to the respective task every p_i time units. If the task cannot accept the event - that is, if it has not completed its execution - the dispatcher deadlocks.

The complete specification is shown below. In the specification of a task, i is the task number and j is the accumulated execution time. We assume that the tasks have distinct periods and are ordered by decreasing periods, so we can use the task number as the priority for processor access.

$$
\begin{array}{rcll}
System & \stackrel{\mathrm{def}}{=} & [(Dispatch_1|Dispatch_2|Task_1 & \\
 & & |Task_2)\backslash\{start_1, start_2\}]_{\{cpu\}} & \\
Dispatch_i & \stackrel{\mathrm{def}}{=} & (start_i!, i).D_{i,0} & i = \{1,2\} \\
D_{i,k} & \stackrel{\mathrm{def}}{=} & k < p_i \rightarrow \{\} : D_{i,k+1} & \\
 & + & k = p_i \rightarrow Dispatch_i & i = \{1,2\}, k = \{0, p_i\} \\
Task_i & \stackrel{\mathrm{def}}{=} & (start_i?, 0).P_{i,0} + \{\} : Task_i & i = \{1,2\} \\
P_{i,j} & \stackrel{\mathrm{def}}{=} & j < e_i \rightarrow (\{\} : P_{i,j} & \\
 & & \quad +\{(cpu, i)\} : P_{i,j+1}) & \\
 & + & j = e_i \rightarrow Task_i & i = \{1,2\}, j = \{0, e_i\}
\end{array}
$$

ACSR analysis techniques allow us to verify the schedulability of a system of tasks for fixed values of parameters e_i and p_i. The correctness criterion being that a resulting process does not deadlock can be checked either by deciding the behavioral equivalence of the process to the process that idles forever, or by performing reachability analysis on the state space of the process to search for deadlock states. For example, we considered two sets of tasks. The task set with parameters $e_1 = 2$, $p_1 = 5$, $e_2 = 1$, $p_2 = 2$ does not exhibit any deadlock, while the set $e_1 = 2$, $p_1 = 3$, $e_2 = 1$, $p_2 = 2$ has a deadlock and thus is not schedulable.

4. PROBABILISTIC ACSR

PACSR (Probabilistic ACSR) extends the process algebra ACSR by associating with each resource a probability. This probability captures the rate at which the resource may fail. Since instantaneous events in PACSR are identical to those of ACSR, we only discuss timed actions, which now can account for resource failure.

Timed Actions. As in ACSR, we assume that a system contains a finite set of serially reusable resources drawn from the set $\mathcal{R}$. We also consider set $\overline{\mathcal{R}}$ that contains, for each $r \in \mathcal{R}$, an element $\overline{r}$, representing the *failed* resource r. We write R for $\mathcal{R} \cup \overline{\mathcal{R}}$. Actions are constructed as in ACSR, but now can contain both normal and failed resources. So now the action $\{(r,p)\}$, $r \in \mathcal{R}$, cannot happen if r has failed. On the other hand, action $\{(\overline{r},q)\}$ takes place with priority q given that resource r has failed. This construct is useful for specifying recovery from failures.

Resource Probabilities. In PACSR we associate each resource with a probability at which the resource may fail. In particular, for all $r \in \mathcal{R}$ we denote by $\mathsf{p}(r) \in [0,1]$ the probability of resource r being up, while $\mathsf{p}(\overline{r}) = 1 - \mathsf{p}(r)$ denotes the probability of r failing. Thus, the behavior of a resource-consuming process has certain probabilistic aspects to it which are reflected in the operational semantics of PACSR. For example, consider the process $\{(cpu,1)\}$: NIL, where resource *cpu* has probability of failure $1/3$, i.e., $\mathsf{p}(\overline{cpu}) = 1/3$. Then, with probability $2/3$, resource *cpu* is available and thus the process may consume it and become inactive, while with probability $1/3$ the resource fails and the process deadlocks.

Probabilistic Processes. The syntax of PACSR processes is the same as that of ACSR. The only extension concerns the appearance of failed resources in timed actions. Thus, it is possible on one hand to assign failure probabilities to resources of existing ACSR specifications and perform probabilistic analysis on them, and, on the other hand, to ignore failure probabilities and apply non-probabilistic analysis of PACSR specifications.

As with ACSR, the semantics of PACSR processes is given in two steps. At the first level, a transition system captures the nondeterministic and probabilistic behavior of processes, ignoring the presence of priorities. Subsequently, this is refined via a second transition system which takes action priorities into account.

The unprioritized semantics is based on the notion of a *world*, which keeps information about the state of the resources of a process. Given a set of resources $Z \subset \mathcal{R}$, the set of possible worlds involving Z is given by $\mathcal{W}(Z) = \{Z' \subseteq Z \cup \overline{Z} \mid x \in Z' \text{ iff } \overline{x} \notin Z'\}$, that is, it contains all possible combinations of the resources in Z being up or down. Given a world $W \in \mathcal{W}(Z)$, we can calculate the probability of W by multiplying the probabilities of every resource in W.

Behavior of a given process P can be given only with respect to the world P is in. A *configuration* is a pair of the form $(P,W) \in \mathsf{Proc} \times 2^{\mathsf{R}}$, representing a PACSR process P in world W. The semantics is given in terms of a labeled transition system whose states are configurations and whose transitions are either probabilistic or nondeterministic. We write S for the set of configurations.

The intuition for the semantics is as follows: for a PACSR process P, we begin with the configuration $(P, \emptyset)$. As computation proceeds, probabilistic transitions are performed to determine the status of resources which are immediately relevant for execution but for which there is no knowledge in the configuration's world. Once the status of a resource is determined by some probabilistic transition, it cannot change until the next timed action occurs. Once a timed action occurs, the state of resources has to be determined anew, since in each time unit resources can fail independently from any previous failures. Nondeterministic transitions (which can be events or actions) may be performed from configurations that contain all necessary knowledge regarding the state of resources.

We partition the set S into probabilistic configurations S_p and nondeterministic configurations S_n. A configuration (P, W) is included in S_n if every resource that can be used in a first step of P is included in W. Transitions for a configuration $(P, W) \in S_n$ are determined in the same way as in ACSR for P, except that a transition labeled by an action A can be taken if every resource $r \in \mathsf{R}$ that appears in A is also contained in W. The probabilistic transition relation takes probabilistic configurations into non-deterministic configurations.

We illustrate the rules of the semantics with the following example. Consider the process $P = \{(r_1, 1), (r_2, 1)\} : P_1 + (e?, 1).P_2$. The immediately relevant resources of P are $\{r_1, r_2\}$. From the probabilistic configuration $(P, \{r_1\})$, where we know that r_1 is up, but have no information about r_2, we have two probabilistic transitions that determine the state of r_2: $(P, \{r_1\}) \overset{\mathsf{p}(r_2)}{\longmapsto} (P, \{r_1, r_2\})$ and $(P, \{r_1\}) \overset{\mathsf{p}(\overline{r_2})}{\longmapsto} (P, \{r_1, \overline{r_2}\})$. Both of these configurations are nondeterministic since we have full information about the relevant resources. Further, $(P, \{r_1, r_2\})$ has two nondeterministic transitions: $(P, \{r_1, r_2\}) \overset{\{(r_1,1),(r_2,1)\}}{\longrightarrow} (P_1, \emptyset)$ and $(P, \{r_1, r_2\}) \overset{(e?,1)}{\longrightarrow} (P_2, \{r_1, r_2\})$. The other configuration allows only one transition: $(P, \{r_1, \overline{r_2}\}) \overset{(e?,1)}{\longrightarrow} (P_2, \{r_1, \overline{r_2}\})$, since r_2 is failed. Note that a probabilistic transition always leads to a nondeterministic configuration. A nondeterministic transition may lead to either nondeterministic configuration or a probabilistic one.

Probabilistic Analysis Techniques. We have defined a probabilistic weak bisimulation [16], which allows us to compare observable behaviors of PACSR processes similar to the case of ACSR. In addition, probabilistic information embedded in the probabilistic transitions allows us to perform quantitative analysis of PACSR specifications. In particular, we can compute the probability of reaching a given state or a deadlocked state.

Example. We illustrate the utility of PACSR in the analysis of fault-tolerance properties by slightly extending the example of Section 3. We consider the same set of tasks running on a processor with an intermittent fault. At any time unit, the processor may be running, in which case the higher-priority task executes normally, or it may be down, in which case none of the tasks execute. We modify the specification of a task to add the alternative behavior where a task can perform an action that contains the failed *cpu* resource and does not increase its execution time.

$$\begin{array}{rcll}
System & = & [(Dispatch_1|Dispatch_2|Task_1 & \\
 & & |\,Task_2)\backslash\{start_1, start_2\}]_{\{cpu\}} & \\
Dispatch_i & = & (start_i!, i).D_{i,0} & i = \{1,2\} \\
D_{i,k} & = & k < p_i \rightarrow \{\} : D_{i,k+1} & \\
 & + & k = p_i \rightarrow Dispatch_i & i = \{1,2\}, k = \{0, p_i\} \\
Task_i & = & (start_i?, 0).P_{i,0} + \{\} : Task_i & i = \{1,2\} \\
P_{i,j} & = & j < e_i \rightarrow (\{\} : P_{i,j} & \\
 & & \quad +\{(cpu, i)\} : P_{i,j+1} & \\
 & & \quad +\{(\overline{cpu}, i)\} : P_{i,j}) & \\
 & + & j = e_i \rightarrow Task_i & i = \{1,2\}, j = \{0, e_i\}
\end{array}$$

We apply the probabilistic analysis to the task set we considered in Section 3: $e_1 = 2$, $p_1 = 5$, $e_2 = 1$, $p_2 = 2$. Even though the task set is schedulable under perfect conditions, in the presence of failures the tasks may still miss their deadlines. Given the probability of a processor failure, we can compute the probability that a deadline is missed. The following list of pairs show results of the experiments we ran. The first element of each pair is the cpu failure probability and the second is the probability of a missed deadline: {(0,0), (0.005, 0.025), (0.01, 0.050), (0.02, 0.100), (0.05, 0.250), (0.075, 0.367), (0.1, 0.473), (0.15, 0.650), (0.2, 0.780), (0.3, 0.926)}.

5. ACSR-VP

ACSR-VP (ACSR with Value Passing) extends the process algebra ACSR described in Section 3 by allowing values to be communicated along communication channels. In this section we present ACSR-VP concentrating on its value-passing capabilities.

We assume a set of variables X ranged over by x, y and a set of values V ranged over by v. Moreover, we assume a set *Expr* of expressions (which includes arithmetic expressions) and we let $BExpr \subset Expr$ be the subset containing boolean expressions. We let e and b range over $Expr$ and $BExpr$, respectively, and we write $\vec{z}$ for a tuple $z_1, \ldots z_n$ of syntactic entities.

As in ACSR, ACSR-VP also has two types of actions: instantaneous events and timed actions. The notion of timed action is identical to that of ACSR. However, instantaneous events are extended to provide value passing in addition to synchronization. An event is denoted as a pair (i, e_p) representing execution of action i at priority e_p, where i ranges over the internal τ, the input event $c?x$, and the output event $c!e$.

The syntax of ACSR-VP processes is similar to that of ACSR except for $C(\vec{x})$.

$$\begin{array}{rcl} P & ::= & \text{NIL} \mid (a,n).P \mid A:P \mid P+P \mid P\|P \mid \\ & & P\,\Delta^{a}_{t}\,(P,P,P) \mid P\backslash F \mid [P]_I \mid P\backslash\backslash I \mid b \rightarrow P \mid C(\vec{x}). \end{array}$$

In the input-prefixed process $(c?x, e).P$ the occurrences of variable x is bound. We write $\mathsf{fv}(P)$ for the set of free variables of P. Each process constant C has an associated definition $C(\vec{x}) \stackrel{\text{def}}{=} P$ where $\mathsf{fv}(P) \subseteq \vec{x}$ and $\vec{x}$ are pairwise distinct. We note that in an input prefix $(c?x, e).P$, e should not contain the bound variable x, although x may occur in P.

An informal explanation of ACSR-VP constructs is similar to that ACSR. The semantics of ACSR-VP process is also defined as a labeled transition system, similarly to that of ACSR. It additionally makes use of the following ideas: Process $(c!e_1, e_2).P$ transmits the value obtained by evaluating expression e_1 along channel c, with priority the value of expression e_2, and then behaves like P. Process $(c?x, p).P$ receives a value v from communication channel c and then behaves like $P[v/x]$, that is, P with v substituted for variable x. In the concurrent composition $(c?x, p_1).P_1 \| (c!v, p_2).P_2$, the two components of the parallel composition may synchronize with each other on channel c resulting in the transmission of value v and producing an event $(\tau, p_1 + p_2)$.

Symbolic Transition System. Consider the simple ACSR-VP process $P \stackrel{\text{def}}{=} (in?x, 1).(out!x, 1).\text{NIL}$ that receives a value along channel in and then outputs it on channel out, and where x ranges over integers. According to traditional methods for providing semantic models for concurrent processes using transition graphs, process P is infinite branching, as it can engage in the transition $(in?n, 1)$ for every integer value n. Thus, standard techniques for analysis and verification of finite state systems cannot be applied to such processes. Several approaches, such as symbolic transition graphs and transition graphs with assignment, have been proposed to deal with this problem for various subclasses of value-passing processes [6, 11, 13, 8]. We now briefly explain how to represent symbolic graphs with assignment for ACSR-VP processes. We only give an overview of the model and we refer to [8] for a complete discussion.

An SGA (Symbolic Graph with Assignment) is a rooted directed graph where each node n has an associated finite set of free variables $\mathsf{fv}(n)$ and each edge is labeled by a guarded action with assignment [11, 17]. Note that a node in SGA is an ACSR-VP term.

The notion of a ***substitution***, which we also call ***assignment***, is defined as follows. A ***substitution*** is any function $\theta: X \to Expr$, such that $\theta(x) \neq x$ for a finite number of $x \in X$. Given a substitution θ, the ***support*** (or ***domain***) of θ is the set of variables $D(\theta) = \{x \mid \theta(x) \neq x\}$. A substitution whose support is empty is called the ***identity substitution***, and is denoted by Id. When $|D(\theta)| = 1$, we use $[\theta(x)/x]$ for the substitution θ.

An SGA for ACSR-VP is a rooted directed graph where each node n has an associated ACSR-VP term and each edge is labeled by a boolean predicate, an action, and an assignment, (b, α, θ). Here we illustrate how to construct an SGA by an example. A set of rules for generating an SGA from an ACSR-VP term can be found in [8]. We use a transition $P \overset{b,\alpha,\theta}{\longmapsto} P'$ to denote that given the truth of boolean expression b, P can evolve to P' by performing actions α and putting into effect the assignment θ. Consider the following process.

$$\begin{array}{lcl} P(x) & \stackrel{\text{def}}{=} & (a?y, 1).P'(x+1, y) \\ P'(x, y) & \stackrel{\text{def}}{=} & (y \leq 2) \to (a!(x+y), 2).NIL \end{array}$$

The SGA for this process is shown below. Note how the value $x + 1$ is assigned along the edge from P to P'.

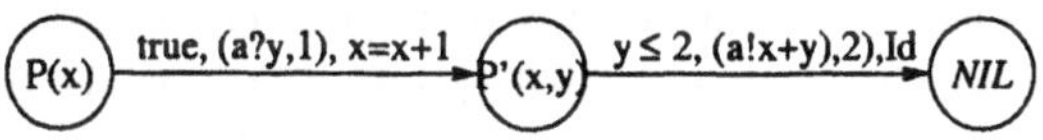

An informal interpretation of the above SGA is to view each process node as a procedure with its respective formal parameters. An edge coming into a node corresponds to a call to the procedure with the actual parameters supplied by the assignment labeling the edge and input events. If some of the variables are missing from the assignment, they are taken to be the same as the variable of the same name in the source node of the edge. After the process has been "called" in this way, it evaluates the guards on the outgoing transitions, applies the preemption relation to the enabled transitions and selects an one of the remaining transitions non-deterministically for the next step.

Symbolic Weak Bisimulation. The bisimulation relation for symbolic transition graphs is defined in terms of relations parametrized on boolean expressions, of the form $\simeq^b$, where $p \simeq^b q$ if and only if, for each interpretation satisfying boolean b, p and q are bisimilar in the traditional notion [8].

Let us compare the process P described above with the following process R:

$$
\begin{array}{lcl}
P(x) & \stackrel{\text{def}}{=} & (a?y,1).P'(x+1,y) \\
P'(x,y) & \stackrel{\text{def}}{=} & (y \le 2) \rightarrow (a!(x+y),2).NIL \\
R(x') & \stackrel{\text{def}}{=} & (a?y',1).R'(x,y') \\
R'(x',y') & \stackrel{\text{def}}{=} & (y' \le 2) \rightarrow (a!(x'+y'+1),2).NIL
\end{array}
$$

The prioritized SGA for R is a minor variation of the SGA for P. Applying the symbolic bisimulation algorithm for processes P and R, we obtain the following predicate equation system.

$$
\begin{array}{lcl}
X_{00}(x,x') & = & \forall z \forall z' X_{11}(z,z',x+1,x') \\
X_{11}(z,z',x,x') & = & z \le 2 \rightarrow z' \le 2 \wedge x+z = x'+z'+1 \\
 & \wedge & z' \le 2 \rightarrow z \le 2 \wedge x'+z'+1 = x+z
\end{array}
$$

This equation system can easily be reduced to the equation $X_{00}(x,x') \equiv x = x'+1$, which allows us to conclude that $P(x)$ and $R(x')$ are bisimilar if and only if $x = x'+1$ holds. In general, if we restrict to the domain of linear expressions, predicate equations obtained from the bisimulation algorithm can be solved using constraint logic programming and integer programming techniques [18].

Example. We revisit the example of Section 3 in order to conduct a more sophisticated, compared to ACSR, analysis allowed by ACSR-VP. With ACSR, the execution time and period of each task had to be fixed in order for the analysis to be performed. The symbolic semantics of ACSR-VP allows us to perform *parametric* analysis of processes. When we treat parameters of tasks as free variables of the specification, the scheduling problem can be restated in ACSR-VP as follows:

$$
\begin{array}{rcll}
System(e_1,e_2,p_1,p_2) & \stackrel{\text{def}}{=} & [(Dispatch_1(p_1)|Dispatch_2(p_2)|Task_1(e_1) & \\
 & & |Task_2(e_2))\backslash\{start_1, start_2\}]_{\{cpu\}} & \\
Dispatch_i(p) & \stackrel{\text{def}}{=} & (start_i!, i).D_i(0,p) & i=\{1,2\} \\
D_i(k,p) & \stackrel{\text{def}}{=} & k < p \rightarrow \{\} : D_i(k+1,p) & \\
 & + & k = p \rightarrow Dispatch_i(p) & i=\{1,2\} \\
Task_i(e) & \stackrel{\text{def}}{=} & (start_i?,0).P_i(0,e) + \{\} : Task_i(e) & i=\{1,2\} \\
P_i(j,e) & \stackrel{\text{def}}{=} & j < e \rightarrow (\{\} : P_i(j,e) & \\
 & & \quad +\{(cpu,i)\} : P_i(j+1,e)) & \\
 & + & j = e \rightarrow Task_i(e) & i=\{1,2\}
\end{array}
$$

We can now perform parametric analysis of *System* for some or all of its parameters. Here we consider the process $System(2, 1, p_1, p_2)$ Omitting the SGA for the example, we show the predicate equation system:

$$
\begin{aligned}
&X_1(p_1, p_2) = X_2(0, 0, p_1, p_2)\\
&X_2(j_{21}, j_{22}, p_1, p_2) = X_3(0, 0, j_{21}, j_{22}, p_1, p_2)\\
&X_3(j_{11}, j_{12}, j_{21}, j_{22}, p_1, p_2) =\\
&\quad (j_{12} < p_1) \wedge (j_{21} < 1) \wedge (j_{22} < p_2) \wedge X_3(j_{11}, j_{12}+1, j_{21}+1, j_{22}+1, p_1, p_2)\\
&\quad +(j_{12} < p_1) \wedge (j_{11} < 2) \wedge (j_{22} < p_2) \wedge X_3(j_{11}+1, j_{12}+1, j_{21}, j_{22}+1, p_1, p_2)\\
&\quad +(j_{12} < p_1) \wedge (j_{11} = 2) \wedge (j_{22} < p_2) \wedge (j_{21} = 1) \wedge X_3(j_{11}, j_{12}+1, j_{21}, j_{22}+1, p_1, p_2)\\
&\quad +(j_{12} = p_1) \wedge (j_{11} = 2) \wedge X_3(0, 0, j_{21}, j_{22}, p_1, p_2)\\
&\quad +(j_{22} = p_2) \wedge (j_{21} = 1) \wedge X_3(j_{11}, j_{12}, 0, 0, p_1, p_2)
\end{aligned}
$$

We have the additional condition $p_2 < p_1$, which comes from the assumption made in Section 3 that processes are sorted by decreasing execution time. When we solve the system of equations under this condition, the set of values for the parameters is given by the pairs of integers (p_1, p_2) satisfying $p_1 > 3, p_2 > 1, p_1 > p_2$.

6. SUMMARY AND CURRENT WORK

We have presented three resource-bound real-time process algebras: ACSR, PACSR and ACSR-VP. ACSR employees a synchronous semantics for resource-consuming actions that take time and an asynchronous semantics for events that are instantaneous. ACSR was developed to handle schedulability analysis in a process-algebraic setting. PACSR supports the notion of probabilistic resource failures, whereas ACSR-VP extends ACSR with value-passing capability during communication and parameterized process definition. To illustrate their features, we have described and analyzed simple real-time systems using ACSR, PACSR, and ACSR-VP.

As mentioned in the introduction section, there are two more formalisms in the family of resource-bound real-time process algebras. One formalism is for visual specification of ACSR and the other is ACSR with dense time. We are currently developing a resource-aware process algebra to capture the notion of power consumption and resource constraints of embedded systems.

Acknowledgments. The authors gratefully acknowledge that the work described in this paper was influenced by the former members of Real-Time Systems Group at the University of Pennsylvania, including Hanêne Ben-Abdallah, Patrice Brémond-Grégoire, Duncan Clarke, Rich Gerber, and Hong-Liang Xie. This research was supported in part by ARO DAAG55-98-1-0393, ARO DAAG55-98-1-0466, DARPA ITO MO-

BIES F33615-00-C-1707, NSF CCR-9988409, NSF CCR-0086147, NSF CISE-9703220, and ONR N00014-97-1-0505.

REFERENCES

[1] H. Ben-Abdallah. *GCSR: A Graphical Language for the Specification, Refinement and Analysis of Real-Time Systems.* PhD thesis, Department of Computer and Information Science, University of Pennsylvania, 1996. Tech Rep IRCS-96-18.

[2] J.A. Bergstra and J.W. Klop. Algebra of Communicating Processes with Abstraction. *Journal of Theoretical Computer Science*, 37:77–121, 1985.

[3] P. Brémond-Grégoire, J.Y. Choi, and I. Lee. A complete axiomatization of finite-state acsr processes. *Information and Computation*, 138(2):124–159, Nov 1997.

[4] P. Brémond-Grégoire and I. Lee. Process Algebra of Communicating Shared Resources with Dense Time and Priorities. *Theoretical Computer Science*, 189:179–219, 1997.

[5] M. Hennessy. *Algebraic Theory of Processes.* MIT Press Series in the Foundations of Computing. MIT Press, 1988.

[6] M. Hennessy and H. Lin. Symbolic bisimulations. *Theoretical Computer Science*, 138:353–389, 1995.

[7] C.A.R. Hoare. *Communicating Sequential Processes.* Prentice-Hall, 1985.

[8] Hee-Hwan Kwak, Jin-Young Choi, Insup Lee, and Anna Philippou. Symbolic weak bisimulation for value-passing calculi. Technical Report MS-CIS-98-22, University of Pennsylvania, Department of Computer and Information Science, 1998.

[9] H.H. Kwak, I. Lee, A. Philippou, J.Y. Choi, and O. Sokolsky. Symbolic schedulability analysis of real-time systems. In *Proceedings of IEEE Real-Time Systems Symposium.* IEEE Computer Society Press, 1998.

[10] I. Lee, P. Brémond-Grégoire, and R. Gerber. A Process Algebraic Approach to the Specification and Analysis of Resource-Bound Real-Time Systems. *Proceedings of the IEEE*, pages 158–171, Jan 1994.

[11] H. Lin. Symbolic graphs with assignment. In V.Sassone U.Montanari, editor, *Proceedings CONCUR 96*, volume 1119 of *Lecture Notes in Computer Science*, pages 50–65. Springer-Verlag, 1996.

[12] R. Milner. *Communication and Concurrency.* Prentice-Hall, 1989.

[13] P. Paczkowski. Characterizing bisimilarity of value-passing parameterised processes. In *Proceedings of the Infinity Workshop on Verification of Infinite State Systems*, pages 47–55, 1996.

[14] D. Park. Concurrency and Automata on Infinite Sequences. In *Proc. of 5th GI Conference.* LNCS 104, Springer Verlag, 1981.

[15] A. Philippou, R. Cleaveland, I. Lee, S. Smolka, and O. Sokolsky. Probabilistic resource failure in real-time process algebra. In *Proc. of CONCUR'98*, pages 389–404. LNCS 1466, Springer Verlag, September 1998.

[16] A. Philippou, O. Sokolsky, and I. Lee. Weak bisimulation for probabilistic systems. In *Proc. of CONCUR'00*, pages 334–349. LNCS 1877, Springer Verlag, August 2000.

[17] J. Rathke. *Symbolic Techniques for Value-passing Calculi.* PhD thesis, University of Sussex, 1997.

[18] R. Saigal. *Linear Programming : A Modern Integrated Analysis.* Kluwer Academic Publishers, 1995.

SURVIVABILITY ANALYSIS OF NETWORKED SYSTEMS

Jeannette M. Wing
Computer Science Department
Carnegie Mellon University
Pittsburgh, Pennsylvania 15213
U.S.A.
wing@cs.cmu.edu

Abstract Survivability is the ability of a system to continue operating despite the presence of abnormal events such as accidental failures and malicious intrusions. Ensuring system survivability has increased in importance as critical infrastructures have become heavily dependent on computers. Examples of these infrastructures are utility, transportation, communication, and financial networks. Complicating the analysis of these networked systems is their inter-dependencies: a failure in one may trigger a failure in another.

In this talk I present a two-phase method for performing survivability analysis of networked systems. First, we inject failure and intrusion events into a system model, use model checking to verify it for fault- and service-related properties, and visually display the model's effects with respect to a given property as a *scenario graph*. Then, we annotate the graphs with symbolic or numeric probabilities to enable reliability analysis using standard Markov Decision Process policy iteration algorithms. We use similar modeling and analysis techniques to do latency and cost-benefit analyses of these networked systems. We model dependencies among events using Bayesian Networks. We applied our two-phase method to two large cases studies from the financial industry and are currently applying it to a case study on intrusion detection systems.

This work is jointly done with Somesh Jha of the University of Wisconsin and Oleg Sheyner of Carnegie Mellon University. A full paper describing this work is in *the Proceedings of the International Conference on Software Engineering,* May 2001.

Keywords: survivability, fault-tolerance, security, reliability, model checking, Markov Decision Processes, Bayesian Networks, scenario graphs.